Wei Wu • Hani Choudhry
Editors

Next Generation Sequencing in Cancer Research

Volume 1: Decoding the Cancer Genome

Springer

Editors
Wei Wu
Department of Pathology
and Laboratory Medicine
University of Calgary
Calgary, AB, Canada

Hani Choudhry
Biochemistry Department
Faculty of Science
King Abdulaziz University
Jeddah, KSA

The Wellcome Trust Centre
for Human Genetics
University of Oxford
Oxford, UK

ISBN 978-1-4614-7644-3 ISBN 978-1-4614-7645-0 (eBook)
DOI 10.1007/978-1-4614-7645-0
Springer New York Heidelberg Dordrecht London

Library of Congress Control Number: 2013943582

Printed on acid-free paper

Springer is part of Springer Science+Business Media (www.springer.com)

Preface

This book is to introduce state-of-the-art Next Generation Sequencing (NGS) technologies for cancer genome research and to show a broad audience, including cancer biologists and oncologists, the technology revolution that is occurring in high-throughput DNA sequencing and the paradigm shifts that it is causing in the cancer research community.

Following the discovery of the double-stranded structure of DNA, different methods to read the nucleotide sequence of DNA placed important milestones in the life sciences and changed the direction of biomedical research, including cancer. Prior to 2005, DNA sequencing relied on the Maxam-Gilbert and the Sanger dideoxy-chain-termination methods. They are considered to be first-generation sequencing technology, but they played a historic role in the decade-long, multibillion-dollar human genome project that was completed in 2003. The demand grew for new technologies to deliver faster, cost-effective, and more accurate genomic information from the complexity of gene regulatory networks in normal and diseased cells. The effort led to the development of next-generation sequencing technologies and the production of massive amounts of sequencing data compared to conventional methods. There has been an astonishing tenfold decline in the cost for sequencing per year, which far outpaces Moore's Law. Numerous commercialized technologies, including Roche/454, ABI/SoLiD, Illumina/Solexa, Helicos, and Pacific Biosciences, have been widely used for diverse applications ranging from biology to medicine and extending to agriculture and fundamental research in evolution.

Cancer is a complex genetic disease and, during the last half-century, we have made great progress in understanding the mechanism of cancer initiation, development, and metastasis at the cellular and molecular levels. As a result, targeted therapy has indeed improved treatment for certain cancers using drugs such as Gleevec (imatinib mesylate) for chronic myelogenous leukaemia, Erlotinib for non-small cell lung cancer with EGFR mutation, Herceptin (trastuzumab) for a subset of breast cancer with HER2/neu gene amplification, and recent BRAF inhibitors for metastatic melanoma.

Scientists around the world are attempting to find the root causes of cancer and they are looking for more direct and effective means to cure cancer. This journey to conquer cancer is more optimistic now with the unfolding of the cancer genome. The cancer genome project was initiated a few years after the completion of human genome project and, through the use of high-throughput mutation detection techniques, it aimed to identify somatically acquired sequence variants and mutations and, hence, to identify the genes that are critical in the development of human cancers. Subsequently, the international cancer genome consortium (ICGC) was developed to depict all the mutations, including copy number changes, insertions and deletions, in 50 types of cancers. To date, numerous cancer genomes have been sequenced, which are helping us to gain an unprecedented understanding of molecular mechanisms underlying the complexity of cancer biology. The translation of cancer genome sequencing is starting to guide diagnosis and effective treatment. The cancer initiatome is the collection of all causes of malignant transformation; it extends beyond the ~2 % protein-coding content of the genome and resides in the non-coding RNA molecules (e.g., piRNAs, microRNAs, long non-coding RNAs) in the actively transcribed regions of the genome. Unravelling the complexity of the cancer initiatome is revealing new players in cancer cells and providing contemporary challenges in a new era of cancer research.

The cancer genome research that has emerged so far from advances in NGS technology is thrilling. It is imperative to bring this frontier research to graduate students, postdoctoral fellows, and junior scientists in basic and translational cancer research. We invited scientists who carry out their research with NGS technology to share their experience, enthusiasm, and passion for research on the cancer genome. The authors present the principles of NGS technologies and massively parallel DNA sequencing and their application of the whole genome sequences (WGS), exome-seq, RNA-seq, miRNA-seq, and ChIP-seq in their research programs. Moreover, the authors discuss their experimental designs and the analysis of their biological data with the latest bioinformatics tools. NGS produces a massive amount of data, and the challenge of managing the data is beyond the scope of this book; nevertheless, the basic concepts of computational analyses are well covered.

We thank all the authors for their enthusiastic contributions to meet the goals of our book. We are grateful to the Springer staff, especially Fiona Sarne, who have been enthusiastic partners in the publication of this frontier cancer genome research and who have ensured that the series is produced in an efficient and timely fashion. Heartfelt gratitude goes to our own families, who continue to patiently support us as we put forward our efforts related to the publication.

Calgary, AB, Canada — Wei Wu
Oxford, UK — Hani Choudhry

Contents

Contributors

Russell Bonneville Department of Biomedical Informatics, The Ohio State University, Columbus, OH, USA

Jennifer A. Chan Department of Pathology & Laboratory Medicine, University of Calgary, Calgary, AB, Canada

Department of Clinical Neurosciences, University of Calgary, Calgary, AB, Canada

Department of Oncology, University of Calgary, Calgary, AB, Canada

Han Chang Research and Development, Bristol-Myers Squibb Company, Princeton, NJ, USA

David N. Cooper Institute of Medical Genetics, School of Medicine, Cardiff University, Cardiff, UK

Aarti N. Desai Persistent Labs, Persistent Systems Ltd., Pune, India

Rajiv Dixit Department of Pathology & Laboratory Medicine, University of Calgary, Calgary, AB, Canada

Francisco J. Esteva Division of Cancer Medicine, Department of Breast Medical Oncology, The University of Texas MD Anderson Cancer Center, Houston, TX, USA

Thalia A. Farazi Laboratory of RNA Molecular Biology, Howard Hughes Medical Institute, The Rockefeller University, New York, NY, USA

Stephen P. Fink Division of Hematology and Oncology, Case Western Reserve University Comprehensive Cancer Center, School of Medicine, Cleveland, OH, USA

Kishore Guda Division of General Medical Sciences (Oncology), Case Western Reserve University Comprehensive Cancer Center, School of Medicine, Cleveland, OH, USA

Barry Iacopetta School of Surgery, University of Western Australia, Crawley, WA, Australia

Elizabeth Iorns Science Exchange, Inc., Palo Alto, CA, USA

Donald G. Jackson Research and Development, Bristol-Myers Squibb Company, Princeton, NJ, USA

Anna-Maria Jegg University of Miami Miller School of Medicine, Miami, FL, USA

Abhay Jere Persistent Labs, Persistent Systems Ltd., Pune, India

Hanlee P. Ji Stanford Genome Technology Center, Stanford University School of Medicine, Stanford, CA, USA

Division of Oncology, Department of Medicine, Stanford University School of Medicine, Stanford, CA, USA

Victor X. Jin Department of Biomedical Informatics, The Ohio State University, Columbus, OH, USA

Paul S. Kayne Research and Development, Bristol-Myers Squibb Company, Princeton, NJ, USA

Chee-Seng Ku Department of Medical Epidemiology and Biostatistics, Karolinska Institutet, Stockholm, Sweden

Saw Swee Hock School of Public Health, National University of Singapore, Singapore

Billy T. Lau Stanford Genome Technology Center, Stanford University School of Medicine, CA, USA

HoJoon Lee Division of Oncology, Department of Medicine, Stanford University School of Medicine, Stanford, CA, USA

Carl S. Leonhardt Laboratory of RNA Molecular Biology, Howard Hughes Medical Institute, The Rockefeller University, New York, NY, USA

Yilu Lu Division of Morbid Genomics, Department of Medical genetics, State Key Laboratory of Biotherapy, West China Hospital, Sichuan University, Chengdu, China

Yongxin Ma Division of Morbid Genomics, Department of Medical genetics, State Key Laboratory of Biotherapy, West China Hospital, Sichuan University, Chengdu, China

Gabriel G. Malouf Department of Medical Oncology, Groupe Hospitalier Pitie-Salpetriere University Pierre and Marie Curie (Paris VI), Paris, France

David R. Mole Henry Wellcome Building for Molecular Physiology, University of Oxford, Oxford, UK

Kenneth Nephew Department of Cellular and Integrative Physiology, Indiana University School of Medicine, Bloomington, IN, USA

Department of Obstetrics and Gynecology, Indiana University School of Medicine, Bloomington, IN, USA

Department of Molecular and Cellular Biochemistry, Indiana University School of Medicine, Bloomington, IN, USA

Hong Kiat Ng Cancer Science Institute of Singapore, National University of Singapore, Singapore

Yudi Pawitan Department of Medical Epidemiology and Biostatistics, Karolinska Institutet, Stockholm, Sweden

Min Qu Department of Urology, Shanghai Changhai Hospital, The Second Military Medical University, Shanghai, China

ShanCheng Ren Department of Urology, Shanghai Changhai Hospital, The Second Military Medical University, Shanghai, China

Petra B. Ross-Macdonald Research and Development, Bristol-Myers Squibb Company, Princeton, NJ, USA

Dimitrios H. Roukos Centre for Biosystems and Genomic Network Medicine, Ioannina University, Ioannina, Greece

Department of Surgery, Ioannina University School of Medicine, Ioannina, Greece

Rolf-Peter Ryseck Research and Development, Bristol-Myers Squibb Company, Princeton, NJ, USA

Johannes Schödel Department of Nephrology and Hypertension, Friedrich-Alexander-University Erlangen-Nuremberg, Erlangen, Germany

Russell Schwartz Department of Biological Sciences, Carnegie Mellon University, Pittsburgh, PA, USA

Lane Center for Computational Biology, Carnegie Mellon University, Pittsburgh, PA, USA

Stanley Shackney Intelligent Oncotherapeutics, LLP, Pittsburgh, PA, USA

Nathan O. Siemers Research and Development, Bristol-Myers Squibb Company, Princeton, NJ, USA

Arvind K. Singla Department of Biochemistry and Molecular Biology, McCaig Institute for Bone and Joint Health, University of Calgary, Calgary, AB, Canada

Xiaoping Su Department of Bioinformatics and Computational Biology, The University of Texas MD Anderson Cancer Center, Houston, TX, USA

Ayshwarya Subramanian Department of Biological Sciences, Carnegie Mellon University, Pittsburgh, PA, USA

Yinghao Sun Department of Urology, Shanghai Changhai Hospital, The Second Military Medical University, Shanghai, China

Chandini Thirukkumaran Translational Research Laboratories, Department of Oncology, Tom Baker Cancer Centre, University of Calgary, Calgary, AB, Canada

Thomas Tuschl Laboratory of RNA Molecular Biology, Howard Hughes Medical Institute, The Rockefeller University, New York, NY, USA

Weixin Wang Department of Biochemistry, LKS Faculty of Medicine, The University of Hong Kong, Hong Kong, China

Shenzhen Institute of Research and Innovation, The University of Hong Kong, Shenzhen, China

Junwen Wang Department of Biochemistry, LKS Faculty of Medicine, The University of Hong Kong, Hong Kong, China

Shenzhen Institute of Research and Innovation, The University of Hong Kong, Shenzhen, China

LKS Faculty of Medicine, Centre for Genomic Sciences, The University of Hong Kong, Hong Kong, China

Toby M. Ward Stanford Cancer Institute, Stanford University School of Medicine, Palo Alto, CA, USA

Henry M. Wood Leeds Institute of Molecular Medicine, University of Leeds, Leeds, UK

Mengchu Wu Cancer Science Institute of Singapore, National University of Singapore, Singapore

Wei Wu Department of Pathology & Laboratory Medicine, University of Calgary, Calgary, AB, Canada

Feng Xu Department of Biochemistry, LKS Faculty of Medicine, The University of Hong Kong, Hong Kong, China

Shenzhen Institute of Research and Innovation, The University of Hong Kong, Shenzhen, China

Chapter 1
The Evolution of High-Throughput Sequencing Technologies: From Sanger to Single-Molecule Sequencing

Chee-Seng Ku, Yudi Pawitan, Mengchu Wu, Dimitrios H. Roukos, and David N. Cooper

Abstract High-throughput sequencing technologies have been advancing rapidly over the last few years, transitioning from the first-generation Sanger sequencing to fourth generation. Each of the new generation sequencing technologies is characterized by new features. In contrast to Sanger sequencing, next-generation sequencing technologies (including the bench-top platforms) are characterized by their ability to perform massively parallel sequencing of up to hundred millions of sequence reads. This has significantly increased the throughput of the sequencing data by several orders of magnitude compared to Sanger sequencing; more importantly, it has also reduced the cost of sequencing quite substantially. These advances have

C.-S. Ku (✉)
Department of Medical Epidemiology and Biostatistics,
Karolinska Institutet, Stockholm, Sweden

Saw Swee Hock School of Public Health,
National University of Singapore, Singapore
e-mail: cheeseng.ku@ki.se

Y. Pawitan
Department of Medical Epidemiology and Biostatistics,
Karolinska Institutet, Stockholm, Sweden

M. Wu
Cancer Science Institute of Singapore, National University of Singapor,
Singapore

D.H. Roukos
Centre for Biosystems and Genomic Network Medicine,
Ioannina University, Ioannina, Greece

Department of Surgery, Ioannina University School of Medicine, Ioannina, Greece

D.N. Cooper
Institute of Medical Genetics, School of Medicine, Cardiff University, Cardiff, UK

W. Wu and H. Choudhry (eds.), *Next Generation Sequencing in Cancer Research: Volume 1: Decoding the Cancer Genome*, DOI 10.1007/978-1-4614-7645-0_1,

made sequencing of the entire human diploid genome and other large-scale omics studies both technically feasible and affordable, which was unachievable using Sanger sequencing. In addition, the sequencing of an entire human genome can be completed within weeks or days, which was unimaginable only a few years ago when the Human Genome Project was completed. This chapter reviews the technologies of these high-throughput sequencing platforms and discusses the evolution of these technologies over the past decade.

Keywords Next-generation sequencing • Third-generation sequencing • Bench-top sequencing • Genome sequencing • Exome sequencing • Microarray • Omics studies • Diagnostics

1 Introduction

Recent rapid developments in high-throughput sequencing technologies [1–3] have not only revolutionized our approach to omics studies (including structural and functional genomics) [4–9] but have also heralded a paradigm shift in genomic medicine [10–12]. Whereas structural genomics studies have focused on identifying genetic variations or polymorphisms in the genome (whether germ line or somatic), functional genomics studies have aimed to interrogate and annotate those elements or sequences that are of direct functional significance. This is evident, for example, with the molecular genomics studies that have previously relied on microarray technologies such as gene expression microarrays and the ChIP-chip methods (i.e., chromatin immunoprecipitation coupled with microarray) for genome-wide interrogation. However, these microarray-based methods were swiftly substituted by high-throughput sequencing methods, namely, RNA-Seq or transcriptome-Seq (to measure transcript or RNA expression levels) and ChIP-Seq (to study protein–DNA interactions involving, e.g., transcription factor binding sites and histone modification patterns), respectively [13].

There are a number of limitations with respect to using microarrays as compared to high-throughput sequencing methods. "Conventional" microarrays do not allow a thorough interrogation of the entire genome, because the synthesis of probes to be immobilized on the solid surface of microarrays (for hybridization) requires prior knowledge of DNA sequence and therefore a reference genome sequence is essential. The probes are needed to detect and measure the abundance of DNA or RNA targets through hybridization. Thus, microarrays are limited to interrogating those genomic regions which are targeted by the probes. It is clear that conventional gene expression (mRNA) microarray studies cannot measure gene expression levels unless probes are available to capture transcripts from these loci; probes are usually designed to capture known (i.e., annotated) protein-coding genes. Therefore, unknown (or yet to be annotated) mRNAs or transcripts from noncoding sequences in the transcriptome cannot be assessed. This also applies to ChIP-chip experiments, where the DNA fragments enriched by immunoprecipitation will go undetected if no complementary probes have been designed to capture them. By contrast, from a theoretical (and also practical) standpoint, high-throughput sequencing approaches are able to capture all the DNA fragments isolated by immunoprecipitation (ChIP-Seq) and all the transcripts (coding and noncoding transcripts) that are available in the transcriptome (RNA-Seq), including low abundance transcripts, if the sequencing depth is sufficient [4, 5, 14, 15].

In structural genomics studies, microarray-based methods such as comparative genomics hybridization (CGH) and single-nucleotide polymorphism (SNP) arrays have been commonly used to detect copy number variations (CNVs) genome-wide [16, 17]. However, these methods have poor sensitivity in the context of detecting smaller CNVs (<10 kb) and are unable to detect copy neutral variations such as balanced translocations and inversions, and complex chromosomal rearrangements. Further, microarrays have limited resolution to define the breakpoints of CNVs and structural variations [18–20]. Such limitations have now been overcome by high-throughput sequencing methods such as paired-end and mate-pair sequencing of the sequence reads and read depth of coverage approach [21–24]. These new and innovative sequencing-based approaches would not have been technically feasible without the high-throughput production of sequencing data, which is unattainable by means of the traditional Sanger sequencing methods. Thus, high-throughput sequencing technologies, since their arrival, have largely replaced traditional Sanger sequencing methods in many large-scale or genome-wide sequencing studies. The high-throughput sequencing technologies are characterized by their ability to simultaneously sequence hundred millions of DNA fragments and generate up to hundreds of gigabases of sequencing data per instrument run. More importantly, the widespread adoption or application of high-throughput sequencing technologies has been made possible by the steep reduction in the cost (per nucleotide) as compared with traditional sequencing methods.

Technological advances in sequencing have also been harnessed to explore genomics, transcriptomics, epigenomics, and metagenomics in international projects. These large-scale omics projects include the 1000 Genomes Project (to identify germ line genetic variants and polymorphisms in human populations from various geographical locations), the International Cancer Genome Consortium (to interrogate the somatic mutational landscape of at least 50 different types of cancers and to perform integrative analysis with transcriptomics and epigenomics data), the NIH Roadmap Epigenomics Program (to generate new research tools, technologies, datasets, and infrastructure to accelerate our understanding of the role of epigenetics), and the Human Microbiome Project (to characterize the microbial communities found at several different sites in the human body) [9]. The Centers for Mendelian Genomics was also established to accelerate the discovery of genes that underlie Mendelian disorders [25]. These new approaches have yielded important new information on the structure and function of our genome (see other chapters in the book).

Although there is currently no consensus in terms of classification, the high-throughput sequencing technologies can be broadly divided into two main categories, namely, next-generation sequencing (NGS) or second-generation sequencing (which encompasses both the conventional and bench-top instruments) and third-generation sequencing (TGS or single DNA molecule sequencing) technologies. There are several distinguishing features for each category (discussed below). The conventional NGS technologies are represented by the Roche 454 GS FLX, Illumina GA/HiSeq, and Life Technologies SOLiD/5500 sequencing platforms. Although these platforms are termed high-throughput sequencing technologies, the throughput for Roche 454 GS FLX (~700 Mb) is substantially lower than that of the other two sequencing platforms that generate up to hundreds of gigabases of sequence data per run. As such, these conventional NGS technologies are very suitable and cost effective for large-scale sequencing studies such as whole-genome sequencing (WGS) and whole-exome sequencing (WES). WGS has been widely applied to sequence the genomes of healthy individuals as well as tumor genomes [26–29]. On the other hand, WES has been instrumental in identifying the causal mutations underlying a range of Mendelian disorders [30, 31] and de novo mutations for both rare and complex diseases [32, 33].

However, in the situation where only the coding regions of small panels of genes (e.g., ranging from several to hundreds of genes rather than the whole genome or exome) are required for sequencing, then the capacity of these high-throughput NGS technologies, especially the Illumina HiSeq and Life Technologies SOLiD/5500 platforms, greatly exceeds requirements. Sequencing of small panels of genes in several patient samples is routine in the clinical diagnostic setting. Indeed, recent studies have already demonstrated the promising application of NGS (coupled with target enrichment methods) as a diagnostic tool for Mendelian disorders such as hereditary breast and colorectal cancers [34–37]. On the other hand, Sanger sequencing is unequal to the task of sequencing on this scale, both in terms of efficiency and cost-effectiveness. For example, the total length of genomic DNA commonly ranges from tens to hundreds of kilobases for the combined coding regions of only a handful of genes [38–40]. Thus, the development of several bench-top NGS instruments (i.e., Roche 454 GS Junior, Life Technologies Ion Torrent, and Illumina MiSeq) has filled a niche in the market for targeted gene sequencing [41, 42]. The throughput of these bench-top NGS instruments ranges from 10 Mb to >1 Gb and the performance of these platforms has been evaluated, each exhibiting its own strengths as well as its limitations [42]. The targeted gene sequencing employed on bench-top NGS instruments holds out great promise as a diagnostic tool in the clinical arena.

Although NGS technologies are characterized by a significantly increased sequencing throughput as compared to Sanger sequencing, they are not without their limitations. For example, NGS technologies (both conventional and bench-top platforms) are still reliant upon amplification steps (similar to Sanger sequencing) to generate clonally amplified DNA copies or fragments (templates) for sequencing. This is to ensure sufficient signal detection for each nucleotide incorporation during the cycle sequencing, i.e., repeated cycles of sequencing comprising "wash" (to remove excess reagents after nucleotide incorporation) and "scan" (imaging and signal detection) steps for each cycle of sequencing before the next cycle begins. These amplification methods are emulsion PCR in solution for Roche 454 and Life Technologies sequencing platforms (i.e., Roche 454 GS FLX, GS Junior, Life Technologies SOLiD, and Ion Torrent) and bridge amplification on solid for Illumina sequencing technologies (i.e., Illumina GA, HiSeq, and MiSeq). The amplification steps can introduce errors and lead to amplification bias or uneven amplification [1, 2]. By contrast, the single DNA molecule sequencing such as Pacific Biosciences Single-Molecule Real-Time (SMRT) sequencing obviates the need for amplification steps and the sequencing occurs in real time (i.e., without cyclic sequencing) [43]. Although the Helicos BioSciences true single-molecule sequencing (tSMS) or

Heliscope Single-Molecule Sequencer also does not need amplification steps, it is still based on cyclic sequencing where each cycle is paused for imaging or signal detection before the next cycle begins [44, 45]. Since the single-molecule sequencing technologies are not reliant on amplification (to sequence the clonally amplified templates), they avoid or at least minimize the errors (or noise) caused by dephasing, i.e., uneven sequencing in the clonally amplified templates.

Currently, there is no clear definition of the different generations of sequencing technologies. However, in this chapter, we distinguish TGS from NGS based on the prominent feature of single DNA molecule sequencing [46]. As such, both the SMRT and tSMS technologies are considered to be TGS. However, if the absence of cycle sequencing is also considered to be a prominent distinguishing feature for TGS, then the Helicos BioSciences tSMS platform would lie in between the transition of NGS to TGS. In addition, some have classified the Life Technologies Ion Torrent as a fourth-generation sequencing technology, because it is the first post-light sequencing instrument on the market, one in which the detection of nucleotide incorporation is not reliant on light emission in contrast to other NGS and TGS technologies [47]. However, we still consider this post-light sequencing technology as an NGS platform because it is still reliant on amplification and cycle sequencing of a large number of individual sequence reads in parallel, similar to the core features of other NGS technologies. For further classification, nanopore sequencing could be considered as a fourth-generation sequencing technology, because it is characterized by (a) single-molecule sequencing (without amplification), (b) real-time sequencing (without repeated cycles), (c) post-light sequencing (without light-based detection), and (d) no sequencing by synthesis (without reliance on the incorporation of nucleotides to synthesize the complementary strand of the DNA template) [48]. The availability of multiple different high-throughput sequencing platforms with significant differences, such as in the throughput, sequencing time, read length, and error rate and type, as well as the cost, has offered a variety of options and the flexibility for researchers to choose the platform(s) best suited to their applications. The aim of this chapter is to describe and compare the technologies of these high-throughput sequencing platforms and to discuss the evolution and development of these technologies over the past decade. However, owing to rapid and ongoing development in the individual sequencing technologies, the reader is encouraged to visit the vendors' homepages for the latest sequencing information. Other chapters in the book cover applications to various omics studies and recent developments and new discoveries from these studies.

2 From Sanger Sequencing to NGS Technologies

2.1 Sanger Sequencing

Sanger dideoxynucleotide or chain termination sequencing was the most widely used sequencing method over the last three decades from the time it was invented in the late 1970s until the first NGS platform was marketed in 2005 [49]. Sanger sequencing has been used for various applications such as mutation discovery, genotyping, and serial analysis of gene expression (SAGE) for measuring gene expression levels; more importantly, it was used to complete the Human Genome Project [50]. Currently, even with the availability of multiple high-throughput sequencing technologies, Sanger sequencing is still considered to be the gold standard method (with the highest raw base accuracy) and hence is still commonly used to validate variants or mutations identified using high-throughput sequencing technologies. As such, Sanger sequencing is far from being an obsolete technology; indeed, it is still commonly used to screen disease genes identified in studies that have applied high-throughput sequencing technologies [34]. For example, Ng et al. (2010) employed Sanger sequencing to identify frameshift indels in the *MLL2* gene in two of the three Kabuki syndrome cases subjected to WES, but who lacked deleterious point mutations and to sequence all 54 exons of the *MLL2* gene in additional cases to identify single base-pair substitutions. Thus, Sanger sequencing may be used to detect indels missed by NGS methods [51].

Although the chemistry of Sanger sequencing is already well known, for the purpose of comparison with the high-throughput sequencing technologies, we briefly summarize it here. Sanger sequencing involves both dNTPs and ddNTPs (nonreversible terminators) to enable the extension and termination of complementary strand synthesis when ddNTPs are incorporated. As a result, DNA fragments of different sizes (e.g., from a PCR amplicon) are generated, each starting from the same location, but each of the different-sized fragments ends with a particular base (ddNTP) that is labeled with one of four fluorescent dyes corresponding to that particular base. These differently terminated fragments are then separated in order of their lengths through capillary electrophoresis, and the information on the last or terminated ddNTP is used to determine the original DNA sequence. This chain termination method is able to generate a sequence length of up to 1 kb. However, the major limitations of Sanger sequencing are low throughput production, i.e., <100 kb (96-capillary fully automated machine), as well as high cost. As a consequence, the sequencing of the haploid human genome in the Human Genome Project employed on Sanger sequencing took ~13 years and cost ~$3 billion dollars.

2.2 Next-Generation Sequencing Technologies (Massive Parallel Sequencing)

The development of commercial NGS technologies since 2005 overcame the low throughput and relatively high cost of Sanger sequencing methods. Shortly after the first NGS instrument was introduced by Roche 454, i.e., the GS 20 System (subsequently replaced by GS FLX System with further improvements in terms of higher throughput and longer-sequence read length) [52], another two biotechnology companies also marketed their sequencing platforms, i.e., Illumina GA and Life Technologies SOLiD [1]. The simultaneous arrival of several NGS sequencers created considerable competition in the sequencing market although, with each technology having its own strengths and limitations. The technologies are also racing towards the goal of a $1,000 genome [53]. In what follows, we compare Sanger sequencing to NGS technologies with regard to the following aspects: (a) sequencing throughput, (b) sequencing chemistry, and (c) sequence read length and error.

2.2.1 Sequencing Throughput

Next-generation sequencing instruments have now largely supplanted Sanger sequencing machines (e.g., 3730xl sequencer) for large-scale omics/sequencing studies. For example, the Beijing Genomics Institute (BGI), which has the world's largest powerhouse of sequencing, possesses 137 Illumina HiSeq2000 and 27 Life Technologies SOLiD machines [54]. This was clearly attributed to the ultrahigh-throughput production of NGS technologies (up to hundreds gigabases of sequencing data per run), which is several orders of magnitude greater than Sanger sequencing. This is achieved through the ability of the NGS technologies to sequence up to billions of DNA fragments (or sequence reads) simultaneously in contrast to the 96-capillary Sanger sequencer. Therefore, they are also known as "massively parallel" sequencing technologies. For example, Illumina HiSeq2000 instrument can produce 6 billion single reads and generate up to >600 Gb of sequence data in a single run. The latest developments in Illumina HiSeq (http://www.illumina.com/systems/hiseq_systems.ilmn) and Life Technologies SOLiD platforms and 5500 Genetic Analyzer systems (see "Next-Generation Sequencing" in https://products.appliedbiosystems.com) have further enhanced the technical performance. However, one major limitation is that these technologies require a much longer run time of several days compared to several hours for Roche GS FLX, which can only generate several hundred megabases (~700 Mb). By contrast, a Sanger sequencing machine such as the 3730xl, which is commonly used in many laboratories, can only produce ~100 kb per run in 3 h [55]. Table 1.1 summarizes the key technical features of NGS and TGS technologies.

2.2.2 Sequencing Chemistry and Approach

One of the major advances introduced by NGS technologies was in enabling the sequencing of complete diploid human genomes [9, 56–58]. On the other hand, Sanger sequencing was used to sequence the diploid genome of Craig Venter [59]. In addition to normal healthy genomes, hundreds of cancer genomes have also been sequenced since 2008 [60]. Nonetheless, one of the major limitations of Sanger sequencing in WGS has been its requirement for in vivo amplification of DNA fragments using bacterial cloning. This differs from the targeted sequencing of candidate regions, where conventional PCR can be used to amplify the regions of interest before they are subjected to Sanger sequencing. The bacterial cloning procedures can introduce a variety of host cloning-related biases; for example, it can adversely affect genome representation because some DNA fragments cannot be cloned and some genomic regions cannot be sequenced. Moreover, these cloning steps are

Table 1.1 Distinguishing technical features between different sequencing technologies

Feature	Roche 454 GS FLX and GS Junior	Illumina GA, HiSeq and MiSeq	Life Technologies SOLiD and 5500	Life Technologies Ion Torrent and Proton	Helicos BioSciences tSMS	Pacific Biosciences SMRT
Generation	Next-generation sequencing	Next-generation sequencing	Next-generation sequencing	Next-generation sequencing	Third-generation sequencing	Third-generation sequencing
Conventional NGS	Roche 454 GS FLX	Illumina GA and HiSeq	Life Technologies SOLiD and 5500	NA	NA	NA
Bench-top NGS	Roche 454 GS Junior	Illumina MiSeq	NA	Life Technologies Ion Torrent and Proton	NA	NA
Synthesizing of complementary strand of DNA template	Yes	Yes	Yes	Yes	Yes	Yes
Sequencing approach	Sequencing by synthesis mediated through DNA polymerase—pyrosequencing	Sequencing by synthesis mediated through DNA polymerase—reversible terminator sequencing	Sequencing by synthesis mediated through DNA ligase	Sequencing by synthesis mediated through DNA polymerase—semiconductor sequencing	Sequencing by synthesis mediated through DNA polymerase—true single-molecule sequencing	Sequencing by synthesis mediated through DNA polymerase—single-molecule real-time sequencing
Nucleotides	Unlabeled nucleotides	Fluorescent labeled nucleotides	Fluorescent labeled nucleotides/probes	Unlabeled nucleotides	Fluorescent labeled nucleotides	Fluorescent labeled nucleotides
Reversible terminator (nucleotide)	No	Yes	No	No	Yes	No

The flow of nucleotides	Only one type of nucleotide is present in each cycle of sequencing	All 4 types of nucleotide are present in each cycle of sequencing	All 16 dinucleotide probes are present in each cycle of sequencing	Only one type of nucleotide is present in each cycle of sequencing	Only one type of nucleotide is present in each cycle of sequencing	No cycle sequencing
Amplification	Emulsion PCR	Bridge amplification	Emulsion PCR	Emulsion PCR	No amplification	No application
Single DNA molecule sequencing	No	No	No	No	Yes	Yes
Cycle versus real-time sequencing	Cycle sequencing	Cycle sequencing	Cycle sequencing	Cycle sequencing	Cycle sequencing	Real-time sequencing
Detection of nucleotide incorporation	Chemiluminescent light emission	Fluorescent light emission	Fluorescent light emission	pH/voltage changes	Fluorescent light emission	Fluorescent light emission
Sequence read length	"Long" read length technology (up to 1 kb)	"Short" read length technology	"Short" read length technology	"Short" read length technology	"Short" read length technology	"Long" read length technology (up to several kilobases)
Throughput	"Low" Roche 454 GS FLX (~700 Mb) and GS Junior (>35 Mb)	"High" up to hundreds of gigabases sequencing data per run	"High" up to hundreds of gigabases sequencing data per run	"Low" Ion Torrent (>1 Gb) Ion Proton chip I (two human exomes) and chip II (one human genome)	"High" up to hundreds of gigabases sequencing data per run	"Low" approximately 75 Mb (75,000 ZMWs × 1 kb)
Sequencing time	"Slow" light detection	"Slow" light detection	"Slow" light detection	"Fast" no light detection	"Slow" light detection	"Fast" real-time sequencing without halting
Additional features	NA	NA	NA	NA	Direct RNA sequencing	Direct methylation detection

tedious and labor intensive [61–63]. However, bacterial cloning amplification has been superceded by the in vitro amplification of millions of DNA fragments simultaneously by NGS technologies using emulsion PCR in solution for Roche 454 and Life Technologies sequencing technologies (which included the bench-top sequencing platforms, i.e., 454 GS Junior and Ion Torrent) and bridge amplification on a solid surface for Illumina sequencing technologies. Figures 1.1, 1.2, and 1.3 illustrate the amplification, sequencing chemistry, and principle for Roche 454 pyrosequencing, Illumina reversible terminator sequencing, and Life Technologies SOLiD sequencing by ligation of dinucleotide probes.

In emulsion PCR, the single-stranded DNA fragments or templates are attached to the surface of beads using adaptors or linkers, each bead being attached to a single individual DNA fragment (monoclonal bead). This is in contrast to polyclonal bead where two or more different individual DNA fragments may be attached. The DNA library is generated through random fragmentation or shearing of the genomic DNA. The surface of the beads contains oligonucleotide probes with sequences that are complementary to the adaptors binding the DNA fragments. The beads are then compartmentalized into separate water–oil emulsion droplets. Each droplet capturing one bead serves as a PCR microreactor for amplification steps to take place and produces clonally amplified copies from a single individual DNA fragment. On the other hand, for bridge amplification on solid surface for Illumina sequencing technologies, the single-stranded DNA fragments are first attached to a solid surface known as a flowcell using adaptors with complementary probes on the flowcell. The other unattached end of the DNA fragments then creates a "bridge-like structure" by bending over and hybridizing to the probes on the flowcell, which form the template for amplification to generate clusters comprising clonally amplified copies of the DNA fragments on the surface of the flowcell. By contrast, TGS technologies are characterized by single-molecule sequencing without the need for amplification steps [1, 2, 61–63].

The amplification process can introduce errors into the template sequence (clonally amplified copies) as well as amplification bias (or uneven amplification), resulting in uneven sequencing coverage (i.e., some regions may be under- or over-sequenced). Under-sequencing can result in the inaccurate detection of variants, while over-sequencing is redundant and not cost effective, as a minimum of tenfold coverage (or sequencing depth) is already considered sufficient for the accurate detection of germ line single-nucleotide variants (SNVs). However, a much higher sequencing depth is generally needed to detect somatic SNVs or point mutations in cancer tissues owing to the inherent genetic heterogeneity (mutations are only present in a subclone of the cancer tissue) and tissue heterogeneity (contamination from adjacent normal tissue) [28, 64]. Currently, the detection of rare mutations using NGS has also been enhanced through the development of "Duplex Sequencing," where this approach can independently tag and sequence each of the two strands of a DNA duplex, distinguishing from sequencing errors [65]. Further, a higher sequencing depth is also required if the sequencing data are also to be used to detect deletions and duplications using the depth of sequence read coverage approach [22, 23]. In addition to the inherent amplification error and bias, from the

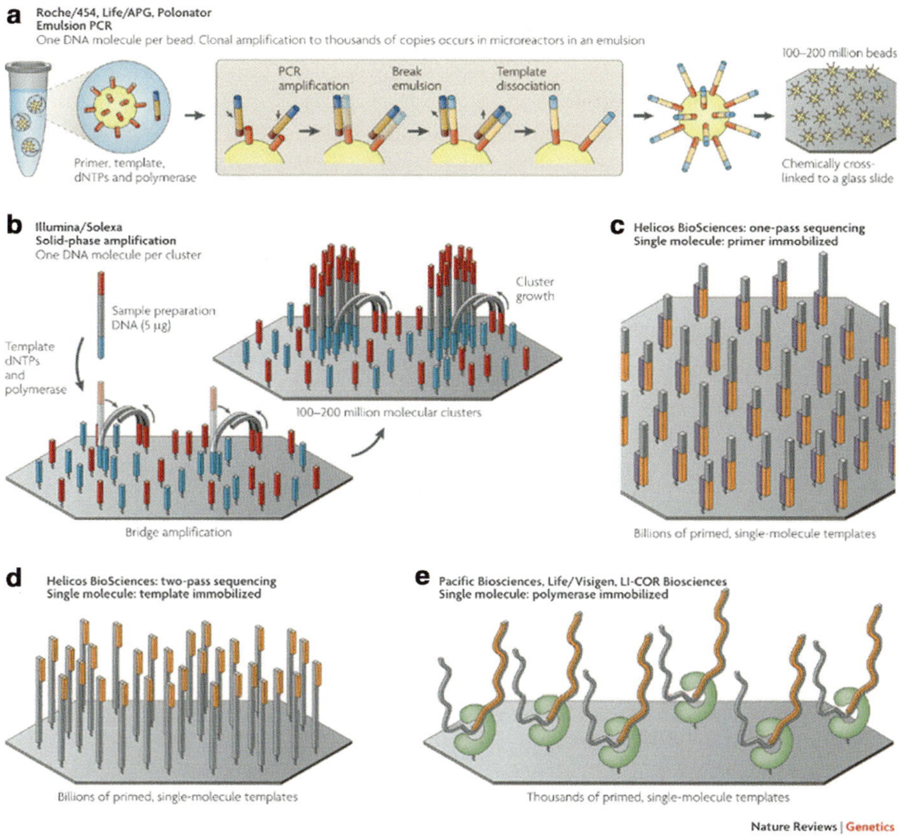

Fig. 1.1 Differences in the amplification between NGS and TGS technologies. (**a**) Emulsion PCR is employed by Roche 454 and Life Technologies SOLiD sequencing technologies. In emulsion PCR, oil–aqueous emulsion droplets are created to encapsulate beads attaching to DNA templates. Each droplet (acts as a PCR microreactor) encapsulates one bead where the PCR amplification is performed to generate clonally amplified copies from the same DNA template. After emulsion PCR, the beads can be deposited onto a glass slide (Life Technologies SOLiD) or into PicoTiterPlate wells (Roche 454 sequencing technologies). (**b**) Bridge amplification on solid surface is employed by the Illumina sequencing technologies, where single-stranded DNA fragments are first attached to a solid surface known as a flowcell using adaptors with complementary probes on the flowcell. The other unattached end of the DNA fragments then creates a "bridge-like structure" by bending over and hybridizing to the probes on the flowcell, which form the template for amplification to generate clusters comprising clonally amplified copies of the DNA fragments on the surface of the flowcell. (**c**) In single DNA molecule sequencing, single-molecule templates are immobilized on solid surface. Individual primer molecules which are spatially distributed are covalently attached to the solid surface. The DNA template (with common adaptor added) is then hybridized to the immobilized primer. (**d**) In this approach, spatially distributed single-molecule templates are covalently attached to the solid support by priming and extending single-stranded, single-molecule templates from immobilized primers. A common primer is then hybridized to the template. (**e**) In SMRT, single polymerase molecules are attached to the bottom of ZMWs where DNA templates are bound. Adapted by permission from Macmillan Publishers Ltd: Nature Reviews Genetics (Nat Rev Genet 11:31–46), copyright (2010)

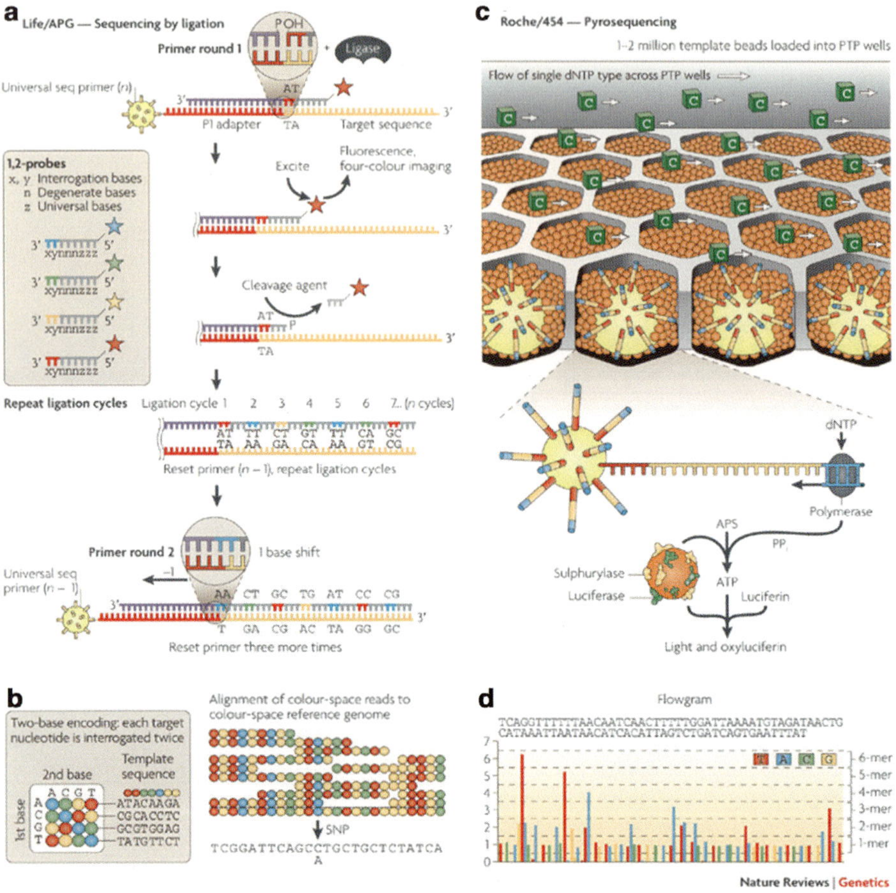

Fig. 1.2 Sequencing chemistry and principle of Life Technologies SOLiD and Roche 454 pyrosequencing. (**a**, **b**) Life Technologies SOLiD employed sequencing by synthesis mediated by ligase enzymes using 16 different dinucleotide probes labeled by four different colors. Primers are hybridized to the P1 adapter sequence on the template and the differently labeled dinucleotide probes compete for ligation to the sequencing primer. Ligation of one probe will interrogate two nucleotides. Following a series of ligation cycles, the synthesizing strand is removed and the template is reset with a primer complementary to the $n-1$ position for a second round of ligation cycles, and five rounds of primer reset are completed for each sequence read. Through the primer reset process, every nucleotide is interrogated in two independent ligation reactions by two different primers. (**c**, **d**) Roche 454 sequencing technologies employed pyrosequencing (sequencing by synthesis mediated by polymerase enzymes). Following loading of the DNA-amplified beads into individual wells on the plate, then a single type of dNTP and reagents are supplied to the plate across the wells. The dNTPs are incorporated into synthesizing strand triggering pyrosequencing reactions, and the emission of chemiluminescent light is detected. The cycles of sequencing repeat with another type of dNTP flowing to the wells sequentially. Adapted by permission from Macmillan Publishers Ltd: Nature Reviews Genetics (Nat Rev Genet 11:31–46), copyright (2010)

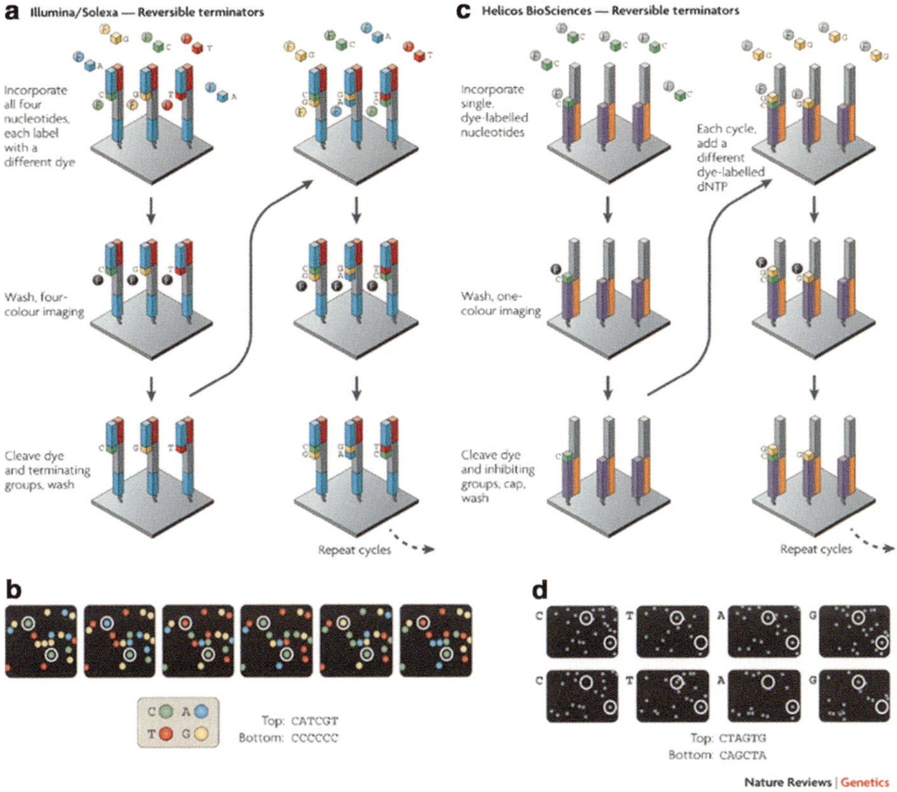

Fig. 1.3 Sequencing chemistry and principle of Illumina reversible terminator sequencing and Helicos tSMS. (**a**, **b**) Illumina reversible terminator sequencing employed sequencing by synthesis mediated by polymerase enzymes using four different reversible nucleotides labeled by four different fluorescent colors. The nucleotides are reversible terminators, allowing for the synthesis of complementary strand to the DNA template in the next cycle of sequencing for the incorporation of other nucleotides. In this reversible terminator sequencing approach, one complementary terminator nucleotide will be incorporated into the synthesis of complementary strand at any one time, followed by washing steps to remove the excess nucleotides and reagents. The imaging of the fluorescence signals is then followed across the whole flowcell. After imaging, the fluorescent moieties are removed and the 3′ blocking group of the reversible terminator nucleotide is also cleaved. These steps are then repeated. (**c**, **d**) In contrast to the Illumina reversible terminators, the Helicos Virtual Terminators are labeled by only one fluorescent dye; as such, the addition of a labeled nucleotide and reagents is controlled, where only one type of nucleotide is present per cycle of sequencing followed by other types of nucleotide consecutively in the next cycles. The sequencing process is also based on the "wash-and-scan" steps in which a labeled nucleotide and reagents flow onto the system, to allow nucleotide incorporation into the synthesizing complementary strand; the excess reagents are then washed and scanned to detect the fluorescent label. The dye–nucleotide linker is then cleaved to release the dye, and this process is repeated. Adapted by permission from Macmillan Publishers Ltd: Nature Reviews Genetics (Nat Rev Genet 11:31–46), copyright (2010)

perspectives of cost and time effectiveness, amplification also has disadvantages in terms of the reagents and time required for sequencing library generation.

The sequencing approaches adopted by NGS technologies can be broadly divided into sequencing by synthesis mediated by DNA polymerase enzymes to incorporate nucleotides (i.e., pyrosequencing for Roche 454 GS FLX and 454 GS Junior (*see* Fig. 1.2c, d) and sequencing by reversible terminator chemistry for Illumina GA/HiSeq/MiSeq (*see* Fig. 1.3a, b)) or sequencing mediated by DNA ligase enzymes through the ligation of dinucleotide probes to synthesize the complementary strand to the DNA fragment (*see* Fig. 1.2a, b, employed by Life Technologies SOLiD, but not by the Ion Torrent sequencing platform). The sequencing chemistry and approach of Ion Torrent is discussed in the following section on bench-top sequencing technologies. However, emulsion PCR amplification is common to both the SOLiD and Ion Torrent sequencing platforms.

In pyrosequencing illustrated in Fig. 1.2a, b, the addition of dNTPs and reagents for repeated cycles of sequencing is controlled. Each type of dNTP flows through the PicoTiterPlate (containing wells where the beads bound to clonally amplified DNA fragments are situated, one bead per well) one at a time, sequentially per cycle of sequencing, followed by a different nucleotide in the next cycle, and so on. This also means that only one type of dNTP is present per cycle of sequencing, followed by another type of dNTP in the next cycle. This is different from the reversible terminator chemistry sequencing for Illumina sequencing technologies where all four types of reversible terminator nucleotides labeled by different fluorescent colors are present in each cycle of sequencing. The incorporation of the complementary nucleotides into the synthesizing strand of the DNA template results in the release of inorganic pyrophosphate, thereby triggering a series of downstream chemiluminescent reactions. The intensity of the chemiluminescent light emitted from each well, corresponding to a single individual DNA template/fragment, is recorded by the detection system and is proportional to the number of nucleotides incorporated. Thus, during pyrosequencing, several nucleotides can be incorporated per cycle of sequencing when there are consecutive identical nucleotides in the sequences (i.e., homopolymer nucleotides). This is in contrast to sequencing involving reversible terminator chemistry where only one reversible terminator nucleotide is incorporated in the synthesis of complementary strand per cycle of sequencing. The intensity of chemiluminescence is proportional to the amount of inorganic pyrophosphate released and hence the number of dNTPs incorporated to the synthesizing complementary strand. For this reason, pyrosequencing is more susceptible to insertion/deletion (indel) errors in homopolymeric sequences longer than six bases. In summary, each cycle of pyrosequencing comprises dNTP incorporation, pyrosequencing reactions, emission of chemiluminescent light, and measurement of the light intensity. The sequencing reagents of the previous cycle are washed away before the next cycle of sequencing takes place [1, 2, 61–63].

Similarly, the Illumina sequencing technologies also employ sequencing by synthesis mediated by DNA polymerase (*see* Fig. 1.3a, b), but it is totally different from pyrosequencing. In reversible terminator chemistry sequencing, all four types of reversible terminator nucleotides and sequencing reagents are added onto the flowcell, and these nucleotides are labeled by four different fluorescent colors. One flowcell has several hundred million clusters, each of which contains clonally amplified copies derived from a single DNA fragment. Similar to the PicoTiterPlate, the format of the flowcell allows simultaneous sequencing of an enormous number of DNA fragments. The nucleotides are reversible terminators (where the blocking group of the ddNTPs can be cleaved off), allowing for the synthesis of complementary strand to the DNA template in the next cycle of sequencing for the incorporation of other nucleotides. In this reversible terminator sequencing approach, one complementary terminator nucleotide will be incorporated into the synthesis of complementary strand at any one time, followed by washing steps to remove the excess nucleotides and reagents (*see* Fig. 1.3a). The imaging of the fluorescence signals is then followed across the whole flowcell (Fig. 1.3b). After imaging, the fluorescent moieties are removed and the 3′ blocking group of the reversible terminator nucleotide is also cleaved. These steps are then repeated (Fig. 1.3a). Since only one nucleotide is incorporated at any one time and since the base calling is not proportional to light intensity but is rather dependent on the fluorescent colors, the reversible terminator chemistry does not have problems in sequencing homopolymeric sequences. However, it is more prone to substitution errors [1, 2, 61–63].

The sequencing-by-synthesis approach adopted by Life Technologies SOLiD is mediated by DNA ligase (Fig. 1.2a, b). As with Roche 454, Life Technologies SOLiD employs emulsion PCR for amplification. The beads containing clonally amplified copies of DNA fragments are then deposited on a glass slide. In brief, the sequencing is based on the sequential ligation of dinucleotide probes, which are labeled by four different fluorescent colors. There are 16 possible combinations of two nucleotides, and these dinucleotide probes will compete for incorporation into the synthesis of complementary strands. As such, ligation of one probe will interrogate two nucleotides in the DNA templates. Using this proprietary sequencing approach, every single position or base in the DNA template is interrogated twice, thereby allowing for distinction between true genetic variations and sequencing errors [1, 2, 61–63].

2.2.3 Sequence Read Length and Error Rate

The NGS technologies have several advantages over Sanger sequencing, but they are not without their limitations. These sequencing technologies are characterized by shorter sequence read lengths as compared to Sanger sequencing, i.e., 150 bases or less for Illumina and Life Technologies sequencing technologies. As a result, these NGS technologies are less suitable for the de novo sequencing of large and complex genomes such as the human genome because the assembly of billions of short sequence reads into large contigs would be most challenging [66, 67]. However, improvements are constantly being made to increase the read length. Longer-sequence read lengths are needed to obtain larger contigs with fewer gaps in between in the assembled consensus sequence. By contrast, the latest improvements in sequencing chemistry have enabled Roche 454 GS FLX to achieve sequence read lengths up to 1 kb, which is comparable to Sanger sequencing (http://454.com/products/gs-flx-system/).

In addition to the limitation in sequence read length, the NGS technologies have also been reported to have higher raw base or base-calling error rates. Life Technologies SOLiD achieved the highest accuracy with a <0.1 % error rate among the NGS technologies, whereas the read base error rates for Illumina and Roche sequencing technologies are 0.2–2 % and 0.5–1.5 %, respectively [68]. However, the error rate has been improved with newer versions of instruments and calling algorithms of the NGS technologies. Although the NGS technologies have higher raw base-calling error rates, this can be remedied by their high-throughput production to increase sequencing depth or coverage to obtain higher consensus accuracy. Whole-genome sequencing studies of human genome have shown that an average sequencing depth of 30-fold is sufficient for the accurate detection of most germ line SNVs [57].

2.3 Bench-Top NGS Platforms

As discussed, owing to their ultrahigh throughput, conventional NGS technologies might be unsuitable (even with barcoding to pool and multiplex the samples) for studies such as the targeted sequencing of panels of genes. The targeted gene sequencing approach has been widely assessed for clinical applications and has been increasingly shown to be a promising diagnostic tool [34]. Thus, conventional NGS technologies might not offer the platform of choice for the clinical diagnostic laboratory. On the other hand, the three bench-top NGS instruments have throughputs ranging from 10 Mb to >1 Gb. For example, three different Ion Torrent sequencing chips are available with minimum throughputs of >10 Mb (chip314), >100 Mb (chip316), and >1 Gb (chip318) per chip, respectively, and with a read length ranging from 100 to 200 bp. A longer read length of up to 400 bases is expected in the near future. Similarly, Illumina MiSeq is expected to generate sequencing data ranging from >120 Mb to >1 Gb depending upon the read length (which ranges from 35 to 150 bases) and whether it is a single-end or paired-end sequencing. By contrast, the Roche 454 GS Junior has a much lower throughput (>35 Mb) per instrument run but has a longer read length of 400 bases on average compared to the other two compatible platforms [41]. These varying throughputs provide multiple options to clinical diagnostic laboratories depending upon their sequencing turnaround time or sample volumes, e.g., an individual hospital-based versus a centralized, state-level, or national-level diagnostic laboratory. The availability of different throughputs from Ion Torrent and MiSeq (>10 Mb to >1 Gb) coupled with sample barcoding offers flexibility for different diagnostic or screening tests that vary with respect to both genomic size and sample volumes. Thus, the major application of these bench-top sequencing machines is intended for targeted sequencing of candidate genes, as, for example, in a diagnostic test. Indeed, none are appropriate for other applications, e.g., human omics studies such as ChIP-Seq and RNA-Seq (requiring up to hundred million reads) owing to the paucity of sequence reads. They are also unsuitable for WES and WGS. However, the introduction of second-generation Ion Torrent sequencing technology, known as Ion Proton, has enabled WES and WGS to be performed efficiently on bench-top machines for small sample sizes (see "Semiconductor Sequencing" in http://www.invitrogen.com/site/us/en/home/Products-and-Services/Applications/Sequencing.html). As with Ion Torrent, Ion Proton Sequencer is a semiconductor sequencing technology (discussed below) and, coupled with two different chips, can generate sequencing data sufficient for two human exomes (Proton I chip) and one human genome (Proton II chip). The higher throughput of Ion Proton is because of the number of wells in the Ion Proton chips which is up to 1,000-fold higher than the chips for Ion Torrent.

The Roche 454 GS Junior and Illumina MiSeq are based on well-established sequencing chemistries, i.e., pyrosequencing and reversible terminator sequencing, respectively (as discussed earlier). By contrast, Ion Torrent and Ion Proton sequencing represent new technology. This is considered to be the world's first "post-light sequencing technology" because this technology does not rely on light emission during cycle sequencing of nucleotide incorporation. Other sequencing technologies are reliant on either fluorescent emission (Illumina GA/HiSeq/MiSeq and Life Technologies SOLiD) or chemiluminescent light emission (i.e., pyrosequencing chemistry used by the Roche 454 sequencing platforms). Similarly, the two TGS technologies already on the market (Helicos BioSciences and Pacific Biosciences) are also reliant on fluorescent emission to detect nucleotide incorporation.

The Ion Torrent and Ion Proton sequencing platforms comprise a sequencer and different semiconductor sequencing chips (a high-density array of wells or micro-machined wells) that perform the sequencing process in a massively parallel way [47]. Figure 1.4 illustrates the sequencing chemistry and approach of the semiconductor sequencing technology (Ion Torrent and Ion Proton). Each well contains a different DNA template (of clonally amplified copies) and beneath the wells is an ion-sensitive layer overlaying a proprietary ion sensor for each well. The Ion Torrent and Ion Proton sequencer sequentially provide the chip with one type of nucleotide after another. When DNA polymerase incorporates a nucleotide into the synthesizing complementary strand of a DNA template, a hydrogen ion is released which causes a voltage change. If the next nucleotide is not complementary, no voltage change will be recorded and no base will be called. In concordance, if there are two identical bases on the DNA template, the voltage will be doubled, and the chip will record two identical bases as called. As such, this sequencing technology does not require light, scanning, and cameras to detect the nucleotide incorporation, thereby simplifying the overall sequencing process and dramatically accelerating the sequencing time.

The number of sequence reads generated by Ion Torrent and Ion Proton depends on the number of wells per chip and the proportion of wells loaded with beads attached to DNA fragments. For example, the number of wells are >1 million (chip314), >6 million (chip316), and >11 million (chip318) for Ion Torrent, and the loading protocols usually fill approximately 30–40 % of the available wells. Thus, the number of reads produced by chip314 is between 0.3 and 0.4 million. As such,

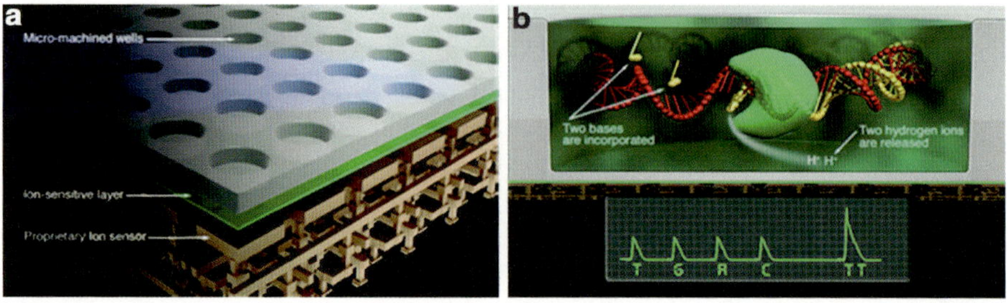

Fig. 1.4 Sequencing chemistry and principle of Ion Torrent/Proton semiconductor sequencing. The Ion Torrent and Ion Proton sequencing platforms comprise a sequencer and different semiconductor sequencing chips (a high-density array of wells or micromachined wells). Each well contains a different DNA template (of clonally amplified copies) and beneath the wells is an ion-sensitive layer overlaying a proprietary ion sensor for each well. The Ion Torrent and Ion Proton sequencer sequentially provide the chip with one type of nucleotide after another. When DNA polymerase incorporates a nucleotide into the synthesizing complementary strand of a DNA template, a hydrogen ion is released which causes a voltage change. If the next nucleotide is not complementary, no voltage change will be recorded and no base will be called. In concordance, if there are two identical bases on the DNA template, the voltage will be doubled, and the chip will record two identical bases as called. Reprinted (adapted) with permission from (Anal Chem 83:4327–4341). Copyright (2011) American Chemical Society

the minimum throughput is estimated to be varied from >10 Mb (chip314) to >100 Mb (chip316) and >1 Gb (chip318). By contrast, the Ion Proton I chip and Ion Proton II chip have up to 1,000-fold higher densities containing 165 million wells and 660 million wells, respectively. As a result, the Ion Proton I chip generates up to 10 Gb data and up to 200 bases read length. On the other hand, it is expected that the Ion Proton II chip, when available, will enable sample-to-variant analysis of an entire human genome in a single day, at up to 20-fold coverage (see "Semiconductor Sequencing" in http://www.invitrogen.com/site/us/en/home/Products-and-Services/Applications/Sequencing.html). This will then represent the fastest ever sequencing of a whole human genome.

Currently, the Ion Torrent platform achieves the fastest sequencing time per chip or per instrument run (i.e., <2 h) compared to the other two competing platforms. The shortest sequencing time is achieved because nucleotide incorporation is directly detected through pH and voltage changes; as such, each nucleotide incorporation event is recorded within seconds [47]. However, this sequencing time excludes the time needed for library preparation. Indeed, the time to diagnosis is likely to be the key factor to consider in a diagnostic context, especially when the accurate and timely molecular diagnosis of a disorder can lead to dramatic improvements in patient treatment and/or care [69].

Several features of Ion Torrent and Ion Proton sequencing chemistry resemble pyrosequencing, i.e., (a) the amplification of adapter-ligated DNA fragments is based on emulsion PCR, (b) only one type of unlabeled nucleotide is present per cycle of sequencing, and (c) the number of nucleotide incorporations is proportional to the voltage change (Ion Torrent and Ion Proton sequencing) or intensity of chemiluminescent light (pyrosequencing). The Ion Torrent and Ion Proton sequencing and 454 GS Junior are based on emulsion PCR; they therefore have a greater level of technical difficulty in sequencing library preparation than MiSeq employed on the cluster generation. This is because DNA loading during the emulsion PCR, the size of the emulsion PCR product, and the uniformity of DNA loading onto the chip or plate have to be optimized to obtain the maximal sequencing output [41]. Thus, for example, the DNA and bead ratio have to be optimized in order to maximize the generation of monoclonal particles/beads. By contrast, polyclonal particles are wasteful, i.e., clonal amplification from more than one DNA template attached to the same single bead during emulsion PCR. Further, the sequencing throughput for Ion Torrent, Ion Proton, and 454 GS Junior is also dependent on the DNA loading (after emulsion PCR) into the wells on the chip or plate. Higher throughput would be expected if the DNA loading step is optimized to increase the proportion of the wells occupied by beads. Compared with other platforms, the strength of the Roche 454 system lies in its longer-sequence reads.

In summary, the sequencing chemistry of all the high-throughput and bench-top NGS technologies comprises successive washing and scanning steps. The wash-and-scan sequencing steps (per cycle) involve the sequential inflow of reagents, such as fluorescent-labeled or unlabeled nucleotides, the incorporation of nucleotides into the synthesizing DNA strands, the stopping of the incorporation reaction, washing out the excess reagent, scanning the light emission to identify the incorporated nucleotides or detecting the pH/voltage changes (for Ion Torrent and Ion Proton), and finally treating the newly incorporated nucleotides (e.g., by removing the fluorescent label and blocking group for Illumina reversible terminator sequencing), to prepare the templates for the next cycle. This is also applicable in single-molecule sequencing, i.e., tSMS (Helicos BioSciences), but not the SMRT (Pacific Biosciences).

3 Single-Molecule Sequencing Technologies

Despite the recent and rapid advances in the omics studies enabled by NGS technologies, a new generation of single (DNA)-molecule sequencing technologies is emerging [3, 46, 70, 71]. In single-molecule sequencing, DNA amplification is bypassed, thereby overcoming issues related to errors and biases introduced by PCR amplification and dephasing. Single-molecule sequencing differs from the NGS technologies that rely on amplification such as bridge amplification to generate clusters (of clonal copies) of a given DNA template, attaching the clusters of DNA templates to a solid surface (flowcell) that is then imaged as the clusters are sequenced by synthesis using a "phased" or "synchronized" approach. As such, dephasing can happen as a result of uneven sequencing of the clonal copies of DNA templates within the clusters, giving rise to sequencing errors especially towards the end of the sequence read. The dephasing causes an increase in noise and sequencing errors as the read extends; it also imposes a constraint on the read length that can be achieved, e.g., the short read lengths (<150 bp) that characterize the Illumina sequencing technologies.

By contrast, TGS technologies (discussed in detail below) possess several important advantages over current NGS technologies, i.e., (a) higher throughput (Helicos BioSciences tSMS), (b) faster turnaround time (real-time sequencing), (c) longer read lengths to enhance de novo assembly and enable direct detection of haplotypes and even whole-chromosome phasing (Pacific Biosciences SMRT), (d) higher consensus accuracy to enable rare variant detection (as a result of high-throughput data), and (e) small amounts of starting material (theoretically only a single molecule may be required for sequencing), and this is particularly important when the clinical material is limited and could be critical for the emerging single cell sequencing methods.

3.1 Helicos BioSciences: True Single-Molecule Sequencing (tSMS)

Helicos BioSciences—true single-molecule sequencing (tSMS)—is the first commercially available sequencing instrument designed to carry out single-molecule sequencing [44, 45]. The Helicos sequencing instrument works by imaging individual DNA molecules (without amplification) attached to a planar surface as they are extended using a defined primer and a modified polymerase as well as proprietary fluorescently labeled nucleotides known as Virtual Terminator nucleotides. The dye is attached to the nucleotide via a chemically cleavable group that allows subsequent steps of sequencing to be performed. Figure 1.3c, d illustrates the tSMS by Helicos BioSciences. More specifically, sample preparation for this tSMS technology involves fragmenting genomic DNA into smaller pieces, adding a 3′ poly(A) tail to the fragments, labeling and blocking by terminal transferase. These templates are then captured onto a surface with covalently bound 5′ dT(50) oligonucleotides through hybridization. The surface is then imaged using CCD sensors, where those templates that have been appropriately captured are identified and then tracked for nucleotide incorporation in cycle sequencing. As with the NGS technologies, the sequencing process is also based on the “wash-and-scan” steps in which a labeled nucleotide and reagents flows onto the system, to allow nucleotide incorporation into the synthesizing complementary strand; the excess reagents are then washed and scanned to detect the fluorescent label. The dye–nucleotide linker is then cleaved to release the dye, and this process is repeated. The controlled addition of a labeled nucleotide and reagents is similar to the Roche 454 pyrosequencing and Ion Torrent/Proton sequencing chemistries, where only one type of nucleotide is present per cycle of sequencing followed by other types of nucleotide consecutively in the next cycles. On the other hand, the Virtual Terminator nucleotides work similarly to Illumina reversible terminator nucleotides, although the former, i.e., Virtual Terminator, nucleotides are only labeled with one fluorescent dye.

Because cycle sequencing (i.e., wash and scan per cycle) is implemented (in a similar way to NGS technologies), the time to sequence a single nucleotide is relatively high, and the read lengths achieved are relatively short (<50 bases). In addition, owing to the nature of single-molecule sequencing, the raw read error rates are much higher than with NGS technologies, i.e., ~5 %. However, this can be remedied by high-throughput production (as it can generate up to one billion short sequence reads) to achieve high fold coverage and a consensus or finished read accuracy of >99 %. Although tSMS is rather newer than NGS technologies, it has now been applied in many studies such as the sequencing of an entire human genome [72], ChIP-Seq [73] and Methyl-Seq [74]. Unlike NGS, the massively parallel sequencing of the huge numbers of sequencing reactions can be carried out asynchronously (i.e., avoiding the issue of dephasing). In addition to the relatively high raw base-calling error rate, deletion and insertion errors can be a significant issue [68]. The deletion and insertion errors could be due to incorporation of unlabeled nucleotides or the signal is too weak to be detected.

In addition to DNA sequencing using DNA polymerase, a further advantage of tSMS technology compared to NGS technologies is that it can perform direct RNA sequencing, by the simple expedient of replacing DNA polymerase by a reverse transcriptase to allow RNA to be sequenced directly [75]. This obviates the need to convert RNA to cDNA or to incorporate ligation/amplification steps. In direct RNA sequencing, each RNA molecule is polyadenylated and 3′-blocked and captured on a surface coated with dT(50) oligonucleotides, similar to the DNA sequencing process. Sequencing is then carried out as described for DNA, but using reverse transcriptase instead of DNA polymerase.

In summary, although Helicos tSMS technology has been successfully applied in many omics studies, representing the first example of true single-molecule sequencing, it still has many of the characteristics of NGS technologies in terms of the sequencing approach and chemistry. While the Helicos technology may struggle to clearly differentiate itself from NGS in some respects, the tSMS and direct RNA-sequencing application constitute the technological advances that have clearly distinguished this technology from NGS.

3.2 Pacific Biosciences: Single-Molecule Real-Time (SMRT) Sequencing

The SMRT sequencing developed by Pacific Biosciences is the first sequencing technology which can directly detect a single molecule of DNA polymerase as it synthesizes the complementary strand of DNA templates in real time [43]. The real-time and continuous (without halting for washing or scanning) detection of incorporation of a single nucleotide taken from a large pool of fluorescent labeled nucleotides during strand synthesis is achieved using zero-mode waveguide (ZMW) technology. The tiny size of the ZMW (deposited on a glass substrate) prevents visible laser

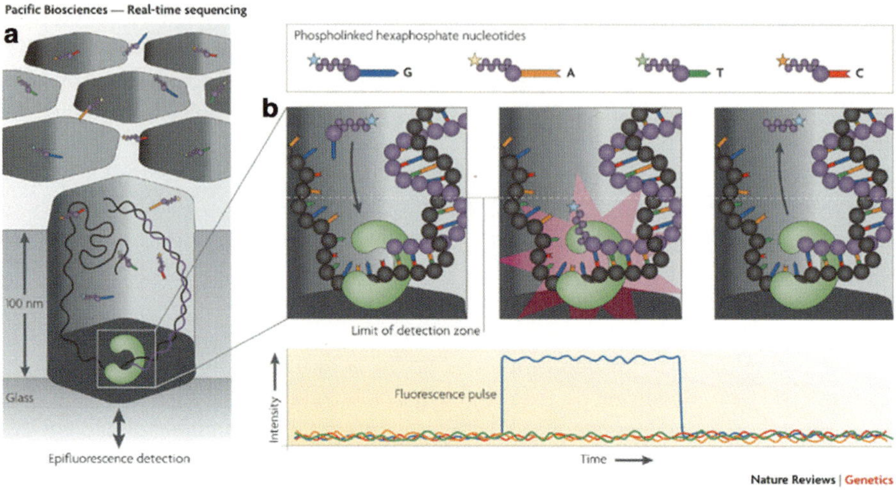

Fig. 1.5 Sequencing chemistry and principle of Pacific Biosciences SMRT. The real-time and continuous (without halting for washing or scanning) detection of incorporation of a single nucleotide during complementary strand synthesis is achieved using ZMW technology. The tiny size of the ZMW (deposited on a glass substrate) prevents visible laser light, from passing entirely through the ZMW. As a result, only the bottom 30 nm of the ZMW becomes illuminated. Within each ZMW, a single DNA polymerase molecule is attached to the bottom glass surface. Four nucleotides, each type labeled with a differently colored fluorophore, are then flooded above an array of ZMWs. When the correct nucleotide is detected by the polymerase, it is incorporated into the synthesizing DNA strand. While held by the polymerase, the fluorescent label of the newly incorporated nucleotide emits fluorescent light and is detected by the sequencing instrument, which corresponds to the nucleotide identity. In SMRT sequencing, the fluorescent dye is attached to the phosphate chain of the nucleotide. As a natural step in the synthesis process, the phosphate chain is cleaved when the nucleotide is incorporated into the DNA strand. Thus, the DNA polymerase naturally frees the dye molecule from the nucleotide when it cleaves the phosphate chain. Adapted by permission from Macmillan Publishers Ltd: Nature Reviews Genetics (Nat Rev Genet 11:31–46), copyright (2010)

light, from passing entirely through the ZMW (i.e., the light decays exponentially as it enters the ZMW). As a result, by shining laser illumination through the glass into the ZMW, only the bottom 30 nm of the ZMW becomes illuminated. This is critical to avoid excitation of the other labeled nucleotides on top of the ZMW in order to minimize the noise. Only the incorporated nucleotide at the bottom of ZMW is excited (Fig. 1.5).

Within each ZMW, a single DNA polymerase molecule is attached to the bottom glass surface. Four nucleotides, each type labeled with a differently colored fluorophore, are then flooded above an array of ZMWs. When the correct nucleotide is detected by the polymerase, it is incorporated into the synthesizing DNA strand. While held by the polymerase, the fluorescent label of the newly incorporated nucleotide emits fluorescent light and is detected by the sequencing instrument, which corresponds to the nucleotide identity. As no laser light penetrates through the ZMWs to excite the fluorescent labels, the labeled nucleotides above the ZMWs do not contribute to the measured signals. Following incorporation, the signal immediately returns to baseline and the process is repeated, with the DNA polymerase continuing to incorporate multiple bases per second. In SMRT sequencing, the fluorescent dye is attached to the phosphate chain of the nucleotide rather than to the base. As a natural step in the synthesis process, the phosphate chain is cleaved when the nucleotide is incorporated into the DNA strand. Thus, upon incorporation of a phospholinked nucleotide, the DNA polymerase naturally frees the dye molecule from the nucleotide when it cleaves the phosphate chain.

Thus, this technology has the ability to detect a single-nucleotide incorporation event against the background of fluorescently labeled nucleotides. The first commercial available SMRT-sequencing instrument consists of an array of approximately 75,000 ZMWs. Each ZMW is capable of containing a DNA polymerase loaded with a different strand of DNA sample. As a result, the array enables the potential detection of approximately 75,000 single-molecule sequencing reactions in parallel. At present, because the DNA polymerase and DNA template to be sequenced are delivered to ZMWs through a random diffusion process, only about a third of the ZMWs in a given array are active for sequencing reactions.

The SMRT-sequencing platform has several important advantages compared to other sequencing technologies in that it requires minimal amounts of reagent and sample preparation to perform sequencing (no PCR amplification is needed) and lacks the time-consuming scanning and washing steps (as the detection of nucleotide incorporation is continuous), resulting in a shorter sequencing time. In SMRT sequencing, the sample preparation consists of fragmenting the DNA into the desired length, blunting the ends, ligating hairpin adaptors, and then sequencing.

The SMRT sequencing achieved longer read lengths than other current technologies, i.e., average read lengths of >1 kb and maximum read lengths greater than 10 kb. In terms of long read length, SMRT is particularly suitable for (and enables) de novo assembly, direct detection of haplotypes (and even provides for the possibility of phasing entire chromosomes if the read length of tens of kilobases is achieved). However, for de novo assembly of human genome, the throughput per run (75,000 ZMWs × 1 kb) for SMRT needs to be further improved. Another interesting feature of SMRT sequencing is the ability to observe and capture kinetic information. The ability to observe the activity of DNA polymerase in real time allows the collection, measurement, and assessment of the dynamics and timing of enzymatic incorporation, referred to as kinetics. As a result, through the SMRT-sequencing process, changes in the kinetics of incorporation associated with the chemical modifications of bases, such as cytosine methylation (i.e., 5-methylcytosine), can be detected in the "normal" sequencing run. In addition, it has also been found to be possible to distinguish 5-methylcytosine from 5-hydroxymethylcytosine (an alternative form of methylation that appears to play critical roles in biological processes) [76, 77]. Despite the many potential advantages of SMRT sequencing, a number of challenges remain. Similar to the Helicos tSMS technology, the raw read error rate is high >5 %, with the errors being dominated by insertions and deletions [78]; this is particularly problematic when aligning sequences and assembling genomes. In addition, the throughput of SMRT sequencing at the current stage is substantially lower than what can be achieved by NGS and Helicos tSMS.

4 Other Sequencing Technologies

Unlike other sequencing technologies on the market (as discussed earlier), the Complete Genomics sequencing platform is not commercially available to end users [79], but providing sequencing service. This sequencing technology is based on a combinatorial approach of probe–anchor hybridization and ligation sequencing. The method uses rolling circle amplification of small DNA sequences into so-called nanoballs. Unchained sequencing by ligation is then used to determine the nucleotide sequence. This method permits large numbers of DNA nanoballs to be sequenced per run and at low consumable cost. The platform has been successfully used in whole-genome sequencing of individuals.

Although single-molecule sequencing such as SMRT is characterized by real-time sequencing, sequencing by synthesis is taking place where complementary strand is synthesized to detect nucleotide incorporation. By contrast, other sequencing technologies are characterized by lacking sequencing by synthesis, for example, nanopore-sequencing technologies in which single molecules of DNA are threaded through a nanopore or positioned in the vicinity of a nanopore and individual bases are detected as they pass through the nanopore [48]. Most nanopore-sequencing technologies rely on the transit of a DNA molecule or its component bases through a hole and the detection of the bases by their eliciting of an electric current or optical signal. Because this type of technology uses single molecules of unmodified DNA, they have the potential to work rapidly on extremely small amounts of input material. Similarly, direct imaging of individual DNA molecules is also possible using advanced microscopy techniques. For example, Halcyon Molecular is developing a single-molecule sequencing approach using transmission electron microscopy to directly image and chemically detect atoms that would uniquely identify the nucleotides comprising a DNA template [3]. As such, no sequencing by synthesis is required; this can therefore be considered as fourth-generation sequencing technologies. However, unlike the tSMS- and SMRT-sequencing technologies which have already been marketed and applied in omics studies, other future sequencing technologies are still in various stages of development and hence might still require some years before they become a mature technology.

5 Summary

All the NGS technologies share some features, such as (a) reliance on in vitro amplification methods, e.g., emulsion PCR and bridge amplification; (b) wash-and-scan steps or cycle sequencing (i.e., nucleotide incorporation cannot be detected in real time except with SMRT-sequencing technology); (c) sequencing by synthesis, i.e., the need to synthesize the complementary strand to decode the nucleotide sequence of the DNA template (except with nanopore-sequencing technologies where the single molecules of DNA are just threaded through a nanopore without synthesis); and (d) reliance on light emission and detection to monitor the nucleotide incorporation (with the exception for Ion Torrent/Proton sequencing chemistry). In addition to their high throughputs, NGS technologies have a major advantage over Sanger sequencing in that individual DNA templates are sequenced. This feature is important in order to detect "rare" mutations, for example, in the case of heterogeneity where mutations that are present only in a subclone of cancer tissue. Therefore, NGS technologies are more sensitive and hence capable of detecting this type of mutation (as long as the sequencing depth is sufficient) compared to Sanger sequencing.

Although NGS technologies have made significant improvements compared to Sanger sequencing, they are not without their limitations. These limitations have been ameliorated by the newer generation technologies. For example, all the NGS technologies are reliant on the amplification steps which can introduce errors and bias, subsequently affecting the accuracy of variant detection, but this has been avoided by single-molecule sequencing (such as tSMS, SMRT, nanopore sequencing, and sequencing by advanced microscopic techniques). In addition, this has bypassed the problems caused by dephasing and increased the sequence read length. Similarly, the need for cycle sequencing (wash-and-scan steps per cycle) has been "avoided" by real-time sequencing. This has significantly shortened the sequencing time required since the "halting" in between cycles of sequencing (or nucleotide detection) can be avoided. Finally, the sequencing by synthesis through enzymatic reactions catalyzed by polymerase or ligase enzymes is eliminated by nanopore-sequencing technologies and sequencing by advanced microscope techniques.

Each of these technologies provides novel approaches to sequencing DNA/RNA and has advantages and disadvantages with respect to specific applications. For example, sequencing technologies that generate a huge number of short sequence reads are very useful in transcriptome sequencing to quantify transcript expression levels and in ChIP-Seq to map protein–DNA interactions. By contrast, sequencing technologies that generate long-sequence read lengths are more suitable for metagenomics studies in investigating bacterial diversity. It is crucial to have longer-sequence read lengths to achieve sufficient discriminatory power of the sequences derived from different bacterial species in a sample. Long-sequence read lengths are also useful for de novo genome assembly. On the other hand, medium throughput bench-top sequencing platforms could be more amenable to the routine clinical diagnostic laboratory situation, where sequencing of panels of genes in small sample sizes of patients is common.

In addition to advances in research, the application of NGS technologies as a diagnostic tool has also become increasingly commonplace. The utility of NGS technologies has been further enhanced by the arrival of bench-top sequencing platforms that make target sequencing of panels of genes technically very feasible and cost effective. More importantly, it has been shown to be a promising diagnostic tool. Substantial advances have now been made by NGS technologies, further advances in research and clinical applications would be anticipated with the developments in the TGS and other future sequencing technologies. For example, high-throughput production of long-sequence read lengths of several kilobases will enhance the de novo assembly of human genomes (rather than re-sequencing through mapping the reads to the reference genome). This is just the beginning of the revolution that high-throughput sequencing technologies are bringing to omics studies. Further, the successful application of NGS in a clinical setting is just the tip of the iceberg. High-throughput sequencing technologies hold out great promise towards personalized genomic medicine. The advances in high-throughput sequencing technologies also pose substantial challenges in bioinformatics and data analysis [80–85]. Finally, the Archon Genomics X PRIZE of $10 million will certainly bring the sequencing technologies to the next higher levels. This prize will be awarded to the first team that can build a whole human genome sequencing device and use it to sequence 100 human genomes within 30 days or less, with an accuracy of no more than one error in every 1,000,000 bases sequenced, with an accuracy rate of at least 98 % of the genome, and at a recurring cost of no more than $1,000 per genome [86, 87].

Financial and Competing Interests Disclosure None to declare.

References

1. Shendure J, Ji H. Next-generation DNA sequencing. Nat Biotechnol. 2008;26(10):1135–45. Epub 2008/10/11.
2. Metzker ML. Sequencing technologies—the next generation. Nat Rev Genet. 2010;11(1):31–46. Epub 2009/12/10.
3. Schadt EE, Turner S, Kasarskis A. A window into third-generation sequencing. Hum Mol Genet. 2010;19(R2):R227–40. Epub 2010/09/23.
4. Wang Z, Gerstein M, Snyder M. RNA-seq: a revolutionary tool for transcriptomics. Nat Rev Genet. 2009;10(1):57–63. Epub 2008/11/19.
5. Park PJ. ChIP-seq: advantages and challenges of a maturing technology. Nat Rev Genet. 2009;10(10):669–80. Epub 2009/09/09.
6. Hirst M, Marra MA. Next generation sequencing based approaches to epigenomics. Brief Funct Genomics. 2010;9(5–6):455–65. Epub 2011/01/27.
7. Hawkins RD, Hon GC, Ren B. Next-generation genomics: an integrative approach. Nat Rev Genet. 2010;11(7):476–86. Epub 2010/06/10.
8. Ku CS, Naidoo N, Wu M, Soong R. Studying the epigenome using next generation sequencing. J Med Genet. 2011;48(11):721–30. Epub 2011/08/10.

9. Naidoo N, Pawitan Y, Soong R, Cooper DN, Ku CS. Human genetics and genomics a decade after the release of the draft sequence of the human genome. Hum Genomics. 2011;5(6):577–622. Epub 2011/12/14.
10. Tran B, Dancey JE, Kamel-Reid S, McPherson JD, Bedard PL, Brown AM, et al. Cancer genomics: technology, discovery, and translation. J Clin Oncol. 2012;30(6):647–60. Epub 2012/01/25.
11. Dixon-Salazar TJ, Silhavy JL, Udpa N, Schroth J, Bielas S, Schaffer AE, et al. Exome sequencing can improve diagnosis and alter patient management. Science translational medicine. 2012;4(138):138ra78. Epub 2012/06/16.
12. Lyon GJ, Wang K. Identifying disease mutations in genomic medicine settings: current challenges and how to accelerate progress. Genome Med. 2012;4(7):58. Epub 2012/07/27.
13. Werner T. Next generation sequencing in functional genomics. Brief Bioinform. 2010;11(5):499–511. Epub 2010/05/27.
14. Hey Y, Pepper SD. Interesting times for microarray expression profiling. Brief Funct Genomic Proteomic. 2009;8(3):170–3. Epub 2009/05/29.
15. Hurd PJ, Nelson CJ. Advantages of next-generation sequencing versus the microarray in epigenetic research. Brief Funct Genomic Proteomic. 2009;8(3):174–83. Epub 2009/06/19.
16. Carter NP. Methods and strategies for analyzing copy number variation using DNA microarrays. Nat Genet. 2007;39 Suppl 7:S16–21. Epub 2007/09/05.
17. Alkan C, Coe BP, Eichler EE. Genome structural variation discovery and genotyping. Nat Rev Genet. 2011;12(5):363–76. Epub 2011/03/02.
18. Redon R, Ishikawa S, Fitch KR, Feuk L, Perry GH, Andrews TD, et al. Global variation in copy number in the human genome. Nature. 2006;444(7118):444–54. Epub 2006/11/24.
19. McCarroll SA, Kuruvilla FG, Korn JM, Cawley S, Nemesh J, Wysoker A, et al. Integrated detection and population-genetic analysis of SNPs and copy number variation. Nat Genet. 2008;40(10):1166–74. Epub 2008/09/09.
20. Ku CS, Pawitan Y, Sim X, Ong RT, Seielstad M, Lee EJ, et al. Genomic copy number variations in three Southeast Asian populations. Hum Mutat. 2010;31(7):851–7. Epub 2010/05/28.
21. Korbel JO, Urban AE, Affourtit JP, Godwin B, Grubert F, Simons JF, et al. Paired-end mapping reveals extensive structural variation in the human genome. Science. 2007;318(5849):420–6. Epub 2007/09/29.
22. Yoon S, Xuan Z, Makarov V, Ye K, Sebat J. Sensitive and accurate detection of copy number variants using read depth of coverage. Genome Res. 2009;19(9):1586–92. Epub 2009/08/07.
23. Medvedev P, Stanciu M, Brudno M. Computational methods for discovering structural variation with next-generation sequencing. Nat Methods. 2009;6 Suppl 11:S13–20. Epub 2009/11/03.
24. Xi R, Kim TM, Park PJ. Detecting structural variations in the human genome using next generation sequencing. Brief Funct Genomics. 2010;9(5–6):405–15. Epub 2011/01/11.
25. Bamshad MJ, Shendure JA, Valle D, Hamosh A, Lupski JR, Gibbs RA, et al. The Centers for Mendelian Genomics: a new large-scale initiative to identify the genes underlying rare Mendelian conditions. Am J Med Genet A. 2012;158A(7):1523–5. Epub 2012/05/26.
26. Mardis ER. A decade's perspective on DNA sequencing technology. Nature. 2011;470(7333): 198–203. Epub 2011/02/11.
27. Lander ES. Initial impact of the sequencing of the human genome. Nature. 2011;470(7333): 187–97. Epub 2011/02/11.
28. Meyerson M, Gabriel S, Getz G. Advances in understanding cancer genomes through second-generation sequencing. Nat Rev Genet. 2010;11(10):685–96. Epub 2010/09/18.
29. Roukos DH, Ku CS. Clinical cancer genome and precision medicine. Ann Surg Oncol. 2012;19(12):3646–50. Epub 2012/08/02.
30. Ku CS, Naidoo N, Pawitan Y. Revisiting Mendelian disorders through exome sequencing. Hum Genet. 2011;129(4):351–70. Epub 2011/02/19.
31. Bamshad MJ, Ng SB, Bigham AW, Tabor HK, Emond MJ, Nickerson DA, et al. Exome sequencing as a tool for Mendelian disease gene discovery. Nat Rev Genet. 2011;12(11): 745–55. Epub 2011/09/29.

32. Ku CS, Polychronakos C, Tan EK, Naidoo N, Pawitan Y, Roukos DH, et al. A new paradigm emerges from the study of de novo mutations in the context of neurodevelopmental disease. Mol Psychiatry. 2013;18(2):141–53. Epub 2012/05/30.
33. Veltman JA, Brunner HG. De novo mutations in human genetic disease. Nat Rev Genet. 2012;13(8):565–75. Epub 2012/07/19.
34. Ku CS, Cooper DN, Polychronakos C, Naidoo N, Wu M, Soong R. Exome sequencing: dual role as a discovery and diagnostic tool. Ann Neurol. 2012;71(1):5–14. Epub 2012/01/26.
35. Walsh T, Lee MK, Casadei S, Thornton AM, Stray SM, Pennil C, et al. Detection of inherited mutations for breast and ovarian cancer using genomic capture and massively parallel sequencing. Proc Natl Acad Sci U S A. 2010;107(28):12629–33. Epub 2010/07/10.
36. Pritchard CC, Smith C, Salipante SJ, Lee MK, Thornton AM, Nord AS, et al. ColoSeq provides comprehensive lynch and polyposis syndrome mutational analysis using massively parallel sequencing. J Mol Diagn. 2012;14(4):357–66. Epub 2012/06/05.
37. Mertes F, Elsharawy A, Sauer S, van Helvoort JM, van der Zaag PJ, Franke A, et al. Targeted enrichment of genomic DNA regions for next-generation sequencing. Brief Funct Genomics. 2011;10(6):374–86. Epub 2011/11/29.
38. Jones MA, Bhide S, Chin E, Ng BG, Rhodenizer D, Zhang VW, et al. Targeted polymerase chain reaction-based enrichment and next generation sequencing for diagnostic testing of congenital disorders of glycosylation. Genet Med. 2011;13(11):921–32. Epub 2011/08/04.
39. Berg JS, Evans JP, Leigh MW, Omran H, Bizon C, Mane K, et al. Next generation massively parallel sequencing of targeted exomes to identify genetic mutations in primary ciliary dyskinesia: implications for application to clinical testing. Genet Med. 2011;13(3):218–29. Epub 2011/01/29.
40. Hoischen A, Gilissen C, Arts P, Wieskamp N, van der Vliet W, Vermeer S, et al. Massively parallel sequencing of ataxia genes after array-based enrichment. Hum Mutat. 2010;31(4):494–9. Epub 2010/02/13.
41. Ku CS, Wu M, Cooper DN, Naidoo N, Pawitan Y, Pang B, et al. Technological advances in DNA sequence enrichment and sequencing for germline genetic diagnosis. Expert Rev Mol Diagn. 2012;12(2):159–73. Epub 2012/03/01.
42. Loman NJ, Misra RV, Dallman TJ, Constantinidou C, Gharbia SE, Wain J, et al. Performance comparison of benchtop high-throughput sequencing platforms. Nat Biotechnol. 2012;30(5):434–9. Epub 2012/04/24.
43. Eid J, Fehr A, Gray J, Luong K, Lyle J, Otto G, et al. Real-time DNA sequencing from single polymerase molecules. Science. 2009;323(5910):133–8. Epub 2008/11/22.
44. Harris TD, Buzby PR, Babcock H, Beer E, Bowers J, Braslavsky I, et al. Single-molecule DNA sequencing of a viral genome. Science. 2008;320(5872):106–9. Epub 2008/04/05.
45. Bowers J, Mitchell J, Beer E, Buzby PR, Causey M, Efcavitch JW, et al. Virtual terminator nucleotides for next-generation DNA sequencing. Nat Methods. 2009;6(8):593–5. Epub 2009/07/22.
46. Thompson JF, Milos PM. The properties and applications of single-molecule DNA sequencing. Genome Biol. 2011;12(2):217. Epub 2011/02/26.
47. Rothberg JM, Hinz W, Rearick TM, Schultz J, Mileski W, Davey M, et al. An integrated semiconductor device enabling non-optical genome sequencing. Nature. 2011;475(7356):348–52. Epub 2011/07/22.
48. Branton D, Deamer DW, Marziali A, Bayley H, Benner SA, Butler T, et al. The potential and challenges of nanopore sequencing. Nat Biotechnol. 2008;26(10):1146–53. Epub 2008/10/11.
49. Sanger F, Air GM, Barrell BG, Brown NL, Coulson AR, Fiddes CA, et al. Nucleotide sequence of bacteriophage phi X174 DNA. Nature. 1977;265(5596):687–95. Epub 1977/02/24.
50. International Human Genome Sequencing C. Finishing the euchromatic sequence of the human genome. Nature. 2004;431(7011):931–45. Epub 2004/10/22.
51. Ng SB, Bigham AW, Buckingham KJ, Hannibal MC, McMillin MJ, Gildersleeve HI, et al. Exome sequencing identifies MLL2 mutations as a cause of Kabuki syndrome. Nat Genet. 2010;42(9):790–3. Epub 2010/08/17.

52. Droege M, Hill B. The Genome Sequencer FLX System—longer reads, more applications, straight forward bioinformatics and more complete data sets. J Biotechnol. 2008;136(1–2):3–10. Epub 2008/07/12.
53. Mardis ER. Anticipating the 1,000 dollar genome. Genome Biol. 2006;7(7):112. Epub 2007/01/17.
54. Liu L, Li Y, Li S, Hu N, He Y, Pong R, et al. Comparison of next-generation sequencing systems. J Biomed Biotechnol. 2012;2012:251364. Epub 2012/07/26.
55. Tucker T, Marra M, Friedman JM. Massively parallel sequencing: the next big thing in genetic medicine. Am J Hum Genet. 2009;85(2):142–54. Epub 2009/08/15.
56. Wheeler DA, Srinivasan M, Egholm M, Shen Y, Chen L, McGuire A, et al. The complete genome of an individual by massively parallel DNA sequencing. Nature. 2008;452(7189):872–6. Epub 2008/04/19.
57. Wang J, Wang W, Li R, Li Y, Tian G, Goodman L, et al. The diploid genome sequence of an Asian individual. Nature. 2008;456(7218):60–5. Epub 2008/11/07.
58. Bentley DR, Balasubramanian S, Swerdlow HP, Smith GP, Milton J, Brown CG, et al. Accurate whole human genome sequencing using reversible terminator chemistry. Nature. 2008;456(7218):53–9. Epub 2008/11/07.
59. Levy S, Sutton G, Ng PC, Feuk L, Halpern AL, Walenz BP, et al. The diploid genome sequence of an individual human. PLoS Biol. 2007;5(10):e254. Epub 2007/09/07.
60. Wong KM, Hudson TJ, McPherson JD. Unraveling the genetics of cancer: genome sequencing and beyond. Annu Rev Genomics Hum Genet. 2011;12:407–30. Epub 2011/06/07.
61. Mardis ER. Next-generation DNA, sequencing methods. Annu Rev Genomics Hum Genet. 2008;9:387–402. Epub 2008/06/26.
62. Strausberg RL, Levy S, Rogers YH. Emerging DNA sequencing technologies for human genomic medicine. Drug Discov Today. 2008;13(13–14):569–77. Epub 2008/07/05.
63. Ansorge WJ. Next-generation DNA, sequencing techniques. New Biotechnol. 2009;25(4):195–203. Epub 2009/05/12.
64. Robison K. Application of second-generation sequencing to cancer genomics. Brief Bioinform. 2010;11(5):524–34. Epub 2010/04/30.
65. Schmitt MW, Kennedy SR, Salk JJ, Fox EJ, Hiatt JB, Loeb LA. Detection of ultra-rare mutations by next-generation sequencing. Proc Natl Acad Sci U S A. 2012;109(36):14508–13. Epub 2012/08/03.
66. Paszkiewicz K, Studholme DJ. De novo assembly of short sequence reads. Brief Bioinform. 2010;11(5):457–72. Epub 2010/08/21.
67. Koren S, Schatz MC, Walenz BP, Martin J, Howard JT, Ganapathy G, et al. Hybrid error correction and de novo assembly of single-molecule sequencing reads. Nat Biotechnol. 2012;30(7):693–700. Epub 2012/07/04.
68. Li Y, Wang J. Faster human genome sequencing. Nat Biotechnol. 2009;27(9):820–1. Epub 2009/09/11.
69. Worthey EA, Mayer AN, Syverson GD, Helbling D, Bonacci BB, Decker B, et al. Making a definitive diagnosis: successful clinical application of whole exome sequencing in a child with intractable inflammatory bowel disease. Genet Med. 2011;13(3):255–62. Epub 2010/12/22.
70. Efcavitch JW, Thompson JF. Single-molecule DNA analysis. Annu Rev Anal Chem. 2010;3:109–28. Epub 2010/07/20.
71. Milos PM. Emergence of single-molecule sequencing and potential for molecular diagnostic applications. Expert Rev Mol Diagn. 2009;9(7):659–66. Epub 2009/10/13.
72. Pushkarev D, Neff NF, Quake SR. Single-molecule sequencing of an individual human genome. Nat Biotechnol. 2009;27(9):847–50. Epub 2009/08/12.
73. Goren A, Ozsolak F, Shoresh N, Ku M, Adli M, Hart C, et al. Chromatin profiling by directly sequencing small quantities of immunoprecipitated DNA. Nat Methods. 2010;7(1):47–9. Epub 2009/12/01.
74. Pastor WA, Pape UJ, Huang Y, Henderson HR, Lister R, Ko M, et al. Genome-wide mapping of 5-hydroxymethylcytosine in embryonic stem cells. Nature. 2011;473(7347):394–7. Epub 2011/05/10.

75. Ozsolak F, Platt AR, Jones DR, Reifenberger JG, Sass LE, McInerney P, et al. Direct RNA sequencing. Nature. 2009;461(7265):814–8. Epub 2009/09/25.
76. Flusberg BA, Webster DR, Lee JH, Travers KJ, Olivares EC, Clark TA, et al. Direct detection of DNA methylation during single-molecule, real-time sequencing. Nat Methods. 2010;7(6):461–5. Epub 2010/05/11.
77. Song CX, Clark TA, Lu XY, Kislyuk A, Dai Q, Turner SW, et al. Sensitive and specific single-molecule sequencing of 5-hydroxymethylcytosine. Nat Methods. 2012;9(1):75–7. Epub 2011/11/22.
78. Carneiro MO, Russ C, Ross MG, Gabriel S, Nusbaum C, Depristo MA. Pacific biosciences sequencing technology for genotyping and variation discovery in human data. BMC Genomics. 2012;13(1):375. Epub 2012/08/07.
79. Drmanac R, Sparks AB, Callow MJ, Halpern AL, Burns NL, Kermani BG, et al. Human genome sequencing using unchained base reads on self-assembling DNA nanoarrays. Science. 2010;327(5961):78–81. Epub 2009/11/07.
80. Horner DS, Pavesi G, Castrignano T, De Meo PD, Liuni S, Sammeth M, et al. Bioinformatics approaches for genomics and post genomics applications of next-generation sequencing. Brief Bioinform. 2010;11(2):181–97. Epub 2009/10/30.
81. Huss M. Introduction into the analysis of high-throughput-sequencing based epigenome data. Brief Bioinform. 2010;11(5):512–23. Epub 2010/05/12.
82. DePristo MA, Banks E, Poplin R, Garimella KV, Maguire JR, Hartl C, et al. A framework for variation discovery and genotyping using next-generation DNA sequencing data. Nat Genet. 2011;43(5):491–8. Epub 2011/04/12.
83. Li H, Homer N. A survey of sequence alignment algorithms for next-generation sequencing. Brief Bioinform. 2010;11(5):473–83. Epub 2010/05/13.
84. Ledergerber C, Dessimoz C. Base-calling for next-generation sequencing platforms. Brief Bioinform. 2011;12(5):489–97. Epub 2011/01/20.
85. Nekrutenko A, Taylor J. Next-generation sequencing data interpretation: enhancing reproducibility and accessibility. Nat Rev Genet. 2012;13(9):667–72.
86. Kedes L, Liu ET. The Archon Genomics X PRIZE for whole human genome sequencing. Nat Genet. 2010;42(11):917–8. Epub 2010/10/29.
87. Kedes L, Campany G. The new date, new format, new goals and new sponsor of the Archon Genomics X PRIZE competition. Nat Genet. 2011;43(11):1055–8. Epub 2011/10/28.

Chapter 2
Next-Generation Sequencing in Cancer Epigenomics and Potential Clinical Applications

Hong Kiat Ng, Dimitrios H. Roukos, Barry Iacopetta, and Chee-Seng Ku

Abstract The arrival of next-generation sequencing (NGS) technologies has greatly empowered research in many fields, including the study of epigenetics in normal biological processes and in cancer. This has transformed epigenetics research from a targeted region approach to the whole epigenome scale at a single-nucleotide resolution. NGS has already been applied in several studies to dissect the cancer epigenome, such as determining the location of different histone marks in various cancer cell lines, whole methylome sequencing of colorectal cancer and adjacent normal tissue and sequencing of small RNAs from different cancer tissues and blood. Epigenetic profiling studies conducted through a common platform (NGS) allow a better understanding of the complementary roles played by each epigenetic mechanism.

Keywords Epigenetics • Epigenomics • Next-generation sequencing • Microarray • DNA methylation • Histone modification • microRNA • Cancer

H.K. Ng
Cancer Science Institute of Singapore, National University of Singapore, Singapore

D.H. Roukos
Centre for Biosystems and Genomic Network Medicine, Ioannina University, Ioannina, Greece

Department of Surgery, Ioannina University School of Medicine, Ioannina, Greece

B. Iacopetta
School of Surgery, University of Western Australia, Crawley, WA, Australia

C.-S. Ku (✉)
Department of Medical Epidemiology and Biostatistics, Karolinska Institutet, Stockholm, Sweden

Saw Swee Hock School of Public Health, National University of Singapore, Singapore
e-mail: cheeseng.ku@ki.se

W. Wu and H. Choudhry (eds.), *Next Generation Sequencing in Cancer Research: Volume 1: Decoding the Cancer Genome*, DOI 10.1007/978-1-4614-7645-0_2,

1 Introduction

Epigenetics is the term used to describe heritable yet reversible changes to the chromatin structure and DNA sequence that result in modulation of gene or transcript expression without direct alteration of the "core" DNA sequence (Fig. 2.1). These changes include modifications of histone tails (different chemical modifications of the amino acids at the histone tail) and DNA methylation (addition of methyl groups to CpG or CHG sites). Histone modifications can alter the chromatin

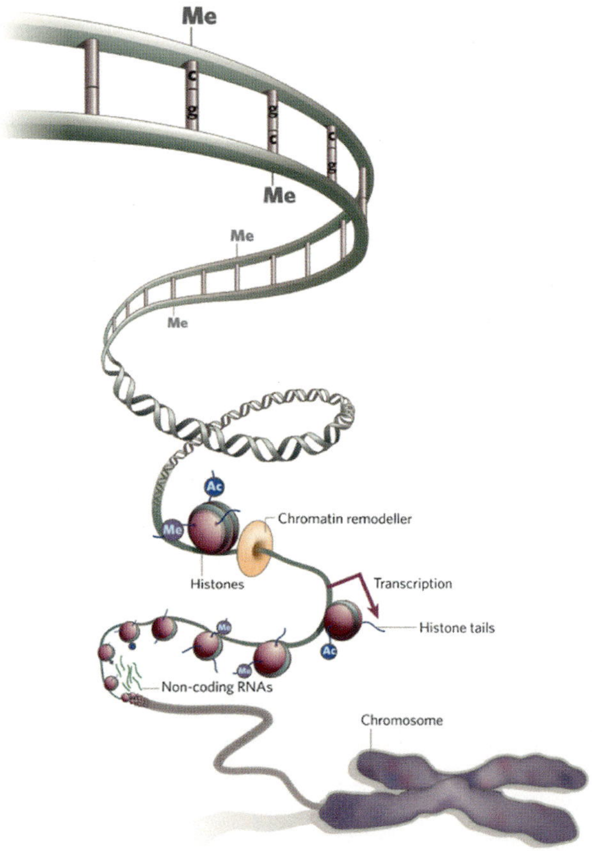

Fig. 2.1 The epigenome. Gene expression is tightly regulated by the interplay between different epigenetic mechanisms such as DNA methylation, histone modifications and miRNAs. DNA methylation is an epigenetic mechanism involving the addition of a methyl group to the C-5 position of the cytosine residue. Chromatin modifications through alterations to histone tails and nucleosome positioning are another important regulatory mechanism for gene transcription. In contrast to the regulation of gene expression at the transcriptional level via DNA methylation and histone modification, miRNAs play a critical role as post-transcriptional regulators of gene expression by impeding protein translation

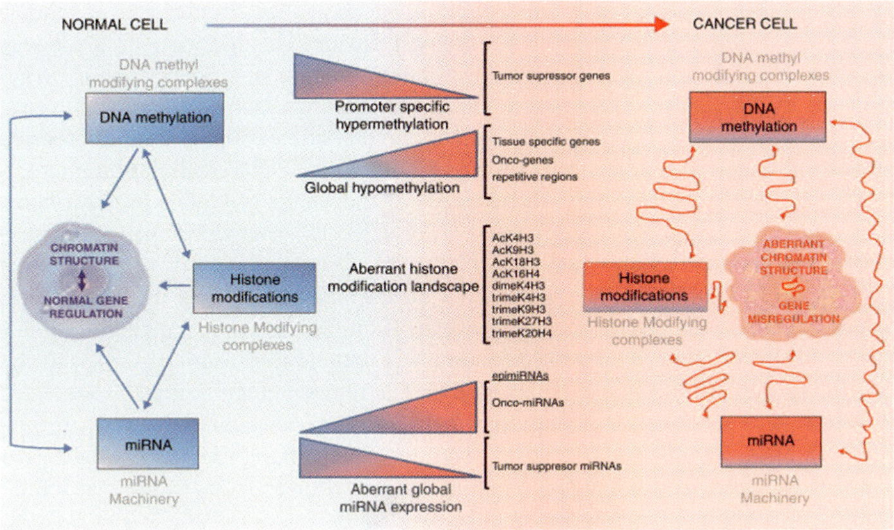

Current Opinion in Genetics & Development

Fig. 2.2 Different epigenetic alterations observed during tumourigenesis. In conjunction with the accumulation of genetic alteration, aberrant DNA methylation pattern, chromatin structure and miRNA expression dysregulate normal gene expression and result in transformations of normal cells to cancer cells

condensation or compaction and thus affect the accessibility of transcription binding proteins, resulting in changes in gene expression. On the other hand, DNA methylation in the CpG islands of gene promoters can also regulate gene expression levels by acting as a physical "obstacle" that prevents binding of transcription machinery proteins to 5′ gene promoter regions and/or the recruitment of additional chromatin remodelling proteins through binding with methyl-CpG-binding proteins [1]. For example, hypermethylation of the gene promoter region and the presence of repressive histone marks (H3K9me3/H3K27me3) are generally associated with gene silencing [2, 3]. Histone modifications and DNA methylation are also involved in genomic imprinting whereby certain genes are selectively silenced in an allelic specific manner, with the genes being expressed from either the paternal or maternal allele [4]. Furthermore, gene expression can also be regulated post-transcriptionally by miRNAs, leading to degradation or translational inhibition of mRNAs [5]. In addition, a subgroup of miRNAs is involved, directly or indirectly, in controlling the expression of epigenetic effectors such as methyl-CpG-binding protein 2 (MeCP2), histone deacetylases (HDACs) and DNA methyltransferases (DNMTs) [6, 7]. The interactions between different epigenetic mechanisms add to the complex nature of the regulation of gene or transcript expression. Epigenetics plays an important role in normal biological and developmental processes, and aberrations have been implicated in various diseases, particularly cancers (Fig. 2.2).

The study of epigenomics (interrogation of epigenetics at the whole-genome level) has advanced considerably over the past decade. The first wave of advance in epigenomics research was heralded by microarray technologies. These arrays allowed whole-genome scale investigation of the distribution of epigenetic events, for example, ChIP-chip (chromatin immunoprecipitation coupled with microarray or chip) to investigate patterns of histone modification. DNA methylation microarrays also allowed the profiling of hundreds of thousands of CpG sites distributed throughout the genome. Although microarrays have also been developed to quantify miRNA expression levels, they suffer from a number of limitations [8] including ascertainment bias in the selection of CpG sites, which have mainly focused on CpG islands. Consequently, a thorough investigation of the entire methylome that included all CpG sites in the whole genome was not possible. Similarly, microarrays used to profile miRNAs have a narrow dynamic range and are limited to annotated miRNAs, making them unsuitable for the discovery of new and biologically important miRNAs. The arrival of next-generation sequencing (NGS) technologies has signalled a paradigm shift in epigenomics and has allowed many of the major limitations associated with microarrays to be overcome.

Currently, NGS has been widely incorporated into epigenomics research for (a) chromatin immunoprecipitation sequencing (ChIP-Seq) to interrogate patterns of histone modification and protein-DNA interactions such as transcription factor binding sites; (b) whole-genome bisulfite sequencing, i.e. MethylC-seq (methylcytosine sequencing) or BS-seq (bisulfite sequencing) to characterise the DNA methylation profile at CpG and CHG sites (where H=A,C or T); and (c) RNA-seq (or transcriptome sequencing) to study both coding and non-coding RNAs [9]. Leveraging on these technological advances in sequencing, several international consortia and initiatives such as the NIH Roadmap Epigenomics Program and the International Human Epigenome Consortium have been launched to study epigenomics in various tissue types and disease models. These projects aim to use NGS-based techniques to produce publicly available, high-quality libraries of quantitative epigenetic reference data for basic biology and disease studies [10]. Due to the high-throughput production of sequencing data at an affordable cost, NGS technologies have already largely replaced microarray-based methods in epigenomic and other omics studies.

In the following sections, we briefly present new developments in the study of epigenomics that have been introduced by NGS technologies. This chapter focuses specifically on DNA methylation, histone tail modifications and miRNAs as these epigenetic factors have been relatively well studied in cancers using microarrays and NGS technologies. We also highlight studies that have used NGS to interrogate the epigenetic profiles of various cancers and discuss the new findings that have emerged in this field. Finally, the potential clinical applications of epigenetics/epigenomics are also discussed.

2 Advances in the Study of Epigenomics

Microarray technologies provide an efficient means of studying epigenetic mechanisms at the genome-wide scale, as opposed to traditional targeted approaches that are hypothesis driven. However, microarray-based methods have several major limitations. Firstly, prior knowledge of the genome is required to design probes for hybridisation with the targets (DNA or RNA), and therefore genomic regions that are not probed cannot be interrogated. This is evident from DNA methylation microarrays such as the Illumina Infinium HumanMethylation 450 Bead array designed to interrogate approximately 450,000 CpG sites but which still represents only a small fraction of the total ~28 million CpG sites in the human genome. Similarly for high-resolution ChIP-chip analysis, at least ten Roche NimbleGen Human ChIP-chip 2.1M whole-genome tiling arrays are required to analyse the DNA fragments immunoprecipitated from a single ChIP experiment. Furthermore, cross-hybridisation with other probes arising from very similar sequences (i.e. repetitive regions, pseudogenes and highly similar miRNA species) are problematic in microarrays. Single-nucleotide polymorphisms or somatic mutations can also affect the binding efficiency of genomic fragments (or targets) to the probes. However, the introduction of NGS has overcome some of the inherent limitations of microarrays. For example, NGS allows analysis of the whole methylome at single-nucleotide resolution. The ability to identify and analyse allele-specific methylation and histone modifications provides new opportunities to study imprinting and gene homologues [11], as well as the differentiation of miRNA species from different but closely related miRNA families. These areas of epigenetics have so far remained largely unexplored. The interrogation of highly repetitive regions also poses a challenge for NGS methods when the sequence reads are being mapped to the reference genome. However, paired-end sequencing (sequencing of both ends of a single sequence read), together with a longer read length, should enhance the ability to interrogate repetitive regions by increasing the specificity of reads aligned uniquely to these regions. The sheer number of sequence reads and the amount of sequencing data produced by NGS technologies means that multiple samples can be barcoded for sequencing per instrument run, particularly for applications that do not require a large amount of sequencing (e.g. profiling of miRNA expression versus sequencing of the entire methylome). This further increases the efficiency of experimentation. In contrast, sample multiplexing is not possible for microarrays.

2.1 DNA Methylation

DNA methylation is an epigenetic mechanism involving the addition of a methyl group to the C-5 position of the cytosine residue. It is mediated by DNA methyltransferase and results in 5-methylcytosine (5mC). The methylation occurs predominantly at CpG dinucleotides in the mammalian genome. Although these CpG sites are distributed throughout the genome, they tend to be enriched in 5′ regions and 2 kb upstream of many genes, thus giving rise to CpG islands and CpG shores, respectively [12]. DNA methylation plays a crucial role in regulating gene transcription, with hypermethylation of promoters generally being associated with gene silencing. Hypermethylation is also involved in genomic imprinting and X chromosome inactivation [4, 13]. In cancer cells, the promoters of tumour suppressor genes are often hypermethylated, resulting in the silencing of gene expression. Furthermore, loss of methylation in repetitive genomic sequences can disrupt the integrity and stability of chromosomes, which is one of the hallmarks of cancer.

Different pretreatment methods such as enzyme digestion, affinity enrichment or sodium bisulfite treatment are used together with NGS to study DNA methylation. Table 2.1 summarises the pros and cons of different NGS-based methods to study DNA methylation, as well as the principle behind this approach. MethylC-seq was used to discover that methylation at non-CpG dinucleotides such as CHG and CHH (where H = A, C or T) occurs in embryonic stem cells and pluripotent stem cells, a finding that would not have been possible using microarray [14]. Approximately 25 % of methylation was detected at non-CpG sites, with enrichment observed in gene bodies but depletion in protein binding sites and enhancers. These findings suggest that an alternative gene regulatory mechanism is used by stem cells to maintain their pluripotency. In contrast, whole-genome bisulfite sequencing of human peripheral blood mononuclear cells and fetal fibroblast cell lines detected only a tiny fraction (<0.2 %) of non-CpG dinucleotide methylation [14, 15].

Although NGS-based methods have been available for epigenomics research, it was not until recently that whole-genome bisulfite sequencing was applied to study the cancer methylome [16, 17]. MethylC-seq was used to examine the DNA methylome of a primary human colorectal tumour sample and its adjacent normal colon tissue at single base pair resolution. After ligating the DNA with fully methylated cytosine sequencing adapter, both libraries were subjected to sodium bisulfite conversion and PCR amplification before being sequenced using the Illumina Genome Analyzer IIx sequencer to generate more than 1.6 billion reads for each of the tumour and normal samples. Reads that were duplicates, contained multiple mismatches or were aligned to multiple locations were removed, and only cytosines covered by three or more C or T nucleotides or with a G on the opposite strand were analysed. Using this approach, approximately 47 million CpGs in each sample were classified as high-confidence CpG sites (>90 % of reads at the position following a cytosine was guanine). Similarly, high-confidence CpHs were also identified in the same manner if >90 % of the adjacent nucleotide was H (A, C, or T). Regions of 10 kb in length with an average methylation of 20–60 % were also classified as partially methylated domains.

From the analysis, methylation at non-CpG cytosines was almost undetectable in both samples (mean CpH of 0.61 % in normal mucosa and 0.65 % in tumour), thus reinforcing the uniqueness of non-CpG methylation in stem cells. Compared to normal mucosa, the tumour CpG methylation profile was found to be less methylated in regions defined as being partially methylated but displaying focal hypermethylation primarily in CpG islands. These regions of hypermethylation are also enriched

Table 2.1 Summary of the pros and cons and the principle behind different NGS-based approaches to study DNA methylation

Approach	Principle	Number of CpGs analysed	Area of interest	Pros	Cons
MethylC-seq	Absolute quantification of DNA methylation in whole DNA methylome through random shearing of DNA followed by bisulfite conversion and NGS or random shearing after bisulfite conversion of DNA	~28 million unique CpGs	Whole DNA methylome including non-CG methylation	Interrogate CpG at repetitive regions	As CpG islands generally lack methylation, CpG coverage at CpG islands is lower Relative quantification of methylation instead of absolute quantification of methylation
MeDIP-seq	Relative methylation level is captured based on affinity enrichment of methylated DNA fragments using an antibody specific to 5-methylcytosine followed by NGS	70–80 % of CpGs	Methylation-rich regions	Increased rate of uniquely mapped sequence, since the four nucleotides are retained Low cost for non-CpG island analysis	As CpG islands generally lack methylation, CpG coverage at CpG islands is lower Relative quantification of methylation instead of absolute quantification of methylation
MBD-seq	Relative methylation level is captured based on affinity enrichment of methylated DNA fragments using methyl-CpG-binding domain protein specific to 5-methylcytosine followed by NGS	70–80 % of CpGs	Methylation-rich regions	Increased rate of uniquely mapped sequence, since the four nucleotides are retained Low cost for non-CpG island analysis	As CpG islands generally lack methylation, CpG coverage at CpG islands is lower

(continued)

Table 2.1 (continued)

Approach	Principle	Number of CpGs analysed	Area of interest	Pros	Cons
RRBS/ERRBS	Absolute DNA methylation level is captured based on enrichment for CpG-rich regions using restriction enzyme fragmentation (e.g. *MspI* digestion), followed by size selection, bisulfite conversion and NGS	10–15 %; 1.2–2 million unique CpGs	High-density CpG regions such as CpG islands/shores and non-CG methylation	Low cost for CpG island analysis	Only CpGs near *MspI* sites are analysed
MethylPlex NGS	Restriction enzyme enrichment without bisulfite conversion	68,508 CpG islands	CpG island methylation	Low cost for CpG island methylation Increased rate of uniquely mapped sequence, since the four nucleotides are retained	Only CpG island methylation is analysed

within regions of long-range (>100 kb) hypomethylation and are associated with replication and attachment to the nuclear lamina in human cell lines. The results suggest that DNA methylation changes contribute to the three-dimensional organisation of chromatin within the nucleus resulting in gene silencing. By employing MethylC-seq, the study further documented substantial DNA methylation differences between tumour and normal tissue and proposed a new mechanism for gene silencing by DNA methylation through its effects on chromatin organisation.

Apart from MethylC-seq, other techniques such as reduced representation bisulfite sequencing (RRBS, endonuclease digestion treatment coupled with NGS) have also been used to characterise the DNA methylation profile in leukaemia [18, 19] and colon cancer [20]. Among the different approaches, methylated DNA immunoprecipitation sequencing (MeDIP-seq) and methyl-CpG-binding domain protein sequencing (MBD-seq) (affinity enrichment coupled with NGS) have been applied in acute myeloid leukaemia [21] and in various cancer types including breast [22], bladder [23], non-small cell lung [24], prostate [25] and ovarian [26].

Using an "enhanced" RRBS procedure that increases the coverage of genomic CpG methylation beyond the traditional CpG islands, Akalin et al. found a completely opposite DNA methylation profile between two primary human bone marrow specimens of genetically distinct acute myeloid leukaemia (AML) subtypes [19]. DNA was extracted from two *IDH* mutant AML samples, two AML cases harbouring t(9;11)(q22,q23) translocations and two normal CD34+ bone marrow cells and then digested by *MspI* restriction enzymes. Libraries containing *MspI* fragments of 70–320 bp in size were then selected for sequencing using Illumina GAII or HiSeq2000, yielding an average coverage of two million CpGs per sample. This "modified" RRBS method achieved a 75 % increase in CpG coverage and a 54 % increase in CpG shore coverage compared to normal RRBS. By comparing the uniquely aligned sequence reads with the base position from at least a tenfold coverage, a set of CpG sites was consistently found to be hypermethylated in both AML subtypes as compared to normal bone marrow. In addition, differences between the two AML subtypes were also noted. Widespread hypermethylation was observed in *IDH* mutant AML, preferentially at promoter regions and in CpG islands surrounding the transcription start sites of genes. This methylation profile was less common in the other AML subtype. In contrast, extensive loss of CpG methylation in introns and in distal intergenic CpG islands and shores was generally observed in the AML harbouring *MLL* gene translocations. The use of this enhanced RRBS method demonstrates the heterogeneity of AML epigenomic profiles. These encompass a complex scenario that includes aberrant DNA methylation of distinct regions and with different mechanisms of gene regulation.

Apart from deciphering the epigenome of different cancers, MeDIP-seq and MBD-seq have been applied to (a) profile changes in the DNA methylation landscape of epithelial to mesenchymal transition (EMT) in breast cancer [22], (b) interrogate drug resistance in ovarian cancer cell lines [26], (c) classify cancer subtypes in AML [21] and (d) discover non-invasive biomarkers for bladder cancer [23]. These studies have made some interesting observations. For example, global DNA methylation landscapes of a breast cancer cell line and of normal mammary epithelium were not significantly altered during EMT induction. This was despite approximately 40 % of the epithelial cell-specific sites being hypomethylated and 40 % of the mesenchymal-specific sites being hypermethylated following induction, thus reinforcing the link between cell-specific DNA methylation alterations and morphological changes [22]. In addition, NGS based on methylation-sensitive enzyme restriction has identified novel genes that might be involved in tumourigenesis, including a cancer-specific, differentially methylated region within the *WFDC2* gene promoter in prostate cancer [25]. Intriguingly, this region was frequently methylated in cancer tissues (17/22 tissues and 6/6 cell lines), but not in benign tissues (0/10) or normal prostate epithelial cells.

Taken together, profiling of global DNA methylation patterns using NGS-based methods have allowed comprehensive interrogation of this epigenetic factor during the pathogenesis of cancer and other important biological processes such as EMT implicated in metastasis. These approaches have also allowed the identification of biomarkers for differentiating cancer subtypes, as well as for disease diagnosis and drug resistance. Table 2.2 summarises the major findings of the studies using NGS-based approaches to study DNA methylation in cancer.

2.2 Histone Modifications

Chromatin modifications through alterations to histone tails and nucleosome positioning are another important regulatory mechanism for gene transcription [27]. Histone proteins compact DNA strands into structures called nucleosomes. However, the histone core and its tail can be subjected to various chemical modifications that change the nucleosome surface or alter the packaging of DNA to regulate the access of DNA-binding protein or other effector complexes. For example, trimethylation of lysine 27 on histone 3 (H3K27me3) or H3K9me3 is often associated with transcriptional silencing, whereas H3K4me3 and H3K4ac are usually associated with transcriptional activation [2].

ChIP-seq was one of the first applications for NGS technologies in mapping the genome-wide distributions of different histone lysine and histone arginine methylations, as well as factors such as H2A.Z, RNA polymerase II and the insulator binding protein CTCF in human T cells [28]. Integration of these data with serial analysis of gene expression has allowed a better understanding of the role played by these modifications in gene regulation. The ENCODE (Encyclopaedia Of DNA Elements) Project had also embarked on producing high-quality maps of several histone modifications, as well as various functional elements using ChIP-seq in multiple cell lines [29].

Similar to DNA methylation, NGS has been used in ChIP-seq to interrogate histone modifications in cancer. It was used to compare the histone modification patterns of H3K4me1, H3K4me3 and H3K9/14ac in human breast cancer cell lines and in a normal mammary epithelial cell line [30]. Using approximately 200 ng of chromatin-immunoprecipitated input DNA, libraries of about 220 bp were enriched

Table 2.2 Summary of NGS-based approaches to the study of DNA methylation in cancer

Name of approach	Cancer type (reference)	Sequencing platform and data	Bioinformatics approach	Major findings
MethylC-seq	Colorectal tumour and adjacent normal [16]	• Illumina GA IIx • 1.6 billion reads generated per sample • 47 million high-confidence CpGs per sample	• Unique reads without duplicates, multiple mismatches or multiple alignment were selected for analysis • Select for high-confidence cytosines covered by three or more C or T nucleotides or with a G on the opposite strand	• Hypermethylated domains enriched within regions of long-range (>100 kb) hypomethylation associated with replication attachment to nuclear lamina • DNA methylation changes contribute to the three-dimensional organisation of chromatin within the nucleus, leading to gene silencing
	Breast cancer cell line and normal mammary epithelial cell line [17]	• Illumina GA II • 889 million unique reads (27-fold coverage) for HCC1954 breast cancer cell line • 20-fold coverage for HMEC human mammary epithelial cells	• Regions of 10 kb with mid–low or mid–high methylation percentage defined as partially methylated domain	• Genes with widespread hypomethylation in partially methylated domains remain silenced • Gene silencing is tightly linked to both DNA hypomethylation and chromatin repressive marks in breast cancer
MeDIP-seq	12 AML patients and 4 normal bone marrow [21]	• Illumina GA II; Generated ~400 millions reads covering 63–87 % for the 16 samples	• Reads mapped to human reference genome hg 18 using high-performance alignment software “Maq” and Bowtie. Reads with Maq quality less than 10 were removed	• SPHKAP and DPP6 have significantly methylated promoters and were repressed in AML • Cytogenetic AML subtype-specific methylome pattern • Cytogenetic AML subtype-specific hypomethylation in some interspersed repeat elements

(continued)

Table 2.2 (continued)

Name of approach	Cancer type (reference)	Sequencing platform and data	Bioinformatics approach	Major findings
	8 breast cancer cell lines and normal mammary epithelial cell line [22]	• Illumina GA II • 97 million uniquely mapped singleton reads and 11 million paired-end reads • 87 % of genomic CpGs were covered in each sample with average 12 times depth	• Reads mapped to human reference genome hg 18 using high-performance alignment software "Maq" • Reads with Maq quality less than 10 were removed. Reads within known regions of genetic amplification or deletion were removed	• Massively reduced methylation level, particularly in CpG-poor regions between breast cancer cell line and normal mammary epithelial cell line • Global DNA methylation level remains unaltered despite 40 % of the epithelial cell-specific sites being hypomethylated and 40 % of the mesenchymal-specific sites being hypermethylated following induction • Reinforces the link between cell-specific DNA methylation and morphological changes
MBD-seq	Cisplatin-sensitive ovarian cell line A2780 and induced cisplatin resistance A2780CP [26]	• Illumina GC II; ~18 million reads generated from cisplatin-sensitive ovarian A2780 line and ~21 million reads from cisplatin-resistant line A2780CP	• Reads with up to 2 mismatches were mapped to hg19 using software SOAP2	• 1,224 hypermethylated and 1,216 hypomethylated DMRs in A2780CP line compared to A2780 line • Lower global CpG methylation in A2780CP cells
	Bladder cancer cell lines (pooled from T24 and 5637) and pooled bladder mucosal tissues [23]	• Illumina GA II; ~6 million reads per sample	• PCR duplicates were removed • BWA alignment with default setting to map reads to hg 19 human reference genome • Hypermethylated peaks were identified using MACS software	• 1,627 hypermethylated promoters in two bladder cancer cell line were identified • Gene signature of 5 genes (VAX1, KCNV1, TAL1, PPOX1 and CFTR) for bladder cancer diagnosis • VAX1 and LMX1A were associated with bladder cancer recurrence

	7 paired lung tumour/normal samples [24]	• Illumina GA IIx; ~15 million unique reads per sample	• Raw mapped reads were computed using Bioconductor project with R statistical computing environment • Reads with quality ≥15 were used to generate genome-wide coverage	• 57 differentially methylated regions were present in all seven NSCLC tumours • Hypermethylated DMRs associated with transcriptional regulators • Subtelomeric regions and satellite repeats were hypomethylated in NSCLC • Adenocarcinoma- and squamous cell carcinoma-specific DMRs were identified
RRBS/ERRBS	CD19+ B cell from 11 chronic lymphocytic leukaemia patients and 3 normal controls [18]	• Illumina GA Iix; • 20–30 million reads per sample • 1.8–2.3 million CpGs in CLL genome	• Bowtie was used to map reads to reference database of human genome (hg18) that were within 100bp from *Msp*I sites at both ends • In-house script was used to compute best alignments for the different loci	• Though global CpG methylation was similar between CLL and normal B cells, 1,746 gene promoters were differentially methylated in at least one CLL when compared to normal • 19 % of differentially methylated genes involved in transcription regulation • Aberrant hypermethylation in all HOX gene clusters
	IDH mutant AML, AML harbouring MLL translocation and normal primary human bone marrow sample [19]	• Illumina GA II or HISeq2000; average 2 million CpGs per sample	• Select for uniquely aligned reads with at least 10× coverage	• Enhanced RRBS has 75 % increase in CpG island coverage and 54 % increase in CpG shore coverage compared to traditional RRBS. Opposite DNA methylation profile between two genetically distinct AML

(continued)

Table 2.2 (continued)

Name of approach	Cancer type (reference)	Sequencing platform and data	Bioinformatics approach	Major findings
MethylPlex NGS	6 benign adjacent, 2 normal, 5 localised prostate cancer, 4 metastatic prostate cancer specimens, prostate cancer (LNCap) and normal prostate (PrEC) cell lines [25]	• Illumina GA II • Average 5 million mappable reads per sample	• Hidden Markov model (HMM)-based algorithm was used to locate peaks from mapped reads	• Overall promoter CpG island methylation increased from ~12.6 % in benign sample to 19.3 % and 21.8 % in localised and metastatic cancer tissue, respectively • 2,481 cancer-specific, differentially methylated regions • Identification of novel prostate cancer-specific methylation in *WFDC2* gene promoter

by running on agarose gels and then sequenced using Illumina GA II. Several million hg18 mapped tags from H3K4me1, H3K4me3, H3K9/14ac immunoprecipitation and input reads were generated for MCF-7 and MCF-A cells.

Enrichment of activating markers (H3K4me3 and H3K9/14ac) at promoter regions was found in this study [30], whereas H3K4me1 was broadly distributed in the proximity of transcription start sites and other genomic regions with enhancer activity. In addition, when gene expression data and histone modification patterns were compared in MCF-7, many differentially expressed genes were found to have divergent histone modification patterns. Intriguingly, breast cancer-related genes such as *ESR1* and 20 other genes were overexpressed while having a high level of H3K9/14ac in their promoter. Thirty-five genomic regions located 3–10 kb upstream of the transcription start site and marked by at least one histone were also tested for transcriptional activity. Of the selected genes, 66 and 60 % showed enhanced transcriptional activity in MCF-7 and MCF-10A, respectively. Most regions marked by H3K4me1 also exhibited enhanced activity, while regions with more than one type of histone modification showed varying activities.

Several studies have investigated the co-distribution of different histone modifications with DNA methylation in cancer cell lines [17, 31–33]. For example, in an integrative analysis of multiple omics data, Hon et al. performed ChIP-seq of several histone modifications, including the repressive H3K9me3 and H3K27me3 marks and the active H3K4me1, H3K4me3, H3K27ac and H3K36me3 marks. They also performed whole-genome bisulfite sequencing to determine DNA methylation in the breast cancer cell line HCC1954 and in the HMEC normal mammary epithelial cell line. Gene expression profiling from RNA sequencing was also employed to determine genome-wide transcript abundance in both cell lines. Interestingly, although widespread DNA hypomethylation was observed in cancer cells, particularly in regions where normal breast epithelial cells were partially methylated, these regions remained largely silenced. The loss of DNA methylation at these domains was often accompanied by the formation of repressive histone marks. A fraction of these regions displayed allelic DNA methylation and allelic specific histone marks H3K9me3/H3K27me3 on the other strand. These results clearly highlight a mutually exclusive relationship between DNA methylation and H3K9me3 and H3K27me3 [17].

When immunoprecipitated DNA enriched for Polycomb H3K27me3 was subjected to sodium bisulfite treatment and high-throughput sequencing [32], DNA methylation and H3K27me3 were not always mutually exclusive. Instead, the data showed that both the methylated and non-methylated allele could be associated with the repressive histone mark H3K27me3 [32]. Furthermore, another study which profiled several histone marks in colorectal cancer cell lines treated with demethylating agent (5′-aza-2′-deoxycytidine) also suggested that other factors such as increased histone 3 acetylation and H3K4me3, as well as simultaneous loss of H3K27me3, were required to achieve long-term transcriptional reactivation of silenced genes [33]. These studies reinforce the complementary yet unique relationship between histone modifications and DNA methylation in different cancer types.

They also highlight the need for co-administration of DNA-demethylating agents with histone deacetylase inhibitor or histone methyltransferase inhibitor in possible future therapies aimed at achieving gene reactivation.

2.3 microRNAs

In contrast to the regulation of gene expression at the transcriptional level via DNA methylation and histone modification, miRNAs play a critical role as post-transcriptional regulators of gene expression by impeding protein translation. Mature miRNAs are a group of small (~22 nucleotides) non-coding RNAs encoded by genomic sequences distributed throughout the genome, including introns of protein-coding genes and intergenic regions. The binding of miRNA to the 3′ untranslated region of mRNA, depending on the degree of complementarity, leads to either mRNA degradation and/or repression of translation [34]. Due to their short length, a single miRNA can target multiple mRNAs, while a single mRNA can also be regulated by multiple miRNAs. To date, 1,921 human mature miRNAs have been discovered and annotated in miRBase Release 18 [35]. Approximately 60 % of mammalian genes are predicted to be targeted or regulated by these miRNAs [36]. Deregulation of miRNAs can occur as a result of genomic aberration (e.g. amplification, deletion, translocation, epigenetic silencing) and has been associated with various diseases, including cancer [37–39].

Small RNA sequencing using NGS was first performed in *Arabidopsis thaliana* [40] before being used to compare the miRNAome of human embryonic stem cells during their differentiation [41]. In the latter study, 334 known and 104 novel miRNAs were identified for two different developmental stages. Of these, 171 of the known and 23 of the novel miRNAs exhibited significantly different expression before and after cell differentiation. NGS technologies have also led to the discovery of mirtrons [42, 43], a class of miRNAs that is generated by splicing from introns and bypassing Drosha processing. Another novel class of miRNAs, known as isomiRs, was also discovered using NGS technologies [41, 44]. IsomiRs have non-template variations in their sequence with respect to reference miRNAs, including 3′ non-template nucleotide additions and nucleotide substitutions. These studies have demonstrated the discovery power of NGS, not only for "conventional" miRNAs but also for new classes of miRNAs. Such discoveries would not have been possible without NGS methods. In addition, direct sequencing of small RNAs allows the detection of variants or mutations that could affect the secondary structure or gene targets for these important regulators [45].

Despite the increasing number of newly discovered miRNAs annotated in miRBase, it is widely believed that many more remain to be discovered and validated. Fortunately, NGS has accelerated the pace of discovery of these small RNAs in various tissues and cancer models [46–49]. For example, when NGS was applied to a pooled fraction of leukaemia cells derived from 89 paediatric cases of seven well-defined genetic subtypes, a total of 554 known, 28 novel and 431 potentially novel and mature miRNAs were identified [49]. Furthermore, by comparing the miRNAome of normal hematopoietic tissue with the different ALL subtypes, a total of 153 known, 16 novel and 170 novel mature miRNAs and miRNA* strands were found to be exclusive to ALL. In contrast, a total of 140 known, 2 novel and 82 novel mature miRNAs and miRNA* strands were expressed only in normal hematopoietic cells.

In addition to identifying different cancer subtypes, miRNA profiling has also been used to predict the clinical outcome of neuroblastoma [47]. Notably, 76 miRNAs were differentially expressed between patients with favourable and unfavourable survival. In this study, the Life Technologies SOLiD system V3 was used to evaluate miRNA expression, and more than 188 million reads were generated. miRNAs in the miR-17-92 cluster were found to be overexpressed in neuroblastoma with poor prognosis. miR-542-5p and miR-628 were expressed in neuroblastoma with good prognosis, but were almost undetectable in cases with unfavourable outcome, thus making them potential prognostic biomarkers. The discovery power of this approach was also demonstrated by the identification of 13 novel miRNAs, of which three were validated in a cohort of 70 neuroblastomas. Widespread post-transcriptional miRNA editing was also observed in the analysis, such as adenosine to inosine (0.7 %), guanosine to adenosine (1.5 %), uracil to cytosine (1.2 %) and uracil to adenosine (1.1 %). These editing events would not have been detected using microarrays and hence demonstrate the multiple discoveries that are made possible by miRNAome sequencing experiments. Furthermore, 29.1 % of all uniquely mapped miRNAs were observed to have a terminal addition, of which 63.1 % were non-template additions. Although single-nucleotide terminal additions predominated (67.2 %), the addition of two or three bases was also observed. The biological relevance of miRNA editing remains to be explored further, but it could be predicted to alter miRNA expression and possibly also the silencing of its targets [50, 51].

miRNAs with potential prognostic value have also been reported in breast cancer [46]. A signature of nine miRNAs was found to differentiate invasive ductal carcinoma from in situ carcinoma. In addition, several miRNAs (miR-210, miR-21, miR-106b*, miR-197 and let-7i) were associated with overall survival and time to metastasis, while miR-210 in particular was implicated in tumour invasion. Expression of the *BRCA1*, *FANCD*, *FANCF*, *PARP1*, *E-cadherin* and *RB1* genes was inversely correlated with miR-210 expression. These genes are activated in in situ carcinoma but repressed in invasive carcinoma.

The presence of circulating miRNAs in blood serum or plasma and other body fluids has led to the sequencing of miRNAs from these samples for the purposes of disease detection and prognostication [52–54]. This has been demonstrated in several cancer types. For example, serum miRNA let-7e, miR-151-5p and miR-222 levels were significantly higher in patients with papillary thyroid carcinoma compared to benign cases [53]. These miRNAs also showed associations with certain clinicopathological variables such as nodal status, tumour size, multifocal lesion status and tumour-node-metastasis stage. Furthermore, the expression of these miRNAs in serum decreased significantly after tumour excision, demonstrating their potential as non-invasive markers for disease detection and monitoring. With evidence growing that deregulation of miRNA expression is a hallmark of cancer, it is important to profile and characterise these small RNA molecules. This should lead to a better understanding of the molecular pathogenesis of diseases or cancer and perhaps also to their use as biomarkers for diagnostic, prognostic and therapeutic purposes.

2.4 Other Epigenetic Mechanisms

In addition to the "conventional" epigenetic mechanisms described above, several new mechanisms were recently reported. DNA hydroxymethylation and RNA methylation have been found to regulate transcription in mammalian systems [55–58]. For example, the recent discovery of N^6-methyladenosine (m^6A) demethylation by the fat mass and obesity-associated (*FTO*) gene [59] has led to the investigation of m^6A sites in the mouse brain at the whole-genome level [56]. These sites occur in highly conserved regions with unique sequence motifs and are enriched near stop codons and in the 3′ UTR of mRNA. They also show association with miRNA-binding sites within the 3′ UTR.

The conversion of 5mC (5 methylcytosine) to 5hmC (5 hydroxymethylcytosine) by TET proteins also suggests a possible mechanism for DNA demethylation. This has led to whole-genome profiling for 5hmC in mouse embryonic stem cells and in human brain tissue [55, 57, 58]. 5hmC was more selectively targeted to genes than was 5mC, with specific enrichment at promoters and gene bodies, but was largely absent in non-genic regions [58]. The presence of 5hmC in gene bodies was also more positively correlated with gene expression than was 5mC. Knock down of TET proteins led to downregulation of the expression of genes related to pluripotency. A significant increase in 5mC level was also observed in the promoters of the same set of genes and induced the cells towards lineage differentiation [55]. The distribution of 5mC and 5hmC was investigated in different normal human tissue types using antibodies specific to α-5hmC and tiling microarrays [60]. 5hmC was primarily but not exclusively found within the bodies of transcribed genes. There are some reports suggesting that loss of 5hmC is associated with cellular transformation [61, 62]; however, a more thorough investigation of 5hmC distribution in human cancer and of its biological role is required.

Apart from gaining an understanding of gene regulation through the mapping of DNA methylation, histone modification and miRNA, it is also important to establish the roles of other genomic regulatory elements such as promoters and enhancers. Several datasets containing 54 unique transcription factors from two cell lines (K562 myelogenous leukaemia and Gm12878 lymphoblastoid) were generated and used to identify different regulatory element clusters [63]. By integrating ChIP-seq data for different histone modifications and open chromatin with various transcription factor binding sites, the methylation of H3K4 was identified as the most specific marker for transcription factors. Clusters located within annotated genes were associated with gene expression and histone modification, while clusters located within intergenic regions were either directly involved in the transcription of miRNAs and long non-coding RNAs or facilitated transcription by long-range interactions. Transcription factors that facilitated long-range interactions were specifically enriched for the activating histone mark H3K4me1 and to a lesser extent H3K27/ac or p300 binding.

There is increasing evidence that other non-coding RNAs such as long intergenic non-coding RNAs (lincRNAs), small nucleolar RNAs (snoRNAs) and PIWI interacting RNAs (piRNAs) are also important regulators of many biological processes [64]. For example, a custom exon tiling array designed for approximately 900 human lincRNA loci identified DNA immunoprecipitated with antibodies against SUZ12 and EZH2 (components of Polycomb Repressive Complex 2, or PRC2). The study revealed that approximately 20 % of lincRNAs expressed in HeLa cells, human lung fibroblasts and foot cells interact with PRC2 and with other chromatin-modifying complexes, suggesting that some lincRNAs recruit chromatin complexes to specific loci in order to regulate gene expression [65]. Furthermore, characterisation of the lincRNA *PCAT-1* in prostate cancer cell lines showed it to be a regulator of cell proliferation and a target of PRC2 [66]. Further characterisation of these non-coding RNAs is required to better understand their biological importance and their role in regulating gene expression.

3 Conclusions

The advent of NGS technologies has greatly enhanced the study of epigenetics and its role in many biological and disease processes, including tumourigenesis. Due to their high resolution at single-nucleotide level, NGS methods have mostly replaced microarrays for investigations at the whole-genome level into DNA methylation, histone modification and other epigenetic mechanisms. The profiling of various histone marks, DNA-binding proteins and transcription factor binding sites in different cancer types, together with whole-genome bisulfite sequencing of colorectal tissue samples and a breast cancer cell line, has demonstrated the technical feasibility of studying the entire epigenome. Similarly, miRNA research has also benefited greatly from NGS technologies, allowing the discovery of novel miRNA species and their derivatives as well as the validation of novel miRNAs. Further advances in bioinformatics tools and a better interpretation of the data should lead to additional biological insights.

The study of different epigenetic mechanisms using NGS technologies has also facilitated the integration of various omics data (integrative analysis) to explore the interaction of these mechanisms in the regulation of transcription. This is apparent from the discovery of a mutually exclusive role between repressive histone marks (H3K9me3 and H3K27me3) and DNA methylation, with the former but not the latter being responsible for gene silencing [17]. However, a fraction of repressed genes are regulated by both allelic DNA methylation and histone repressive marks.

In summary, the studies discussed above offer a glimpse into the possibilities offered by NGS for the investigation of different epigenetic mechanisms and how the results obtained can provide novel biological insights. The application of these new technologies has not only improved our understanding of various epigenetic mechanisms, but also highlights the interactions between them. While NGS has already opened new opportunities and led to important discoveries, these are likely to represent only the tip of the iceberg. Further progress in technology and bioinformatics should advance epigenetics research to an even higher level.

Financial and Competing Interests Disclosure None to declare.

References

1. Nan X, Meehan RR, Bird A. Dissection of the methyl-CpG binding domain from the chromosomal protein MeCP2. Nucleic Acids Res. 1993;21(21):4886–92.
2. Bernstein BE, Mikkelsen TS, Xie X, Kamal M, Huebert DJ, Cuff J, et al. A bivalent chromatin structure marks key developmental genes in embryonic stem cells. Cell. 2006;125(2):315–26.
3. Bird A. DNA methylation patterns and epigenetic memory. Genes Dev. 2002;16(1):6–21.
4. Fang F, Hodges E, Molaro A, Dean M, Hannon GJ, Smith AD. Genomic landscape of human allele-specific DNA methylation. Proc Natl Acad Sci U S A. 2012;109(19):7332–7.
5. Olsen PH, Ambros V. The lin-4 regulatory RNA controls developmental timing in Caenorhabditis elegans by blocking LIN-14 protein synthesis after the initiation of translation. Dev Biol. 1999;216(2):671–80.
6. Varambally S, Cao Q, Mani RS, Shankar S, Wang X, Ateeq B, et al. Genomic loss of microRNA-101 leads to overexpression of histone methyltransferase EZH2 in cancer. Science. 2008;322(5908):1695–9.
7. Garzon R, Liu S, Fabbri M, Liu Z, Heaphy CE, Callegari E, et al. MicroRNA-29b induces global DNA hypomethylation and tumor suppressor gene reexpression in acute myeloid leukemia by targeting directly DNMT3A and 3B and indirectly DNMT1. Blood. 2009;113(25): 6411–8.
8. Hurd PJ, Nelson CJ. Advantages of next-generation sequencing versus the microarray in epigenetic research. Brief Funct Genomic Proteomic. 2009;8(3):174–83.
9. Mardis ER. The impact of next-generation sequencing technology on genetics. Trends Genet. 2008;24(3):133–41.
10. Satterlee JS, Schubeler D, Ng HH. Tackling the epigenome: challenges and opportunities for collaboration. Nat Biotechnol. 2010;28(10):1039–44.
11. Gertz J, Varley KE, Reddy TE, Bowling KM, Pauli F, Parker SL, et al. Analysis of DNA methylation in a three-generation family reveals widespread genetic influence on epigenetic regulation. PLoS Genet. 2011;7(8):e1002228.
12. Irizarry RA, Ladd-Acosta C, Wen B, Wu Z, Montano C, Onyango P, et al. The human colon cancer methylome shows similar hypo- and hypermethylation at conserved tissue-specific CpG island shores. Nat Genet. 2009;41(2):178–86.
13. Sharp AJ, Stathaki E, Migliavacca E, Brahmachary M, Montgomery SB, Dupre Y, et al. DNA methylation profiles of human active and inactive X chromosomes. Genome Res. 2011; 21(10):1592–600.
14. Lister R, Pelizzola M, Dowen RH, Hawkins RD, Hon G, Tonti-Filippini J, et al. Human DNA methylomes at base resolution show widespread epigenomic differences. Nature. 2009; 462(7271):315–22.

15. Li Y, Zhu J, Tian G, Li N, Li Q, Ye M, et al. The DNA methylome of human peripheral blood mononuclear cells. PLoS Biol. 2010;8(11):e1000533.
16. Berman BP, Weisenberger DJ, Aman JF, Hinoue T, Ramjan Z, Liu Y, et al. Regions of focal DNA hypermethylation and long-range hypomethylation in colorectal cancer coincide with nuclear lamina-associated domains. Nat Genet. 2012;44(1):40–6.
17. Hon GC, Hawkins RD, Caballero OL, Lo C, Lister R, Pelizzola M, et al. Global DNA hypomethylation coupled to repressive chromatin domain formation and gene silencing in breast cancer. Genome Res. 2012;22(2):246–58.
18. Pei L, Choi JH, Liu J, Lee EJ, McCarthy B, Wilson JM, et al. Genome-wide DNA methylation analysis reveals novel epigenetic changes in chronic lymphocytic leukemia. Epigenetics. 2012;7(6):567–78.
19. Akalin A, Garrett-Bakelman FE, Kormaksson M, Busuttil J, Zhang L, Khrebtukova I, et al. Base-pair resolution DNA methylation sequencing reveals profoundly divergent epigenetic landscapes in acute myeloid leukemia. PLoS Genet. 2012;8(6):e1002781.
20. Gu H, Bock C, Mikkelsen TS, Jager N, Smith ZD, Tomazou E, et al. Genome-scale DNA methylation mapping of clinical samples at single-nucleotide resolution. Nat Methods. 2010;7(2):133–6.
21. Saied MH, Marzec J, Khalid S, Smith P, Down TA, Rakyan VK, et al. Genome wide analysis of acute myeloid leukemia reveal leukemia specific methylome and subtype specific hypomethylation of repeats. PLoS One. 2012;7(3):e33213.
22. Ruike Y, Imanaka Y, Sato F, Shimizu K, Tsujimoto G. Genome-wide analysis of aberrant methylation in human breast cancer cells using methyl-DNA immunoprecipitation combined with high-throughput sequencing. BMC Genomics. 2010;11:137.
23. Zhao Y, Guo S, Sun J, Huang Z, Zhu T, Zhang H, et al. Methylcap-seq reveals novel DNA methylation markers for the diagnosis and recurrence prediction of bladder cancer in a Chinese population. PLoS One. 2012;7(4):e35175.
24. Hughes Carvalho R, Haberle V, Hou J, van Gent T, Thongjuea S, van Ijcken W, et al. Genome-wide DNA methylation profiling of non-small cell lung carcinomas. Epigenetics Chromatin. 2012;5(1):9.
25. Kim JH, Dhanasekaran SM, Prensner JR, Cao X, Robinson D, Kalyana-Sundaram S, et al. Deep sequencing reveals distinct patterns of DNA methylation in prostate cancer. Genome Res. 2011;21(7):1028–41.
26. Yu W, Jin C, Lou X, Han X, Li L, He Y, et al. Global analysis of DNA methylation by methyl-capture sequencing reveals epigenetic control of cisplatin resistance in ovarian cancer cell. PLoS One. 2011;6(12):e29450.
27. Milosavljevic A. Emerging patterns of epigenomic variation. Trends Genet. 2011;27(6): 242–50.
28. Barski A, Cuddapah S, Cui K, Roh TY, Schones DE, Wang Z, et al. High-resolution profiling of histone methylations in the human genome. Cell. 2007;129(4):823–37.
29. Rosenbloom KR, Dreszer TR, Long JC, Malladi VS, Sloan CA, Raney BJ, et al. ENCODE whole-genome data in the UCSC Genome Browser: update 2012. Nucleic Acids Res. 2012;40(Database issue):D912–7.
30. Choe MK, Hong CP, Park J, Seo SH, Roh TY. Functional elements demarcated by histone modifications in breast cancer cells. Biochem Biophys Res Commun. 2012;418(3):475–82.
31. Dudziec E, Gogol-Doring A, Cookson V, Chen W, Catto J. Integrated epigenome profiling of repressive histone modifications, DNA methylation and gene expression in normal and malignant urothelial cells. PLoS One. 2012;7(3):e32750.
32. Statham AL, Robinson MD, Song JZ, Coolen MW, Stirzaker C, Clark SJ. Bisulfite sequencing of chromatin immunoprecipitated DNA (BisChIP-seq) directly informs methylation status of histone-modified DNA. Genome Res. 2012;22(6):1120–7.
33. Mossman D, Scott RJ. Long term transcriptional reactivation of epigenetically silenced genes in colorectal cancer cells requires DNA hypomethylation and histone acetylation. PLoS One. 2011;6(8):e23127.

34. Huntzinger E, Izaurralde E. Gene silencing by microRNAs: contributions of translational repression and mRNA decay. Nat Rev Genet. 2011;12(2):99–110.
35. Kozomara A, Griffiths-Jones S. miRBase: integrating microRNA annotation and deep-sequencing data. Nucleic Acids Res. 2011;39(Database issue):D152–7.
36. Friedman RC, Farh KK, Burge CB, Bartel DP. Most mammalian mRNAs are conserved targets of microRNAs. Genome Res. 2009;19(1):92–105.
37. Calin GA, Dumitru CD, Shimizu M, Bichi R, Zupo S, Noch E, et al. Frequent deletions and down-regulation of micro-RNA genes miR15 and miR16 at 13q14 in chronic lymphocytic leukemia. Proc Natl Acad Sci U S A. 2002;99(24):15524–9.
38. Zhang L, Huang J, Yang N, Greshock J, Megraw MS, Giannakakis A, et al. microRNAs exhibit high frequency genomic alterations in human cancer. Proc Natl Acad Sci U S A. 2006;103(24):9136–41.
39. Agirre X, Martinez-Climent JA, Odero MD, Prosper F. Epigenetic regulation of miRNA genes in acute leukemia. Leukemia. 2012;26(3):395–403.
40. Rajagopalan R, Vaucheret H, Trejo J, Bartel DP. A diverse and evolutionarily fluid set of microRNAs in Arabidopsis thaliana. Genes Dev. 2006;20(24):3407–25.
41. Morin RD, O'Connor MD, Griffith M, Kuchenbauer F, Delaney A, Prabhu AL, et al. Application of massively parallel sequencing to microRNA profiling and discovery in human embryonic stem cells. Genome Res. 2008;18(4):610–21.
42. Ruby JG, Jan CH, Bartel DP. Intronic microRNA precursors that bypass Drosha processing. Nature. 2007;448(7149):83–6.
43. Babiarz JE, Ruby JG, Wang Y, Bartel DP, Blelloch R. Mouse ES cells express endogenous shRNAs, siRNAs, and other Microprocessor-independent, Dicer-dependent small RNAs. Genes Dev. 2008;22(20):2773–85.
44. Burroughs AM, Ando Y, de Hoon MJ, Tomaru Y, Nishibu T, Ukekawa R, et al. A comprehensive survey of 3′ animal miRNA modification events and a possible role for 3′ adenylation in modulating miRNA targeting effectiveness. Genome Res. 2010;20(10):1398–410.
45. Ryland GL, Bearfoot JL, Doyle MA, Boyle SE, Choong DY, Rowley SM, et al. MicroRNA genes and their target 3′-untranslated regions are infrequently somatically mutated in ovarian cancers. PLoS One. 2012;7(4):e35805.
46. Volinia S, Galasso M, Sana ME, Wise TF, Palatini J, Huebner K, et al. Breast cancer signatures for invasiveness and prognosis defined by deep sequencing of microRNA. Proc Natl Acad Sci U S A. 2012;109(8):3024–9.
47. Schulte JH, Marschall T, Martin M, Rosenstiel P, Mestdagh P, Schlierf S, et al. Deep sequencing reveals differential expression of microRNAs in favorable versus unfavorable neuroblastoma. Nucleic Acids Res. 2010;38(17):5919–28.
48. Persson H, Kvist A, Rego N, Staaf J, Vallon-Christersson J, Luts L, et al. Identification of new microRNAs in paired normal and tumor breast tissue suggests a dual role for the ERBB2/Her2 gene. Cancer Res. 2011;71(1):78–86.
49. Schotte D, Moqadam FA, Lange-Turenhout EA, Chen C, van Ijcken WF, Pieters R, et al. Discovery of new microRNAs by small RNAome deep sequencing in childhood acute lymphoblastic leukemia. Leukemia. 2011;25(9):1389–99.
50. Blow MJ, Grocock RJ, van Dongen S, Enright AJ, Dicks E, Futreal PA, et al. RNA editing of human microRNAs. Genome Biol. 2006;7(4):R27.
51. Kawahara Y, Zinshteyn B, Sethupathy P, Iizasa H, Hatzigeorgiou AG, Nishikura K. Redirection of silencing targets by adenosine-to-inosine editing of miRNAs. Science. 2007;315(5815):1137–40.
52. Liu R, Zhang C, Hu Z, Li G, Wang C, Yang C, et al. A five-microRNA signature identified from genome-wide serum microRNA expression profiling serves as a fingerprint for gastric cancer diagnosis. Eur J Cancer. 2011;47(5):784–91.
53. Yu S, Liu Y, Wang J, Guo Z, Zhang Q, Yu F, et al. Circulating microRNA profiles as potential biomarkers for diagnosis of papillary thyroid carcinoma. J Clin Endocrinol Metab. 2012;97(6):2084–92.

54. Liu R, Chen X, Du Y, Yao W, Shen L, Wang C, et al. Serum microRNA expression profile as a biomarker in the diagnosis and prognosis of pancreatic cancer. Clin Chem. 2012;58(3):610–8.
55. Ficz G, Branco MR, Seisenberger S, Santos F, Krueger F, Hore TA, et al. Dynamic regulation of 5-hydroxymethylcytosine in mouse ES cells and during differentiation. Nature. 2011;473(7347):398–402.
56. Meyer KD, Saletore Y, Zumbo P, Elemento O, Mason CE, Jaffrey SR. Comprehensive analysis of mRNA methylation reveals enrichment in 3′ UTRs and near stop codons. Cell. 2012;149(7):1635–46.
57. Booth MJ, Branco MR, Ficz G, Oxley D, Krueger F, Reik W, et al. Quantitative sequencing of 5-methylcytosine and 5-hydroxymethylcytosine at single-base resolution. Science. 2012;336(6083):934–7.
58. Jin SG, Wu X, Li AX, Pfeifer GP. Genomic mapping of 5-hydroxymethylcytosine in the human brain. Nucleic Acids Res. 2011;39(12):5015–24.
59. Jia G, Fu Y, Zhao X, Dai Q, Zheng G, Yang Y, et al. N6-methyladenosine in nuclear RNA is a major substrate of the obesity-associated FTO. Nat Chem Biol. 2011;7(12):885–7.
60. Nestor CE, Ottaviano R, Reddington J, Sproul D, Reinhardt D, Dunican D, et al. Tissue type is a major modifier of the 5-hydroxymethylcytosine content of human genes. Genome Res. 2012;22(3):467–77.
61. Haffner MC, Chaux A, Meeker AK, Esopi DM, Gerber J, Pellakuru LG, et al. Global 5-hydroxymethylcytosine content is significantly reduced in tissue stem/progenitor cell compartments and in human cancers. Oncotarget. 2011;2(8):627–37.
62. Kudo Y, Tateishi K, Yamamoto K, Yamamoto S, Asaoka Y, Ijichi H, et al. Loss of 5-hydroxymethylcytosine is accompanied with malignant cellular transformation. Cancer Sci. 2012;103(4):670–6.
63. Rye M, Saetrom P, Handstad T, Drablos F. Clustered ChIP-Seq-defined transcription factor binding sites and histone modifications map distinct classes of regulatory elements. BMC Biol. 2011;9:80.
64. Ghildiyal M, Zamore PD. Small silencing RNAs: an expanding universe. Nat Rev Genet. 2009;10(2):94–108.
65. Khalil AM, Guttman M, Huarte M, Garber M, Raj A, Rivea Morales D, et al. Many human large intergenic noncoding RNAs associate with chromatin-modifying complexes and affect gene expression. Proc Natl Acad Sci U S A. 2009;106(28):11667–72.
66. Prensner JR, Iyer MK, Balbin OA, Dhanasekaran SM, Cao Q, Brenner JC, et al. Transcriptome sequencing across a prostate cancer cohort identifies PCAT-1, an unannotated lincRNA implicated in disease progression. Nat Biotechnol. 2011;29(8):742–9.

Chapter 3
Next-Generation Sequencing for Cancer Genomics

Aarti N. Desai and Abhay Jere

Abstract In the last couple of decades, availability of high-throughput genomic technologies such as microarrays and next-generation sequencing (NGS) has provided unprecedented insights into the complexity of cancer genomics. In particular, NGS with its ability to provide an unbiased view of the genome is a very useful tool in studying the cancer genome which is characterized by de novo genetic aberrations. Using NGS, gene expression signatures, copy number variations, mutations, and epigenetic changes such as methylation as well as histone modifications can be identified which could point towards novel diagnostic and/or prognostic biomarkers. Comprehensive understanding of the cancer genomics could also provide mechanistic insights into cancer susceptibility, development, and progression. This chapter provides an overview of the studies that have applied NGS technologies to further our understanding of cancer.

Keywords Next-generation sequencing • Cancer • Genetics • Genomics • Carcinomas • Transcriptome • Methylation • SNP

1 Introduction

Cancer is a complex and heterogeneous disease that results from either inherited or somatic genetic aberrations such as single base substitutions, insertions, deletions, copy number variations, and epigenetic modifications (Fig. 3.1). Since the genetic basis of cancer has long been acknowledged, cancer research has seen the application of some of the latest advances in genomic technologies [1–5]. Next-generation

A.N. Desai (✉) • A. Jere
Persistent Labs, Persistent Systems Ltd., Pingala – Aryabhata, 12A/12,
Off Karve Road, Erandwane, 411004 Pune, India
e-mail: aarti_desai@persistent.co.in

W. Wu and H. Choudhry (eds.), *Next Generation Sequencing in Cancer Research: Volume 1: Decoding the Cancer Genome*, DOI 10.1007/978-1-4614-7645-0_3,

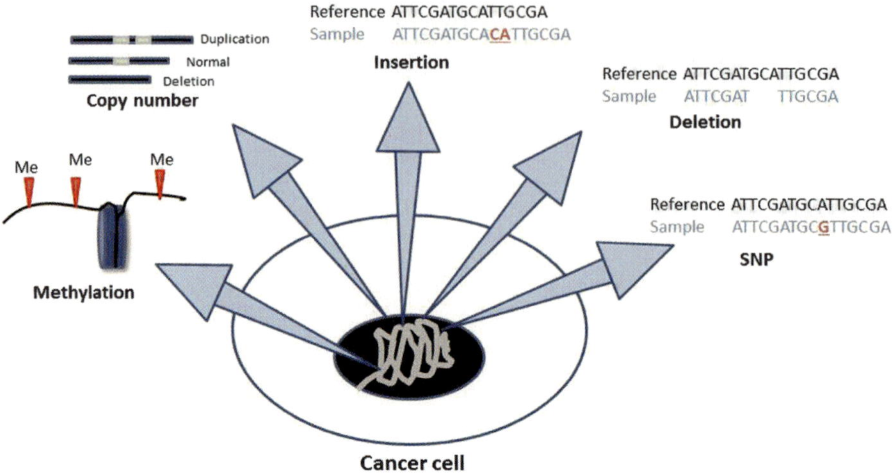

Fig. 3.1 Genetic aberrations in cancer cells. The genetic aberrations (germ line or somatic) found in cancer cells are single base substitutions (SNPs), insertions, deletions, copy number variations, and epigenetic modifications such as hypo- or hypermethylation

sequencing (NGS) technologies are the latest sets of advanced genomic tools to be applied to cancer. Over the past few years, the advances made in NGS technologies have enabled a more thorough understanding of the genetics underlying cancer.

NGS are high-throughput sequencing technologies that can perform millions of sequencing reactions in parallel [6, 7]. Illumina HiSeq, Roche 454, and Ion torrent PGM are the three leading NGS platforms currently in the market. In principal, sequence generation using an NGS platforms has the following common steps: [8–11] (1) DNA to be sequenced is sheared and universal adaptors are ligated to DNA fragments of a selected size, (2) size-selected fragments are clonally amplified (emulsion PCR or bridge PCR) to generate a few thousand copies of each of the fragment, (3) amplified fragments are sequenced by synthesizing the complementary strand (either sequencing by synthesis or sequencing by ligation), and (4) base incorporation event is detected (e.g., image capture of fluorescent dye or light emission signal or change in current). The sequence data is generated in the form of millions of short reads (75–400 bp), and based on the platform used, the throughput can range from 1 to 600 GB in a single run (Table 3.1). A detailed discussion of the sequencing chemistries for each of these platforms as well as their comparison is not in the scope of this chapter, and several excellent reviews are already available [7, 12–16]. Briefly, the high scale of throughput and low cost of sequencing are the biggest enabler for the undertaking large-scale genomic projects which can provide a better understanding of the heterogeneity and complexity of genetic alterations in cancer.

Several international, multicentric projects such as the International Cancer Genome Consortium (ICGC; [17]) and The Cancer Genome Atlas (TCGA; http://cancergenome.nih.gov/) are currently using NGS technology to unravel the genomic and epigenomic landscape of some of the most prevalent forms of cancer.

Table 3.1 Comparison of sequencing platforms

Sequencer name	Read length	Throughput	Cost per million bases[a]	Advantages	Disadvantages
Illumina HiSeq series	100–150 bp	60–600 GB	$ 0.07	High throughput High-quality data Multiplexing	Complex data analysis Short read length
Illumina MiSeq	35–150 bp	Up to 1 GB per run	$ 0.5	Low-cost bench top instrument High-quality data Multiplexing	Complex data analysis Low throughput
Life Technologies' SOLiD	35–75 bp	Up to 120 GB per run	$ 0.13	Low error rate Proven sequencing technology Multiplexing	Complex data analysis
Ion Torrent Personal Genome Machine (PGM)	100–200 bp	Up to 1 GB with 318 chip	$ 0.63	Low-cost, bench top instrument Short run time Direct signal detection Easy sample preparation	Low throughput High cost of sequence generation
Roche 454 GS FLX	700 bp	700 Mb per run	$ 10	Long read length Low error rate Proven sequencing technology	Low throughput High cost of sequence generation
Roche 454 Junior	400 bp	35 MB per run	$ 0.42[b]	Low cost, bench top instrument Short run time Long read length Low error rate	Low throughput High cost of sequence generation

[a]Cost per base information (except for Roche 454 Junior) is obtained from Liu et al. [16]

[b]Estimated from the data provided in [99]

Additionally, numerous whole genome, exome, and targeted resequencing studies have helped identify inherited as well as somatic mutations in cancers such as diffuse large cell lymphoma [18], pancreatic cysts [19], chronic myelocytic leukemia [20], acute myeloid leukemia [21], multiple myeloma [22], and breast cancer [23]. In this chapter, we will discuss some prominent studies that have used NGS technology and contributed significantly to further our understanding of cancer genomics.

2 Understanding Cancer Genomics Using the Whole Genome Sequencing Approach

The human genome is made of approximately 3 billion base pairs, divided in 24 chromosomes, and only about 5 % of the genome contains genes with even lesser portion (1–2 %) coding for proteins [24, 25]. Genome of every individual is unique and when compared to other individuals harbors genetic differences in the form of single base substitutions or single nucleotide polymorphisms (SNPs), copy number variations (multiplications, deletions), or translocations. Single base substitutions or SNPs are the most common type of genetic variation in the human genome. As of the latest release of dbSNP (build 137), there are a little over 38 million validated SNPs identified in humans (http://www.ncbi.nlm.nih.gov/projects/SNP/snp_summary.cgi). Additionally, a large number of copy number variations (CNVs) have been identified in the human genome [26–29]. Copy number variations are DNA aberrations that result in an abnormal number of copies (greater than or less than 2 in case of the diploid human genome) of small regions or entire chromosomes. Copy number alterations or gross chromosomal abnormalities are some of the most commonly identified genetic aberrations in cancer genomes and are believed to play an important role in cancer [30, 31].

Genetic variations present in germ line cells such as the gametes are called germ line variations and could be inherited from the parents, whereas variations acquired by an individual during their lifetime are called somatic variations. Majority of these variations are benign, i.e., they result in differences such as hair color, eye color, height, and skin color. However, a few variations can be pathogenic and contribute to increased susceptibility to diseases such as diabetes [32–35], hypertension [36–39], autism [40–44], and cancer [21, 45–47].

As mentioned earlier in the chapter, NGS technologies provide an unbiased view of the alterations in the genome, and since many a times cancer arises from novel somatic mutations, it provides a very useful tool to study the alterations in cancer genome. Additionally, since a very high throughput can be achieved on the NGS platforms, it is possible to sequence entire human genome at a depth that can provide meaningful information regarding the genomic aberrations. Hence, in cases where there is no prior knowledge of genetic aberrations underlying the cancer phenotype, the approach taken is to sequence the entire genome.

Link et al. [21] used the whole genome sequencing approach to identify mutations in the genome of a patient with early onset breast and ovarian cancer who also developed therapy-related acute myeloid leukemia. The patient had no family history of breast or ovarian cancer and tested negative for the conventional BRCA1 and BRCA2 mutations tested by commercial tests. Performing whole genome sequencing on the skin (normal) and bone marrow (leukemia) genome of the patient revealed a novel 3 kb heterozygous deletion in the TP53 gene (exons 7–9) of the normal genome and a 17.6 MB region of uniparental disomy on chromosome 17 with resultant homozygous deletion of the same region (exons 7–9 of TP53 gene) of the leukemia genome. The loss of exons 7–9 results in loss of DNA binding domain in the TP53 protein and thus produces a functionally defective protein. The authors concluded that this deletion mutation in the TP53 gene was most likely contributing to high cancer susceptibility in the patient. Similarly, Berger et al. [48] performed whole genome sequencing on tumor and matched normal genomic DNA of seven patients with "high-risk" primary prostate cancer. Among the significant findings of this study was the discovery of structural rearrangements in PTEN tumor suppressor gene and a PTEN-interacting protein MAGI (membrane-associated guanylate kinase, WW and PDZ domain containing 2). The authors suggest that the rearrangement events that disrupt the functionality of PTEN and associated proteins may (e.g., through loss of *MAGI2*) adversely affect the PI3 kinase signaling in prostate cancer.

Another example is the whole genome sequencing of tumor/normal samples from three pancreatic adenocarcinoma (PAC) patients performed by Liang et al. [49]. According to the authors, even though mutations in genes such as *KRAS* (v-Ki-ras2 Kirsten rat sarcoma viral oncogene homolog), *TP53* (tumor protein p53), *SMAD4*/*DPC4* (SMAD family member 4/deletion target in pancreatic carcinoma 4 homolog), *CDKN2A* (cyclin-dependent kinase inhibitor 2A; p16), and *BRCA2* (breast cancer 2, early onset) are heavily implicated in PAC, there is a dearth of effective treatment. With the aim of gaining a comprehensive understanding of the PAC genome, they performed transcriptome profiling and whole genome sequencing (WGS) of tumor biopsy DNA as well as matched normal (blood) DNA from three separate PAC patients. Using this integrated approach, the authors identified that the KRAS pathway was the most significantly impacted pathway in the three tumor samples they studied.

Aforementioned studies successfully employed whole genome sequencing in order to identify mutations underlying increased susceptibility to cancer as well as potentially new therapeutic targets. However, performing whole genome sequencing routinely for clinical diagnostics is impractical and often not necessary. In majority cancers, candidate genes are known, and only sequencing the coding region (exome) or smaller regions is a much more feasible strategy.

3 Understanding Cancer Genomics Using the Targeted Resequencing Approach

Targeted sequencing approach involves sequencing only the regions of interest (exome or a few candidate regions) and provides the benefit of cost as well as ease of analysis [50, 51]. In order to select genomic region to sequence, probes or "baits" are designed which specifically hybridize to the region of interest. DNA to be sequenced is first fragmented and then hybridization reaction is carried out either in solution or on a solid surface to enrich for regions to be sequenced. Sample library is prepared in a manner similar to whole genome sequencing, except that barcode sequences can also be included to facilitate multiplexing. Both whole exome and targeted sequencing methods have been extensively employed in cancer genomics.

Walsh et al. [46] have designed a genetic test "BROCA" to identify mutations in women diagnosed with primary ovarian, peritoneal, or fallopian tube carcinoma. The BROCA test includes a panel of 21 tumor suppressor genes, BRCA1, BRCA2, CHEK2, PALB2, BRIP1, p53, PTEN, STK11, CDH1, ATM, BARD1, MLH1, MRE11, MSH2, MSH6, MUTYH 1, NBN, PMS1, PMS2, RAD50, and RAD51C, that the same group had previously successfully tested in breast cancer patients [23]. In the ovarian cancer study, Walsh et al. [46] reported that the 360 subjects that they sequenced for the 21 genes contained germ line loss of mutations in 12 genes (BRCA1, BRCA2, TP53, RAD51C, RAD50, PALB2, NBN, MSH6, MRE11, CHEK2, BRIP1, and BARD1). Out of these genes, BARD1, BRIP1, CHEK2, MRE11A, NBN, and RAD50 were not previously implicated in inherited ovarian carcinoma. According to the authors, BROCA test is a highly sensitive and cost-effective test that can be used in diagnosing breast, ovarian, and fallopian tube carcinoma. Now, Ambry Genetics has made available a breast cancer screening test called BreastNext™ (http://ambrygen.com/tests/breastnext) based on the findings from Walsh et al. [23], Heikkinen et al. [52], and Ghimenti et al. [53].

Similarly, Pritchard et al. [47] have reported a comprehensive targeted assay panel called "ColoSeq" to identify all classes of mutations in Lynch and polyposis syndrome genes. They used this test for colon cancer specimens and colon cancer cell lines with defined mutations to correctly identify 100 % pathogenic mutations in MLH1, MSH2, MSH6, PMS2, EPCAM, APC, and MUTYH, including single nucleotide variants (SNVs), small insertions and deletions, and large copy number variants. In order to validate the assay further, they used the panel in HapMap samples and identified 222 of 224 heterozygous SNVs (99.4 %). The authors suggest that ColoSeq is a sensitive and cost-effective test for Lynch and polyposis syndromes. Moreover, Baylor Medical College, Houston, TX, has a comprehensive cancer gene mutation panel (https://www.bcm.edu/geneticlabs/test_detail.cfm?testcode=9700) that uses targeted resequencing approach to sequence DNA from tumor and normal samples.

In addition to targeted resequencing approach, exome sequencing has been applied to differentiate between subtypes of pancreatic cysts by Wu et al. [19]. They performed exome sequencing on genetic material obtained from the pancreatic cyst fluid to identify mutations that are unique to each of the four malignant cysts: intraductal papillary mucinous neoplasms (IPMNs), mucinous cystic neoplasms (MCNs), serous cystadenomas (SCAs), and solid pseudopapillary neoplasms (SPNs). Based on their analysis, the authors short-listed five genes (VHL, RNF43, CTNNB1, GNAS, and KRAS) that are mutated uniquely in each of the cyst types and can be used as biomarkers for classification. They found that in 4 out of 8 SCAs analyzed, there was a mutation in the von Hippel–Lindau gene (VHL) which is a key component of the VHL ubiquitin ligase complex. The remaining four SCAs had Loss of Heterozygosity (LOH) in or adjacent to VHL. None of the SCAs had mutations in the other four genes short listed by the authors. Mutations were observed on RNF3, GNAS, or KRAS, but not in CTNNB1 or VHL in all the eight IPMN exomes sequenced. Similarly, mutations were only found in RNF4 and KRAS in the MCN cyst type, and the SPN cysts only contained CTNNB1 mutations. The authors make a case that uniqueness of these mutations in each of the cyst types along with clinical and radiological data could provide an accurate classification of the cyst type.

Aforementioned are just some examples of targeted and exome sequencing to highlight the utility of this approach in diagnostics and classification. There are hundreds of other reports available in which targeted sequencing approach is applied for different forms of cancer such as clear cell renal carcinoma [54], pediatric medulloblastoma [31], diffuse large B cell lymphoma [18], and glioblastoma [55].

4 Understanding Cancer Genomics Using Transcriptome Profiling Approach

A transcriptome can be simplistically defined as the entire collection of transcripts present in a cell at a given point in time. The transcripts serve as link between the genotype of an individual and the observed phenotype. The human genome has 3 billion bases, out of which only about 1–2 % contains protein coding genes. Genes are composed of exons, introns, and regulatory region. The process of transcription involves making a copy of the gene (exons and introns) to produce a precursor mRNA and processing the precursor mRNA to remove the introns while fusing the exons to produce a mature RNA or the transcript. The precursor mRNA can be processed in multiple ways to produce multiple transcripts by a process called alternate splicing (Fig. 3.2). Thus, a transcriptome does not follow one gene–one transcript relationship. Every gene can give rise to multiple transcripts, and it has now been acknowledged that a large number (between 35–75 %) of genes have multiple transcripts [24, 25, 56]. The complement of transcripts present in a cell at any given time depends on the function and the physiological state of the cell. Hence, profiling the transcriptome of a particular cell or tissue can provide important mechanistic insights in the observed phenotype of the cells.

Microarrays have been used extensively to generate gene expression profiles that enable classification of tumors and predict survival as well as response to therapy. Van de Vijver et al. [57] used microarrays to generate gene expression profiles for RNA extracted from 295 female patients with primary breast carcinoma. All patients in the study had stage I or II breast cancer and were younger than 53 years old. Out of the 295 patients, 151 had lymph node- negative disease, and 144 had lymph node-positive disease. The authors examined the gene expression ratio of 70 marker genes that they have previously established as signature of prognosis [58] in these 295 patients and were able to show a very good correlation between the prognosis signature and the absence of (early) distant metastases or death.

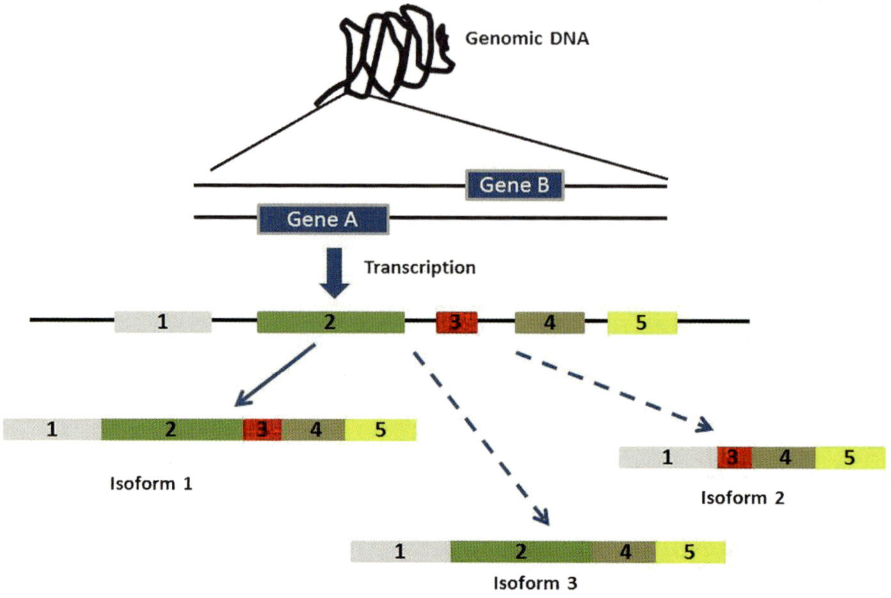

Fig. 3.2 Alternate splicing of mRNA to produce multiple isoforms. Genes are composed of exons, introns, and regulatory region. The *solid boxes* 1–5 are the five exons present in the gene and the *solid lines* are introns. The process of transcription involves making a copy of the gene (exons and introns) to produce a precursor mRNA and processing the precursor mRNA to remove the introns while fusing the exons to produce a mature RNA or transcript. In this example, the native isoform (Isoform 1) has all the five exons in the transcript, whereas in case of isoform 2, exon 2 is spliced out, and in case of ifiveoform 3, exon 3 is spliced out

Commercial microarray-based assay such as the MammaPrint® is an FDA approved test that is available as a predictive test for women who are under the age of 61 years and have either estrogen receptor-positive or -negative, lymph node-negative breast cancer [59]. Even though microarrays have been very useful in gene expression profiling, they are being slowly replaced by next-generation sequencing approaches as the latter provides an opportunity to drill down till isoform level and identify novel exon–exon junctions, novel isoforms, and allele-specific expression and thus provides a complete picture of the transcriptome.

Seo et al. [60] applied a comprehensive exome and transcriptome sequencing approach in 87 lung adenocarcinoma and 77 matched control samples from Korean patients. They found 45 in-frame fusion transcripts in total from the 87 cancer samples. According to the authors, the eight chimeric tyrosine kinases that they found in the cancer samples are likely to play an important role in cancer development. According to the literature cited by the authors, EML4-ALK has been identified in non-small-cell lung cancer [61, 62], KIF5B-RET in lung adenocarcinoma [63, 64], and CD74-ROS1and SLC34A2-ROS1 in lung cancer [65, 66]. However, four out of the eight fusion genes (*CCDC6–ROS1*, *FGFR2–CIT*, *AXL–MBIP*, and

SCAF11–PDGFRA) identified in their study were novel and carry protein tyrosine kinase domains and dimerization, features that are essential to activate chimeric tyrosine kinases. Moreover, the authors also mention that the other fusion genes identified in the same study, such as *MAP4K3–PRKCE*, *BCAS3–MAP3K3*, *ERBB2IP–MAST4*, and *APLP2–TNFSF11*, may also have functional importance since the genes are serine–threonine kinases or involved in signaling pathways.

Eswaran et al. [67] have reported a comprehensive transcriptome profile of the triple negative breast cancer (TNBC), non-TNBC, and HER2-positive breast cancers. In each of the three breast cancer types, they were able to identify an average of 80,279 known transcripts (p value and FDR less than 0.05), which seem to arise from diverse array of isoforms present in the cancer samples since the number of genes were only 16,245. Along with defining the native transcriptome of each of the cancer type, the authors performed gene expression analysis across the three cancer types in order to identify genes that were differentially expressed. Their analysis identified 2,617 transcripts that were differentially expressed between the TNBC and non-TNBC groups, out of which 962 transcripts were upregulated and 1,655 transcripts were downregulated.

The studies mentioned here demonstrate the power of NGS technologies in identifying novel gene fusion events, which are very common, in cancer tissues. The technology can also be employed to elucidate in very minute detail the complete repertoire of transcripts expressed in cancer tissues. This information can be extremely useful in accurate classification and diagnosis of cancer.

5 Understanding Cancer Genomics Using Methylation Profiling

The phrase epigenetic modifications is used to describe heritable modifications which do not involve change in DNA sequence but include changes in DNA methylation pattern and histone modification (e.g., acetylation). DNA methylation or histone modifications change the chromatin structure such that gene transcription is affected. DNA methylation is a process by which a methyl group is added to cytosine (CpG) in the DNA sequence, and usually this acts as a suppressor of transcription. DNA methylation is very important in the process of cell development and differentiation [68, 69], and there is a very specific change in the pattern of DNA methylation during development and differentiation [70]. In the early stages of embryogenesis, the genome is demethylated with subsequent reestablishment of methylation. This orchestrated methylation–demethylation forms a very important mechanism by which gene expression, and thereby a cellular phenotype is regulated [71]. Importantly, the base pairing of the genome facilitates reciprocal maintenance of methylation during the subsequent replication cycles, and hence, epigenetic modifications have implications not only for the individual in which the modification arose, but also for subsequent generations. As DNA methylation is very important in regulating gene expression, any perturbation of the methylation state of the genome can result in altered gene expression. A number of studies have reported a relationship between altered methylation of genome and gene expression variability in different types of cancers [72–75].

Bisulfite sequencing was used by Tanaka et al. [76] to determine the methylation status of trefoil factor 1 (TFF1) in cancer cells from gastric cancer patients. Prior to sequencing, the authors performed expression analysis and found that the TFF1 gene was downregulated in the cancer cells and that low expression of TFF1 was associated with poor survival rate on the 108 patients that were treated with surgery alone, which could be because cells that have decreased expression of TFF1 are highly invasive. They were further able to establish that the TFF1 expression and lymph node metastasis as predictive factors for survival. The bisulfite sequencing, along with expression profiling, revealed that there is a strong relationship between DNA methylation status of TFF1 gene and its expression level. Specifically, methylation status of two CpG sites, one located close to a TATA box and another located close to hypoxia response element, was a strong indicator of TFF1 expression level. Thus, the authors were able to demonstrate that altered DNA methylation of the TFF1 gene alters its expression and can act as a useful marker for predicting survival rate of gastric cancer patients.

A novel NGS-based methodology, MethylPlex–next-generation sequencing (M-NGS), which enriches methylated DNA using restriction enzymes and requires minimal input genomic DNA was applied by Kim et al. [77] to decipher the genome-wide methylation pattern in prostate tissue. They studied 17 prostate tissues: six benign adjacent, two normal, five localized prostate cancer, and four metastatic prostate cancer specimens. Their analysis revealed significantly differentially methylated regions (DMRs) and methylation patterns on comparing with different prostate tissues. The methylation levels observed in the intergenic and intragenic region were comparable across the four prostate tissues; however, the methylation pattern in the promoter region showed marked differences. The normal tissue had lowest level of methylation in promoter region, while the prostate cancer and metastatic prostate samples had high levels of methylation. More importantly, their results were able to reproduce all of the previously published 56 prostate cancer methylated regions. In order to test the significance of the observed methylation differences, the authors also studied the transcription profiles of the prostate samples and found that hypermethylation of promoters in the cancer tissue were significantly associated with gene repression. Thus, the authors used a comprehensive approach to establish the utility of next-generation sequencing-based methylation analysis to further our understanding of methylation state of the cancer genome.

In addition to DNA methylation, histone modifications are also demonstrated to play an important role in cancer susceptibility and progression [78, 79]. Histones are proteins around which the DNA strand is wound to create a highly compact chromatin structure. Histones undergo various kinds of chemical modifications such as methylation, acetylation, phosphorylation, and ubiquitination at various residues to either activate or suppress gene expression. Some of the posttranslational histone modifications that are known to play a role in transcription regulation are histone H3 lysine 4 trimethylation (H3K4me3) and histone acetylation (e.g., H3K9/16ac) that are usually found at promoter regions or transcription start sites (TSSs).

Choe et al. [80] used ChIPSeq analysis to identify gene regulatory elements in human breast cancer cell lines. They tested for histone H3 lysine 4 trimethylation (H3K4me3), histone H3 lysine 4 methylation (H3K4me1), and histone 3 lysine 9 and 14 acetylation (H3K9/14ac) in the normal and breast cancer cell line. Their analysis showed that the H3K4me3 and H3K9/14ac were significantly enriched at promoter regions, and H3K4me1 has a relatively broad distribution over proximity of TSSs as well as other genomic regions.

6 Bioinformatic Analysis Approach for Resequencing, Transcriptome Profiling, and Methylation Analysis

Analysis of NGS data presents significant challenge because the amount of data that needs to be processed is high (a few hundred GB in case of whole genome sequencing), and the algorithms used to process the data require high-end computational resources. Almost all of the algorithms that are available for NGS data analysis are open-source algorithms developed at various academic labs during the course of projects undertaken by them. NGS data is produced from the sequencers in the form of short read data and the analysis is a multistep process (Figs. 3.3, 3.4 and 3.5). Performing quality check on the raw read data is an optional, but highly recommended, step in NGS data analysis (http://www.bioinformatics.babraham.ac.uk/projects/fastqc/). If needed, the data can be preprocessed to trim low quality and adaptor bases using tools such as the FastX toolkit (http://hannonlab.cshl.edu/fastx_toolkit/index.html). After preprocessing, the first common step in resequencing, transcriptome profiling, and methylation analysis is "mapping" of the read data to a reference genome. Tools such as BWA [81], Bowtie [82], and Tophat [83] can be used for mapping. In case of resequencing, the second step is variant calling which can be done using tools such as GATK [84] and SAMTools [85]. The third step is variant filtration and annotation using SNPEff [86].

In case of transcriptome profiling, post mapping the second step is quantification, and isoform detection usually is done using tools such as Cufflinks [87, 88] or MISO [89], followed by differential expression analysis using Cuffdiff [87, 88] and/or DESeq [90]. Once a list of differentially expressed genes is available, traditional methods such as pathway analysis, GSEA enrichment, and GO analysis are applied to derive biological significance of the genes.

In case of methylome analysis, the second step of the workflow is base-by-base calling of methylated bases using tools such as BisSNP [91], followed by quantification of % methylation of genome and identifying genes in the vicinity of abnormally methylated regions.

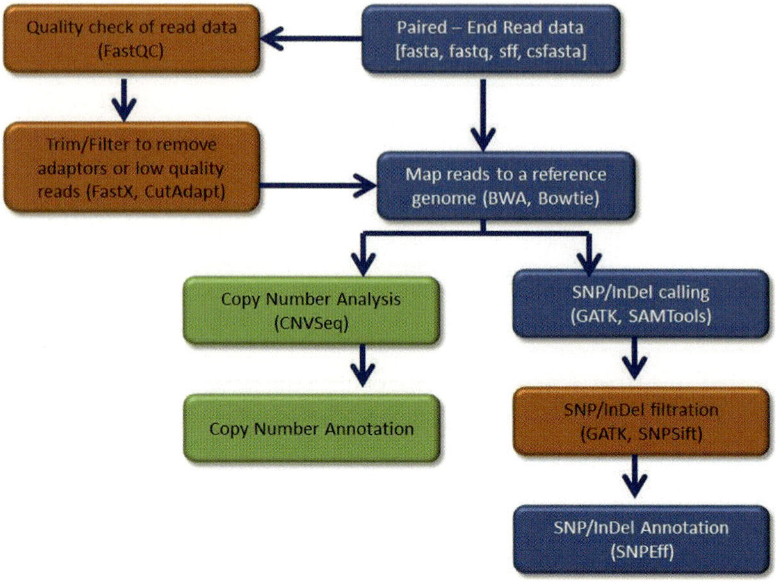

Fig. 3.3 Data analysis protocol for genetic variation detection. The genetic variation detection data analysis pipeline has the following steps: (1) *Quality check*: The step provides information about the average base quality per position of a read, % duplication, and adaptor contamination (if any). FastQC is the most commonly used tool for performing quality check on read data. (2) *Trimming/filtering*: Read data is filtered using tools such as FastX and Cutadapt to remove duplicate reads and trimmed to remove low quality and adaptor bases from the reads. (3) *Map reads to reference genome*: The raw read data (most commonly paired-end reads) are aligned against the reference genome using tools such as BWA and Bowtie. (4) *SNP/InDel/copy number calling*: The aligned read data is used by variant callers such as GATK, SamTools, and CNVSeq to call SNPs, InDel, and large copy number variations in the sample. (5) *SNP filtering*: SNPs are filtered using GATK or SNPSift to remove low-quality SNPs or SNPs with low coverage. (6) *SNP and copy number annotation*: SNPs and copy number variants are annotated with gene, impact, and database (e.g., dbSNP) information using tools like SNPEff. Quality check is an optional, but highly recommended step. Trimming/filtering and SNP filtering are also optional steps

7 Advantages and Disadvantages of NGS Technology

7.1 Advantages

(a) High throughput of NGS technologies enables sequencing of regions of interest at very high depth. This enables detection of alleles that occur at much lower frequency (MAF = 1 %; [92, 93]).
(b) NGS enables detection of novel genetic abnormalities since sequence of every base in the region of interest is detected.

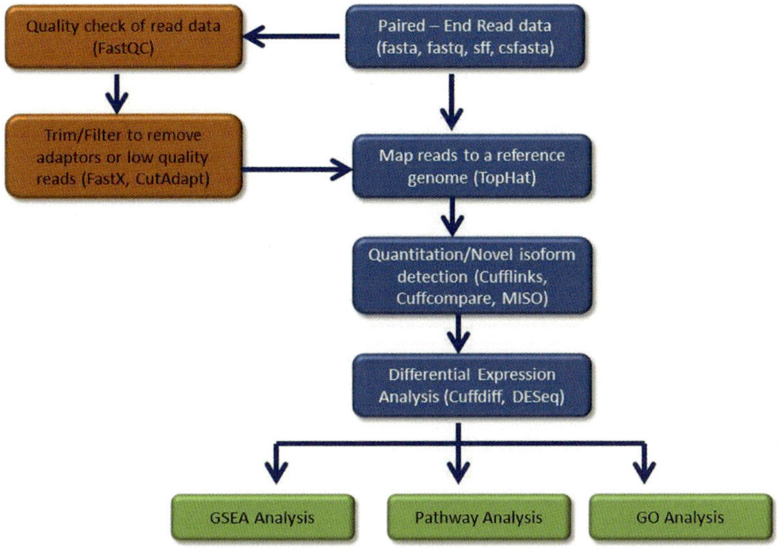

Fig. 3.4 Data analysis protocol for transcriptome profiling. The transcriptome profiling data analysis pipeline has the following steps: (1) *Quality check*: The step provides information about the average base quality per position of a read, % duplication, and adaptor contamination (if any). FastQC is the most commonly used tool for performing quality check on read data. (2) *Trimming/filtering*: Read data is filtered using tools such as FastX and Cutadapt to remove duplicate reads and trimmed to remove low quality and adaptor bases from the reads. (3) *Map reads to reference genome*: The raw read data (most commonly paired-end reads) are aligned against the reference genome using tools such as TopHat. (4) *Quantification/Novel isoform detection*: The aligned read data is then used to quantify the abundance at gene and isoform level using Cufflinks or MISO. This step can also provide information about novel genes/isoforms present in the sample. (5) *Differential expression analysis*: Perform statistical tests using Cuffdiff or DESeq (*t*-test, ANOVA) on the list of genes and isoforms to identify entities that are differentially expressed in conditions of the study. Once the list of differentially expressed genes is identified, the data can be analyzed for pathway and gene set enrichment. Quality check is an optional, but highly recommended step. Trimming/filtering is also an optional step

(c) Novel transcripts, exon–exon junctions, and allele-specific gene expression can be detected using the NGS approach [94, 95].
(d) A targeted resequencing and multiplexing allows processing of multiple samples in a single instrument run, thereby reducing the cost of per sample analysis [50, 96].
(e) Unlike in Sanger sequencing, a large number of loci of interest (multiple genes) can be analyzed in a single panel, thereby simplifying the tests while reducing the cost and turnaround time [46].
(f) Small amount of genetic material (as low as 50 ng) can be used to generate the sequence data [10].
(g) Sequence data can also be generated from FFPE [97] and partially degraded DNA samples (e.g., ancient DNA; [98]).

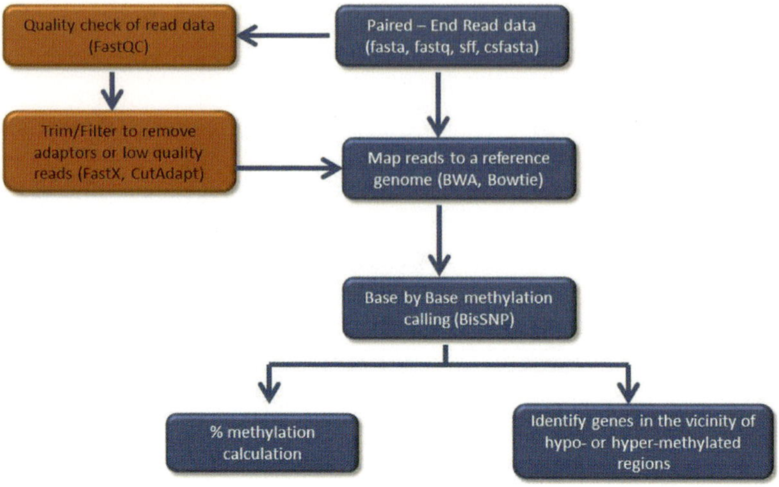

Fig. 3.5 Data analysis protocol for methylation analysis. The methylation data analysis pipeline has the following steps: (1) *Quality check*: The step provides information about the average base quality per position of a read, % duplication, and adaptor contamination (if any). FastQC is the most commonly used tool for performing quality check on read data. (2) *Trimming/filtering*: Read data is filtered using tools such as FastX and Cutadapt to remove duplicate reads and trimmed to remove low quality and adaptor bases from the reads. (3) *Map reads to reference genome*: The raw read data (most commonly paired-end reads) are aligned against the reference genome using tools such as BWA and Bowtie. (4) *Base-by-base methylation calling*: Methylation state of each base is called using tools such as BisSNP, and % methylation level is calculated. (5) *Identify genes that are in the vicinity of hypo- or hypermethylated regions* to short-list candidates with gene expression differences. Quality check is an optional, but highly recommended step. Trimming/filtering is also an optional step

7.2 Disadvantages

(a) A large amount of data is generated for every run on NGS systems. Analysis of the data to generate a meaningful output (annotated variants in case of genome sequencing) is a multistep process and requires high-end computing hardware (e.g., analyzing whole genome human data for variant analysis requires as much as 32 GB RAM and up to 500 GB storage space depending on the depth to which the genome is sequenced).
(b) Storage and distribution of the data requires significant investment in hardware and IT support.
(c) NGS data tends to have slightly higher error rate (0.1–1 %) than traditional Sanger sequencing (0.1 %). Hence, there is a high probability of obtaining false positive data subsequent to analysis.
(d) Due to the high error rate, any novel information (variants, transcripts, methylation sites) obtained from NGS data needs to be thoroughly validated.
(e) Though the cost of sequence generation is low, the cost associated with the analysis, storage, and validation of data can cause a significant increase in the overall cost of the test.

(f) Since NGS provides information about all the bases being sequenced, variants associated with other pathological conditions or unknown clinical significance could be identified in the patient. This could have implications not only for the patient but also immediate family members. There could be cases where other variants identified may predispose the patient and family members to late onset diseases such as Alzheimer's or Parkinson's or diseases with no cure. In such cases, physicians and genetic counselors need to make decision about whether all information should be shared with the patient and family members.
(g) Information about whether a person carries predisposing variants could be used to discriminate against them by insurance companies and future employers. The Genetic Information Nondiscrimination Act (GINA) was passed in 2008 in the United States to prevent against this happening. However, as sequencing becomes cheaper and more widespread, more comprehensive measures need to be in place to prevent misuse of genetic information.

8 Summary and Future Directions

With the rapid improvements in the sequencing technology and decrease in cost, it is likely that all cancer samples will be sequenced not only at the whole genome level to document all somatic and inherited mutations, but also at the transcriptome and epigenome level. Complete knowledge of an individual's cancer genome will facilitate accurate diagnosis, subtyping, and appropriate treatment choice, thereby improving the outcome for the patient.

References

1. Graham K, Ge X, de Las MA, Tripathi A, Rosenberg CL. Gene expression profiles of estrogen receptor-positive and estrogen receptor-negative breast cancers are detectable in histologically normal breast epithelium. Clin Cancer Res. 2011;17(2):236–46.
2. Uddin S, Ahmed M, Hussain A, Abubaker J, Al-Sanea N, AbdulJabbar A, et al. Genome-wide expression analysis of Middle Eastern colorectal cancer reveals FOXM1 as a novel target for cancer therapy. Am J Pathol. 2011;178(2):537–47.
3. Kim HK, Choi IJ, Kim CG, Kim HS, Oshima A, Michalowski A, et al. A gene expression signature of acquired chemoresistance to cisplatin and fluorouracil combination chemotherapy in gastric cancer patients. PLoS One. 2011;6(2):e16694.
4. Kadota M, Yang HH, Gomez B, Sato M, Clifford RJ, Meerzaman D, et al. Delineating genetic alterations for tumor progression in the MCF10A series of breast cancer cell lines. PLoS One. 2010;5(2):e9201.
5. Bessarabova M, Kirillov E, Shi W, Bugrim A, Nikolsky Y, Nikolskaya T. Bimodal gene expression patterns in breast cancer. BMC Genomics. 2010;11 Suppl 1:S8.
6. Shendure J, Ji H. Next-generation DNA sequencing. Nat Biotechnol. 2008;10:1135–45.
7. Metzker ML. Sequencing technologies – the next generation. Nat Rev Genet. 2010;11:31–46.
8. Margulies M, Egholm M, Altman WE, Attiya S, Bader JS, Bemben LA, et al. Genome sequencing in microfabricated high-density picolitre reactors. Nature. 2005;437(7057):376–80.

9. Shendure J, Porreca GJ, Reppas NB, Lin X, McCutcheon JP, Rosenbaum AM, et al. Accurate multiplex polony sequencing of an evolved bacterial genome. Science. 2005;309(5741):1728–32.
10. Bentley DR, Balasubramanian S, Swerdlow HP, Smith GP, Milton J, Brown CG, et al. Accurate whole human genome sequencing using reversible terminator chemistry. Nature. 2008;456(7218):53–9.
11. Rothberg JM, Hinz W, Rearick TM, Schultz J, Mileski W, Davey M, et al. An integrated semiconductor device enabling non-optical genome sequencing. Nature. 2011;475(7356):348–52.
12. Mardis ER. Next-generation DNA, sequencing methods. Annu Rev Genomics Hum Genet. 2008;9:387–402.
13. Tucker T, Marra M, Friedman JM. Massively parallel sequencing: the next big thing in genetic medicine. Am J Hum Genet. 2009;85(2):142–54.
14. Pareek CS, Smoczynski R, Tretyn A. Sequencing technologies and genome sequencing. J Appl Genet. 2011;52(4):413–35.
15. Desai AN, Jere A. Next generation sequencing: ready for the clinics? Clin Genet. 2012;81(6):503–10.
16. Liu L, Li Y, Li S, Hu N, He Y, Pong R, et al. Comparison of next generation sequencing systems. J Biomed Biotechnol. 2012;2012:251364.
17. Hudson TJ, Anderson W, Artez A, Barker AD, Bell C, Bernabé RR, et al; International Cancer Genome Consortium. International network of cancer genome projects. Nature. 2010; 464(7291):993–998
18. Lohr JG, Stojanov P, Lawrence MS, Auclair D, Chapuy B, Sougnez C, et al. Discovery and prioritization of somatic mutations in diffuse large B-cell lymphoma (DLBCL) by whole-exome sequencing. Proc Natl Acad Sci U S A. 2012;109(10):3879–84.
19. Wu J, Jiao Y, Dal Molin M, Maitra A, de Wilde RF, Wood LD, et al. Whole-exome sequencing of neoplastic cysts of the pancreas reveals recurrent mutations in components of ubiquitin-dependent pathways. Proc Natl Acad Sci U S A. 2011;108(52):21188–93.
20. Puente XS, Pinyol M, Quesada V, Conde L, Ordóñez GR, Villamor N, et al. Whole-genome sequencing identifies recurrent mutations in chronic lymphocytic leukemia. Nature. 2011;475(7354):101–5.
21. Link DC, Schuettpelz LG, Shen D, Wang J, Walter MJ, Kulkarni S, et al. Identification of a novel TP53 cancer susceptibility mutation through whole-genome sequencing of a patient with therapy-related AML. JAMA. 2011;305(15):1568–76.
22. Chapman PB, Hauschild A, Robert C, Haanen JB, Ascierto P, Larkin J, et al. Improved survival with vemurafenib in melanoma with BRAF V600E mutation. N Engl J Med. 2011;364(26):2507–16.
23. Walsh T, Lee MK, Casadei S, Thornton AM, Stray SM, Pennil C, et al. Detection of inherited mutations for breast and ovarian cancer using genomic capture and massively parallel sequencing. Proc Natl Acad Sci U S A. 2010;107(28):12629–33.
24. Lander ES, Linton LM, Birren B, Nusbaum C, Zody MC, Baldwin J, et al; International Human Genome Sequencing Consortium. Initial sequencing and analysis of the human genome. Nature. 2001; 409(6822):860–921.
25. Venter JC, Adams MD, Myers EW, Li PW, Mural RJ, Sutton GG, et al. The sequence of the human genome. Science. 2001;291:1304–51.
26. Levy S, Sutton G, Ng PC, Feuk L, Halpern AL, Walenz BP, et al. The diploid genome sequence of an individual human. PLoS Biol. 2007;5(10):e254.
27. Zogopoulos G, Ha KC, Naqib F, Moore S, Kim H, Montpetit A, et al. Germ-line DNA copy number variation frequencies in a large North American population. Hum Genet. 2007;122(3–4):345–53.
28. Cooper GM, Zerr T, Kidd JM, Eichler EE, Nickerson DA. Systematic assessment of copy number variant detection via genome-wide SNP genotyping. Nat Genet. 2008;40(10):1199–203.
29. Yim SH, Kim TM, Hu HJ, Kim JH, Kim BJ, Lee JY, et al. Copy number variations in East-Asian population and their evolutionary and functional implications. Hum Mol Genet. 2010;19(6):1001–8.

30. Stratton MR, Campbell PJ, Futreal PA. The cancer genome. Nature. 2009;458(7239):719–24.
31. Parsons DW, Li M, Zhang X, Jones S, Leary RJ, Lin JC, et al. The genetic landscape of the childhood cancer medulloblastoma. Science. 2011;331(6016):435–9.
32. Reinbothe TM, Alkayyali S, Ahlqvist E, Tuomi T, Isomaa B, Lyssenko V, et al. The human L-type calcium channel Ca(v)1.3 regulates insulin release and polymorphisms in CACNA1D associate with type 2 diabetes. Diabetologia. 2013;56(2):340–9.
33. Landman GW, van Vliet-Ostaptchouk JV, Kleefstra N, van Hateren KJ, Drion I, Groenier KH, et al. Association between 9p21 genetic variants and mortality risk in a prospective cohort of patients with type 2 diabetes (ZODIAC-15). Cardiovasc Diabetol. 2012;11:138.
34. Janipalli CS, Kumar MV, Vinay DG, Sandeep MN, Bhaskar S, Kulkarni SR, et al. Analysis of 32 common susceptibility genetic variants and their combined effect in predicting risk of Type 2 diabetes and related traits in Indians. Diabet Med. 2012;29(1):121–7.
35. Wang Y, Nie M, Li W, Ping F, Hu Y, Ma L, et al. Association of six single nucleotide polymorphisms with gestational diabetes mellitus in a Chinese population. PLoS One. 2011;6(11):e26953.
36. Gong Y, McDonough CW, Wang Z, Hou W, Cooper-Dehoff RM, Langaee TY, et al. Hypertension susceptibility Loci and blood pressure response to antihypertensives: results from the pharmacogenomic evaluation of antihypertensive responses study. Circ Cardiovasc Genet. 2012;5(6):686–91.
37. Saade S, Cazier JB, Ghassibe-Sabbagh M, Youhanna S, Badro DA, Kamatani Y, et al. Large scale association analysis identifies three susceptibility loci for coronary artery disease. PLoS One. 2011;6(12):e29427.
38. Wain LV, Verwoert GC, O'Reilly PF, Shi G, Johnson T, Johnson AD, et al. Genome-wide association study identifies six new loci influencing pulse pressure and mean arterial pressure. Nat Genet. 2011;43(10):1005–11.
39. Ehret GB, Munroe PB, Rice KM, Bochud M, Johnson AD, Chasman DI. International Consortium for Blood Pressure Genome-Wide Association Studies Genetic variants in novel pathways influence blood pressure and cardiovascular disease risk. Nature. 2011;478(7367):103–9.
40. Steinberg KM, Ramachandran D, Patel VC, Shetty AC, Cutler DJ, Zwick ME. Identification of rare X-linked neuroligin variants by massively parallel sequencing in males with autism spectrum disorder. Mol Autism. 2012;3(1):8.
41. Prandini P, Pasquali A, Malerba G, Marostica A, Zusi C, Xumerle L, et al. The association of rs4307059 and rs35678 markers with autism spectrum disorders is replicated in Italian families. Psychiatr Genet. 2012;22(4):177–81.
42. Williams NM, Franke B, Mick E, Anney RJ, Freitag CM, Gill M, et al. Genome-wide analysis of copy number variants in attention deficit hyperactivity disorder: the role of rare variants and duplications at 15q13.3. Am J Psychiatry. 2012;169(2):195–204.
43. Chung RH, Ma D, Wang K, Hedges DJ, Jaworski JM, Gilbert JR, et al. An X chromosome-wide association study in autism families identifies TBL1X as a novel autism spectrum disorder candidate gene in males. Mol Autism. 2011;2(1):18.
44. Casey JP, Magalhaes T, Conroy JM, Regan R, Shah N, Anney R, et al. A novel approach of homozygous haplotype sharing identifies candidate genes in autism spectrum disorder. Hum Genet. 2012;131(4):565–79.
45. Beroukhim R, Mermel CH, Porter D, Wei G, Raychaudhuri S, Donovan J, et al. The landscape of somatic copy-number alteration across human cancers. Nature. 2010;463(7283):899–905.
46. Walsh T, Casadei S, Lee MK, Pennil CC, Nord AS, Thornton AM, et al. Mutations in 12 genes for inherited ovarian, fallopian tube, and peritoneal carcinoma identified by massively parallel sequencing. Proc Natl Acad Sci U S A. 2011;108(44):18032–7.
47. Pritchard CC, Smith C, Salipante SJ, Lee MK, Thornton AM, Nord AS, et al. ColoSeq provides comprehensive lynch and polyposis syndrome mutational analysis using massively parallel sequencing. J Mol Diagn. 2012;14(4):357–66.
48. Berger MF, Lawrence MS, Demichelis F, Drier Y, Cibulskis K, Sivachenko AY, et al. The genomic complexity of primary human prostate cancer. Nature. 2011;470(7333):214–20.

49. Liang WS, Craig DW, Carpten J, Borad MJ, Demeure MJ, Weiss GJ, et al. Genome-wide characterization of pancreatic adenocarcinoma patients using next generation sequencing. PLoS One. 2012;7(10):e43192.
50. Teer JK, Bonnycastle LL, Chines PS, Hansen NF, Aoyama N, Swift AJ, et al. Systematic comparison of three genomic enrichment methods for massively parallel DNA sequencing. Genome Res. 2010;20(10):1420–31.
51. Tewhey R, Nakano M, Wang X, Pabón-Peña C, Novak B, Giuffre A, et al. Enrichment of sequencing targets from the human genome by solution hybridization. Genome Biol. 2009;10(10):R116.
52. Heikkinen K, Rapakko K, Karppinen SM, Erkko H, Knuutila S, Lundán T. RAD50 and NBS1 are breast cancer susceptibility genes associated with genomic instability. Carcinogenesis. 2006;27(8):1593–9.
53. Ghimenti C, Sensi E, Presciuttini S, Brunetti IM, Conte P, Bevilacqua G, et al. Germline mutations of the BRCA1-associated ring domain (BARD1) gene in breast and breast/ovarian families negative for BRCA1 and BRCA2 alterations. Genes Chromosomes Cancer. 2002;33(3):235–42.
54. Varela I, Tarpey P, Raine K, Huang D, Ong CK, Stephens P, et al. Exome sequencing identifies frequent mutation of the SWI/SNF complex gene PBRM1 in renal carcinoma. Nature. 2011;469(7331):539–42.
55. Keller A, Harz C, Matzas M, Meder B, Katus HA, Ludwig N, et al. Identification of novel SNPs in glioblastoma using targeted resequencing. PLoS One. 2011;6(6):e18158.
56. Neverov AD, Artamonova II, Nurtdinov RN, Frishman D, Gelfand MS, Mironov AA. Alternative splicing and protein function. BMC Bioinformatics. 2005;6:266–74.
57. van de Vijver MJ, He YD, van't Veer LJ, Dai H, Hart AA, Voskuil DW, et al. A gene-expression signature as a predictor of survival in breast cancer. N Engl J Med. 2002;347(25):1999–2009.
58. van't Veer LJ, Dai H, van de Vijver MJ, He YD, Hart AA, Bernards R, et al. Expression profiling predicts clinical outcome of breast cancer. Nature. 2002;415:530–6.
59. Slodkowska EA, Ross JS. MammaPrint 70-gene signature: another milestone in personalized medical care for breast cancer patients. Expert Rev Mol Diagn. 2009;9(5):417–22.
60. Seo JS, Ju YS, Lee WC, Shin JY, Lee JK, Bleazard T, et al. The transcriptional landscape and mutational profile of lung adenocarcinoma. Genome Res. 2012;22(11):2109–19.
61. Soda M, Choi YL, Enomoto M, Takada S, Yamashita Y, Ishikawa S, et al. Identification of the transforming EML4-ALK fusion gene in non-small-cell lung cancer. Nature. 2007;448:561–6.
62. Choi YL, Takeuchi K, Soda M, Inamura K, Togashi Y, Hatano S, et al. Identification of novel isoforms of the EML4-ALK transforming gene in non-small cell lung cancer. Cancer Res. 2008;68(13):4971–6.
63. Ju YS, Lee WC, Shin JY, Lee S, Bleazard T, Won JK, Kim YT, Kim JI, Kang JH, Seo JS. A transforming KIF5B and RET gene fusion in lung adenocarcinoma revealed from whole-genome and transcriptome sequencing. Genome Res. 2012;22:436–45.
64. Kohno T, Ichikawa H, Totoki Y, Yasuda K, Hiramoto M, Nammo T, Sakamoto H, Tsuta K, Furuta K, Shimada Y, et al. KIF5B-RET fusions in lung adenocarcinoma. Nat Med. 2012;18:375–7.
65. Rikova K, Guo A, Zeng Q, Possemato A, Yu J, Haack H, Nardone J, Lee K, Reeves C, Li Y, et al. Global survey of phosphotyrosine signaling identifies oncogenic kinases in lung cancer. Cell. 2007;131:1190–203.
66. Takeuchi K, Soda M, Togashi Y, Suzuki R, Sakata S, Hatano S, Asaka R, Hamanaka W, Ninomiya H, Uehara H, et al. RET, ROS1 and ALK fusions in lung cancer. Nat Med. 2012;18:378–81.
67. Eswaran J, Cyanam D, Mudvari P, Reddy SD, Pakala SB, Nair SS, et al. Transcriptomic landscape of breast cancers through mRNA sequencing. Sci Rep. 2012;2:264.
68. Schmidt CS, Bultmann S, Meilinger D, Zacher B, Tresch A, Maier KC, et al. Global DNA hypomethylation prevents consolidation of differentiation programs and allows reversion to the embryonic stem cell state. PLoS One. 2012;7(12):e52629.

69. Fonseca MB, Nunes AF, Morgado AL, Solá S, Rodrigues CM. TAp63γ Demethylation Regulates Protein Stability and Cellular Distribution during Neural Stem Cell Differentiation. PLoS One. 2012;7(12):e52417.
70. Novak P, Stampfer MR, Munoz-Rodriguez JL, Garbe JC, Ehrich M, Futscher BW, et al. Cell-type specific DNA methylation patterns define human breast cellular identity. PLoS One. 2012;7(12):e52299.
71. Lee ST, Xiao Y, Muench MO, Xiao J, Fomin ME, Wiencke JK, et al. A global DNA methylation and gene expression analysis of early human B-cell development reveals a demethylation signature and transcription factor network. Nucleic Acids Res. 2012;40(22):11339–51.
72. Hansen KD, Timp W, Bravo HC, Sabunciyan S, Langmead B, McDonald OG, et al. Increased methylation variation in epigenetic domains across cancer types. Nat Genet. 2011;43(8):768–75.
73. Feber A, Wilson GA, Zhang L, Presneau N, Idowu B, Down TA, et al. Comparative methylome analysis of benign and malignant peripheral nerve sheath tumors. Genome Res. 2011;21(4):515–24.
74. Brenet F, Moh M, Funk P, Feierstein E, Viale AJ, Socci ND, et al. DNA methylation of the first exon is tightly linked to transcriptional silencing. PLoS One. 2011;6(1):e14524. Epub 2011/01/27.
75. Taylor BS, DeCarolis PL, Angeles CV, Brenet F, Schultz N, Antonescu CR. Frequent alterations and epigenetic silencing of differentiation pathway genes in structurally rearranged liposarcomas. Cancer Discov. 2011;1(7):587–97.
76. Tanaka T, Nakamura J, Kitajima Y, Kai K, Miyake S, Hiraki M, et al. Loss of trefoil factor 1 is regulated by DNA methylation and is an independent predictive factor for poor survival in advanced gastric cancer. Int J Oncol. 2013;42:894–902. doi:10.3892/ijo.2013.1759.
77. Kim JH, Dhanasekaran SM, Prensner JR, Cao X, Robinson D, Kalyana-Sundaram S, et al. Deep sequencing reveals distinct patterns of DNA methylation in prostate cancer. Genome Res. 2011;21(7):1028–41.
78. Wang GG, Allis CD, Chi P. Chromatin remodeling and cancer, Part I: covalent histone modifications. Trends Mol Med. 2007;13(9):363–72.
79. Füllgrabe J, Kavanagh E, Joseph B. Histone onco-modifications. Oncogene. 2011;30(31):3391–403.
80. Choe MK, Hong CP, Park J, Seo SH, Roh TY. Functional elements demarcated by histone modifications in breast cancer cells. Biochem Biophys Res Commun. 2012;418(3):475–82.
81. Li H, Durbin R. Fast and accurate short read alignment with Burrows-Wheeler transform. Bioinformatics. 2009;25(14):1754–60.
82. Langmead B, Trapnell C, Pop M, Salzberg SL. Ultrafast and memory-efficient alignment of short DNA sequences to the human genome. Genome Biol. 2009;10(3):R25.
83. Trapnell C, Patcher L, Salzberg SL. TopHat: discovering splice junctions with RNASeq. Bioinformatics. 2009;25(9):1105–11.
84. McKenna A, Hanna M, Banks E, Sivachenko A, Cibulskis K, Kernytsky A, et al. The genome analysis toolkit: a MapReduce framework for analyzing next-generation DNA sequencing data. Genome Res. 2010;20(9):1297–303.
85. Li H, Handsaker B, Wysoker A, Fennell T, Ruan J, Homer N, et al. The sequence alignment/map format and SAMtools. Bioinformatics. 2009;25(16):2078–9.
86. Cingolani P, Platts A, le Wang L, Coon M, Nguyen T, Wang L, et al. A program for annotating and predicting the effects of single nucleotide polymorphisms, SnpEff: SNPs in the genome of Drosophila melanogaster strain w1118; iso-2; iso-3. Fly (Austin). 2012;6(2):80–92. doi:10.4161/fly.19695.
87. Mortazavi A, Williams BA, McCue K, Schaeffer L, Wold B. Mapping and quantifying mammalian transcriptomes by RNA-Seq. Nat Methods. 2008;5(7):621–8.
88. Trapnell C, Williams BA, Pertea G, Mortazavi A, Kwan G, van Baren MJ, et al. Transcript assembly and quantification by RNA-Seq reveals unannotated transcripts and isoform switching during cell differentiation. Nat Biotechnol. 2010;28(5):511–5.

89. Katz Y, Wang ET, Airoldi EM. Analysis and design of RNA sequencing experiments for identifying isoform regulation. Nat Methods. 2010;7(12):1009–15.
90. Anders S, Huber W. Differential expression analysis for sequence count data. Genome Biol. 2010;11:R106.
91. Liu Y, Siegmund KD, Laird PW, Berman BP. Bis-SNP: combined DNA methylation and SNP calling for Bisulfite-seq data. Genome Biol. 2012;13(7):R61.
92. Ingman M, Glyllensten U. SNP frequency estimation using massively parallel sequencing of pooled DNA. Eur J Hum Genet. 2009;17:383–6.
93. Out AA, van Minderhout IJ, Goeman JJ. Deep sequencing to reveal new variants in pooled DNA samples. Hum Mutat. 2009;30(12):1703–12.
94. Sultan M, Schulz MH, Richard H, Magen A, Klingenhoff A, Scherf M, et al. A global view of gene activity and alternative splicing by deep sequencing of the human transcriptome. Science. 2008;321(5891):956–60.
95. Costa V, Angelini C, D'Apice L. Massive-scale RNA-Seq analysis of non ribosomal transcriptome in human trisomy 21. PLoS One. 2011;6(4):e18493.
96. Elsharawy A, Warner J, Olson J, Forster M, Schilhabel MB, Link DR, et al. Accurate variant detection across non-amplified and whole genome amplified DNA using targeted next generation sequencing. BMC Genomics. 2012;13:500.
97. Kerick M, Isau M, Timmermann B, Sültmann H, Herwig R, Krobitsch S, et al. Targeted high throughput sequencing in clinical cancer settings: formaldehyde fixed-paraffin embedded (FFPE) tumor tissues, input amount and tumor heterogeneity. BMC Med Genomics. 2011;4:68.
98. Meyer M, Kircher M, Gansauge MT, Li H, Racimo F, Mallick S, et al. A high-coverage genome sequence from an archaic Denisovan individual. Science. 2012;338(6104):222–6.
99. Dudley DM, Chin EN, Bimber BN, Sanabani SS, Tarosso LF, Costa PR, et al. Low-cost ultra-wide genotyping using Roche/454 pyrosequencing for surveillance of HIV drug resistance. PLoS One. 2012;7(5):e36494.

Chapter 4
Genomic Landscape of Cancer Metastasis

Arvind K. Singla, Chandini Thirukkumaran, Rajiv Dixit, and Wei Wu

Abstract Cancer metastasis is the most critical determinant of overall survival in cancer patients. With the significant advancement in next-generation sequencing technologies, our understanding of mechanism of cancer metastasis has been considerably changed especially after the discovery of noncoding RNAs (ncRNAs). The ncRNAs play critical roles in a variety of cellular processes, including development, differentiation, and tumorigenesis. Remarkably, discovery of a class of ncRNAs named as miRNAs led to a paradigm change in our understanding of gene regulation and function. Moreover, long noncoding RNAs (lncRNAs), another type of ncRNAs, have recently depicted their role as drivers of tumorigenesis by regulating tumor suppressive and oncogenic pathways. In this chapter, we describe metastasis and introduce ncRNAs with major focus on lncRNAs and outline a detailed description of few cancer metastasis-associated lncRNAs (MALAT-1, HOTAIR, BC200, and SRA RNA). Overall, a complete understanding of the genomic landscape of metastasis will generate new opportunities in early detection and therapeutic intervention.

A.K. Singla, Ph.D. (✉)
Department of Biochemistry and Molecular Biology, McCaig Institute for Bone and Joint Health, University of Calgary, 3280 Hospital Drive NW, Calgary, AB, Canada T2N 4Z6
e-mail: aksingla@ucalgary.ca

C. Thirukkumaran
Translational Research Laboratories, Department of Oncology, Tom Baker Cancer Centre, University of Calgary, Calgary, AB, Canada

R. Dixit • W. Wu
Department of Pathology & Laboratory Medicine, University of Calgary, Calgary, AB, Canada

W. Wu and H. Choudhry (eds.), *Next Generation Sequencing in Cancer Research: Volume 1: Decoding the Cancer Genome*, DOI 10.1007/978-1-4614-7645-0_4,

Keywords Genomic landscape • Cancer • Metastasis • ncRNA • miRNA • lncRNA • HOTAIR • MALAT-1 • BC200 • SRA RNA

1 Introduction

Despite advances in early detection and improvements in the treatment of cancer, mortality from this disease remains high and in most cases invariably associated with the development of metastatic disease [1, 2]. More than 90 % of cancer deaths are due to metastasis, an incomplete understanding of genomic changes associated with metastasis has hindered the development of effective therapies against it. For the past 60–70 years, major efforts in understanding the genome were focused on protein-coding genes, which represent ~2 % of the total genome. Recent studies from the ENCODE project estimated that up to 70 % of the genome is actively transcribed [3]. Unfortunately, the vast majority of the nonprotein-coding genome was thought to be a "junk DNA." Although, the earliest-discovered ncRNAs such as ribosomal RNA (rRNA), transfer RNA (tRNA), and small nuclear RNAs are well known to play an important role in regulation of translation [4, 5]. Recent advancements in next-generation sequencing technologies led to the discovery of an array of ncRNAs with specific expression, unique biological function, and mechanism of action [5, 6]. As the discovery of ncRNAs led to a paradigm change in our understanding of gene regulation and function, a big question that needs to be answered at this point of time is how can we utilize this new piece of knowledge towards developing new diagnostic tools and therapeutic drugs especially in terminal diseases like advance stage metastasis. In this chapter, we will first describe metastasis, steps of the metastatic cascade, and the role of next-generation sequencing technologies in cancer metastasis and then describe ncRNAs with a major focus on lncRNAs.

2 Metastasis

Metastasis is the spread of cancer cells from the primary tumor site to different organs of the body. For example, in lung cancer patients, metastatic founder cells originate from primary tumor in the lung and metastasize to lymph nodes, brain, bone, liver, and adrenal glands (Fig. 4.1). Similarly in breast cancer patients, metastatic founder cells originate from breast tissue and metastasize to lung, brain, bone, and liver (Fig. 4.1). A prominent feature of metastasis is the capability of metastatic founder cells to colonize multiple organs (multiorgan metastasis or late stage metastasis). The complete understanding of the genetic alterations associated with initiation and progression of late stage metastasis will provide new opportunities for diagnosis and treatment.

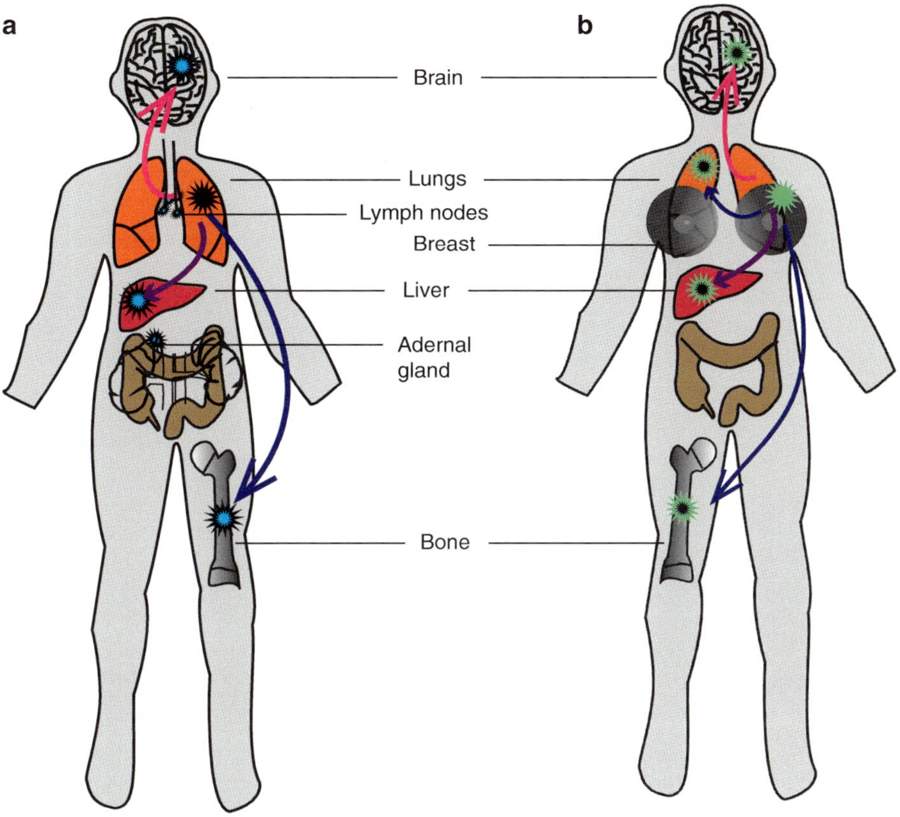

Fig. 4.1 Breast and lung cancer metastasis. (**a**) Lung cancer metastasis; primary tumor originated in lung tissue, metastasize to lymph nodes, contralateral lung, liver, bone, brain, and other sites. (**b**) Breast cancer metastasis; primary tumor originated in breast tissue metastasize to lung, liver, bone, brain, and other sites

2.1 Events of the Metastasis Cascade

Metastasis is a multistep process with a sequential cascade of events such as local invasion, intravasation, survival in the circulation, extravasation, micrometastasis and macrometastasis (Fig. 4.2) [7]. In brief, the main features of each step are described below.

2.1.1 Local Invasion

Cancer cells within primary tumor invade into the surrounding stroma and adjacent normal tissue through mechanisms like epithelial to mesenchymal transition (EMT) and degradation of basement membrane by secretion of matrix metalloproteinases (MMPs), commonly, activated during the course of metastasis initiation [7–9].

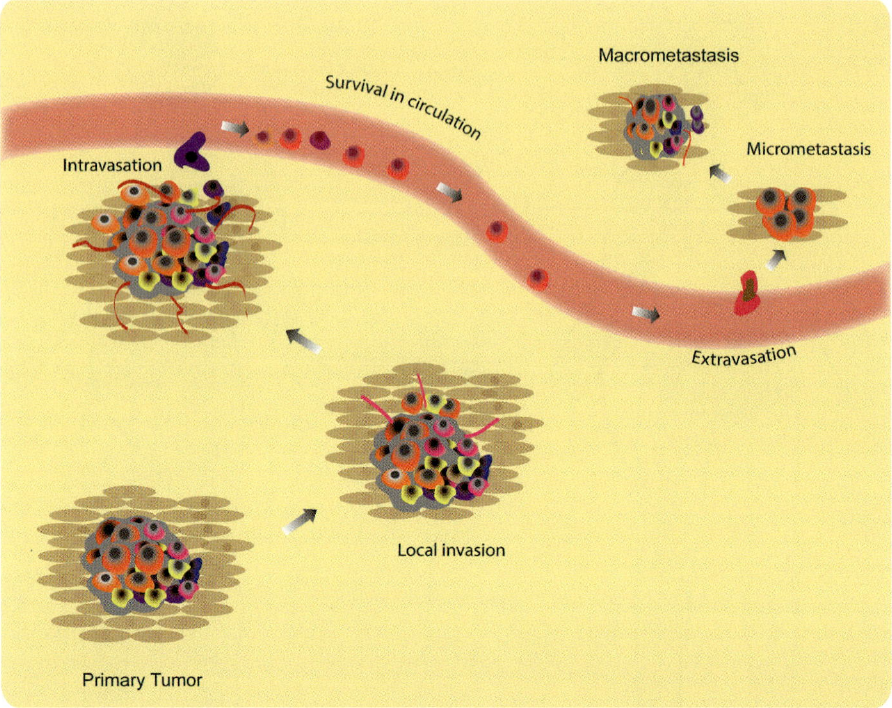

Fig. 4.2 Events of metastatic cascade. (**a**) Local invasion: the metastatic process is initiated when cancer cells within primary tumors invade into the surrounding stroma. (**b**) Intravasation: cancer cells entering into the circulation through blood or lymphatic vessels. (**c**) Survival in circulation: cancer cells traveling through the circulation called as circulating tumor cells (CTCs). CTCs must overcome several stresses in order to reach distant metastatic sites. (**d**) Extravasation: CTC able to survive in circulation could infiltrate distant organs. During extravasation process, the cancer cells emigrate from the vessels into the parenchyma of a distant organ tissue. (**e**) Micrometastasis and macrometastasis formation: newly invaded cancer cells capable of adapting to the new microenvironment evade the immune responses and survive, reinitiate proliferation and angiogenesis to form micrometastasis. Micrometastatic tumor with time develop into a macrometastatic tumor

2.1.2 Intravasation

Intravasation is a set of cellular and molecular events whereby invasive cancer cells enter into the lumina of blood vessels (or lymphatic vessels) (Fig. 4.2). These vessels are lined by endothelial and pericyte cell layers, and cancer cells enter into circulation by penetrating through these layers [7, 9].

2.1.3 Survival in the Circulation

Cancer cells that have successfully intravasated into the circulation must overcome several stresses in order to reach distant metastatic sites. Hemodynamic shear forces can threaten cell membrane integrity, detachment from neighboring stroma can induce cell death (anoikis), and attack from host's immune cells is easier once the cells are removed from the primary site [7–9]. Multiple advanced technologies have eased detection of circulating tumor cells in cancer patients; evidences support their role as prognostic indicator of disease [10–12].

2.1.4 Extravasation

During extravasation, the cancer cells emigrate from the vessels by penetrating through the endothelial and pericyte layers and into the parenchyma of a distant organ tissue (Fig. 4.2). Although extravasation is opposite to the process of intravasation, both these processes are quite different from one another in terms of difference in vasculature and host-associated stromal cell at a primary and metastatic tumor site [7–9, 13].

2.1.5 Micrometastasis and Macrometastasis

Once the cancer cells are invaded into a new metastatic organ site, they face different types of microenvironmental challenges. For First, as there is a dramatic difference in the microenvironment of the metastatic organ site compared to primary tumor site; in terms of primary tumors ability to produce high amounts of cytokines/growth factors/MMPs, that facilitate cancer cell growth, and a leaky microvasculature, which may be lacking in a newly colonized metastatic site [7–9]. Secondly, microenvironmental differences within each metastatic organ, for example, bone versus brain or liver vs adrenal gland, can be dramatically different, therefore cancer cells after colonizing metastatic sites must learn to adapt to each of these microenvironmental variations. Due to these challenges, a large number of disseminated tumor cells in new metastatic niches suffer slow rates of death or persist as single or a small number of cells in a dormant stage [14].

Metastatic cells that are able to adapt to the local microenvironment evade the host immune responses and survive, reinitiate proliferation, and induce neoangiogenesis to successfully form micrometastatic tumors. These micrometastatic tumors thereby generate macrometastasis that could be detected by current imaging technologies [7].

Currently, therapeutic options are very limited for patients with multiorgan metastatic tumors, so there is an urgent need to completely understand late stage of metastasis. Few researchers have attempted to study multiorgan metastasis by isolating and characterizing organ-specific metastatic variants from human metastatic tumors or cell lines [15–17]. Comparison of these organ-specific metastatic variants gene and ncRNA expression with their respective parental line led to the identification of novel genes and ncRNAs involved in organ-specific metastasis. Clinical relevance of these genes and ncRNAs is obtained by examining their expression in patient tumor tissues or by comparing with available clinical gene or ncRNA expression data sets. Notably, Joan Massague and colleagues characterized a set of genes associated with breast cancer brain metastasis (*COX2*, *HBEGF*, and *ST6GALNAC5*) and lung metastasis (*CXCL1*, *ROBO1*, *MMP1*, *COX2*, and *ID1*) [15, 16]. Similarly, we characterized adrenal and bone-specific metastatic sublines under H1299 (parental) background; comparison with parental cells led to identification of ANG2 and miR-21's role in organ- specific metastasis ([17], Singla Ak et al., unpublished results).

2.2 *Models of Metastasis*

Metastasis is a complex process; based on our current understanding, few models exist to define dissemination of the metastatic cells; in brief, main features of each model are described below.

2.2.1 Unidirectional or Multidirectional Metastasis

As our understanding of how cancer spreads is still emerging, it is important to note that initially, the process of metastasis has been viewed as a unidirectional process, in which cancer cells move in one direction from primary tumor to distant metastatic sites (Fig. 4.3). In contrast, cancer cells are viewed as fluid seeds, they are capable of seeding not only metastatic sites but also their tumor of origin (self-seeding), and this process is proposed as multidirectional process, based on preclinical and clinical evidences (Fig. 4.3) [18–20].

2.2.2 Linear and Parallel Model of Metastasis

Based on the time of metastatic dissemination of cancer cells, there are two widely used models, linear and parallel. According to the linear model of metastasis, the metastatic dissemination occurs at a later stage, when the primary tumor has already grown to a large volume. Here, the primary tumor volume represents the risk factor for progression to metastasis, greater the volume, higher the chances of metastasis [21, 22]. This model was supported from studies showed similar molecular signatures for primary and their corresponding metastatic tumors [21, 22]. In contrast, the parallel model of metastasis is based on that the metastatic founder cells disseminate early (before primary tumor is fully malignant or before clinically detectable disease) from the primary lesion. Based on this model, metastatic lesions and primary tumors evolve independently; this model is supported by autopsy studies and molecular genetic analyses of primary and disseminated tumor cells [22–24].

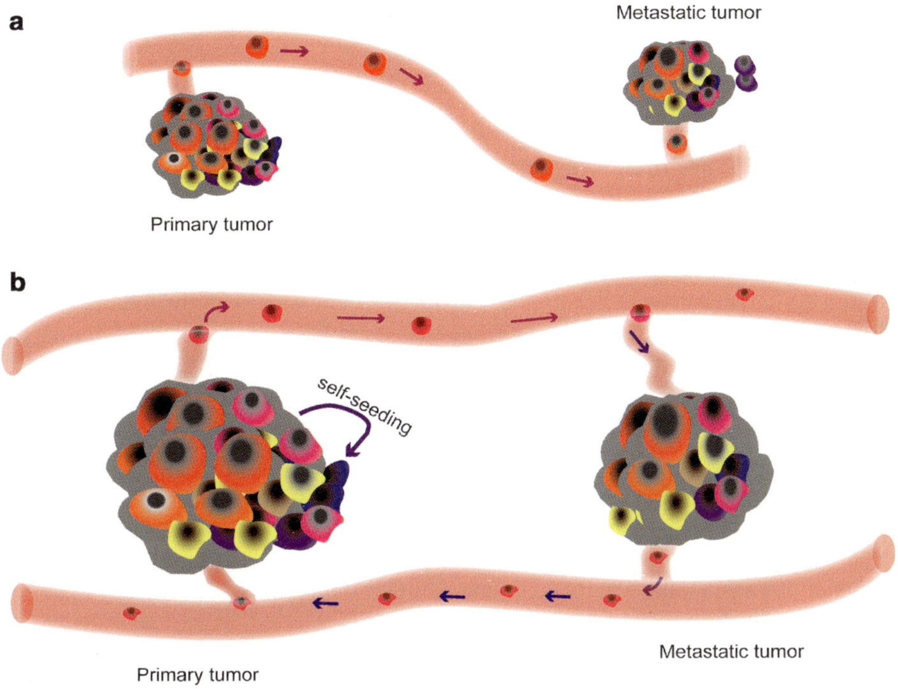

Fig. 4.3 Unidirectional and multidirectional metastasis. (**a**) Unidirectional metastasis: cancer cells originated from the primary tumor move only in one direction, from primary tumor to distant metastatic sites. (**b**) Multidirectional metastasis: cancer cells originated from primary tumor after entering into the circulation, can seed the original tumor (self-seeding), or enter into distant metastatic sites such as the brain, bone, lung and are capable of extravasating from a distant metastatic site and can return to the original primary tumor

3 Genomic Signature of Metastasis Revealed by Next Generation Sequencing (NGS)

Cancer progression is caused by the accumulation of genetic alterations; complete understanding of genetic variations exist within primary, and metastatic tumor might significantly enhance our understanding of cancer metastasis. Recent advancements in whole-genome sequencing technologies have provided us with powerful and cost-effective tools to study gene copy number variations (CNVs), single nucleotide variations (SNVs), loss of heterozygosity (LOH), and mutational and structural aberrations within primary and metastatic tumors [25–27]. Using these next generation technologies, few groups have attempted to characterize genetic changes associated with cancer metastasis. Utilizing peripheral blood sample as a control, Ding et al. performed a whole-genomic analysis on primary breast tumor and brain metastasis samples, obtained from a basal-like (ERBB2 and ER negative) breast cancer patient [25]. They observed that most of the genomic aberrations including mutations, deletions, inversions, and translocations were common between primary and metastatic tumor. However, an increased prevalence of 26 mutations was observed in the metastatic tumor. In contrast to primary tumor, metastatic tumor contained two de novo mutations: a missense mutation (T708I) in *SNED1* and a silent mutation (N2483) in the *FLNC* gene plus a large deletion [25]. Interestingly, the extensive variety of mutant allele frequencies in the cellular population of primary tumor was also reported suggestive of significant genetic heterogeneity. Similarly, Turajlic et al. performed whole-genome sequencing on primary acral melanoma and its lymph node metastasis tumor obtained from a patient with no prior treatment [26]. They observed that the primary and the metastatic tumor were mostly similar for gene copy number alterations, loss of heterozygosity, and single nucleotide variations. However, two mutations appeared to arise de novo in the metastasis samples: one single nucleotide variation affecting coding region of *WNT1* (C369F) and the other affected a splice site in *SUPT5H*. In addition, a substantial genetic heterogeneity was also reported in both primary and metastatic tumor [26]. Notably, most of these above mentioned genomic studies were carried out using single biopsy sample from primary and metastatic tumor. An elegant study performed by Gerlinger et al. using multiple spatially separated biopsy samples obtained from a primary and metastatic tumor of a renal cell carcinoma patient [27]. The whole exome sequencing, chromosome aberration, and DNA ploidy profiling were employed on these samples. Remarkably, they observed that the majority of somatic mutations found in single biopsy were not consistently detectable throughout all the tested regions of the same patient's tumor. In addition, ploidy profiling and allelic composition analysis depicted widespread intratumoral heterogeneity [27].

Overall, next-generation sequencing technologies have characterized genetic changes within primary and metastatic tumors highlighted the presence of significant intratumoral heterogeneity within an individual tumor. But this new piece of information has posed intriguing questions to clinicians and researchers on issues like: Is one tumor biopsy sample sufficient or not? How many biopsy samples might accurately represent the genomic landscape of tumor? How this huge amount of genomic information can be used to enhance the overall treatment outcome? etc.

4 The Role of ncRNAs in Cancer Metastasis

The discovery of ncRNAs has dramatically altered our understanding of cancer biology, as deregulations of ncRNAs are involved in cancer initiation, progression, and metastasis. The ncRNAs are divided into two main groups based on the size of the transcript: small ncRNAs (SncRNAs) and long ncRNAs (lncRNAs). In the following sections, we will further describe ncRNAs associated with cancer metastasis.

4.1 Small Noncoding RNAs (SncRNAs)

Small ncRNAs are 18–200 nucleotides in size with a wide variety of gene regulatory functions [28]. The major types in this category are microRNAs (miRNAs), small interfering RNAs (siRNA), piwi-interacting RNAs (piRNAs), small nucleolar RNAs (snoRNAs), and transcription initiation RNAs [6, 28]. Remarkably, the discovery of miRNAs has significantly enhanced the importance of small ncRNAs in cancer metastasis, as many elegant studies have depicted regulation of metastatic events by specific miRNAs [29]. In brief, miR-9, 10b, 200, 205, 31, and 29b are involved in local invasion, miR-21 is associated with intravasation, miR-7 and miR-31 are related to survival in circulation, miR-31 and 214 are associated with extravasation step, and miR-15/16, 31, 34, 335, and 126 and let7 play a role in micro- and macro-metastasis formation [29, 30]. The detailed information about their function and mechanism of action can be found in these reviews [29–31].

4.2 Long Noncoding RNAs (lncRNAs)

Long ncRNAs are 200 bases to >100 kilobase (kb) in size; the major kinds are large intergenic ncRNAs (lincRNAs), long intronic ncRNAs, antisense RNAs (aRNAs), transcribed ultraconserved regions (T-UCRs), and pseudogenes [32, 33]. Depending upon their location of transcription in genome, long intronic, lincRNA, and aRNAs are described as: (a) long intronic ncRNAs are transcribed from an intron of another transcript, (b) large intergenic ncRNAs lie between sets of protein-coding genes, and (c) antisense RNAs are transcribed on the opposite strand from a protein-coding gene [32]. Similarly, based on their evolutionary conserved nature, another type of lncRNA is T-UCR, found in intergenic, intronic, and exonic regions of the genome. These comprise a subset of 481 sequences, completely conserved between orthologous regions of human, mouse, and rat genomes [32].

The majority of the known lncRNAs are transcribed by RNA polymerase II and are polyadenylated; however, an increasing number of reports are showing lncRNAs transcribed by RNA polymerase III and non-polyadenylated [32, 34–36]. For example lncRNA BC200 described later in this chapter is transcribed by RNA polymerase III and is not polyadenylated [32, 60]. LncRNAs dysregulated expression was observed in various cancer types including breast, lung, prostate, colon, and hepatocellular carcinoma [32, 37–39]. LncRNAs have been shown to play an important role in diverse biological processes such as splicing, epigenetics, development, stem cell pluripotency, differentiation, localization, survival, transcription, cell cycle, and metabolism [32, 35, 36]. LncRNAs function through *cis* (genetic elements located near the same genomic locus) or *trans* (genetic elements located on distant genomic locus) regulatory mechanisms and regulate gene expression both at transcriptional and posttranscriptional levels [35, 40]. LncRNAs regulate gene expression through various mechanisms such as acting as precursor to small RNAs, direct interaction with chromatin-modifying proteins and transcription factors, as regulators of mRNA decay, through lncRNA transcription, and by regulation of chromatin structure [36, 41–43]. The detailed information about lncRNAs biogenesis, function, and mechanism of action can be seen in Chap. 10 of this book (Wei Wu et al.).

In contrast to well-defined nomenclature for microRNAs like miR-1, miR-21, miR-31, and miR-222, lncRNA nomenclature is less refined, based in part on their discovery in a particular cancer type or tissue expression pattern and function. For example, lncRNA PCAT-1 was identified in prostate cancer thereby named as prostate cancer-associated transcript 1. Similarly, lncRNA HULC was named as highly upregulated in liver cancer. Others include MALAT-1 (metastasis-associated lung adenocarcinoma transcript-1), XIST (X inactive-specific transcript), and HOTAIR (HOX antisense intergenic RNA) [32, 35, 37].

4.3 Advancement in Technology for the Discovery of lncRNAs

In 1990, the first lncRNA, H19, was discovered, followed by the discovery of another lncRNA, XIST in 1991 [68, 44]. The majority of lncRNAs discovered during that time were using traditional mapping approaches like large-scale sequencing of full-length cDNA libraries, screening of cDNA libraries, and cloning methods [32, 43, 44]. As these methods were very time consuming, were slowly replaced by lncRNA microarrays. However, the major drawback of lncRNA microarray platforms is their incapability to identify new lncRNAs, as arrays are based on known lncRNA probes. To overcome these problems, techniques like DNA tiling array was used, which is based on tiling of non-overlapping or partially overlapping probes to cover the whole genome. The lncRNA HOTAIR was discovered using the HOX tiling array technique [53]. Currently, with the advancement of high-throughput RNA sequencing technologies (whole transcriptome RNA-seq, series analysis of gene expression, and 3′ end sequencing for expression quantification) and validation techniques (immunohistochemistry, northern blot, in situ hybridization, and rapid amplification of cDNA ends), new lncRNAs are being discovered and characterized.

Taken together, current evidences support that lncRNAs are evolutionary conserved with unique mode of action and are dysregulated in multiple diseases including cancer. In contrast to small ncRNAs, the field of lncRNAs is yet emerging, with some lncRNAs depicting a role in cancer metastasis. A detailed account of a few lncRNAs known to play a role in cancer metastasis is described in the following section.

5 LncRNAs Associated with Cancer Metastasis

5.1 MALAT-1

MALAT-1 (metastasis-associated lung adenocarcinoma transcript 1) also referred as NEAT2 (Nuclear-enriched abundant transcript 2) is 8.7-kb transcript and is located on chromosome 11q13 [45, 46]. This lncRNA was discovered in 2003, using subtractive hybridization approach in stage I non-small cell lung cancer (NSCLC) tumors that either metastasized or not [45]. Subsequent studies have reported its overexpression in other cancer types like bladder, colorectal, hepatocellular, lung squamous cell carcinoma, and laryngeal squamous cell cancer (LSCC) [46–50].

In vitro studies utilizing ectopic up- or downregulation of MALAT-1 expression in A549, HepG2, and NIH 3T3 cell lines have demonstrated its role in regulating cell motility, migration, invasion, and viability [46, 49, 51, 52]. MALAT-1 regulates the expression of motility-related genes at transcriptional or posttranscriptional levels [51]. In preclinical in vivo studies using experimental xenograft models (intravenously inoculated A549 cells or subcutaneously established EBC-1 tumor), MALAT-1's downregulation or loss of function led to decrease in metastasis formation [52]. Moreover, downregulation of MALAT-1 reduced the growth of A549 subcutaneous tumors [46, 52]. MALAT-1 has also been shown to play a role in bladder cancer metastasis where downregulation of MALAT-1 reduces the levels of epithelial to mesenchymal transition-associated markers (ZEB1, ZEB2, and Slug) and enhances E-cadherin [48]. Epithelial to mesenchymal transition (EMT) plays an important role in metastasis initiation and progression [7]. Overall, MALAT-1 was initially discovered in NSCLC patients; subsequent studies have depicted its involvement in other cancer types, currently an emerging candidate in cancer metastasis.

5.2 *HOTAIR*

HOTAIR (HOX antisense intergenic RNA) is located in the HoxC cluster on chromosome 12q13.13 and regulates the HoxD cluster genes in a *trans*-regulatory mechanism [32, 53]. Elevated HOTAIR levels have been observed in breast, colon, pancreatic, laryngeal squamous cell cancer (LSCC), gastrointestinal stromal tumors (GIST), and hepatocellular carcinoma [53–58]. In human breast cancer tumors, HOTAIR expression is upregulated and positively correlates with metastasis and overall survival [53]. Similarly, stage IV colorectal cancer patient's upregulated HOTAIR correlated with the presence of liver metastasis [54]. In addition, upregulated HOTAIR expression in HCC patients treated with liver transplantation showed an increasing risk of HCC recurrence and shorter survival [55].

Evidences generated from in vitro studies indicate that the ectopic overexpression of HOTAIR in breast cancer cell lines (MCF-10A, MCF-7, SK-BR3, and MDA-MB-231) enhances cancer cell invasion, whereas knockdown of HOTAIR in MCF-7, Hep-2, HepG2, Panc1, and L3.6pL cell lines decreases invasion [53–58]. The in vivo studies were carried out in preclinical orthotopic breast cancer mouse model; overexpression of HOTAIR in highly invasive MDA-MB-231 cells enhanced the rate of primary tumors and promoted lung metastasis [53]. The proposed mechanism of action for overexpressed HOTAIR in breast cancer is through the enhanced recruitment of PRC2 complex to the genomic locations of target genes, leading to H3K27 trimethylation and epigenetic silencing of metastasis suppressor genes that results in enhancement of tumor invasion and metastasis [53]. Taken together, these studies provide preclinical and clinical evidences supporting the role of HOTAIR in cancer metastasis.

5.3 BC200

BC200 also known as BCYRN1 (brain cytoplasmic RNA 1) was previously described as neuron-specific RNA [59, 60]. Chen et al. in 1997 first reported the presence of BC200 in nonneuronal tissues such as breast, lung, ovary, parotid, cervix, and esophagus cancers [59]. Although the tumor tissue samples for each tumor type were very low in this study, high BC200 expression was reported in lung metastasis of a melanoma, infiltrating carcinoma of the breast and in breast cancer tissues from patients with metastatic tumors in comparison to corresponding normal tissues with no BC200 expression [59]. A following study performed by the same group in 2004 compared BC200 RNA expression using normal/benign, ductal carcinoma in situ (DCIS) and invasive carcinoma breast tumor tissues [60]. BC200 RNA was expressed significantly higher in invasive breast cancer tissues compared to normal tissues or benign tumors. In addition, BC200 RNA expression was upregulated in high-grade ductal carcinoma in situ (DCIS) breast tumor tissues compared to non-high-grade DCIS tissues [60]. Importantly, the higher the tumor grade, the faster the growth and invasion compared to no or low-grade DCIS tumors (where cancer cells resemble normal breast cells). Also, of potential relevance BC1 RNA, the rodent analogue of human BC200 RNA, was also found to be deregulated in murine tumor tissues [61]. Although upregulated expression of BC200 RNA was reported in invasive or high-grade/stage tumors, future preclinical research is clearly needed to completely define its role in cancer metastasis.

5.4 SRA RNA

SRA RNA (steroid receptor RNA activator) is also known as RNA with dual function, was first discovered in 1999 [62]. The SRA gene encodes for both coding and noncoding transcripts; the coding SRA transcript produces an endogenous protein (SRAP) expressed in prostate, breast, and muscle tissues [63, 64]. Noncoding SRA transcript comprise either a full or a partial intron-1 sequence, generated by alternative splicing [63, 64].

In vitro studies using breast cancer cell lines have shown that both fully spliced SRAP-coding and noncoding SRA RNA can coexist. Interestingly, more invasive breast cancer cell lines such as MDA-MB-231 and MDA-MB-468 express greater relative levels of noncoding SRA RNA than less/noninvasive cell lines (MCF-10A1, MCF-7, and T47D) [65]. A relative increase in SRA RNA levels through reprogramming of endogenous SRA splicing in T5 breast cancer cells led to enhanced plasminogen urokinase activator (PLAU) gene expression, a well-known gene important for metastasis progression [63]. This depicts a change in balance of non-coding and coding transcripts capable of defining a specific phenotype.

Notably, knockdown of SRA RNA in highly invasive breast cancer MDA-MB-231 cells significantly reduced their in vitro invasion and invasion/metastasis-related genes such as CAV1, TMPRSS2, MMP-1, and MMP-9 [66]. Clinical studies using human tumor tissues showed SRA RNA is overexpressed in human tumors of the breast, ovary, and uterus. Also of potential relevance, SRA RNA levels are augmented in infiltrating ductal tumors, breast fibroadenomas, high-grade breast adenocarcinomas, and infiltrating carcinomas, compared to normal breast tissue [64, 66, 67]. Overall, these findings reveal that SRA RNA plays a role in cancer metastasis, but future research will be necessary in order to more exhaustively examine its role in regulating particular event of the metastatic cascade.

6 Summary

Next-generation sequencing technologies have provided novel insights into human genome and transcriptome, including the discovery of ncRNAs (SncRNAs and lncRNAs). Even though the current list of known ncRNAs is short (especially lncRNAs), as the field is emerging rapidly, novel ncRNAs with supportive preclinical and clinical evidence are likely to be obtained in the future. Due to their unique mechanism of action and expression, it could be anticipated that ncRNAs would bring new therapeutic options to the clinic. However, comprehensive analysis of their function is necessary before their application in humans. Currently, most of the patients with advance stage metastasis are incurable; a paradigm change in our understanding of cancer metastasis will provide new opportunities for next-generation drugs to target cancer metastasis differently compared to existing methods.

Acknowledgements The authors acknowledge limitations in citing important contributions of many other scientists due to space constraints. We thank Franz Zemp for providing critical reviews and suggestions for this manuscript. The research for this manuscript was supported by the Canadian Cancer Society Research Institute (CCSRI).

Disclosure of potential conflicts of interest: No potential conflicts of interest were disclosed.

References

1. Jemal A, Bray F, Center MM, Ferlay J, Ward E, Forman D. Global cancer statistics. CA Cancer J Clin. 2011;61(2):69–90.
2. Siegel R, Ward E, Brawley O, Jemal A. Cancer statistics, 2011: the impact of eliminating socioeconomic and racial disparities on premature cancer deaths. CA Cancer J Clin. 2011;61(4):212–36.

3. Dunham I, Kundaje A, Aldred SF, Collins PJ, Davis CA, Doyle F, et al. An integrated encyclopedia of DNA elements in the human genome. Nature. 2012;489(7414):57–74.
4. Costa FF. Non-coding RNAs: meet thy masters. Bioessays. 2010;32(7):599–608.
5. Kugel JF, Goodrich JA. Non-coding RNAs: key regulators of mammalian transcription. Trends Biochem Sci. 2012;37(4):144–51.
6. Iorio MV, Croce CM. microRNA involvement in human cancer. Carcinogenesis. 2012;33(6):1126–33.
7. Valastyan S, Weinberg RA. Tumor metastasis: molecular insights and evolving paradigms. Cell. 2011;147(2):275–92.
8. Spano D, Heck C, De Antonellis P, Christofori G, Zollo M. Molecular networks that regulate cancer metastasis. Semin Cancer Biol. 2012;22(3):234–49.
9. Nguyen DX, Bos PD, Massagué J. Metastasis: from dissemination to organ-specific colonization. Nat Rev Cancer. 2009;9(4):274–84.
10. Stott SL, Hsu CH, Tsukrov DI, Yu M, Miyamoto DT, Waltman BA, et al. Isolation of circulating tumor cells using a microvortex-generating herringbone-chip. Proc Natl Acad Sci U S A. 2010;107(43):18392–7.
11. Nagrath S, Sequist LV, Maheswaran S, Bell DW, Irimia D, Ulkus L, et al. Isolation of rare circulating tumour cells in cancer patients by microchip technology. Nature. 2007;450(7173):1235–9.
12. Alix-Panabières C, Schwarzenbach H, Pantel K. Circulating tumor cells and circulating tumor DNA. Annu Rev Med. 2012;63:199–215.
13. Downey CM, Singla AK, Villemaire ML, Buie HR, Boyd SK, Jirik FR. Quantitative ex-vivo micro-computed tomographic imaging of blood vessels and necrotic regions within tumors. PLoS One. 2012;7(7):e41685.
14. Chambers AF, Groom AC, MacDonald IC. Dissemination and growth of cancer cells in metastatic sites. Nat Rev Cancer. 2002;2(8):563–72.
15. Minn AJ, Gupta GP, Siegel PM, Bos PD, Shu W, Giri DD, et al. Genes that mediate breast cancer metastasis to lung. Nature. 2005;436:518–24.
16. Bos PD, Zhang XH, Nadal C, Shu W, Gomis RR, Nguyen DX, et al. Genes that mediate breast cancer metastasis to the brain. Nature. 2009;459:1005–9.
17. Rana S, Singla AK, Bajaj A, Elci SG, Miranda OR, Mout R, et al. Array-based sensing of metastatic cells and tissues using nanoparticle-fluorescent protein conjugates. ACS Nano. 2012;6(9):8233–40.
18. Comen E, Norton L. Self-seeding in cancer. Recent Results Cancer Res. 2012;195:13–23.
19. Comen E, Norton L, Massagué J. Clinical implications of cancer self-seeding. Nat Rev Clin Oncol. 2011;8(6):369–77.
20. Kim MY, Oskarsson T, Acharyya S, Nguyen DX, Zhang XH, Norton L, et al. Tumor self-seeding by circulating cancer cells. Cell. 2009;139(7):1315–26.
21. Lorusso G, Rüegg C. New insights into the mechanisms of organ-specific breast cancer metastasis. Semin Cancer Biol. 2012;22(3):226–33.
22. Klein CA. Parallel progression of primary tumours and metastases. Nat Rev Cancer. 2009;9(4):302–12.
23. Stoecklein NH, Klein CA. Genetic disparity between primary tumours, disseminated tumour cells, and manifest metastasis. Int J Cancer. 2010;126(3):589–98.
24. Pantel K, Brakenhoff RH, Brandt B. Detection, clinical relevance and specific biological properties of disseminating tumour cells. Nat Rev Cancer. 2008;8:329–40.
25. Ding L, Ellis MJ, Li S, Larson DE, Chen K, Wallis JW, et al. Genome remodelling in a basal-like breast cancer metastasis and xenograft. Nature. 2010;464(7291):999–1005.
26. Turajlic S, Furney SJ, Lambros MB, Mitsopoulos C, Kozarewa I, Geyer FC, et al. Whole genome sequencing of matched primary and metastatic acral melanomas. Genome Res. 2012;22(2):196–207.
27. Gerlinger M, Rowan AJ, Horswell S, Larkin J, Endesfelder D, Gronroos E, et al. Intratumor heterogeneity and branched evolution revealed by multiregion sequencing. N Engl J Med. 2012;366(10):883–92.

28. Holley CL, Topkara VK. An introduction to small non-coding RNAs: miRNA and snoRNA. Cardiovasc Drugs Ther. 2011;25(2):151–9.
29. Valastyan S. Roles of microRNAs and other non-coding RNAs in breast cancer metastasis. J Mammary Gland Biol Neoplasia. 2012;17(1):23–32.
30. Valastyan S, Weinberg RA. MicroRNAs: crucial multi-tasking components in the complex circuitry of tumor metastasis. Cell Cycle. 2009;8(21):3506–12.
31. Garofalo M, Croce CM. microRNAs: master regulators as potential therapeutics in cancer. Annu Rev Pharmacol Toxicol. 2011;51:25–43.
32. Gibb EA, Brown CJ, Lam WL. The functional role of long non-coding RNA in human carcinomas. Mol Cancer. 2011;10:38.
33. Sana J, Faltejskova P, Svoboda M, Slaby O. Novel classes of non-coding RNAs and cancer. J Transl Med. 2012;10:103.
34. Gutschner T, Diederichs S. The hallmarks of cancer: a long non-coding RNA point of view. RNA Biol. 2012;9(6):703–19.
35. Prensner JR, Chinnaiyan AM. The emergence of lncRNAs in cancer biology. Cancer Discov. 2011;1(5):391–407.
36. Rinn JL, Chang HY. Genome regulation by long noncoding RNAs. Annu Rev Biochem. 2012;81:145–66.
37. Wapinski O, Chang HY. Long noncoding RNAs and human disease. Trends Cell Biol. 2011;21(6):354–61.
38. Chen LL, Carmichael GG. Long noncoding RNAs in mammalian cells: what, where, and why? Wiley Interdiscip Rev RNA. 2010;1(1):2–21.
39. Brunner AL, Beck AH, Edris B, Sweeney RT, Zhu SX, Li R, et al. Transcriptional profiling of lncRNAs and novel transcribed regions across a diverse panel of archived human cancers. Genome Biol. 2012;13(8):R75.
40. Guil S, Esteller M. Cis-acting noncoding RNAs: friends and foes. Nat Struct Mol Biol. 2012;19(11):1068–75.
41. Tsai MC, Manor O, Wan Y, Mosammaparast N, Wang JK, Lan F, et al. Long noncoding RNA as modular scaffold of histone modification complexes. Science. 2010;329(5992):689–93.
42. Khalil AM, Guttman M, Huarte M, Garber M, Raj A, Rivea Morales D, et al. Many human large intergenic noncoding RNAs associate with chromatin-modifying complexes and affect gene expression. Proc Natl Acad Sci U S A. 2009;106(28):11667–72.
43. Tsai MC, Spitale RC, Chang HY. Long intergenic noncoding RNAs: new links in cancer progression. Cancer Res. 2011;71(1):3–7.
44. Brown CJ, Ballabio A, Rupert JL, Lafreniere RG, Grompe M, Tonlorenzi R, et al. A gene from the region of the human X inactivation centre is expressed exclusively from the inactive X chromosome. Nature. 1991;349(6304):38–44.
45. Ji P, Diederichs S, Wang W, Böing S, Metzger R, Schneider PM, et al. MALAT-1, a novel noncoding RNA, and thymosin beta4 predict metastasis and survival in early-stage non-small cell lung cancer. Oncogene. 2003;22(39):8031–41.
46. Schmidt LH, Spieker T, Koschmieder S, Schäffers S, Humberg J, Jungen D, et al. The long noncoding MALAT-1 RNA indicates a poor prognosis in non-small cell lung cancer and induces migration and tumor growth. J Thorac Oncol. 2011;6(12):1984–92. Erratum in: J Thorac Oncol. 2012;7(7):1206.
47. Feng J, Tian L, Sun Y, Li D, Wu T, Wang Y, Liu M. Expression of long non-coding ribonucleic acid metastasis-associated lung adenocarcinoma transcript-1 is correlated with progress and apoptosis of laryngeal squamous cell carcinoma. Head Neck Oncol. 2012;4(2):46.
48. Ying L, Chen Q, Wang Y, Zhou Z, Huang Y, Qiu F. Upregulated MALAT-1 contributes to bladder cancer cell migration by inducing epithelial-to-mesenchymal transition. Mol Biosyst. 2012;8(9):2289–94.
49. Lai MC, Yang Z, Zhou L, Zhu QQ, Xie HY, Zhang F, et al. Long non-coding RNA MALAT-1 overexpression predicts tumor recurrence of hepatocellular carcinoma after liver transplantation. Med Oncol. 2012;29(3):1810–6.

50. Xu C, Yang M, Tian J, Wang X, Li Z. MALAT-1: a long non-coding RNA and its important 3′ end functional motif in colorectal cancer metastasis. Int J Oncol. 2011;39(1):169–75.
51. Tano K, Mizuno R, Okada T, Rakwal R, Shibato J, Masuo Y, Ijiri K, Akimitsu N. MALAT-1 enhances cell motility of lung adenocarcinoma cells by influencing the expression of motility-related genes. FEBS Lett. 2010;584(22):4575–80.
52. Gutschner T, Hämmerle M, Eißmann M, Hsu J, Kim Y, Hung G, et al. The non-coding RNA MALAT1 is a critical regulator of the metastasis phenotype of lung cancer cells. Cancer Res. 2013;73(3):1180–9.
53. Gupta RA, Shah N, Wang KC, Kim J, Horlings HM, Wong DJ, et al. Long non-coding RNA HOTAIR reprograms chromatin state to promote cancer metastasis. Nature. 2010; 464(7291):1071–6.
54. Kogo R, Shimamura T, Mimori K, Kawahara K, Imoto S, Sudo T, et al. Long noncoding RNA HOTAIR regulates polycomb-dependent chromatin modification and is associated with poor prognosis in colorectal cancers. Cancer Res. 2011;71(20):6320–6.
55. Yang Z, Zhou L, Wu LM, Lai MC, Xie HY, Zhang F, Zheng SS. Overexpression of long non-coding RNA HOTAIR predicts tumor recurrence in hepatocellular carcinoma patients following liver transplantation. Ann Surg Oncol. 2011;18(5):1243–50.
56. Li D, Feng J, Wu T, Wang Y, Sun Y, Ren J, Liu M. Long intergenic noncoding RNA HOTAIR is overexpressed and regulates PTEN methylation in laryngeal squamous cell carcinoma. Am J Pathol. 2013;182(1):64–70.
57. Kim K, Jutooru I, Chadalapaka G, Johnson G, Frank J, Burghardt R, et al. HOTAIR is a negative prognostic factor and exhibits pro-oncogenic activity in pancreatic cancer. Oncogene. 2012;21:1–10.
58. Niinuma T, Suzuki H, Nojima M, Nosho K, Yamamoto H, Takamaru H, et al. Upregulation of miR-196a and HOTAIR drive malignant character in gastrointestinal stromal tumors. Cancer Res. 2012;72(5):1126–36.
59. Chen W, Böcker W, Brosius J, Tiedge H. Expression of neural BC200 RNA in human tumours. J Pathol. 1997;183(3):345–51.
60. Iacoangeli A, Lin Y, Morley EJ, Muslimov IA, Bianchi R, Reilly J, et al. BC200 RNA in invasive and preinvasive breast cancer. Carcinogenesis. 2004;25(11):2125–33.
61. Chen W, Heierhorst J, Brosius J, Tiedge H. Expression of neural BC1 RNA: induction in murine tumours. Eur J Cancer. 1997;33(2):288–92.
62. Lanz RB, McKenna NJ, Onate SA, Albrecht U, Wong J, Tsai SY, et al. A steroid receptor coactivator, SRA, functions as an RNA and is present in an SRC-1 complex. Cell. 1999;97(1):17–27.
63. Cooper C, Guo J, Yan Y, Chooniedass-Kothari S, Hube F, Hamedani MK, et al. Increasing the relative expression of endogenous non-coding Steroid Receptor RNA Activator (SRA) in human breast cancer cells using modified oligonucleotides. Nucleic Acids Res. 2009; 37(13):4518–31.
64. Ulveling D, Francastel C, Hubé F. When one is better than two: RNA with dual functions. Biochimie. 2011;93(4):633–44.
65. Hube F, Guo J, Chooniedass-Kothari S, Cooper C, Hamedani MK, Dibrov AA, et al. Alternative splicing of the first intron of the steroid receptor RNA activator (SRA) participates in the generation of coding and noncoding RNA isoforms in breast cancer cell lines. DNA Cell Biol. 2006;25(7):418–28.
66. Foulds CE, Tsimelzon A, Long W, Le A, Tsai SY, Tsai MJ, et al. Research resource: expression profiling reveals unexpected targets and functions of the human steroid receptor RNA activator (SRA) gene. Mol Endocrinol. 2010;24(5):1090–105.
67. Lanz RB, Chua SS, Barron N, Söder BM, DeMayo F, O'Malley BW. Steroid receptor RNA activator stimulates proliferation as well as apoptosis in vivo. Mol Cell Biol. 2003; 23(20):7163–76.
68. Brannan CI, Dees EC, Ingram RS, Tilghman SM. The product of the H19 gene may function as an RNA. Mol Cell Biol. 1990;10(1):28–36.

Chapter 5
Mapping the HIF Transcription Factor in Cancer by ChIP-Seq Technology

Johannes Schödel and David R. Mole

Abstract Hypoxia-inducible factor (HIF) is the major transcriptional regulator mediating the cellular physiological response to reduced levels of oxygen (hypoxia). Upregulated as a consequence, both of intra-tumour hypoxia and through activation of oncogenic pathways, HIF has an important role in the pathogenesis of many cancers. However, activation of major physiological pathways in cancer upregulates pathways with both pro- and antitumorigenic actions and therefore confers a co-selection penalty. It is likely that both genetic and epigenetic factors alter the HIF-transcriptional response to favour a more tumorigenic profile. Mapping HIF transcription factor binding in cancer by ChIP-seq technology provides a framework for studying the mechanisms by which both genetic and epigenetic signatures associated with cancer may alter this HIF response.

Keywords HIF • Hypoxia • Hypoxia-inducible factor • Chromatin • Chromatin immunoprecipitation • ChIP • Next-generation sequencing • Cancer

1 Introduction

By September 2011, there were 1,617 published genome-wide association studies reporting significant ($p<5\times10^{-8}$) associations for 249 separate traits [1]. Since the order of amino acids in a peptide chain can be derived from the DNA sequence, the

J. Schödel
Department of Nephrology and Hypertension, Friedrich-Alexander-University Erlangen-Nuremberg, Ulmenweg 18, 91054 Erlangen, Germany
e-mail: johannes.schoedel@uk-erlangen.de

D.R. Mole (✉)
Henry Wellcome Building for Molecular Physiology, University of Oxford, Roosevelt Drive, OX3 7BN Oxford, UK
e-mail: drmole@well.ox.ac.uk

W. Wu and H. Choudhry (eds.), *Next Generation Sequencing in Cancer Research: Volume 1: Decoding the Cancer Genome*, DOI 10.1007/978-1-4614-7645-0_5,

functional consequences of variants in protein-coding regions are more readily understood. However, the Human Genome Project identified as few as 20,000 protein-coding genes spanning only 1.5 % [2, 3] of the human genome and most non-Mendelian disease-associated polymorphisms have so far mapped to nonprotein-coding regions. Consequently, the causative polymorphism and/or the molecular basis of its contribution to disease have been ascertained for only a small proportion of these polymorphisms [4–7].

More recently the ENCODE consortium has indicated that the vast majority (80.4 %) of the human genome that was previously thought to be nonfunctional participates in at least one RNA-coding and/or chromatin-associated event in at least one cell type [8]. Therefore, rather than altering the function of a translated protein, noncoding polymorphisms most likely act in cis to influence the transcriptional output of distant genetic elements (e.g. through regulation of transcription or splicing).

Characterising the regulatory elements (enhancers, promoters, insulators, repressors, silencers, etc.) within at-risk polymorphic regions will be crucial in elucidating the mechanisms underlying disease association. Studying the primary DNA sequence is of limited utility in defining such elements. With the exception of core promoters, regulatory elements are frequently cell-type specific and may be unconstrained across mammalian evolution [4]. Therefore, sequence conservation is relatively insensitive in detecting such regions. Similarly, transcription factor-binding motifs are relatively promiscuous and therefore the number of potential binding sites across the genome too numerous to accurately predict actual binding.

Active regulatory elements are characterised by cell-type-specific patterns of epigenetic marks, such as DNA accessibility (as characterised by DNAse hypersensitivity or formaldehyde-assisted identification of regulatory elements—FAIRE) or histone modifications [9, 10]. Analysis of such marks has proven to be a powerful approach in annotating tissue-specific regulatory elements. Identification of these regulatory elements at sites that are genetically linked to disease-associated polymorphisms may help identify putative disease-causing polymorphism(s). However, such regions may encompass a number of such polymorphisms. The more precise identification of transcription factor-binding sites, facilitated by chromatin immunoprecipitation (ChIP) coupled with motif analysis, may permit further refinement of polymorphisms in close genetic linkage to disease-associated alleles as well as being more functionally informative [5–7].

2 The HIF Pathway

The transcription pathway mediated by hypoxia-inducible factor (HIF) is one such transcriptional regulator that is of great importance in many types of cancer. Conserved across all known metazoan species [11, 12], HIF is composed of a heterodimer of α- and β-subunits [13]. Each subunit contains N-terminal beta helix–loop–helix (bHLH) and PAS domains responsible for dimerisation and

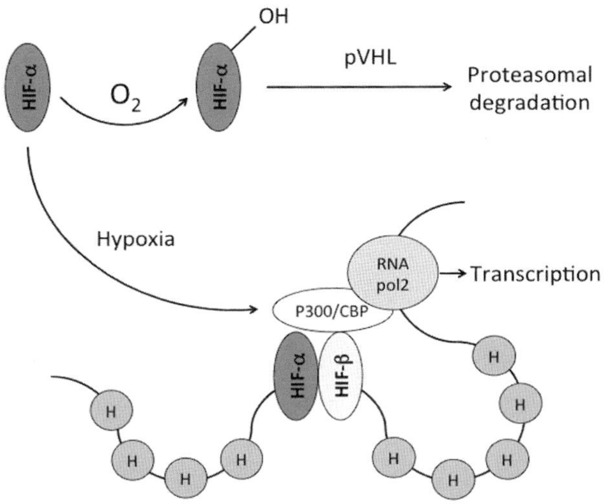

Fig 5.1 Overview of HIF transcription factor. HIF is composed of a heterodimer of α- and β-subunits. In the presence of oxygen, the HIF-α subunit is hydroxylated by prolyl hydroxylase domain (PHD) enzymes. The modified HIF-α subunits are then recognised by the von Hippel-Lindau tumour suppressor protein (pVHL), ubiquitylated and targeted for degradation in the proteasome. Under low levels of oxygen or when HIF-α is stabilised by oncogenic factors such as the loss of active pVHL, it heterodimerises with HIF-β, binds to hypoxia-response elements (HREs) on accessible regions of chromatin and interacts with the basal transcription machinery to upregulate RNA transcription

DNA binding. The β-subunit (HIF-1β), also known as aryl hydrocarbon nuclear receptor translocator (ARNT), is constitutively expressed and together with the aryl hydrocarbon receptor (AHR) is also essential for the xenobiotic response to certain environmental hydrocarbons. The HIF-α subunit exists as three isoforms [13–15], with the best studied being the HIF-1α subunit and the HIF-2α subunit, also known as endothelial PAS protein-1 (EPAS-1). These HIF-α subunits are highly regulated through oxygen-dependent domains in the C-terminal portion of the protein that in turn control degradation of the protein and interaction with transcriptional co-activators, most notably p300/CBP [12]. Regulation is conferred through hydroxylation of key residues within these domains by a series of oxygen-sensitive 2-oxoglutarate-dependent dioxygenase enzymes [16, 17]. In the presence of oxygen, hydroxylation of these residues targets the α-subunits for degradation by the ubiquitin-proteasomal pathway and inhibits binding to p300/CBP and hence their transactivating ability. When oxygen is limited, HIF-α is stabilised, dimerises with HIF-1β, binds to DNA and transactivates the expression of many hundreds of genes involved in both restoring oxygen supply and reducing oxygen consumption (Fig. 5.1).

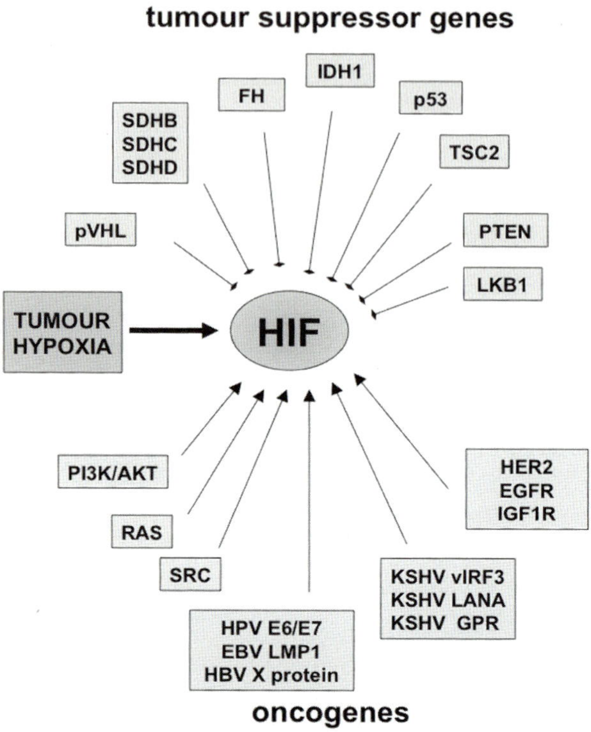

Fig. 5.2 Tumour suppressor and oncogenic pathways impacting the regulation of HIF. *pVHL* von Hippel-Lindau tumour suppressor, *SDH* succinate dehydrogenase, *FH* fumarate hydratase, *IDH1* isocitrate dehydrogenase-1, *TSC* tuberose sclerosis gene, *PTEN* phosphatase and tensin homologue, *LKB1* liver kinase B1, *PI3K* phosphoinositide 3-kinase, *RAS* rat sarcoma, *HPV* human papilloma virus, *EBV* Epstein-Barr virus, *KSHV* Kaposi sarcoma herpesvirus, *vIRF3* viral interferon regulatory factor-3, *LANA* latency-associated nuclear antigen, *GPR* G-protein-coupled receptor, *HER2* human epidermal growth factor receptor 2, *EGFR* epidermal growth factor receptor, *IGF1R* insulin-like growth factor 1 receptor

2.1 HIF and Cancer

The rapid proliferation of cells that is characteristic of cancer generates increased oxygen demand, which frequently outstrips that provided by the blood supply, resulting in regions of, often profound, hypoxia (reduced oxygen levels) within the tumours. For example, in breast cancers, the median pO_2 was only 10 mmHg compared to 65 mmHg in normal breast tissue [18]. This hypoxia contributes to the stabilisation of HIF-α subunits and the activation of HIF-transcriptional responses that is seen within the majority of primary tumours and their metastases [19, 20]. In addition, activation of oncogenic pathways may be directly coupled to activation of HIF [21] (Fig. 5.2). For example, HIF-α synthesis is increased through mTOR by the activation of phosphatidylinositol-3-kinase (PI-3-K) and MAP-kinase as a result of oncogenic gain-of-function mutations or increased signalling from tyrosine kinase receptors or G-protein-coupled receptors [22–25]. The protein products of several oncogenic viruses associated with human cancers (e.g. the Kaposi's sarcoma herpesvirus) may also activate the HIF-transcriptional response by various mechanisms [26–31]. Inactivation of tumour suppressor pathways can also lead to HIF activation [23–25, 32–37]. Most notably, the von Hippel-Lindau (VHL) tumour suppressor acts as the ubiquitin E3 ligase that is central to the oxygen-dependent degradation of HIF-α subunits. Loss or inactivation of this tumour suppressor (seen in the majority of renal carcinomas) leads to constitutive activation of HIF pathways mimicking hypoxia [34]. Loss of function of other tumour suppressor genes such as fumarate hydratase (FH) or succinate dehydrogenase (SDH) subunits generates high levels of metabolic inhibitors that interfere with the 2-oxoglutarate-dependent dioxygenase enzymes that regulate HIF [32, 33]. The convergence of intratumoral hypoxia, oncogenic and tumour suppressor pathways on the HIF-transcriptional cascade suggest a causal role for this HIF activation in cancer pathogenesis. Additional lines of evidence further support this hypothesis.

Extensive observational studies in many types of human cancers show a correlation between high levels of HIF-1α and HIF-2α protein detected by immunohistochemical staining and poor prognosis as assessed by patient mortality, which was particularly striking in early breast, cervical and endometrial cancers [21]. Interventional studies, predominantly in animal xenograft models, show that HIF gain of function generally accelerates tumour growth, whilst HIF loss of function retards progression [21]. Many of the newer classes of anticancer drugs that target angiogenic pathways have effects either on the HIF pathway or on HIF-transcriptional targets.

Finally, although not exhaustive, the number of genes that are known to be regulated by HIF in cancer cells is gradually increasing [38, 39]. Many of these genes have key roles in cancer-associated pathways, including angiogenesis, stem cell maintenance, metabolic reprogramming, autocrine growth factor signalling, epithelial to mesenchymal transitioning, invasion, metastasis and resistance to radio- and chemotherapy (Fig. 5.3).

2.2 Historical Identification of HIF-Transcriptional Targets

Historically, identification of the transcriptional targets of HIF in cancer has been a two-stage process (see Wenger [40] for review of early HIF-target genes). Firstly, regulation of gene expression has been identified, either by candidate-based approaches such as RNAse protection assays or quantitative PCR or more systematically using expression array analysis. Such microarray experiments have identified several thousand HIF-regulated genes [41–50] with many being regulated in a highly cell-specific manner together with a smaller "core" group of genes that are regulated by HIF across a broad range of cell types [38]. However, it is likely that many of these genes are not regulated by direct binding of HIF to their regulatory regions, but indirectly through additional HIF-dependent transcription and/or

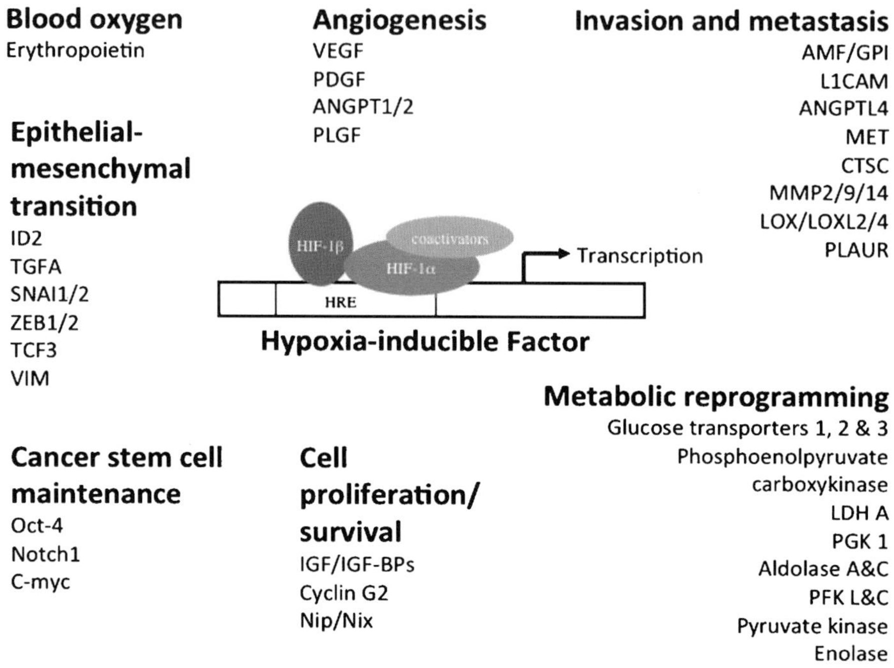

Fig. 5.3 Key HIF-target genes involved in oncogenic processes. *VEGF* vascular endothelial growth factor, *PDGF* platelet-derived growth factor, *ANGPT1/2* angiopoietin 1/2, *PLGF* placental growth factor, *AMF/GPI* autocrine motility factor/glucose-6-phosphate isomerase, *L1CAM* L1 cell adhesion molecule, *ANGPTL4* angiopoietin-like 4, *MET* met proto-oncogene, *CTSC* cathepsin C, *MMP* matrix metallopeptidase, *LOX(L)* lysyl oxidase (like), *PLAUR* plasminogen activator, urokinase receptor, *ID2* inhibitor of DNA binding 2, *TGFA* transforming growth factor alpha, *SNAI1/2* snail homologue 1/2, *ZEB1/2* zinc finger E-box binding homeobox 1/2, *TCF3* transcription factor 3, *VIM* vimentin, Oct-4 octamer-binding protein 4, *IGF* insulin-like growth factor, *IGF-BP* IGF binding protein, *LDH* lactate dehydrogenase, *PGK* phosphoglycerate kinase, *PFK* phosphofructokinase

mRNA stability factors [51–57]. Therefore, in attempts to show direct regulation by HIF, many groups have attempted to determine the cis-acting regulatory elements that are responsible for conferring HIF/hypoxic regulation on gene expression, either by a deletional approach in mouse models [58, 59] or by fusing potential regulatory elements to reporter genes [60–64]. DNA-binding activity could then be determined using DNAse footprinting and/or electrophoretic mobility shift assays (EMSA) [63–65]. Both labour intensive and inherently biased towards the identification of promoter-proximal regulatory elements, the analysis of non-chromatinised DNA in transient transfection reporter assays and electrophoretic mobility shift assays may not accurately reflect behaviour at the endogenous gene locus. Nevertheless, this approach has identified sufficient numbers of regulatory elements that it has been possible to define a consensus motif termed the hypoxia-response element (HRE) [40].

The presence of an HRE motif in a defined regulatory element of an HIF-regulated gene is highly indicative of direct transcriptional regulation of that gene by HIF. However, since the core-binding motif (RCGTG) is very short (only 5 bp), it occurs with high frequency (>1 million) across the genome (particularly in GC-rich promoter regions—given the high GC content of the core motif) [66]. Many of these core motifs will not lie in functional regulatory regions, so that the presence of such a motif within a regulated gene is not of itself sufficient to identify direct transcriptional targets of HIF, rendering bioinformatic approaches at distinguishing the direct HIF-transcriptional response ineffective. This has lead to alternative approaches to define functional HREs by demonstrating HIF binding at the endogenous gene locus.

3 Chromatin Immunoprecipitation

The presence of HIF binding to a candidate site in native chromatin can be examined in cells using chromatin immunoprecipitation (ChIP) [67–69]. HIF is first stabilised using either hypoxia or a chemical mimic of hypoxia such as the prolyl hydroxylase inhibitor dimethyloxalylglycine (DMOG). The cells are then treated with a cross-linking agent such as formaldehyde that generates covalent bonds between interacting proteins and between proteins and interacting DNA sequences. The cells are then lysed and the chromatin is sheared or digested into manageable sized fragments using sonication or DNAse enzyme. Antibodies specific to each HIF subunit recover not only the transcription factor but also DNA sequences that are bound to HIF. After reversal of cross linking and digestion of proteins, the resulting DNA is assayed by quantitative PCR (ChIP-qPCR). Primers spanning the putative binding site are used and results are usually expressed as fold enrichment compared to a control immunoprecipitation or input chromatin and normalised to a noninteracting region of the genome.

Since each candidate site needs to be assayed by a separate PCR reaction, this approach is effective where the potential number of core HIF-binding motifs at a regulated gene locus is limited. In practice it is possible to assay candidate binding sites that are close to the promoter, whereas given the high frequency of HRE motifs across the genome, the more distant candidate sites may be too numerous to test individually. Thus, ChIP-qPCR is still biased towards detecting promoter-proximal HIF binding. Therefore, in order to determine the full extent of the genome-wide direct transcriptional response to HIF, an unbiased pan-genomic assessment of HIF binding using nontarget-based approaches must be used.

3.1 ChIP-Seq

Early approaches to the pan-genomic assessment of HIF chromatin immunoprecipitation experiments utilised tiled microarrays to interrogate the material for the presence of interacting sequences [70–73]. However, whole-genome tiled microarrays are large and expensive, requiring multiple chips, and so more limited approaches using tiled microarrays spanning just the promoter regions of annotated genes have generally been employed [70, 71, 73]. These remain inherently biased towards promoter-proximal HIF binding and therefore dilute the advantages of using a genome-wide strategy. In addition, array-based strategies have a relatively low signal-to-noise ratio rendering the statistical determination of binding sites less powerful.

More recently, next-generation, high-throughput sequencing strategies to determine the DNA sequences present in HIF chromatin immunoprecipitation material have been employed [66, 74]. These approaches allow the sequence of the terminal (50–100) bases on each DNA fragment in the chromatin immunoprecipitate to be determined. These sequence "tags" can then be mapped (by sequence comparison) to the reference genome. The relative abundance of these tags reflects the amount of DNA from each genomic region in the immunoprecipitate (Fig. 5.4). Thus, genomic regions to which multiple sequence tags align reflect high chromatin abundance in the immunoprecipitate and thus HIF-interacting regions of the chromatin. Since, immunoprecipitated chromatin contains double-stranded DNA, tags mapping to the 5′ end of each (positive or negative) DNA strand will be separated by the average length of the sheared chromatin fragments. Thus, computer algorithms to determine peaks of mapped sequence read generally employ filters not only for read density but also for the biphasic nature of reads in each frame.

This approach generally gives very good genome-wide coverage, although there may be some bias as a result of difficulty mapping sequences to regions containing multiple repeats, regions of genomic duplication or regions of divergence from the reference genome. The signal-to-noise ratio is generally much higher than for array-based technologies. This not only increases confidence in peak calling but also allows for clearer delineation of the density profile of the mapped sequences and therefore results in a more precise definition of the position of HIF-interacting sites. The main advantage of this technique, however, is its ability to detect HIF binding at sites remote from gene promoters that may lie within the body of the gene, 3′ to the gene, or in intergenic regions remote from any annotated gene.

3.2 How Many HIF-Binding Sites?

Various computer algorithms exist to identify binding sites when analysing ChIP-seq data (e.g. MACS [75], CisGenome [76]). Generally these rely on the maximum read density in a read cluster or peak, an estimate of the distribution of reads between the Watson and Crick reading frames and a statistical measure of the false discovery rate based on the distribution of reads across the genome. A minimum threshold is usually set to distinguish "true" peaks from random (background) noise. For many purposes a set of well-defined "high-stringency" binding sites with a low false-positive rate is required for further analysis (e.g. motif searching) and this threshold is often set deliberately high. However, the number of binding sites identified

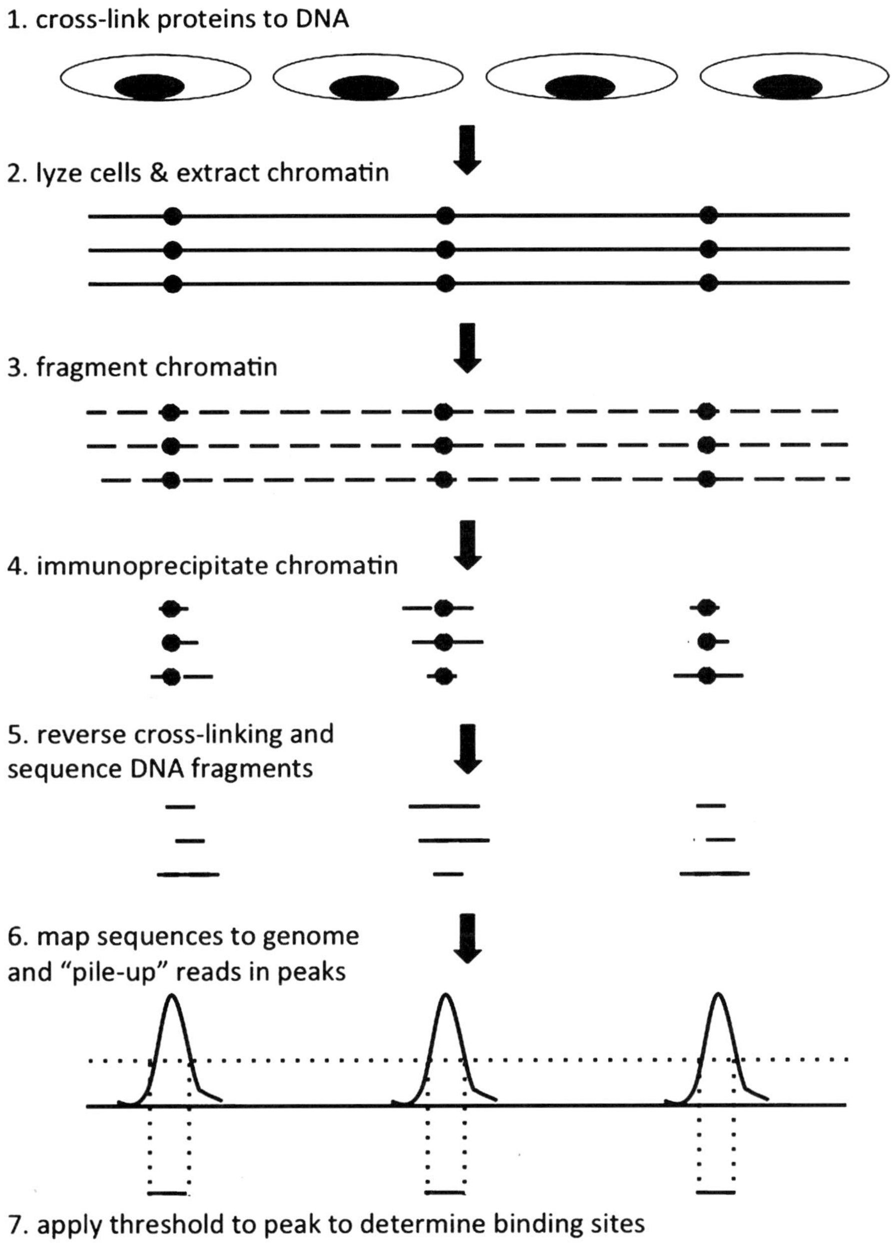

Fig. 5.4 Overview of ChIP-seq methodology. (1) Cells are first treated with formaldehyde to covalently cross-link proteins to DNA. (2) Cells are lysed and the chromatin isolated. (3) The chromatin is fragmented by sonication or enzymatic digestion. (4) The DNA-binding protein of interest is immunoprecipitated using antibodies that recognise formaldehyde-treated protein. (5) Cross linking is reversed and the protein component digested. (6) The immunoprecipitated chromatin fragments are sequenced using next-generation sequencing. The reads are then "mapped" to the reference genome by sequence comparison. (7) Immunoprecipitated regions of chromatin produce multiple overlapping reads and thresholds are applied to determine binding sites

depends greatly on the threshold used, so it is difficult to estimate the number of true binding sites and even harder to exclude HIF binding at a given gene locus.

One method of determining an appropriate threshold is to use a "seed" group of binding sites or gene loci at which the frequency of HIF binding is judged or expected to be high [66]. For example, this might be a set of HIF-regulated gene loci or a set of binding sites defined in prior analyses or in additional cell types. When peaks identified in the ChIP-seq dataset are ranked by peak height, the frequency of overlap between this "seed" dataset and the strongest peaks will be great (i.e. a high rate of true positivity). As the peak height decreases, there comes a point when random noise is gradually reached and the overlap between the seed dataset and the ChIP-seq peaks falls to a low background level defined by the limited size of the human genome. For example, when binding sites defined by ChIP-chip analysis are used as the "seed" dataset, the frequency of overlap with HIF-a ChIP-seq peaks falls to background levels after the 1,000 strongest peaks [66]. This provides an upper estimate of the number of HIF-binding sites and indicates an appropriate threshold for defining positive peaks. This type of approach is generally robust to the size of the "seed" group used so the "seed" dataset need not be exhaustive (e.g. ChIP-chip analysis does not cover the entire genome).

Since the rate of overlap gradually (rather than abruptly) declines as the ChIP-seq peak height falls, a threshold corresponding to the top 1,000 peaks will include some false-positive sites, and so additional filters are generally employed to define a set of high-stringency binding sites, a random sample of which can be confirmed by ChIP-qPCR. In MCF-7 cells, this method has defined 400 high-stringency HIF-1α-binding sites at 356 gene loci and 425 high-stringency HIF-2α-binding sites at 357 gene loci. This approach will likely exclude some true binding sites and therefore provides a lower estimate of the number of HIF-α-binding sites in MCF-7 cells.

Since the group of HIF-α-binding sites has somewhat blurred margins, an analysis of the overlap between HIF-1α- and HIF-2α-binding sites does not lend itself to conventional analysis by Venn diagram. Instead a similar analysis to that outlined above has been used to define overlapping HIF-1α and HIF-2α sites. For example, when the high-stringency HIF-1α-binding genes are used as the seed dataset and HIF-2α binding genes are ranked by peak read density, approximately 250 out of 356 HIF-1α binding genes also bind HIF-2α. This overlap is robust to whether HIF-1α or HIF-2α high-stringency binding genes are used as the "seed" dataset and indicates that HIF-1α and HIF-2α commonly bind at the same gene loci.

3.3 HIF-1β Binding

Although HIF-α proteins were initially identified as part of a DNA-binding complex with HIF-1β[13], interaction with other transcription factors that might result in DNA binding of one subunit without the other has been defined for both HIF-α and HIF-1β subunits [77, 78]. Parallel analysis of HIF-α and HIF-1β binding by

ChIP-seq allows this question to be addressed by using similar analyses to those outlined above to determine the degree of overlap between HIF-α- and HIF-1β-binding sites. This indicates very high concordance between the top 200 HIF-1α- and HIF-2α-binding sites and HIF-1β binding (95 % and 94.5 %, respectively) [66]. Conversely the top 200 HIF-1β-binding sites exhibited 95 % overlap with HIF-α sites. On further testing by ChIP-qPCR, the majority of HIF-α only sites were found to be composed of false-positive HIF-α signal in the ChIP-seq analysis. Thus, HIF-α binds almost exclusively in combination with HIF-1β, and therefore, the presence of HIF-1β can be used as a further filter when defining HIF-binding sites. Conversely, in the absence of stimuli to induce other HIF-1β dimerisation partners such as aryl hydrocarbon receptor (Ahr) [78], HIF-1β is almost exclusively found on chromatin at sites that bind HIF-α.

The preparation of material for next-generation sequencing involves amplification during library preparation and always results in comparable numbers of reads irrespective of the amount of input DNA. Therefore, in the absence of chromatin binding, the background signal will be exaggerated, making it difficult to exclude the possibility of DNA binding. For example, the ChIP-seq analysis of HIF-1β binding in normoxia does give apparent signal, although many of the peaks lack the usual shape resulting from the random distribution of DNA fragment length arising from sonication. However, these signals bear no correlation with transcriptional start sites, DNA hypersensitivity sites or with hypoxic HIF-1β-binding sites, suggesting that they are false-positive signals arising from amplified background noise and again indicating a lack of HIF-1β DNA binding in the absence of stimuli to stabilise heterodimerisation partners (Fig. 5.5).

3.4 *Distribution of HIF-Binding Sites*

Across the genome, nearest neighbour distance analysis demonstrated some clustering of HIF-binding gene loci. However, this merely reflected the nonrandom distribution of genes across the genome and was comparable to that observed for a random selection of gene loci. This suggests that if genes are localised into hypoxia-inducible "transcriptional factories", this is not reflected in the one-dimensional organisation of the genome.

Since HIF ChIP-seq identifies all HIF-binding sites irrespective of distance from gene promoters, it permits a complete description of their distribution. When mapped to the closest transcriptional start site (TSS), both HIF-1 and HIF-2 demonstrated most frequent binding approximately 100-bp upstream (5′) of the nearest gene. However, many HIF-binding sites were more distant from the transitional start site with approximately 60 % of HIF-1 sites and 80 % of HIF-2 sites lying more than 2.5 kb and in some cases more than 100 kb from the nearest TSS. In this respect HIF-2 was more likely to bind distant from the TSS and HIF-1 was more frequently seen in association with the promoter.

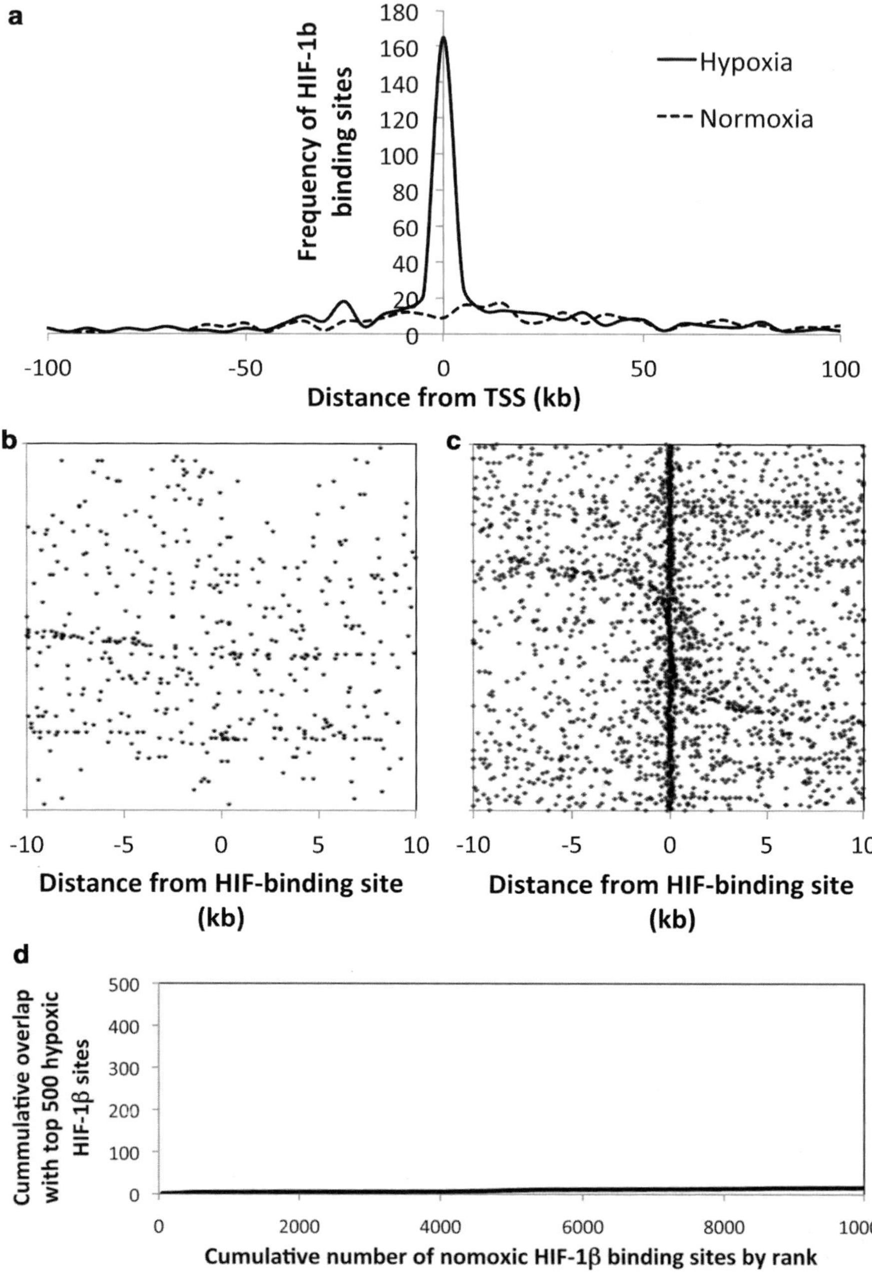

Fig. 5.5 HIF-1β binding in normoxia and hypoxia. (**a**) Frequency distribution showing the number of HIF-1β-binding sites according to distance from the nearest transcription start site (TSS). Hypoxic HIF-1β sites cluster close to promoters, whilst normoxic sites show no association with promoter regions. (**b** and **c**) Association between normoxic DNAse1 hypersensitivity sites—HSS (ENCODE Digital DNAse1 hypersensitivity data, http://genome.ucsc.edu/cgi-bin/hgTables).

3.5 Motif Analysis

The identification of many hundreds of HIF-binding sites permits the analysis of binding sequences for underlying recognition motifs. This has confirmed the core hypoxia-response element (HRE) recognition motif (RCGTG) defined by previous analyses of more limited numbers of binding sites and by mutation analysis. Additional preferences for a more extended recognition motif and for more distant motifs have also been identified, but these are not an absolute requirement for HIF binding. Thus, the core recognition motif for HIF binding is short (just 5 bp) and therefore relatively abundant (>1.1 million) across the genome. Since analysis of the primary sequence at binding and nonbinding HRE motifs does not predict binding, other "epigenetic" factors must play a role in directing HIF binding. Key amongst these is DNA accessibility.

Correlating HIF ChIP-seq datasets with publically available DNAse-seq datasets, in the same cell line, from the ENCODE consortium demonstrated that the concurrence of an HRE with a DNAse hypersensitive site is approximately 20 times more likely to bind HIF than an HRE that is in a DNAse-insensitive region. Importantly, DNAse sensitivity was defined in normoxia and therefore suggests that these sites are constitutively accessible and that accessibility itself is not regulated by hypoxia. However, only a small proportion of DNAse hypersensitive HRE motifs bound to HIF, suggesting that further epigenetic factors, in addition to DNA accessibility, are required to permit HIF binding.

3.6 HIF-Binding Patterns in Different Cancers

Different cell types not only have differing transcriptional repertoires but also regulate different sets of genes in response to hypoxia. Therefore, it is not surprising that different cell types also have differing patterns of HIF binding. Comparison of HIF-binding patterns in MCF-7 breast cancer cells and 786-O renal cancer cells indicated concordant binding in both cell lines at approximately 40 % of HIF-binding

Fig. 5.5 (continued) HIF-1β-binding regions were ranked according to distance upstream/downstream of the nearest TSS (*vertical axis*). For each HIF-1β-binding region, the position of DNASE1 HSS, relative to the centre of the HIF-binding region, was plotted (*horizontal axis*). (**b**) Normoxic HIF-1b sites show no association with DNA accessibility, whilst (**c**) hypoxic-binding sites are strongly associated with regions of open chromatin. (**d**) The degree of overlap between normoxic and hypoxic HIF-1β-binding sites. The top 10,000 normoxic sites were ranked according to peak height and the number overlapping with the 500 strongest hypoxic sites plotted against this rank. No significant overlap was observed between normoxic and hypoxic HIF-1β-binding sites

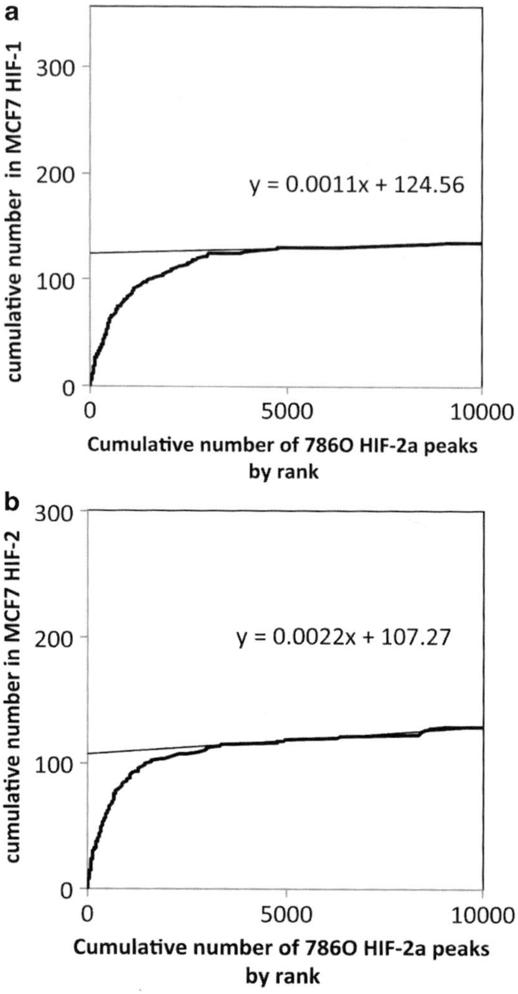

Fig. 5.6 Comparison of HIF-binding sites in MCF-7 breast cancer cells and 786-O kidney cancer cells by ChIP-seq. HIF-2a ChIP-seq peaks from 786-O cells were ranked by peak height and plotted on the *x* axis. The number overlapping with (**a**) high-stringency HIF-1 sites and (**b**) high-stringency HIF-2 sites were plotted on the *y* axis. The degree of overlap was highest for the strongest 786-O peaks falling to low background levels after 2,000–3,000 peaks. Extrapolation of these background levels indicates approximately 125/356 and 107/301 (35 %) overlap between 786-O and MCF-7 HIF-1β-binding sites

sites (Fig. 5.6). Analysis of common sites and those unique to 786-O cells showed that whilst the former were accessible in MCF-7 cells, the 786-O unique sites were inaccessible as defined by DNAse hypersensitivity. This was the case even when the neighbouring promoter was accessible. Thus, local accessibility at the HIF-binding site helps define cell-type-specific patterns of binding to HREs. The factors governing local accessibility and the other epigenetic events involved in directing HIF binding remain unclear. However, there is emerging evidence that these processes are important in the pathogenesis of cancer with the description of cancer-associated mutations in increasing numbers of epigenetic modifiers.

3.7 HIF Binding and Gene Regulation

The diffuse distribution of HIF binding sites can make it difficult to identify the functional transcriptional target for that binding site, particularly where the binding site is remote and/or equidistant from two promoters. To date most groups have used the nearest promoter as a first approximation to identify the potential transcriptional target of a given transcription factor. The juxtaposition of ChIP-seq data with pre-existing pan-genomic expression data from microarray studies has allowed this assumption to be explored in greater detail.

Gene Set Enrichment Analysis [79, 80] (GSEA) is a computational method that defines whether an a priori defined set of genes (e.g. the closest genes to the defined HIF-binding sites) is differentially expressed between two biological states (e.g. normoxia versus hypoxia or hypoxia versus hypoxia and HIF suppression). Essentially genes are ranked according to the differential expression between the two states. Each occurrence of a gene, in the predefined subset, is marked by a positive score, and each occurrence of a gene absent from the subset is marked by a negative score. A running total of the metric is plotted and used to derive a statistical measure of the enrichment of the subset of genes amongst differentially expressed (regulated) genes. Genes that neighbour HIF-binding sites are highly enriched amongst genes regulated in hypoxia and by the presence of HIF, indicating that a large proportion of HIF-binding sites regulate the closest gene. Importantly, HIF-binding sites are only enriched amongst genes upregulated in hypoxia or downregulated by HIF suppression, indicating that HIF is predominantly a transcriptional activator. This association remains true even when the HIF-binding site is remote (e.g. >100 kb) from the nearest promoter, indicating that HIF can exhibit transcriptional effects over long genomic intervals.

However, not all genes neighbouring HIF-binding sites are regulated by hypoxia/HIF. When these binding sites are examined, analysis of the second nearest promoter again shows enrichment amongst genes upregulated by HIF/hypoxia, indicating that HIF binding is also capable of regulating non-neighbouring genes. Thus, whilst the annotation of genes neighbouring HIF-binding sites as HIF-target genes provides a first approximation, it does result in false-positive and false-negative calls. Newer bioinformatic methods using machine learning algorithms to identify transcription factor regulatory targets from pan-genomic binding and expression data are being developed (e.g. EMBER—Expectation Maximization of Binding and Expression Profiles [81]) but to date have not been applied to HIF. Ultimately, experimental validation of predicted binding-regulatory interactions requires that a physical interaction between the binding site and the gene promoter be demonstrated using techniques such as chromatin-conformation capture (3C) assays, which are both time consuming and technically demanding.

The correlation of HIF-binding ChIP-seq data with expression data from microarray studies assumes that the transcriptional targets of HIF binding are represented on the microarray chip used. However, it is now recognised that the transcriptional output includes many classes of RNA that may not be included in traditional microarrays, which typically only cover protein-coding RNAs. Furthermore, some types of RNA may not be polyadenylated and may be excluded from the cDNA library through selective PCR or size exclusion. Newer techniques of pan-genomic expression analysis such as next-generation sequencing of RNA (RNA-seq), particularly involving total RNA, have the capability of analysing all classes of RNA in parallel and providing a much more complete picture of the transcriptional output [82, 83]. The possibility of combining ChIP-seq data with these types of analysis holds the exiting possibility of defining the entire transcriptional output of the HIF pathway.

4 HIF-Target Genes and Renal Cancer

Despite the constraints of expression microarrays, the analysis of ChIP-seq data in the light of these datasets represents the most comprehensive description of the direct transcriptional output of HIF in cancer to date. This has identified several hundred HIF binding and regulated genes that are both common and unique to each cancer and helped delineate some of the processes contributing to the tissue specificity. These HIF-target genes are frequently upregulated in cancer. For example, when HIF-target genes are examined for differential expression using publically accessible microarray studies of renal cancer and normal renal tissue, GSEA analysis reveals that a high proportion of HIF-target genes are upregulated in the cancers compared to the normal tissue [84] (Fig. 5.7).

Many of these transcriptional targets are well-recognised HIF targets with important oncogenic roles. However, others are previously less well described, some with unclear roles in cancer and some with potentially tumour suppressing functions. The upregulation of HIF in cancer, whether it is as a consequence of physiological hypoxia or as a result of oncogenic mutation such as inactivation of the VHL tumour suppressor, will entrain global upregulation of all of these target genes. Distinguishing which activated genes are "driver" genes promoting tumorigenesis and which are merely "passenger" genes acquired as a consequence of co-selection with the "driver" genes is critical to understanding the pathogenesis of cancer and to the effective design of new therapeutic strategies.

The acquisition of HIF binding during progression from non-malignant to malignant cells might indicate important HIF targets with a direct role in promoting tumorigenesis. Therefore, HIF-binding sites that occur commonly and specifically in certain types of cancer, but not in noncancerous tissue, may help distinguish "driver" from "passenger" HIF-target genes.

However, another strategy for determining which HIF-target genes are important to the pathogenesis of cancer is to cross-reference these HIF ChIP-seq and expression datasets with data arising from cancer genetic studies, either germ line studies of cancer predisposition or somatic studies of cancer-associated mutation. The presence of cancer-susceptibility polymorphisms, close to a HIF-target gene, will prioritise those gene loci for further analysis. Conversely, the identification of HIF-binding sites close to intergenic cancer-susceptibility loci may identify potential mechanisms by which polymorphic variants might lead to cancer predisposition.

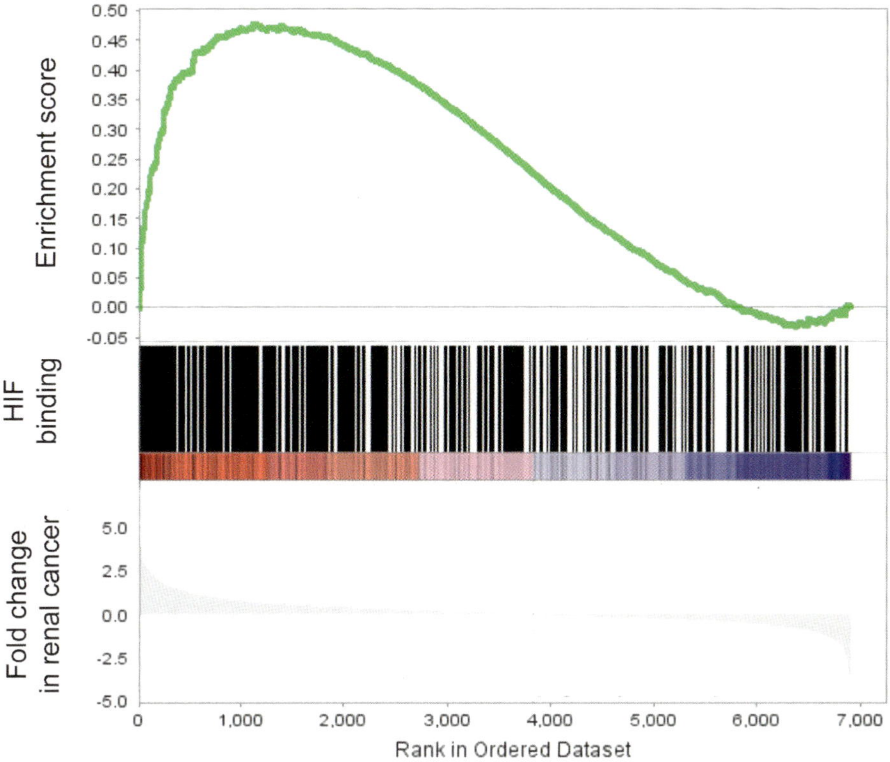

Fig. 5.7 786-O, HIF-2 binding genes are upregulated in renal tumours. Gene Set Enrichment Analysis (GSEA) was performed using microarray data (Gumz et al. [84], GSE6344) comparing gene expression levels in ten renal tumours with surrounding normal renal tissue. Genes were ranked according to fold upregulation in renal tumours. Each gene that binds HIF-2 in 786-O cells is marked with a *black bar*. The running enrichment score is shown in *green*

Polymorphisms that are in high linkage disequilibrium with cancer-associated polymorphisms will be coinherited and therefore will have candidacy as potential disease-causing polymorphisms. Those at HIF-binding sites have the potential to disrupt HIF binding and so affect gene expression. However, HIF-target gene function might also be influenced by polymorphisms/mutations in the coding region and expression might be further affected by variation at additional transcriptional enhancer sites that cooperate with the HIF-binding site. Therefore, any polymorphic variants close to an HIF-binding site might disrupt its transcriptional target whether or not the region of high linkage disequilibrium extends to the HIF-binding site itself.

This strategy combining pan-genomic expression and ChIP-seq analysis of regulatory elements with genetic studies is a novel type of approach to interpreting genetic data in cancer. However, it is already paying dividends in understanding the role of HIF activation, especially in renal cancer in which VHL inactivation leads to constitutive activation of HIF.

4.1 11q13.3 Polymorphisms and Renal Cancer

Recent genome-wide association studies (GWAS) have identified two renal cancer-susceptibility loci on chromosome 2p21 and 11q13.3. The first of these two loci contains polymorphisms that lie within the first intron of the EPAS1 (HIF-2a) gene. The second locus on 11q13.3 lies in a gene-poor region of previously unknown function approximately 150 kb centromeric to the MYEOV1 gene and approximately 220 kb telomeric to the CCND1 gene locus. Importantly, HIF binding was identified close to and in high linkage disequilibrium (LD) with the index disease-associated SNP at this locus. Indeed polymorphisms lying just 10 and 15 bp either side of the HRE at the HIF-binding site are in high LD with the disease-associated polymorphism.

Consistent with a specific role for the 11q13.3 binding site in renal cancer, HIF binding was only observed in VHL-defective clear cell renal cancer cell lines and in renal tumours. Binding was not observed in VHL wild-type cell lines of renal or non-renal origin in which HIF had been temporarily stabilised by hypoxia or the HIF stabiliser, dimethyloxalylglycine (DMOG). It is not known whether this binding site develops as a consequence of VHL loss and chronic HIF activation or whether renal cancers develop in a rare subpopulation of renal cells in which the enhancer predates inactivation of VHL. However, the selective presence of the 11q13.3 HIF-binding site in VHL-defective cancers in which HIF is constitutively expressed strongly suggests that this site provides a survival advantage when HIF levels are high and therefore contributes to the pathogenesis of renal cancer.

Concordant with genome-wide findings that local DNA accessibility determines cell-specific patterns of DNA binding, the presence of HIF at this site mirrors local DNA accessibility as defined by formaldehyde-assisted isolation of regulatory elements (FAIRE). Additional chromatin immunoprecipitation experiments demonstrated the presence of histone methylations (H3K4me1 and H3K4me3) compatible with an active enhancer and interaction of this site with RNApol2 both of which again mirror VHL and HIF status. These properties are not reversed when wild-type VHL is stably reexpressed in VHL-defective clear cell renal cancer cell lines (i.e. the enhancer remains accessible, retains H3K4me1 and H3K4me3 histone methylation and remains able to bind HIF). This indicates that the epigenetic mechanisms that lead to activity of this enhancer in VHL-defective renal cancer are not readily reversible.

Compatible with genome-wide patterns indicating that HIF may regulate genes other than the nearest gene, the more proximal MYEOV gene is only weakly regulated by HIF in the VHL-defective cancer setting, whilst the more distant CCND1 gene is one of the most highly HIF-regulated genes on the entire chromosome 11. This HIF-binding enhancer at the 11q13.3 locus physically interacts with the CCND1 promoter as demonstrated by chromatin-conformation capture (3C) assay and by fluorescent in situ hybridisation (FISH) providing direct experimental evidence that CCND1 is a transcriptional target of HIF binding at the 11q13.3 locus. This promoter is approximately 220 kb from the HIF-binding site and emphasises the importance of experimentally determining the promoter(s) that interact with a cis-acting regulatory region of interest (be it a transcription factor-binding site, a DNAse hypersensitive site or an intergenic disease association locus) when predicting the transcriptional target of that region.

4.2 11q13.3 Allele-Specific Effects on HIF-Binding, Epigenetic Environment and CCND1 Expression

Although this type of analysis can tell us that the 11q13.3 locus is a long-range enhancer of CCND1 expression that is specific to VHL-defective clear cell renal carcinoma cells, it does not tell us whether polymorphisms at this site affect the function of the enhancer. Genotype–phenotype correlations, such as those undertaken by Cao et al. [85], which illustrate a correlation between the genotype at the HIF-binding enhancer and CCND1 expression level provide some evidence that this may be the case. However, such analyses are inherently noisy and may require many tumour samples to determine an effect, particularly (as at the 11q13.3 renal cancer-susceptibility locus) when the frequency of one of the alleles is low. Furthermore, whilst in vitro analyses such as electrophoretic mobility shift assays (EMSA) or transient transfection reporter assays to determine enhancer activity may distinguish differential properties of the different polymorphic variants, these techniques rely on exogenous DNA molecules, which may not accurately reflect the properties of native chromatin.

In order to capture the effects of polymorphic variants on HIF binding in the setting of native chromatin, "allele-specific" chromatin immunoprecipitation can be used. To undertake this type of analysis, a cell line heterozygous for SNPs underlying the HIF-binding peak (i.e. close enough to the HRE that they lie on the same sonication fragments generated during the chromatin preparation) and in high LD with the disease-associated SNP must first be determined. Chromatin immunoprecipitation is then undertaken and the material is analysed using a method (e.g. qPCR using allele-specific TaqMan probes) that can determine the relative abundance of each allele. This can then be compared with the relative abundance of each SNP in the input chromatin (cancer cells are frequently polypoid, which may alter the allelic ratios). Any allelic imbalance in the immunoprecipitated chromatin compared to the native chromatin or control immunoprecipitations will signify preferential binding (e.g. of HIF) to one allele compared with the other. Using this approach in KTCL140 cells, both HIF-2α and HIF-1β preferentially bound to the 11q13.3 allele that predisposes to renal cancer, when compared to the protective (minor) allele.

Since polymorphisms within the disease-associated haplotype are all in high linkage disequilibrium with each other, they are all coinherited, and therefore, it is not possible to distinguish the SNP that is responsible for disrupting HIF binding using this technique. Furthermore, whilst these polymorphic SNPs are close to the HRE, none disrupts the core-binding motif, and therefore, none would be predicted to affect HIF-binding directly. Instead, it is likely that one or more of these SNPs alter other properties of the local chromatin, which in turn governs HIF binding. For example, it is possible to examine "allele-specific" chromatin accessibility using similar methodologies applied to material prepared from the same KTCL140 cells using FAIRE. Sequences from the RCC-predisposing allele were preferentially enriched in the FAIRE material compared to the input chromatin signifying greater DNA accessibility of the HIF-binding site on this allele compared to the RCC-protective allele.

"Allele-specific" analyses can also be extended to determine functional effects of polymorphisms at the 11q13.3 enhancer on CCND1 expression, even though the CCND1 promoter is in low LD with the RCC-associated haplotype. Firstly, ChIP-seq analysis of RNApol2 binding in normoxic and hypoxic MCF-7 cells identifies small peaks of RNApol2 coinciding with HIF binding and presumably reflecting interaction of HIF with the basal transcriptional machinery. In KTCL140 cells, "allele-specific" analysis of RNApol2 binding at the 11q13.3 enhancer indicates preferential binding to the RCC-predisposing allele, suggesting that this allele preferentially drives transcription.

Secondly, the preferential expression of CCND1 from each allele in a cell line can be determined from the relative abundance a heterozygous transcribed SNP in mRNA compared to genomic DNA. KTCL140 cells demonstrate just such an allelic imbalance when heterozygous SNPs within the 3′ UTR of the CCND1 gene were analysed. Although these SNPs are distant from and in low LD with the disease-associated SNPs, these cells are all derived from the same source and have not been through subsequent meioses and so the SNPs at the two loci will all have the same phase. Thus, it can be inferred that one allele at the cancer-associated locus is associated with preferential CCND1 expression, although it is not formally possible to determine which.

Finally, transcriptional activity at a gene locus can be inferred from the amount of RNApol2 bound to the transcribed region of the gene. For example, ChIP-seq analysis of RNApol2 binding in normoxic and hypoxic MCF-7 cells indicates that HIF-dependent gene transcription is associated with increased "run through" of RNApol2 across the body of the gene. Analysing RNApol2 immunoprecipitates from KTCL140 cells using the same heterozygous transcribed SNPs (within the 3′ UTR) that were used to establish differential mRNA abundance demonstrated a similar imbalance in RNApol2 binding. This indicates an allelic imbalance in transcriptional activity at the CCND1 gene locus that mirrors mRNA abundance.

5 Future Directions

This example illustrates the utility of chromatin immunoprecipitation-based strategies in the study of cancer-associated genetic signals of undetermined function. Firstly, ChIP-seq approaches directed at transcription factors with a known role in cancer can be used to determine potential regulatory regions that are in high linkage disequilibrium with disease-associated polymorphisms. Analysis of public databases of ChIP-seq signals demonstrates frequent concordance in binding sites for many different transcription factors across assorted cell types, and in the future, this may provide a short cut to identifying sites of interest for further study, particularly when relevant transcriptional pathways have not been identified. However, our analysis of HIF binding in different cell lines has demonstrated significant cell-type-specific patterns of HIF binding. Furthermore, it is these differences that may be particularly important in driving cancer progression. Indeed, the 11q13.3 CCND1 enhancer (important in renal cancer predisposition) would not have been identified by examining currently available public databases.

Secondly, at the 11q13.3 RCC-susceptibility locus, ChIP for epigenetic marks and for RNApol2 was used to confirm the regulatory role of transcription factor-binding sites identified by ChIP-seq. An alternative, more generic strategy might involve reversing the order of these analyses. For example, ChIP-seq analysis of histone modifications associated with active enhancers together with pan-genomic assessment of DNA accessibility using FAIRE-seq or DNAse-seq may provide a way of identifying all active enhancers within a region of interest, irrespective of transcription factor. These prioritised sites could then be examined for binding to a panel of potentially relevant transcription factors in simpler ChIP-qPCR experiments.

This approach is particularly relevant to those cancers (including RCC) in which mutations in epigenetic modifiers are being linked to its pathogenesis (e.g. PBRM1, SETD2, KDM6A, KDM5C, MLL2 [86–91]). Studying pan-genomic patterns of binding of these modifiers, the epigenetic marks that they generate and/or the effect of their mutation on patterns of transcription factor binding such as HIF binding will be important in determining the mechanisms and target involved.

Finally, an alternate way of identifying regulatory regions is to determine sites that interact in cis with known promoters using chromatin-conformation capture (3C) assays. This has been used to identify a distant enhancer of DEXI that lies within a region of chromosome 16p13 that has been linked by GWAS with autoimmune diseases (including type 1 diabetes mellitus and multiple sclerosis) but lies intronic to the CLEC16A gene. Newer techniques such as 5C have the potential to identify DNA-DNA interactions across the genome. Interaction between sites within a disease-associated locus and a gene promoter may provide valuable clues to the function of polymorphism at the former site.

6 Summary

The major transcriptional pathway orchestrated by HIF is commonly upregulated in cancer and is generally associated with a poor prognosis. However, the physiological responses induced by HIF posses both pro- and antitumorigenic actions. Therefore, in cancer, it is likely that both genetic and epigenetic events modulate the HIF-transcriptional response to favour a more tumorigenic profile. By studying pan-genomic patterns of HIF-binding in cancer in combination with genetic and epigenetic signatures associated with cancer, it is now possible to dissect some of these mechanisms.

References

1. Hindorff LA, MacArthur J, Wise A, Junkins HA, Hall PN, Klemm AK, et al. A Catalog of Published Genome-Wide Association Studies. Available from: http://www.genome.gov/gwastudies.
2. Lander ES, Linton LM, Birren B, Nusbaum C, Zody MC, Baldwin J, et al. Initial sequencing and analysis of the human genome. Nature. 2001;409(6822):860–921. PubMed PMID: 11237011.
3. IHGS Consortium. Finishing the euchromatic sequence of the human genome. Nature. 2004;431(7011):931–45. PubMed PMID: 15496913. Epub 2004/10/22. eng.
4. Freedman ML, Monteiro AN, Gayther SA, Coetzee GA, Risch A, Plass C, et al. Principles for the post-GWAS functional characterization of cancer risk loci. Nat Genet. 2011;43(6):513–8. PubMed PMID: 21614091. Pubmed Central PMCID: 3325768. Epub 2011/05/27. eng.
5. Pomerantz MM, Ahmadiyeh N, Jia L, Herman P, Verzi MP, Doddapaneni H, et al. The 8q24 cancer risk variant rs6983267 shows long-range interaction with MYC in colorectal cancer. Nat Genet. 2009;41(8):882–4. PubMed PMID: 19561607. Pubmed Central PMCID: 2763485. Epub 2009/06/30. eng.
6. Jia L, Landan G, Pomerantz M, Jaschek R, Herman P, Reich D, et al. Functional enhancers at the gene-poor 8q24 cancer-linked locus. PLoS Genet. 2009;5(8):e1000597. PubMed PMID: 19680443. Pubmed Central PMCID: 2717370. Epub 2009/08/15. eng.
7. Schodel J, Bardella C, Sciesielski LK, Brown JM, Pugh CW, Buckle V, et al. Common genetic variants at the 11q13.3 renal cancer susceptibility locus influence binding of HIF to an enhancer of cyclin D1 expression. Nat Genet. 2012;44(4):420–5. Epub 11 Mar 2012.
8. Consortium EP, Dunham I, Kundaje A, Aldred SF, Collins PJ, Davis CA, et al. An integrated encyclopedia of DNA elements in the human genome. Nature. 2012;489(7414):57–74. PubMed PMID: 22955616. Pubmed Central PMCID: 3439153.
9. Barski A, Cuddapah S, Cui K, Roh TY, Schones DE, Wang Z, et al. High-resolution profiling of histone methylations in the human genome. Cell. 2007;129(4):823–37. PubMed PMID: 17512414. eng.
10. Yip KY, Cheng C, Bhardwaj N, Brown JB, Leng J, Kundaje A, et al. Classification of human genomic regions based on experimentally determined binding sites of more than 100 transcription-related factors. Genome Biol. 2012;13(9):R48. PubMed PMID: 22950945. Pubmed Central PMCID: 3491392.
11. Loenarz C, Coleman ML, Boleininger A, Schierwater B, Holland PW, Ratcliffe PJ, et al. The hypoxia-inducible transcription factor pathway regulates oxygen sensing in the simplest animal, Trichoplax adhaerens. EMBO Rep. 2011;12(1):63–70. PubMed PMID: 21109780. Pubmed Central PMCID: 3024122. Epub 2010/11/27. eng.
12. Kaelin Jr WG, Ratcliffe PJ. Oxygen sensing by metazoans: the central role of the HIF hydroxylase pathway. Mol Cell. 2008;30(4):393–402. PubMed PMID: 18498744. Epub 2008/05/24. eng.
13. Wang GL, Jiang BH, Rue EA, Semenza GL. Hypoxia-inducible factor 1 is a basic-helix-loop-helix-PAS heterodimer regulated by cellular O2 tension. Proc Natl Acad Sci USA. 1995;92(12):5510–4. PubMed PMID: 7539918.
14. Gu YZ, Moran SM, Hogenesch JB, Wartman L, Bradfield CA. Molecular characterization and chromosomal localization of a third alpha-class hypoxia inducible factor subunit, HIF3alpha. Gene Expr. 1998;7(3):205–13. PubMed PMID: 9840812.

15. Tian H, McKnight SL, Russell DW. Endothelial PAS domain protein 1 (EPAS1), a transcription factor selectively expressed in endothelial cells. Genes Dev. 1997;11(1):72–82. PubMed PMID: 9000051.
16. Schofield CJ, Ratcliffe PJ. Oxygen sensing by HIF hydroxylases. Nat Rev Mol Cell Biol. 2004;5(5):343–54. PubMed PMID: 15122348.
17. Lando D, Gorman JJ, Whitelaw ML, Peet DJ. Oxygen-dependent regulation of hypoxia-inducible factors by prolyl and asparaginyl hydroxylation. Eur J Biochem/FEBS. 2003;270(5):781–90. PubMed PMID: 12603311.
18. PubMed PMID: 15063685. Epub 2004/04/06. eng
19. Talks KL, Turley H, Gatter KC, Maxwell PH, Pugh CW, Ratcliffe PJ, et al. The expression and distribution of the hypoxia-inducible factors HIF-1alpha and HIF-2alpha in normal human tissues, cancers, and tumor-associated macrophages. Am J Pathol. 2000;157(2):411–21. PubMed PMID: 10934146.
20. Zhong H, De Marzo AM, Laughner E, Lim M, Hilton DA, Zagzag D, et al. Overexpression of hypoxia-inducible factor 1alpha in common human cancers and their metastases. Cancer Res. 1999;59(22):5830–5. PubMed PMID: 10582706.
21. Semenza GL. Defining the role of hypoxia-inducible factor 1 in cancer biology and therapeutics. Oncogene. 2010;29:625–34. PubMed PMID: 19946328. Epub 2009/12/01. Eng.
22. Laughner E, Taghavi P, Chiles K, Mahon PC, Semenza GL. HER2 (neu) signaling increases the rate of hypoxia-inducible factor 1alpha (HIF-1alpha) synthesis: novel mechanism for HIF-1-mediated vascular endothelial growth factor expression. Mol Cell Biol. 2001;21(12):3995–4004. PubMed PMID: 11359907.
23. Zhong H, Chiles K, Feldser D, Laughner E, Hanrahan C, Georgescu MM, et al. Modulation of hypoxia-inducible factor 1alpha expression by the epidermal growth factor/phosphatidylinositol 3-kinase/PTEN/AKT/FRAP pathway in human prostate cancer cells: implications for tumor angiogenesis and therapeutics. Cancer Res. 2000;60(6):1541–5. PubMed PMID: 10749120.
24. Zundel W, Schindler C, Haas-Kogan D, Koong A, Kaper F, Chen E, et al. Loss of PTEN facilitates HIF-1-mediated gene expression. Genes Dev. 2000;14(4):391–6. PubMed PMID: 10691731Pubmed Central PMCID: 316386. Epub 2000/02/26. eng.
25. Brugarolas JB, Vazquez F, Reddy A, Sellers WR, Kaelin WG. TSC2 regulates VEGF through mTOR-dependent and -independent pathways. Cancer Cell. 2003;4(2):147–58. PubMed PMID: 12957289.
26. Kondo S, Seo SY, Yoshizaki T, Wakisaka N, Furukawa M, Joab I, et al. EBV latent membrane protein 1 up-regulates hypoxia-inducible factor 1alpha through Siah1-mediated down-regulation of prolyl hydroxylases 1 and 3 in nasopharyngeal epithelial cells. Cancer Res. 2006;66(20):9870–7. PubMed PMID: 17047048. Epub 2006/10/19. eng.
27. Yoo YG, Cho S, Park S, Lee MO. The carboxy-terminus of the hepatitis B virus X protein is necessary and sufficient for the activation of hypoxia-inducible factor-1alpha. FEBS Lett. 2004;577(1–2):121–6. PubMed PMID: 15527772. Epub 2004/11/06. eng.
28. Tomita M, Semenza GL, Michiels C, Matsuda T, Uchihara JN, Okudaira T, et al. Activation of hypoxia-inducible factor 1 in human T-cell leukaemia virus type 1-infected cell lines and primary adult T-cell leukaemia cells. Biochem J. 2007;406(2):317–23. PubMed PMID: 17576198. Pubmed Central PMCID: 1948965. Epub 2007/06/20. eng.
29. Sodhi A, Montaner S, Patel V, Zohar M, Bais C, Mesri EA, et al. The Kaposi's sarcoma-associated herpes virus G protein-coupled receptor up-regulates vascular endothelial growth factor expression and secretion through mitogen-activated protein kinase and p38 pathways acting on hypoxia-inducible factor 1alpha. Cancer Res. 2000;60(17):4873–80. PubMed PMID: 10987301. Epub 2000/09/15. eng.
30. Shin YC, Joo CH, Gack MU, Lee HR, Jung JU. Kaposi's sarcoma-associated herpesvirus viral IFN regulatory factor 3 stabilizes hypoxia-inducible factor-1 alpha to induce vascular endothelial growth factor expression. Cancer Res. 2008;68(6):1751–9. PubMed PMID: 18339855. Epub 2008/03/15. eng.

31. Nakamura M, Bodily JM, Beglin M, Kyo S, Inoue M, Laimins LA. Hypoxia-specific stabilization of HIF-1alpha by human papillomaviruses. Virology. 2009;387(2):442–8. PubMed PMID: 19321184. Pubmed Central PMCID: 2674135. Epub 2009/03/27. eng.
32. Selak MA, Armour SM, MacKenzie ED, Boulahbel H, Watson DG, Mansfield KD, et al. Succinate links TCA cycle dysfunction to oncogenesis by inhibiting HIF-alpha prolyl hydroxylase. Cancer Cell. 2005;7(1):77–85. PubMed PMID: 15652751.
33. Isaacs JS, Jung YJ, Mole DR, Lee S, Torres-Cabala C, Chung Y-L, et al. HIF overexpression correlates with biallelic loss of fumarate hydratase in renal cancer: Novel role of fumarate in regulation of HIF stability. Cancer Cell. 2005;8(2):143–53.
34. Maxwell PH, Wiesener MS, Chang GW, Clifford SC, Vaux EC, Cockman ME, et al. The tumour suppressor protein VHL targets hypoxia-inducible factors for oxygen-dependent proteolysis. Nature. 1999;399(6733):271–5. PubMed PMID: 10353251.
35. Ravi R, Mookerjee B, Bhujwalla ZM, Sutter CH, Artemov D, Zeng Q, et al. Regulation of tumor angiogenesis by p53-induced degradation of hypoxia-inducible factor 1alpha. Genes Dev. 2000;14(1):34–44. PubMed PMID: 10640274.
36. Shackelford DB, Vasquez DS, Corbeil J, Wu S, Leblanc M, Wu CL, et al. mTOR and HIF-1alpha-mediated tumor metabolism in an LKB1 mouse model of Peutz-Jeghers syndrome. Proc Natl Acad Sci USA. 2009;106(27):11137–42. PubMed PMID: 19541609. Pubmed Central PMCID: 2708689. Epub 2009/06/23. eng.
37. Zhao S, Lin Y, Xu W, Jiang W, Zha Z, Wang P, et al. Glioma-derived mutations in IDH1 dominantly inhibit IDH1 catalytic activity and induce HIF-1alpha. Science. 2009;324(5924):261–5. Med PMID: 19359588. Pubmed Central PMCID: 3251015. Epub 2009/04/11. eng.
38. PubMed PMID: 20061373. Pubmed Central PMCID: 2853119. Epub 2010/01/12. eng
39. Semenza GL. Hypoxia-inducible factors: mediators of cancer progression and targets for cancer therapy. Trends Pharmacol Sci. 2012;33(4):207–14. PubMed PMID: 22398146. Epub 2012/03/09. Eng.
40. Wenger RH, Stiehl DP, Camenisch G. Integration of oxygen signaling at the consensus HRE. Sci STKE. 2005;2005(306):re12. PubMed PMID: 16234508. eng.
41. Elvidge GP, Glenny L, Appelhoff RJ, Ratcliffe PJ, Ragoussis J, Gleadle JM. Concordant regulation of gene expression by hypoxia and 2-oxoglutarate-dependent dioxygenase inhibition: the role of HIF-1alpha, HIF-2alpha, and other pathways. J Biol Chem. 2006;281(22):15215–26. PubMed PMID: 16565084.
42. Manalo DJ, Rowan A, Lavoie T, Natarajan L, Kelly BD, Ye SQ, et al. Transcriptional regulation of vascular endothelial cell responses to hypoxia by HIF-1. Blood. 2005;105(2):659–69. PubMed PMID: 15374877. eng.
43. Hu CJ, Iyer S, Sataur A, Covello KL, Chodosh LA, Simon MC. Differential regulation of the transcriptional activities of hypoxia-inducible factor 1 alpha (HIF-1alpha) and HIF-2alpha in stem cells. Mol Cell Biol. 2006;26(9):3514–26. PubMed PMID: 16611993. eng.
44. Hu CJ, Wang LY, Chodosh LA, Keith B, Simon MC. Differential roles of hypoxia-inducible factor 1alpha (HIF-1alpha) and HIF-2alpha in hypoxic gene regulation. Mol Cell Biol. 2003;23(24):9361–74. PubMed PMID: 14645546. eng.
45. Greijer AE, van der Groep P, Kemming D, Shvarts A, Semenza GL, Meijer GA, et al. Up-regulation of gene expression by hypoxia is mediated predominantly by hypoxia-inducible factor 1 (HIF-1). J Pathol. 2005;206(3):291–304. PubMed PMID: 15906272. eng.
46. Imamura T, Kikuchi H, Herraiz MT, Park DY, Mizukami Y, Mino-Kenduson M, et al. HIF-1alpha and HIF-2alpha have divergent roles in colon cancer. Int J Cancer. 2009;124(4):763–71. PubMed PMID: 19030186. Epub 2008/11/26. eng.
47. Wang V, Davis DA, Haque M, Huang LE, Yarchoan R. Differential gene up-regulation by hypoxia-inducible factor-1alpha and hypoxia-inducible factor-2alpha in HEK293T cells. Cancer Res. 2005;65(8):3299–306. PubMed PMID: 15833863. eng.
48. Sung FL, Hui EP, Tao Q, Li H, Tsui NB, Dennis Lo YM, et al. Genome-wide expression analysis using microarray identified complex signaling pathways modulated by hypoxia in nasopharyngeal carcinoma. Cancer Lett. 2007;253(1):74–88. PubMed PMID: 17320280. Epub 2007/02/27. eng.

49. discussion 7. PubMed PMID: 17486380. eng
50. Choi SM, Oh H, Park H. Microarray analyses of hypoxia-regulated genes in an aryl hydrocarbon receptor nuclear translocator (Arnt)-dependent manner. FEBS J. 2008;275(22):5618–34. PubMed PMID: 18959748. Epub 2008/10/31. eng.
51. Oikawa M, Abe M, Kurosawa H, Hida W, Shirato K, Sato Y. Hypoxia induces transcription factor ETS-1 via the activity of hypoxia-inducible factor-1. Biochem Biophys Res Commun. 2001;289(1):39–43. PubMed PMID: 11708773. Epub 2001/11/16. eng.
52. Laderoute KR, Calaoagan JM, Gustafson-Brown C, Knapp AM, Li GC, Mendonca HL, et al. The response of c-jun/AP-1 to chronic hypoxia is hypoxia-inducible factor 1 alpha dependent. Mol Cell Biol. 2002;22(8):2515–23. PubMed PMID: 11909946. eng.
53. PubMed PMID: 12354771. Epub 2002/10/02. eng
54. Koshiji M, Kageyama Y, Pete EA, Horikawa I, Barrett JC, Huang LE. HIF-1alpha induces cell cycle arrest by functionally counteracting Myc. EMBO J. 2004;23(9):1949–56. PubMed PMID: 15071503. eng.
55. Gordan JD, Lal P, Dondeti VR, Letrero R, Parekh KN, Oquendo CE, et al. HIF-alpha effects on c-Myc distinguish two subtypes of sporadic VHL-deficient clear cell renal carcinoma. Cancer Cell. 2008;14(6):435–46. PubMed PMID: 19061835. Epub 2008/12/09. eng.
56. Gustafsson MV, Zheng X, Pereira T, Gradin K, Jin S, Lundkvist J, et al. Hypoxia requires notch signaling to maintain the undifferentiated cell state. Dev Cell. 2005;9(5):617–28. PubMed PMID: 16256737. Epub 2005/11/01. eng.
57. Kaidi A, Williams AC, Paraskeva C. Interaction between beta-catenin and HIF-1 promotes cellular adaptation to hypoxia. Nat Cell Biol. 2007;9(2):210–7. PubMed PMID: 17220880. Epub 2007/01/16. eng.
58. Semenza GL, Traystman MD, Gearhart JD, Antonarakis SE. Polycythemia in transgenic mice expressing the human erythropoietin gene. Proc Natl Acad Sci USA. 1989;86(7):2301–5. PubMed PMID: 2928334. eng.
59. Semenza GL, Dureza RC, Traystman MD, Gearhart JD, Antonarakis SE. Human erythropoietin gene expression in transgenic mice: multiple transcription initiation sites and cis-acting regulatory elements. Mol Cell Biol. 1990;10(3):930–8. PubMed PMID: 2304468. eng.
60. Levy NS, Goldberg MA, Levy AP. Sequencing of the human vascular endothelial growth factor (VEGF) 3' untranslated region (UTR): conservation of five hypoxia-inducible RNA-protein binding sites. Biochim Biophys Acta. 1997;1352(2):167–73. PubMed PMID: 9199248.
61. Pugh CW, Tan CC, Jones RW, Ratcliffe PJ. Functional analysis of an oxygen-regulated transcriptional enhancer lying 3' to the mouse erythropoietin gene. Proc Natl Acad Sci USA. 1991;88(23):10553–7. PubMed PMID: 1961720. eng.
62. Pugh CW, Ebert BL, Ebrahim O, Maxwell PH, Ratcliffe PJ. Analysis of cis-acting sequences required for operation of the erythropoietin 3' enhancer in different cell lines. Ann N Y Acad Sci. 1994;718:31–9. discussion 9-40. PubMed PMID: 8185238. eng.
63. Semenza GL, Nejfelt MK, Chi SM, Antonarakis SE. Hypoxia-inducible nuclear factors bind to an enhancer element located 3' to the human erythropoietin gene. Proc Natl Acad Sci USA. 1991;88(13):5680–4. PubMed PMID: 2062846. eng.
64. Firth JD, Ebert BL, Ratcliffe PJ. Hypoxic regulation of lactate dehydrogenase A. Interaction between hypoxia-inducible factor 1 and cAMP response elements. J Biol Chem. 1995;270(36):21021–7. PubMed PMID: 7673128. Epub 1995/09/08. eng.
65. Dimova EY, Moller U, Herzig S, Fink T, Zachar V, Ebbesen P, et al. Transcriptional regulation of plasminogen activator inhibitor-1 expression by insulin-like growth factor-1 via MAP kinases and hypoxia-inducible factor-1 in HepG2 cells. Thromb Haemost. 2005;93(6):1176–84. PubMed PMID: 15968405.
66. Schodel J, Oikonomopoulos S, Ragoussis J, Pugh CW, Ratcliffe PJ, Mole DR. High-resolution genome-wide mapping of HIF-binding sites by ChIP-seq. Blood. 2011;117(23):e207–17. PubMed PMID: 21447827. Epub 2011/03/31. eng.
67. O'Neill LP, Turner BM. Immunoprecipitation of chromatin. Methods Enzymol. 1996;274:189–97. PubMed PMID: 8902805. Epub 1996/01/01. eng.

68. Collas P. The current state of chromatin immunoprecipitation. Mol Biotechnol. 2010;45(1): 87–100. PubMed PMID: 20077036. Epub 2010/01/16. eng.
69. Lau KW, Tian YM, Raval RR, Ratcliffe PJ, Pugh CW. Target gene selectivity of hypoxia-inducible factor-alpha in renal cancer cells is conveyed by post-DNA-binding mechanisms. Br J Cancer. 2007;96(8):1284–92. PubMed PMID: 17387348. eng.
70. Mole DR, Blancher C, Copley RR, Pollard PJ, Gleadle JM, Ragoussis J, et al. Genome-wide association of hypoxia-inducible factor (HIF)-1{alpha} and HIF-2{alpha} DNA binding with expression profiling of hypoxia-inducible transcripts. J Biol Chem. 2009;284(25):16767–75. PubMed PMID: 19386601. Pubmed Central PMCID: 2719312. Epub 2009/04/24. eng.
71. Xia X, Kung AL. Preferential binding of HIF-1 to transcriptionally active loci determines cell-type specific response to hypoxia. Genome Biol. 2009;10(10):R113. PubMed PMID: 19828020. Epub 2009/10/16. Eng.
72. Xia X, Lemieux ME, Li W, Carroll JS, Brown M, Liu XS, et al. Integrative analysis of HIF binding and transactivation reveals its role in maintaining histone methylation homeostasis. Proc Natl Acad Sci USA. 2009;106(11):4260–5. PubMed PMID: 19255431. Eng.
73. Krieg AJ, Rankin EB, Chan D, Razorenova O, Fernandez S, Giaccia AJ. Regulation of the histone demethylase JMJD1A by hypoxia-inducible factor 1 alpha enhances hypoxic gene expression and tumor growth. Mol Cell Biol. 2010;30(1):344–53. PubMed PMID: 19858293. Pubmed Central PMCID: 2798291. Epub 2009/10/28. eng.
74. Tanimoto K, Tsuchihara K, Kanai A, Arauchi T, Esumi H, Suzuki Y, et al. Genome-wide identification and annotation of HIF-1alpha binding sites in two cell lines using massively parallel sequencing. Hugo J. 2010;4(1–4):35–48. PubMed PMID: 22132063. Pubmed Central PMCID: 3051044. Epub 2011/12/02. eng.
75. Zhang Y, Liu T, Meyer CA, Eeckhoute J, Johnson DS, Bernstein BE, et al. Model-based analysis of ChIP-Seq (MACS). Genome Biol. 2008;9(9):R137. PubMed PMID: 18798982. Pubmed Central PMCID: 2592715. Epub 2008/09/19. eng.
76. Ji H, Jiang H, Ma W, Johnson DS, Myers RM, Wong WH. An integrated software system for analyzing ChIP-chip and ChIP-seq data. Nat Biotechnol. 2008;26(11):1293–300. PubMed PMID: 18978777. eng.
77. Koshiji M, To KK, Hammer S, Kumamoto K, Harris AL, Modrich P, et al. HIF-1alpha induces genetic instability by transcriptionally downregulating MutSalpha expression. Mol Cell. 2005;17(6):793–803. PubMed PMID: 15780936. Epub 2005/03/23. eng.
78. Hankinson O. The aryl hydrocarbon receptor complex. Annu Rev Pharmacol Toxicol. 1995;35:307–40. PubMed PMID: 7598497.
79. Subramanian A, Tamayo P, Mootha VK, Mukherjee S, Ebert BL, Gillette MA, et al. Gene set enrichment analysis: a knowledge-based approach for interpreting genome-wide expression profiles. Proc Natl Acad Sci USA. 2005;102(43):15545–50. PubMed PMID: 16199517. Pubmed Central PMCID: 1239896. Epub 2005/10/04. eng.
80. Mootha VK, Lindgren CM, Eriksson KF, Subramanian A, Sihag S, Lehar J, et al. PGC-1alpha-responsive genes involved in oxidative phosphorylation are coordinately downregulated in human diabetes. Nat Genet. 2003;34(3):267–73. PubMed PMID: 12808457. Epub 2003/06/17. eng.
81. Maienschein-Cline M, Zhou J, White KP, Sciammas R, Dinner AR. Discovering transcription factor regulatory targets using gene expression and binding data. Bioinformatics. 2012;28(2):206–13. PubMed PMID: 22084256. Pubmed Central PMCID: 3259433.
82. Wang Z, Gerstein M, Snyder M. RNA-Seq: a revolutionary tool for transcriptomics. Nat Rev Genet. 2009;10(1):57–63. PubMed PMID: 19015660. Pubmed Central PMCID: 2949280.
83. Ozsolak F, Platt AR, Jones DR, Reifenberger JG, Sass LE, McInerney P, et al. Direct RNA sequencing. Nature. 2009;461(7265):814–8. PubMed PMID: 19776739.
84. Gumz ML, Zou H, Kreinest PA, Childs AC, Belmonte LS, LeGrand SN, et al. Secreted frizzled-related protein 1 loss contributes to tumor phenotype of clear cell renal cell carcinoma. Clin Cancer Res. 2007;13(16):4740–9. PubMed PMID: 17699851. Epub 2007/08/19. eng.

85. Cao Q, Qin C, Meng X, Ju X, Ding Q, Wang M, et al. Genetic polymorphisms in APE1 are associated with renal cell carcinoma risk in a Chinese population. Mol Carcinog. 2011;50(11):863–70. PubMed PMID: 21538578. Epub 2011/05/04. eng.
86. COSMIC. Catalogue Of Somatic Mutations In Cancer 2012. Available from: http://www.sanger.ac.uk/genetics/CGP/cosmic/.
87. van Haaften G, Dalgliesh GL, Davies H, Chen L, Bignell G, Greenman C, et al. Somatic mutations of the histone H3K27 demethylase gene UTX in human cancer. Nat Genet. 2009;41(5):521–3. PubMed PMID: 19330029. Epub 2009/03/31. eng.
88. Dalgliesh GL, Furge K, Greenman C, Chen L, Bignell G, Butler A, et al. Systematic sequencing of renal carcinoma reveals inactivation of histone modifying genes. Nature. 2010;463(7279):360–3. PubMed PMID: 20054297. Pubmed Central PMCID: 2820242. Epub 2010/01/08. eng.
89. Varela I, Tarpey P, Raine K, Huang D, Ong CK, Stephens P, et al. Exome sequencing identifies frequent mutation of the SWI/SNF complex gene PBRM1 in renal carcinoma. Nature. 2011;469(7331):539–42. PubMed PMID: 21248752. Pubmed Central PMCID: 3030920. Epub 2011/01/21. eng.
90. Duns G, Hofstra RM, Sietzema JG, Hollema H, van Duivenbode I, Kuik A, et al. Targeted exome sequencing in clear cell renal cell carcinoma tumors suggests aberrant chromatin regulation as a crucial step in ccRCC development. Hum Mutat. 2012;33(7):1059–62. PubMed PMID: 22461374. Epub 2012/03/31. eng.
91. Duns G, van den Berg E, van Duivenbode I, Osinga J, Hollema H, Hofstra RM, et al. Histone methyltransferase gene SETD2 is a novel tumor suppressor gene in clear cell renal cell carcinoma. Cancer Res. 2010;70(11):4287–91. PubMed PMID: 20501857. Epub 2010/05/27. eng.

Chapter 6
Application of Next-Generation Sequencing to Analysis of TGFβ/SMAD4 Targets in Ovarian Cancer

Russell Bonneville, Kenneth Nephew, and Victor X. Jin

Abstract Deregulation of the transforming growth factor-β (TGFβ) signaling pathway in epithelial ovarian cancer has been reported, but the precise mechanism underlying disrupted TGFβ signaling in the disease remains unclear. We performed chromatin immunoprecipitation followed by sequencing (ChIP-seq) to investigate genome-wide screening of TGFβ-induced SMAD4 binding in epithelial ovarian cancer. Following TGFβ stimulation of the A2780 epithelial ovarian cancer cell line, we identified 2,362 SMAD4 binding loci and 318 differentially expressed SMAD4 target genes. Comprehensive examination of SMAD4-bound loci revealed four distinct binding patterns. TGFβ-stimulated SMAD4-bound loci were primarily classified as either Stimulated Only or Shift, indicating that TGFβ stimulation alters SMAD4 binding patterns in epithelial ovarian cancer cells. Furthermore, based on gene regulatory network analysis, we determined that the TGFβ-induced, SMAD4-dependent regulatory network was strikingly different in ovarian cancer compared to normal cells. Importantly, the TGFβ/SMAD4 target genes identified in the A2780 epithelial ovarian cancer cell line were predictive of patient survival, based on in silico mining of publically available patient databases. In conclusion, our data highlight the utility of next-generation sequencing technology to identify genome-wide SMAD4 target genes in epithelial ovarian cancer and link aberrant TGFβ/SMAD

R. Bonneville • V.X. Jin (✉)
Department of Biomedical Informatics, The Ohio State University, Columbus, OH 43210, USA
e-mail: Victor.Jin@osumc.edu

K. Nephew
Department of Cellular and Integrative Physiology, Indiana University School of Medicine, Bloomington, IN 47405, USA

Department of Obstetrics and Gynecology, Indiana University School of Medicine, Bloomington, IN 47405, USA

Department of Molecular and Cellular Biochemistry, Indiana University School of Medicine, Bloomington, IN 47405, USA

W. Wu and H. Choudhry (eds.), *Next Generation Sequencing in Cancer Research: Volume 1: Decoding the Cancer Genome*, DOI 10.1007/978-1-4614-7645-0_6,

signaling to ovarian tumorigenesis. Furthermore, the identified SMAD4 binding loci, combined with gene expression profiling and in silico data mining of patient cohorts, may provide a powerful approach to determine potential gene signatures with biological and future translational research in ovarian and other cancers.

Keywords TGFβ/SMAD • ChIP-seq • Epithelial ovarian cancer • Gene signatures

1 Introduction

The transforming growth factor-β (TGFβ) signaling pathway is important in regulating many cellular processes including growth in ovarian surface epithelial cells (OSE) [1, 2]. TGFβ signaling dysregulation is frequently observed in epithelial ovarian cancer (EOC) and may be critical to EOC development [3, 4]. The effects of TGFβ are mediated by three TGFβ ligands—TGFβ1, TGFβ2, and TGFβ3, acting through TGFβ type 1 and 2 receptors [5–7]. TGFBR2 is the specific receptor for TGFβ ligands. The TGFβ-TGFBR2 complex regulates the activation of downstream Smad and non-Smad pathways [8], and the phosphorylated type 1 receptor recruits and phosphorylates receptor-regulated Smads and R-Smads. Of the five R-Smads in mammals, the TGFBR2-ALK5 complex activates SMAD2 and SMAD3, and the TGFBR2-ALK1 complex activates SMAD1, SMAD5, and SMAD8 [9]. Activated R-Smads form heteromeric complexes with the common partner Smad (SMAD4 in mammals) and translocate into the nucleus [6]. Since the activated Smad complex has insufficient affinity for the Smad binding element to support association with endogenous promoters, Smad complexes associate with other DNA binding transcription factors (TFs) to regulate gene expression [7, 10]. Many studies have shown that various families of TFs can serve as SMAD4 partner proteins to achieve high affinity and selectivity for target promoters with the appropriate binding elements [11–15].

The A2780 human epithelial ovarian cancer cell line provides a useful model for studying drug-sensitive ovarian cancer, as it is sensitive to cisplatin, a platinum-type agent used in ovarian cancer treatment. A2780 cells also exhibit partial TGFβ dysregulation, indicated by only a modest increase in SMAD4 expression and translocation from the cytoplasm to the nucleus following TGFβ stimulation [16]. Therefore, this cell line is also an appropriate model system for genome-wide mapping of SMAD4 target genes and identifying the dysregulated TGFβ/SMAD4 target genes and pathways implicated in ovarian cancer patients.

ChIP-seq (chromatin immunoprecipitation-sequencing) has been shown by many studies to provide mapping of transcription factor binding and histone modifications on a genome-wide scale with much higher resolution, greater depth, and improved accuracy compared to array-based approaches such as ChIP-chip [17–20]. In this study, we used ChIP-seq to study TGFβ/SMAD4 regulation in the platinum-sensitive A2780 ovarian cancer cell line. We profiled SMAD4 binding loci with TGFβ stimulation, and used computational methods to investigate the SMAD4 binding pattern in TGFβ-stimulated A2780 and compare it with the SMAD4 binding patterns in both a normal immortalized ovarian surface epithelial cell (IOSE) line from our previous study [13] and human keratinocytes (HaCaT) from Koinuma et al. [12]. Furthermore, we generated TGFβ/SMAD4-regulated gene signatures and correlated them with clinical outcome data from two publicly available ovarian cancer patient cohorts using in silico mining. This integrative approach revealed significant associations of TGFβ/SMAD4 regulatory networks with both progression-free and overall survival in ovarian cancer patients. Our ChIP-seq data identified thousands of SMAD4 binding loci as well as SMAD4-regulated genes, providing both a new resource for studying the mechanisms of dysregulation of TGFβ signaling in ovarian cancer cells as well as potential prognostic biomarkers for future ovarian cancer translational research.

2 ChIP-seq: Chromatin Immunoprecipitation and Massively Parallel Sequencing

Chromatin immunoprecipitation (ChIP) was performed as previously described [21, 22] with some noteworthy changes. Briefly, cells were rinsed with room temperature PBS before being cross-linked in a 1 % formaldehyde solution. Cells were then harvested and homogenized in the presence of protease inhibitors before DNA was sonicated. Magnetic Dynabeads (Invitrogen) combined with a mixture of antibodies (20 % SMAD4 #9515 (Cell Signaling Technology, Danvers, MA) and 80 % SMAD4 DCS-46 (Santa Cruz Biotechnology, Santa Cruz, CA)) were used to pull down SMAD4 overnight. Purified DNA was used to detect fold enrichment by SYBR Green qRT-PCR.

Sequencing libraries were generated for massive parallel sequencing using standard methods. Briefly, 500 ng of pulldown DNA was subjected to end repair, terminal adenylation, and adapter ligation before fragments ranging from ~175 to 250 were isolated from a 2 % E-gel (Invitrogen). Subsequent to a standardized 12 cycle PCR, DNA quality was evaluated on a DNA 1000 Bioanalyzer chip (Agilent Technologies, Santa Clara, CA) before being submitted for sequencing on an Illumina GAII. All ChIP-seq data is deposited in the Gene Expression Omnibus (GEO) database at the National Center for Biotechnology Information (http://www.ncbi.nlm.nih.gov/geo), accession number GSE27526 (Fig. 6.1).

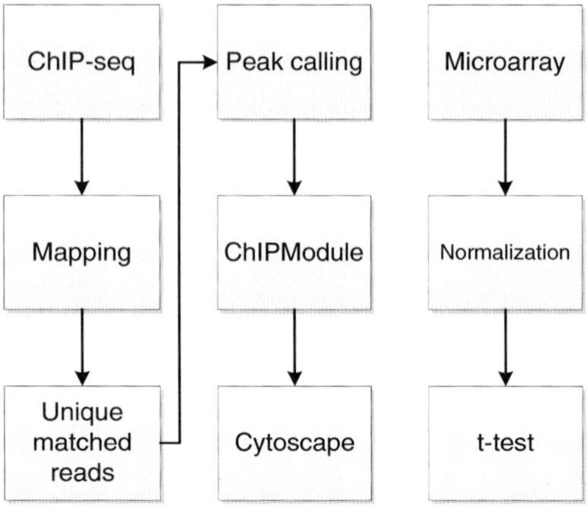

Fig. 6.1 Data analysis workflow. A flowchart of the ChIP-seq analysis pipeline and processing of the A2780 microarray data

3 Processing ChIP-Seq and Microarray Gene Expression Data

A standard procedure for extracting image files, mapping the reads onto human genome, and filtering the mapped reads to unique reads was followed with the Solexa 1.6 pipeline. The TGFβ-stimulated and TGFβ-unstimulated samples were each produced in two lanes of raw reads. The reads from these two lanes were combined in to a single data set. Both samples in the combined data set were processed using BELT [23, 24] developed in our laboratory, with a 300 nt bin size at an acceptance threshold of 0.996 vs. an input sample.

The microarray expression data was normalized using the standard protocol for the MAS5 algorithm implemented by Affymetrix in R, and a student's *t*-test was performed to determine the significance of the difference between the sets of biological triplicates for the stimulated and untreated samples. Significance was liberally defined as $p<0.10$, and a differential fold chance was defined as $\log_2$-fold change >0.50.

4 Gene Regulatory Network Analysis

We apply our computational analytical approach developed in our laboratory [25], which includes a de novo method ChIPModule [26] to identify the Hub TFs for 318 TGFβ/SMAD4 genes in A2780 cells and 92 TGFβ/SMAD4 genes in HaCaT cells, respectively. The Hub TFs for 150 TGFβ/SMAD4 genes in IOSE cells were from our previous study [13] which used the CART model [27], a similar machine-learning approach. The gene regulatory networks were constructed by scanning the binding loci of each gene using the position-weight matrix (PWM) of Hub TFs. The topology and visualization of the resulted hierarchal network is built by Cytoscape [28], where blue nodes represent Hub TFs, while red and green nodes correspond to up- and downregulated genes, respectively. The significance of the network is statistically tested by a permutation test to determine the probability of each edge of the network under random circumstances.

5 Results

Many previous studies [2, 4, 29–31] including ours [13, 16, 32, 33] have attempted to establish and characterize the molecular mechanisms of dysregulated TGFβ-mediated signaling in ovarian cancer cells and acquired cisplatin-resistant ovarian cancer cells. We used ChIP-seq of SMAD4 in A2780 cells with and without TGFβ stimulation to further elucidate the details of these underlying mechanisms.

We performed ChIP-seq for SMAD4 on all samples to generate a set of 36 bp raw reads. The reads were sequenced using the Illumina/Solexa GAII system, yielding ~43 million to ~51 million reads per sample. The reads were mapped to the UCSC human hg18 reference assembly, resulting in ~26 million unique mapped reads (UMRs) in unstimulated A2780 and ~32 million UMRs in TGFβ-stimulated A2780.

We then used our previously developed peak-calling program BELT [23, 24] to identify SMAD4 binding loci in both sets of reads. BELT uses a percentile-scoring method to determine locally enriched regions of the genome and then localizes actual peaks within them. The false discovery rate (FDR) is estimated by comparison with a simulated background randomly generated with similar read frequencies and Monte Carlo synthetic peak generation. BELT confirmed multiple previously known SMAD4 binding sites, including Gadd45A, CTGF, JAG1, LEMD3 [15], MYC [34], EDN1, RYBP, DST, and BCAT1 [12].

5.1 TGFβ-Unstimulated Binding

In unstimulated A2780, 2,009 SMAD4 binding sites were detected (Fig. 6.2); of these, 1,499 (74.6 %) were located within ±100 kb of an annotated RefSeq gene [35]. Only 267 of the 1,499 loci (13.3 %) were within ±8 kb of the promoter region of a gene, while the majority of these loci were within 10 kb upstream of a TSS. These results highlight the advantage of unbiased whole-genome target site analysis through ChIP-seq over ChIP-chip promoter arrays [12, 13, 36], which may only find subsets of SMAD4 target sites.

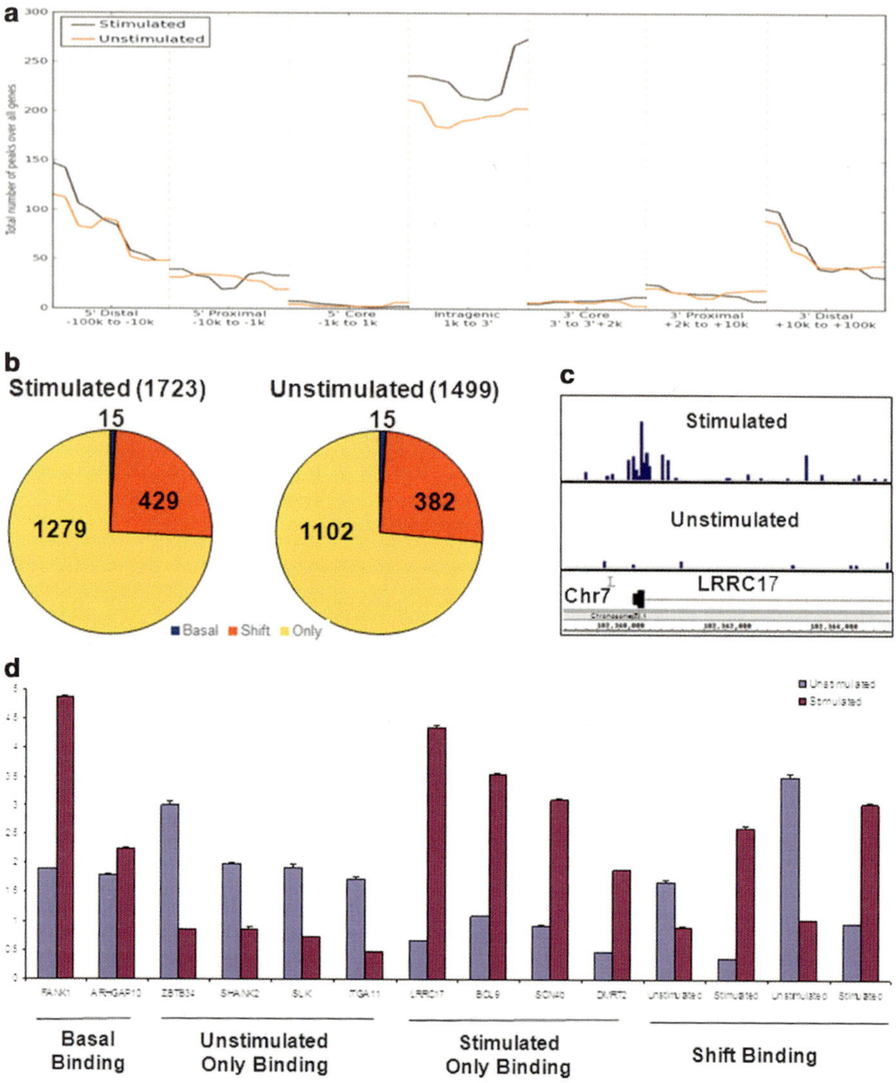

Fig. 6.2 Identification of TGFβ/SMAD4 binding loci. (**a**) The distribution of the location of SMAD4 binding loci in a histogram plot based on their relative to a closest known RefGene 5′TSS. (**b**) Classification of SMAD4 binding loci into four binding patterns. Stimulated Only binding loci are those whose associated RefGene has binding loci only in the stimulated set, likewise for Unstimulated Only. Shift binding loci have a binding loci appearing on the same gene in both conditions, and they are greater than 1,000 nt apart. Basal binding loci appear on the same gene in both conditions, but they are less than 1,000 nt apart. (**c**) A screenshot showing LRRC17 binding pattern, where SMAD4 binds to 5′TSS of LRRC17 after TGFβ stimulation, is categorized to Stimulated Only Binding. (**d**) Abundance of DNA following SMAD4 ChIP pulldown as compared to DNA present following pulldown with nonspecific IgG antibody as determined by quantitative SYBR Green PCR. U and S used to represent the Unstimulated and Stimulated binding regions of SLC40A1, respectively. *Asterisks* represent a *t*-test *p* value of less than 0.05 and denote significant enrichment relative to IgG control (Reprinted from Kennedy BA et al. [54], original figure 1. Licensed by PLOS under Creative Commons Attribution License (CCAL))

5.2 *TGFβ-Stimulated Binding*

2,362 SMAD4 binding sites were detected in TGFβ-stimulated A2780. A similar distribution of SMAD4 binding loci was observed in stimulated A2780 as unstimulated. We then classified the binding loci within ±100 kb of any annotated RefSeq genes (1,723 loci for stimulated and 1,499 loci for unstimulated) into four binding patterns. The binding patterns of SMAD4 were dramatically different between stimulated and unstimulated A2780. 74.2 % (1,279 of 1,723) and 73.5 % (1,102 of 1,499) of the binding loci were in the Stimulated Only Binding and Unstimulated Only Binding categories, respectively. 24.8 % (429 of 1,723) and 25.5 % (382 of 1,499) binding loci were in the shift binding category for stimulated and unstimulated A2780, respectively. Only 15 binding loci in each sample (0.9 % of stimulated and 1.0 % of unstimulated) were classified as basal binding. These results showed that TGFβ stimulation of ovarian cancer cells may alter the landscape of SMAD4 binding patterns (Fig. 6.2d).

In order to confirm these binding changes we observed in the ChIP-seq data, we randomly chose 22 of the targets identified by ChIP-seq and performed ChIP-qPCR with DNA from a separate immunoprecipitation. This ChIP-qPCR validation confirmed these targets identified by ChIP-seq and provided further evidence that exogenous TGFβ signaling is able to dramatically influence SMAD4 binding patterns.

5.2.1 Regulation of TGFβ-Stimulated SMAD4 Target Gene Expression in A2780

We performed gene expression microarray assays to determine the expression status of SMAD4 target genes after TGFβ stimulation. A2780 mRNA from three independent replicates of both before and after 3 h of TGFβ stimulation was prepared and analyzed with the Affymetrix U133 Plus 2 Platform. 3,191 genes were significantly up- or downregulated after TGFβ stimulation, defined as at least a 0.5 $\log_2$-fold change of expression and $p < 0.1$ of significant expression difference. After correlating the differentially expressed genes with 1,443 TGFβ-stimulated SMAD4 target genes (encompassing 1,723 SMAD4 binding loci), we found that 2,873 of the differentially expressed genes lacked SMAD4 binding loci, leaving 318 genes with at least one SMAD4 binding locus and at least a 0.5 $\log_2$-fold expression change with 3 h of TGFβ stimulation (Fig. 6.3).

Gene ontology analysis of these 318 genes showed that they were significantly enriched with genes involved with cell part morphogenesis and developmental proteins (Fig. 6.3c), as is consistent with previous studies in different cell types [13, 37]. Additionally, the 1,125 genes associated with SMAD4 binding, but lacking differential expression were enriched for genes with EGF-like domain and polymorphism, suggesting that there may be other signaling pathways than TGFβ that mediate SMAD4 functions. The 2,873 genes with differential expression but lacking SMAD4 binding were enriched in immune functions and proteinaceous extracellular matrix. We next reduced the p value threshold to 0.05 and obtained a set of 1,763 genes differentially

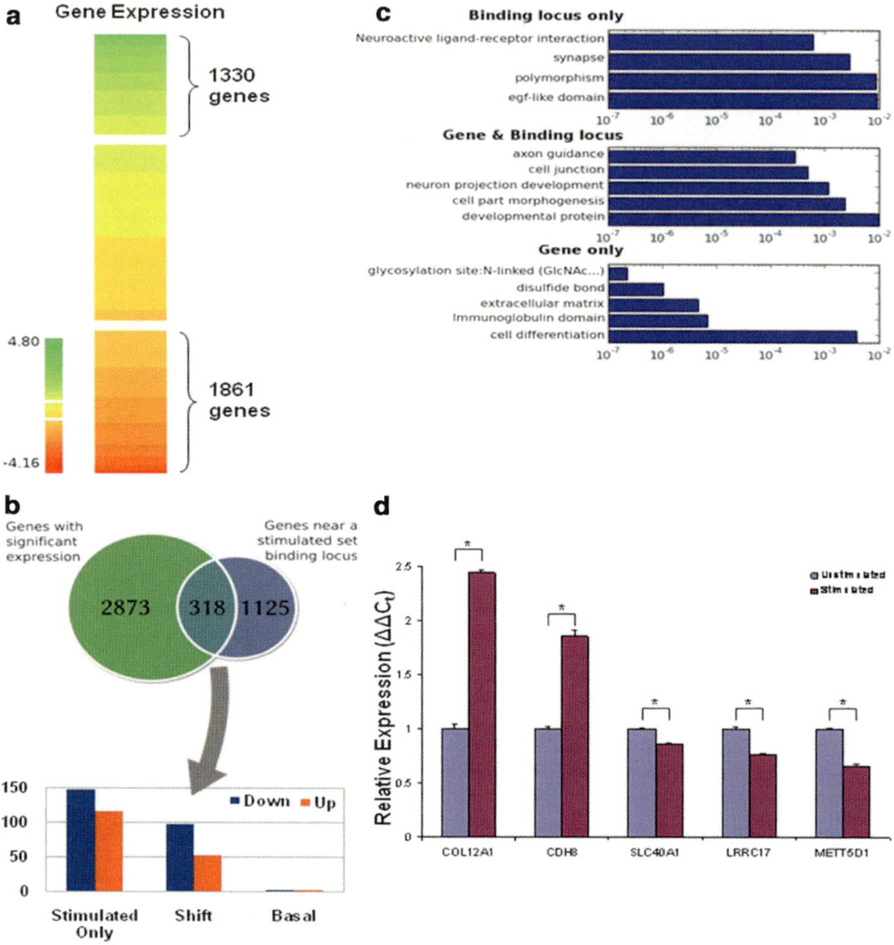

Fig. 6.3 TGFβ/SMAD4-regulated genes. (**a**) A heatmap of the expression fold changes for genes between the unstimulated and the TGFβ-stimulated condition, showing three group of genes, upregulated, no change, and downregulated. Up- and downregulated genes are defined as having a $\log_2$-fold change of greater than 0.5 or less than −0.5, respectively. (**b**) A comparison between the genes with SMAD4 binding loci (1443) in the TGFβ-stimulated condition with all genes showing differential expression (3193), showing three different groups, those with differential expression and no SMAD4 binding loci, those with no differential expression and a SMAD4 binding loci, and those with both. (**c**) GO annotations for the three different group genes showing in the Venn diagram (**b**). (**d**) RNA expression level as determined by qRT-PCR relative to GAPDH expression levels. Experiments were performed in biological triplicate. *Asterisks* represent a *t*-test *p* value of less than 0.05 and denote significant difference in expression between unstimulated and stimulated conditions (Reprinted from Kennedy BA et al. [54], original figure 2. Licensed by PLOS under Creative Commons Attribution License (CCAL))

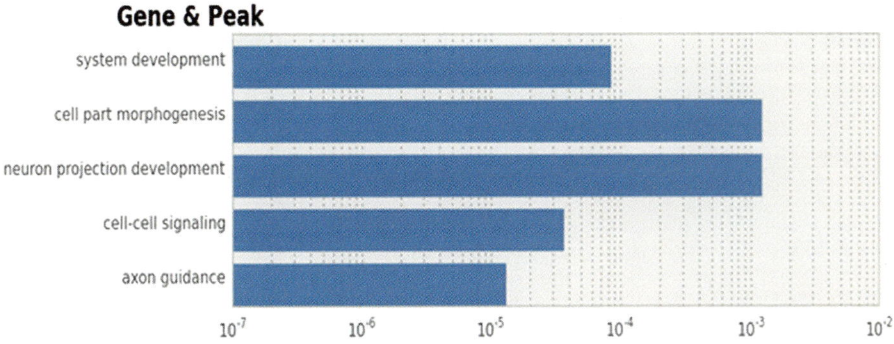

Fig. 6.4 GO for 184 TGFβ/SMAD4 differentially expressed genes with a p value of less than 0.05 showing a similar functional categories with 318 genes. Together with RT-qPCR validations, our results demonstrated that the identified genes (318) in the study are valid for the further downstream analysis (Reprinted from Kennedy BA et al. [54], original figure S2. Licensed by PLOS under Creative Commons Attribution License (CCAL))

expressed with TGFβ stimulation. 184 (10.4 %) of these had at least one TGFβ-stimulated SMAD4 binding site. This is consistent with the observed percentage of differentially observed genes with at least one TGFβ-stimulated SMAD4 site under $p<0.10$ (10.0 %). The top GO results were also similar (Fig. 6.4). To further confirm these differentially expressed SMAD4 genes, we chose a random set of 18 targets identified by this analysis for RT-qPCR, which validated 13 (72.2 %) of them.

5.3 *SMAD4-Dependent Gene Regulatory Networks in TGFβ-Induced Ovarian Cancer Cells*

One of our previous studies [13] and a study from Koinuma et al. [12] identified 150 TGFβ-stimulated SMAD4 target genes in IOSE (an immortalized ovarian surface epithelial cell line) and 92 TGFβ-stimulated SMAD4 target genes in HaCaT (an immortalized keratinocyte cell line). We were not surprised to find very little overlap among the genes in these studies and the 318 SMAD4 target genes we have identified (Fig. 6.5a), as only A2780 is a cancer cell line. Additionally, the other two studies were performed with promoter ChIP-chip arrays, which cannot identify as many targets as the ChIP-seq experiments performed with this study. The SMAD4 target genes in the other two studies were primarily involved in regulation of cell proliferation (or anti-apoptosis) and development process (muscle development) according to GO analysis [38], which was different than the GO results obtained with the A2780 SMAD4 target genes (Fig. 6.5b).

We then applied a computational analysis approach we previously developed [31] to build the SMAD4-dependent regulatory networks in HaCaT, IOSE, and A2780, respectively (Fig. 6.6). With this approach, we first start with ChIP data sets

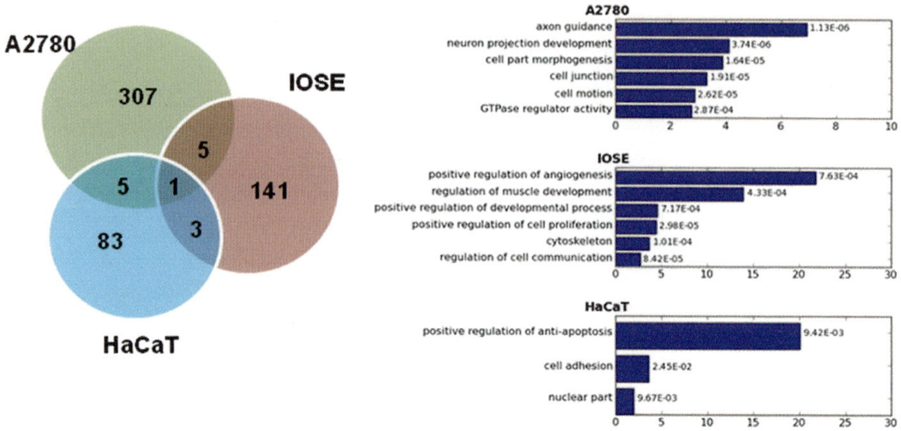

Fig. 6.5 A comparison of TGFβ/SMAD4 target genes. (**a**) A Venn diagram shows the comparison of TGFβ/SMAD4 target genes in three different cell types. (**b**) GO annotations for the unique genes for each cell type (Reprinted from Kennedy BA et al. [54], original figure 3. Licensed by PLOS under Creative Commons Attribution License (CCAL))

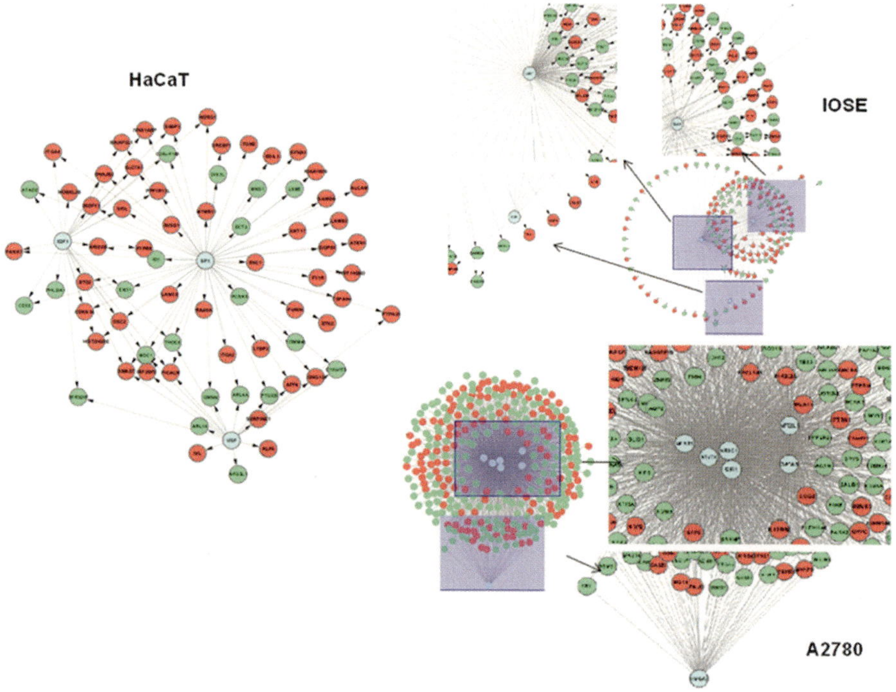

Fig. 6.6 TGFβ-induced SMAD4-dependent gene regulatory networks in (**a**) HaCaT, (**b**) IOSE, (**c**) A2780 cells (Reprinted from Kennedy BA et al. [54], original figure 4. Licensed by PLOS under Creative Commons Attribution License (CCAL))

and gene expression data. Each SMAD4 binding locus is matched to a known RefSeq ID, and the RefSeq IDs are examined for differential gene expression. We used our set of 318 differentially regulated SMAD4 target genes after TGFβ stimulation to find the most significant transcription factor (TF) binding partners according to ChIPMotifs [39] or ChIPModules [26], which we used as Hub TFs. The connection between the Hub TFs and genes was determined by scanning the position-weight matrices of the Hub TFs in all SMAD4 binding loci, and a permutation test was used to assess the reliability of each connection to the network. We visualized the resulting regulatory network with Cytoscape [28].

We identified six Hub TFs from our set of 318 SMAD4-dependent, differentially regulated genes in A2780: GFI1, NR3C1, SOX17, STAT4, ZNF354C, and TCF8. We identified four Hub TFs in IOSE cells with a similar approach (CART model) [13]: LEF1 (TCF), ELK1, COUPTF (NR2F5), and E2F. We identified three Hub TFs for 92 SMAD4-dependent target genes in HaCaT cells: E2F1, SP1, and USF. This was very similar to the TF motifs identified by Koinuma et al. [12]. Our results did not detect the top motif reported by their study (AP1), due to the advanced classification algorithm used by ChIPModules [26] and its ability to eliminate TF motifs that are also enriched in random sets. We found that the Hub TF E2F (E2F1) was shared by IOSE and HaCaT, but not A2780. These results, along with GO function analysis, suggest that E2F may act as a major SMAD4 co-transcription factor partner in mediating cell proliferation that is lost in carcinoma cells. This gene regulatory network analysis strongly indicates that TGFβ stimulates a different SMAD4-dependent regulatory mechanism in ovarian cancer cells than normal cells. Although we did not investigate potential explanations of this difference, a more recent study by Matsumura et al. suggests that alteration of TGFβ signaling in ovarian cancer may be at least partially the result of accumulation of DNA methylation over time [40].

5.4 Gene Signatures of Selection and Clinical Outcome

In order to identify gene signatures potentially capable of providing better prognostic information than standard clinical and pathological parameters [41, 42], we next addressed the relationship of TGFβ-stimulated SMAD4-dependent target genes to the clinical outcomes of ovarian cancer patients. We examined the 307 target genes we identified in A2780 cells that were not identified in previous studies of normal cells, in two different clinical ovarian cancer cohort studies that had reported survival data [37, 43]. We first classified the patients into two different subgroups according to their gene signatures and then correlated our data with the patient survival information. With the 153 patient cohort from Bild et al. [37], we were able to apply hierarchical clustering with distance-based measures from a trial-and-error perspective to 187 of 307 genes identified in the gene expression data set, classifying the genes into four gene groups (Fig. 6.7a). For each of the four gene groups, we further clustered the 153 samples into four patient groups (PGs) and correlated each

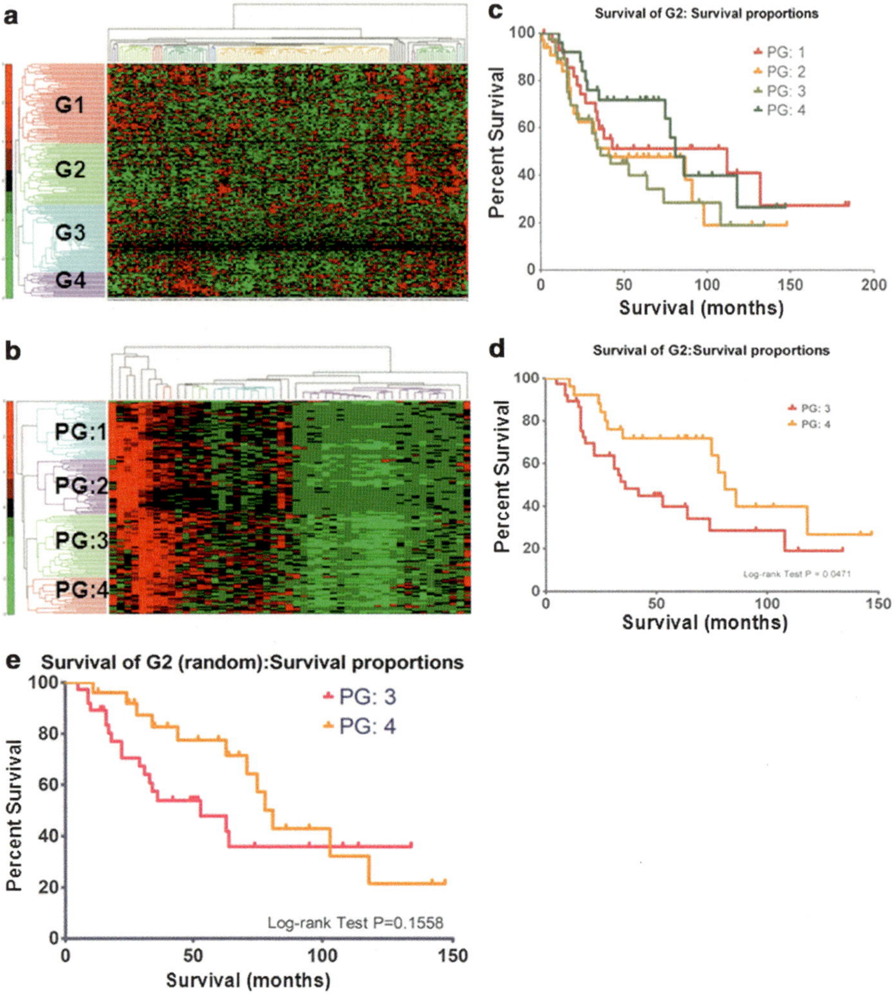

Fig. 6.7 The selection of gene signatures and their associated clinical outcome. (**a**) The hierarchical clustering result of the 187 genes into four gene groups, namely, G1, G2, G3, and G4. The *vertical axis* represents the gene clusters (187 genes), and the *horizontal axis* stands for diverse samples (153 patients). (**b**) The hierarchical clustering result of the 153 patients into four patient groups, namely, PG: 1, PG: 2, PG: 3, and PG: 4, by using the G2 group of 49 genes. (**c**) Survival curve plot for the G2 gene group. The *horizontal axis* represents the survival months and the *vertical* for the percent survival (%) within the corresponding patient group. Totally four patient groups, i.e., PG: 1, PG: 2, PG: 3, and PG: 4, are analyzed for the G2 gene group. (**d**) A detailed survival curve plot for two patient groups, PG: 3 and PG: 4, showing a significant log-rank test p value of 0.0471. (**e**) A survival curve plot for two patient groups, PG: 3 and PG: 4 using a randomly selected 49 genes (where they are not within 49 G2 genes) showing log-rank test p value is 0.1558 (Reprinted from Kennedy BA et al. [54], original figure 5. Licensed by PLOS under Creative Commons Attribution License (CCAL))

PG with its survival information (Fig. 6.7b; 29/153 patients did not have complete survival information and were thus excluded from survival curve plots). We found a signature of 49 genes (G2 gene group) that was able to predict a survival correlation for 62 of the patients with $p=0.0471$ (Fig. 6.7c, d). Specifically, the median survival of PG: 4 (25 patients) was 31 months, compared to PG: 3 [37] patients with a median survival of 63 months. A survival curve plot for PG: 3 and PG: 4 using a random selection of 49 genes (excepting those within the G2 group) showed a log-rank test p value of 0.1558 (Fig. 6.7e). Due to the limited pathological information available for this patient cohort, we were unable to significantly correlate our gene signatures with other clinical outcomes. However, a high percentage of stage IV patients clustered into PG: 3, while all stage IC and two stage IIC patients clustered into PG: 4, despite a similar number of stage IIIC patients in each, perhaps indicating that the TGFβ/SMAD4-regulated genes we have identified could be potentially used to classify a subtype of ovarian cancer patients. Applying the same in silico mining approach to the second patient cohort from Lu et al. [43] (42 patients and 5 normal people) showed that a gene signature of 19 of the 307 genes predicted better survival rates for PG: 4 and Normals than other PGs ($p=0.0078$).

6 Discussion

In this study, we have for the first time applied ChIP-seq to genome-wide mapping of TGFβ-stimulated and SMAD4-dependent regulated genes in the A2780 ovarian cancer cell line. We chose to use ChIP-seq to analyze SMAD4 binding because of its ability to detect many more TF binding loci than ChIP-chip. Additionally, ChIP-seq offers a higher signal-to-noise ratio in most cases [20] as well as much higher resolution (within tens of base pairs rather than probes). This capability of ChIP-seq was essential to classifying binding events as basal or shift binding. ChIP-seq is also capable of interrogating a much greater portion of the genome than ChIP-chip, as it is not limited by the regions spanned by probes. Indeed, only 13.3 % of detected SMAD4 binding loci in unstimulated A2780 cells were in the promoter regions of known genes, the rest of which would not have been detected with a promoter array.

The results show that a majority of SMAD4 binding loci are either newly bound to chromatin (74.2 %) or shifted bound (24.8 %) upon TGFβ stimulation compared to without TGFβ stimulation, suggesting TGFβ-stimulated cancer cells may alter the landscape of SMAD4 binding patterns. Furthermore, our gene ontology analysis revealed striking similarities between the top 10 GO categories for 1,443 and 1,316 SMAD4 target genes in stimulated and unstimulated conditions (data not shown). However, 318 differentially expressed genes containing at least one stimulated SMAD4 binding locus were significantly enriched for more specific GO terms, such as cell part morphogenesis and developmental proteins, indicating that TGFβ signaling may induce SMAD4 to regulate a very specific set of target genes, thereby facilitating specific functions in that cell type. Indeed, in GO analysis of SMAD4 target genes without changes in expression level with TGFβ stimulation, the term

"EGF like signaling" was enriched, providing further evidence that other signaling pathways may modulate SMAD4-dependent regulated genes in ovarian cancer. For example, the bone morphogenic proteins (BMPs) are upstream of SMAD4 and therefore may be capable of regulating some of these SMAD4 target genes. Studies have shown the role of BMPs as key regulators of ovarian physiology and their involvement in ovarian and other cancer development [44–46]. Our future studies will further delineate the regulation of SMAD4 target genes identified in this study by each signaling pathway.

We observed that a majority (>70 %) of SMAD4 binding loci were located more than 8 kb away from the 5′TSS of a known RefSeq gene. This is consistent with other findings for transcription factors, such as estrogen receptor alpha (ERα) [47–49], androgen receptor (AR) [50], and peroxisome proliferator-activated receptor (PPAR) [51]. This might suggest that the TGFβ binding loci become in close proximity to the promoter through chromosomal looping upon TGFβ stimulation. Perhaps this could be investigated with Hi-C, a chromosome conformation capture method using massively parallel sequencing to detect inter- and intra-chromosome interactions within the entire genome [52, 53]. Interestingly, our de novo motif analysis also identified a SMAD-like motif in a set of 5-distal binding loci but not in a set of 5′-promoter loci (data not shown). Our genome-wide location analysis also helped to pinpoint the importance of whole-genome-wide sequencing technologies, revealing that many binding loci are located far ways from the 5′TSS of a known gene, many of which may not be interrogated by ChIP-chip (microarray) probes complementary to a transcription factor. Therefore, to begin to uncover the underlying mechanism of TGFβ/SMAD4-mediated gene regulation, we will conduct ChIP-3C-qPCR to confirm whether these distal binding loci are indeed related to these particular genes.

One of the important aspects of this study was our use of in silico mining of publicly available patient cohort data to identify a subset of TGFβ/SMAD4 target genes as a gene signature for predicting clinical (survival) outcomes. To our knowledge, this is the first study to attempt to classify ovarian cancer patients into different subtypes of patient groups on the basis of TGFβ signaling response by SMAD4-regulated genes, as well as statistically significant prediction of poor survival from good survival populations (Fig. 6.7). Thus, our approach of combining ChIP-seq identification of binding loci, gene expression profiling, and in silico mining of patient cohorts may be useful for identifying potential biologically and clinically important gene signatures.

In conclusion, we provided the first comprehensive map of thousands of TGFβ/SMAD4 targets in an ovarian cancer cell line (A2780), which could also be used for studying SMAD functions in tumorigenesis. To our knowledge, this is the first study to correlate TGFβ/SMAD4-regulated genes with clinical data of ovarian cancer patient survival to identify potential gene signatures for prognoses in ovarian cancer.

References

1. Berchuck A, Rodriguez G, Olt G, Whitaker R, Boente MP, Arrick BA, et al. Regulation of growth of normal ovarian epithelial cells and ovarian cancer cell lines by transforming growth factor-beta. Am J Obstet Gynecol. 1992;166(2):676–84.
2. Wong AS, Leung PC. Role of endocrine and growth factors on the ovarian surface epithelium. J Obstet Gynaecol. 2007;33(1):3–16.
3. Derynck R, Akhurst RJ, Balmain A. TGF-beta signaling in tumor suppression and cancer progression. Nat Genet. 2001;29(2):117–29.
4. Nilsson EE, Skinner MK. Role of transforming growth factor beta in ovarian surface epithelium biology and ovarian cancer. Reprod Biomed Online. 2002;5(3):254–8.
5. Heldin CH, Miyazono K, ten Dijke P. TGF-ß signalling from cell membrane to nucleus through SMAD proteins. Nature. 1997;390(6659):465–71.
6. Shi Y, Massagué J. Mechanisms of TGF-beta signaling from cell membrane to the nucleus. Cell. 2003;113(6):685–700.
7. Feng XH, Derynck R. Specificity and versatility in TGF-beta signaling through Smads. Annu Rev Cell Dev Biol. 2005;21:659–93.
8. Derynck R, Zhang YE. Smad-dependent and Smad-independent pathways in TGF-beta family signalling. Nature. 2003;425(6958):577–84.
9. Miyazawa K, Shinozaki M, Hara T, Furuya T, Miyazono K. Two major Smad pathways in TGF-beta superfamily signalling. Genes Cells. 2002;7(12):1191–204.
10. Wang G, Li C, Wang Y, Chen G. Cooperative assembly of co-Smad4 MH1 with R-Smad1/3 MH1 on DNA: a molecular dynamics simulation study. PLoS One. 2013;8(1):e53841.
11. Koinuma D, Tsutsumi S, Kamimura N, Taniguchi H, Miyazawa K, Sunamura M, et al. Chromatin immunoprecipitation on microarray analysis of Smad2/3 binding sites reveals roles of ETS1 and TFAP2A in transforming growth factor beta signaling. Mol Cell Biol. 2009;29(1):172–86.
12. Koinuma D, Tsutsumi S, Kamimura N, Imamura T, Aburatani H, Miyazono K. Promoter-wide analysis of Smad4 binding sites in human epithelial cells. Cancer Sci. 2009;100(11):2133–42.
13. Qin H, Chan MWY, Liyanarachchi S, Balch C, Potter D, Souriraj IJ, et al. An integrative ChIP-chip and gene expression profiling to model SMAD regulatory modules. BMC Syst Biol. 2009;3:73.
14. Ikushima H, Komuro A, Isogaya K, Shinozaki M, Hellman U, Miyazawa K, et al. An Id-like molecule, HHM, is a synexpression group-restricted regulator of TGF-beta signalling. EMBO J. 2008;27(22):2955–65.
15. Gomis R, Alarcón C, He W, Wang Q, Seoane J, Lash A, et al. A FoxO–Smad synexpression group in human keratinocytes. Proc Natl Acad Sci USA. 2006;103(34):12747–52.
16. Li M, Balch C, Montgomery JS, Jeong M, Chung JH, Yan P, et al. Integrated analysis of DNA methylation and gene expression reveals specific signaling pathways associated with platinum resistance in ovarian cancer. BMC Med Genomics. 2009;2:34.
17. Barski A, Cuddapah S, Cui K, Roh TY, Schones DE, Wang Z, et al. High-resolution profiling of histone methylations in the human genome. Cell. 2007;129(4):823–37.
18. Johnson DS, Mortazavi A, Myers RM, Wold B. Genome-wide mapping of in vivo protein-DNA interactions. Science. 2007;316(5830):1497–502.
19. Mikkelsen TS, Ku M, Jaffe DB, Issac B, Lieberman E, Giannoukos G, et al. Genome-wide maps of chromatin state in pluripotent and lineage-committed cells. Nature. 2007;448(7153):553–60.
20. Ho JW, Bishop E, Karchenko PV, Nègre N, White KP, Park PJ. ChIP-chip versus ChIP-seq: Lessons for experimental design and data analysis. BMC Genomics. 2011;12:134.
21. Cheng AS, Jin VX, Fan M, Smith LT, Liyanarachchi S, Yan PS, et al. Combinatorial analysis of transcription factor partners reveals recruitment of c-MYC to estrogen receptor-alpha responsive promoters. Mol Cell. 2006;21(3):393–404.

22. Lee TI, Johnstone SE, Young RA. Chromatin immunoprecipitation and microarray-based analysis of protein location. Nat Protoc. 2006;1(2):729–48.
23. Lan X, Bonneville R, Apostolos J, Wang W, Jin VX. W-ChIPeaks: a comprehensive web application tool to process ChIP-chip and ChIP-seq data. Bioinformatics. 2011;27(3):428–30.
24. Frietze S, Lan X, Jin VX, Farnham PJ. Genomic targets of the KRAB and SCAN domain-containing zinc finger protein 263. J Biol Chem. 2010;285(2):1393–403.
25. Gu F, Hsu H-K, Hsu P-Y, Wu J, Ma Y, Parvin J, et al. Inference of hierarchal regulatory network of estrogen-dependent breast cancer through ChIP-based data. BMC Syst Biol. 2010;4:170.
26. Jin VX, Rabinovich A, Squazzo SL, Green R, Farnham PJ. A computational genomics approach to identify cis-regulatory modules for chromatin immunoprecipitation microarray data—a case study using E2F1 in cancers. Genome Res. 2006;16(12):1585–95.
27. Breiman L, Friedman JH, Olshen RA, Stone CJ. Classification and regression trees. New York, NY: Chapman & Hall; 1984.
28. Shannon P, Markiel A, Ozier O, Baliga NS, Wang JT, Ramage D, et al. Cytoscape: a software environment for integrated models of biomolecular interaction networks. Genome Res. 2003;13(11):2498–504.
29. Yamada SD, Baldwin RL, Karlan BY. Ovarian carcinoma cell cultures are resistant to TGF-beta1-mediated growth inhibition despite expression of functional receptors. Gynecol Oncol. 1999;75(1):72–7.
30. Baldwin RL, Tran H, Karlan BY. Loss of c-myc repression coincides with ovarian cancer resistance to transforming growth factor beta growth arrest independent of transforming growth factor beta/Smad signaling. Cancer Res. 2003;63(6):1413–9.
31. Tanaka Y, Kobayashi H, Suzuki M, Kanayama N, Terao T. Transforming growth factor-beta1–dependent urokinase up-regulation and promotion of invasion are involved in Src-MAPK–dependent signaling in human ovarian cancer cells. J Biol Chem. 2004;279(10):8567–76.
32. Chan MWY, Huang YW, Hartman-Frey C, Kuo CT, Deatherage D, Qin H, et al. Aberrant transforming growth factor beta1 signaling and SMAD4 nuclear translocation confer epigenetic repression of ADAM19 in ovarian cancer. Neoplasia. 2008;10(9):908–19.
33. Chou JL, Su HY, Chen LY, Liao Y-P, Hartman-Frey C, Lai YH, et al. Promoter hypermethylation of FBXO32, a novel TGF-beta/SMAD4 target gene and tumor suppressor, is associated with poor prognosis in human ovarian cancer. Lab Invest. 2010;90(3):414–25.
34. Lim SK, Hoffmann FM. Smad4 cooperates with lymphoid enhancer-binding factor 1/T cell-specific factor to increase c-myc expression in the absence of TGF- signaling. PNAS. 2006;103(49):18580–5.
35. Kent WJ, Sugnet CW, Furey TS, Roskin KM, Pringle TH, Zahler AM, et al. The human genome browser at UCSC. Genome Res. 2002;12:996–1006.
36. Fei T, Xia K, Li Z, Zhou B, Zhu S, Chen H, et al. Genome-wide mapping of SMAD target genes reveals the role of BMP signaling in embryonic stem cell fate determination. Genome Res. 2010;20(1):36–44.
37. Bild AH, Yao G, Chang JT, Wang Q, Potti A, Chasse D, et al. Oncogenic pathway signatures in human cancers as a guide to targeted therapies. Nature. 2006;439(7074):353–7.
38. Huang DW, Sherman BT, Lempicki RA. Systematic and integrative analysis of large gene lists using DAVID Bioinformatics Resources. Nat Protoc. 2009;4(1):44–57.
39. Jin VX, Apostolos J, Nagisetty NS, Farnham PJ. W-ChIPMotifs: a web application tool for de novo motif discovery from ChIP-based high-throughput data. Bioinformatics. 2009;25(23):3191–3.
40. Matsumura N, Huang Z, Mori S, Baba T, Fujii S, Konishi I, et al. Epigenetic suppression of the TGF-beta pathway revealed by transcriptome profiling in ovarian cancer. Genome Res. 2011;21(1):74–82.
41. van de Vijver MJ, He YD, van't Veer LJ, Dai H, Hart AA, Voskuil DW, et al. A gene-expression signature as a predictor of survival in breast cancer. N Engl J Med. 2002;347(25):1999–2009.

42. Chibon F, Lagarde P, Salas S, Pérot G, Brouste V, Tirode F, et al. Validated prediction of clinical outcome in sarcomas and multiple types of cancer on the basis of a gene expression signature related to genome complexity. Clin Cancer Res. 2004;10(10):3291–300.
43. Lu KH, Patterson AP, Wang L, Marquez RT, Atkinson EN, Baggerly KA, et al. Selection of Potential Markers for Epithelial Ovarian Cancer with Gene Expression Arrays and Recursive Descent Partition Analysis. Clin Cancer Res. 2004;10(10):3291–300.
44. Bleuming SA, He XC, Kodach LL, Hardwick JC, Koopman FA, Ten Kate FJ, et al. Bone morphogenetic protein signaling suppresses tumorigenesis at gastric epithelial transition zones in mice. Cancer Res. 2007;67(17):8149–55.
45. Kodach LL, Wiercinska E, de Miranda NF, Bleuming SA, Musler AR, Peppelenbosch MP, et al. The bone morphogenetic protein pathway is inactivated in the majority of sporadic colorectal cancers. Gastroenterology. 2008;134(5):1332–41.
46. Milano F, van Baal JW, Buttar NS, Rygiel AM, de Kort F, DeMars CJ, et al. Bone morphogenetic protein 4 expressed in esophagitis induces a columnar phenotype in esophageal squamous cells. Gastroenterology. 2007;132(7):2412–21.
47. Carroll JS, Meyer CA, Song J, Li W, Geistlinger TR, Eeckhoute J, et al. Genome-wide analysis of estrogen receptor binding sites. Nat Genet. 2006;38(11):1289–97.
48. Welboren WJ, van Driel MA, Janssen-Megens EM, van Heeringen SJ, Sweep FC, Span PN, et al. ChIP-Seq of ERalpha and RNA polymerase II defines genes differentially responding to ligands. EMBO J. 2009;28(10):1418–28.
49. Fullwood MJ, Liu MH, Pan YF, Liu J, Xu H, Mohamed YB, et al. An oestrogen-receptor-alpha-bound human chromatin interactome. Nature. 2009;462(7269):58–64.
50. Wang Q, Li W, Zhang Y, Yuan X, Xu K, Yu J, et al. Androgen receptor regulates a distinct transcription program in androgen-independent prostate cancer. Cell. 2009;138(2):245–56.
51. Nielsen R, Pedersen TA, Hagenbeek D, Moulos P, Siersbaek R, Megens E, et al. Genome-wide profiling of PPARgamma:RXR and RNA polymerase II occupancy reveals temporal activation of distinct metabolic pathways and changes in RXR dimer composition during adipogenesis. Genes Dev. 2008;22(21):2953–67.
52. Erez LA, van Berkum NL, Williams L, Imakaev M, Ragoczy T, Telling A, et al. Comprehensive mapping of long-range interactions reveals folding principles of the human genome. Science. 2009;326(5950):289–93.
53. Belton JM, McCord RP, Gibcus JH, Naumova N, Zhan Y, Dekker J. Hi-C: A comprehensive technique to capture the conformation of genomes. Methods. 2012;58(3):268–76.
54. Kennedy BA, Deatherage DE, Gu F, Tang B, Chan MWY, Nephew KP, et al. ChIP-seq defined genome-wide map of TGFβ/SMAD4 targets: implications with clinical outcome of ovarian cancer. PLoS One. 2002;6(7):e22606.

Chapter 7
Targeted Sequencing Strategies in Cancer Research

HoJoon Lee, Billy T. Lau, and Hanlee P. Ji

Abstract Despite the rapidly decreasing cost and increasing throughput of next-generation sequencing technologies, deep targeted resequencing has maintained its importance in the sensitive detection of mutations by focusing only on targets of biomedical or clinical relevance. In this chapter, we review current as well as emerging technologies that enable the capture of specific elements from a sample. In particular, we focus on targeted resequencing in cancer research. To date, the cancer research community has sequenced over 2,500 cancer samples from 17 different tissue types by whole-exome targeted resequencing. In addition, clinically relevant targets such as known cancer genes and cancer-associated miRNAs and viruses have been sequenced selectively by targeted next-generation sequencing technologies. We expect that the use of targeted resequencing in cancer research will soon become a standard research and analysis method, and in turn further technological developments that allow for the improved extraction of genetic components will be rapidly adopted.

Keywords Next-generation sequencing (NGS) • Cancer genomics • Targeted sequencing • Whole-exome sequencing

---

HoJoon Lee and Billy T. Lau have contributed equally to this work.

H.J. Lee
Division of Oncology, Department of Medicine, Stanford University School of Medicine,
CCSR 1115, 269 Campus Drive, Stanford, CA 94305, USA

B.T. Lau
Stanford Genome Technology Center, Stanford University School of Medicine,
855 S. California Avenue, Palo Alto, CA 94304, USA

H.P. Ji (✉)
Stanford Genome Technology Center, Stanford University School of Medicine,
855 S. California Avenue, Stanford, CA 94304, USA

Division of Oncology, Department of Medicine, Stanford University School of Medicine,
CCSR 1115, 269 Campus Drive, Stanford, CA 94305, USA
e-mail: genomics_ji@stanford.edu

W. Wu and H. Choudhry (eds.), *Next Generation Sequencing in Cancer Research: Volume 1: Decoding the Cancer Genome*, DOI 10.1007/978-1-4614-7645-0_7,

1 Introduction

Targeted sequencing is a strategy originally developed to reduce the costs of next-generation sequencing by directly capturing desired sequences of interest. However, despite the rapidly decreasing costs of next-generation sequencing, the use of targeted resequencing has continued to increase. For applications involving the detection of rare molecules or mutations in mixed samples, the resolution (i.e., sequencing depth) required is still much higher than what is currently possible with whole-genome "shotgun" approaches. Studies that focus on specific genomic regions but use shotgun sequencing methods will suffer from extraneous reads from outside regions of interest. Post-sequencing software packages have been developed to filter out such contaminating reads; however, the total capacity of a given sequencing run will be underutilized. Targeted sequencing therefore is an attractive tool for researchers to study specific regions of interest at significantly higher resolution. As an example, targeted sequencing of the human exome—about 1–2 % of the human genome—would yield an increase in sequencing depth of approximately 50- to 100-fold compared to shotgun approaches.

Targeted sequencing has been of significant interest to clinical and translational settings; in particular, substantial efforts have been applied to cancer research. As genomic instabilities that lead to cancerous growth can be due to point mutations in coding regions, exome sequencing becomes a natural choice for detecting them. Because mutations in coding regions generally lead to changes in protein structure, exome sequencing is also relevant for developing targeted drug therapies and assessing resistances that may occur during treatment. Similarly, although genomic structural variations possess a clear signature in the analysis pipeline, finding the specific breakpoint is extremely difficult with shotgun sequencing. Targeted sequencing of RNA transcripts therefore becomes an attractive method for detecting these rare events with high sensitivity.

The specific next-generation sequencing technology being used has generally evolved over time, including but not limited to products offered by Illumina, Roche, and Life Technologies. However, none of these sequencing strategies provide a complete DNA capturing platform for the purpose of downstream targeted sequencing. Therefore, researchers generally either have sought alternative DNA capture strategies from commercial sources or developed their own for the purpose of sample preprocessing before the actual sequencing reaction. These capture strategies are generally platform-agnostic and work on any type of next-generation sequencing platform. In this chapter, we will briefly provide an overview of the types of target capture strategies available. As each target capture technologies attempt to occupy a specific niche, it would be inappropriate to provide side-by-side technical comparisons. Furthermore, while other reviews focus on the recent and numerous technical optimizations for each approach [1–4], our intent is to provide an evaluation of when each approach would be appropriate. Lastly, we will provide an overview of current progress that has been made towards targeted sequencing in cancer research.

2 Overview of Targeted DNA Capture Technologies

All targeted sequencing technologies revolve around the specific enrichment of a target sequence of interest through the general method of DNA capture. A complementary probe, whether it is another nucleic acid or a DNA-binding protein, is designed to hybridize against a desired region in a nucleic acid sample and is then extracted for sequencing. A variety of technological platforms have been developed that utilize this principle with varying degrees of efficiency, expandability, and cost.

2.1 *Oligonucleotide Pools*

Hybridization of oligonucleotide probes to their targets in aqueous solutions remains one of the most popular methods of target enrichment due to its simplicity. The experimental procedure consists of the synthesis of an oligonucleotide pool complementary to regions of interest followed by incubation with the source material. This oligonucleotide pool is often modified with selectable end groups so that successfully captured samples can be conveniently extracted. Common modifications include biotinylated oligonucleotides that are then extracted with streptavidin-coated beads. The first implementation of this technique was first demonstrated using RNA capture probes greater than 100 bp [5] but has since been extended to DNA-based probes as well [6–8]. While this approach is simple, its accuracy is limited by nonspecific binding to off-target DNA sequences as well as sequential capture whereby a captured DNA fragment can nonspecifically capture another (Fig. 7.1).

As the number of targets that can be enriched using this strategy is directly proportional to the number of probes synthesized, massively parallel oligonucleotide synthesis strategies must be employed. Its usage either is consequently restricted to

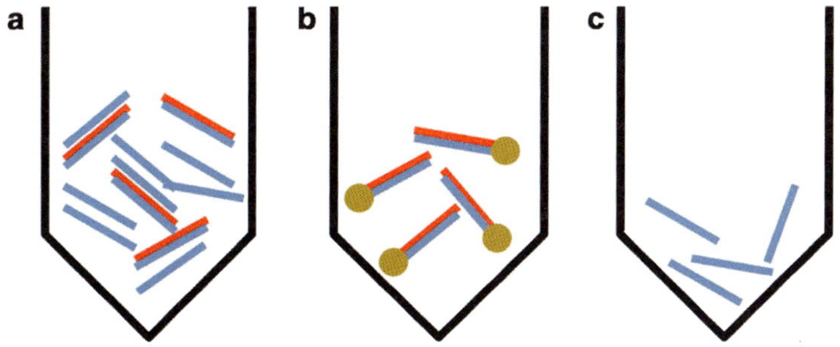

Fig. 7.1 In-solution target capture by hybridization. (**a**) Nucleic acid fragments (*blue*) are captured by oligonucleotide probes (*red*) in a single test tube. (**b**) The oligonucleotide probes can be selectively extracted by the use of affinity beads (*yellow*), and the nonbinding fragments can be washed away. (**c**) The captured fragments are then eluted from the probes and afterward are compatible with typical sequencing workflows

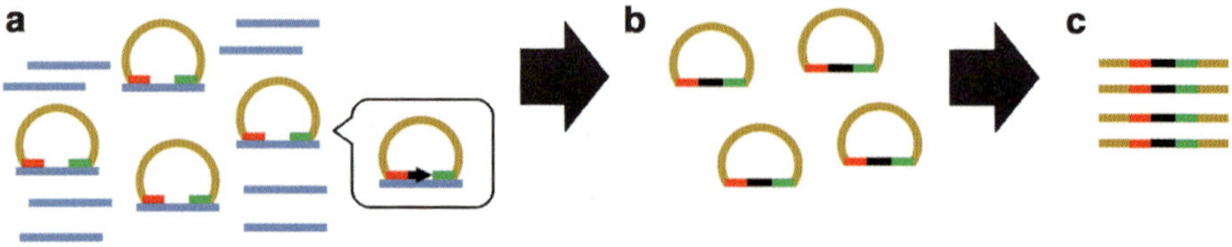

Fig. 7.2 Target capture by molecular inversion probes. (**a**) Probes flanking a region of interest hybridize to their targets (*blue*). *Inset*: The addition of a polymerase and ligase allows the surrounded region to be filled in and effectively circularizes the probe. (**b**) Linear fragments are removed by exonuclease digestion, leaving the circularized probes intact. (**c**) The circularized probes are linearized and become compatible for sequencing workflows

small-scale studies of several hundred customized oligonucleotide probes or is dependent on outsourcing to specialized companies that can perform parallel synthesis in excess of tens of thousands of unique oligonucleotides. Companies such as Agilent and Roche together provide the vast majority of commercial targeted sequencing solutions. As will be discussed later, in-solution hybridization of oligonucleotide probes remains one of the most popular strategies for targeted sequencing in cancers.

2.2 *Molecular Inversion Probes*

Molecular inversion probes (MIPs) are a unique class of single-stranded DNA probes that, upon successful targeting to a desired sequence, undergo a change in conformation that allows the molecules to be directly selected by amplification strategies or protected from cleanup reactions. MIPs differ from previous in-solution hybridization strategies in that they depend on polymerase and ligase activity. Typically, upon successful hybridization to a target, MIPs are extended across the target region of interest and are then closed through a ligation reaction to form a circular molecule that is protected against exonuclease digestion. These molecules can then be linearized by restriction enzyme digest and can then be directly sequenced or can be amplified by PCR or rolling circle amplification. MIP-based targeting strategies on genomic DNA also obviate the need for shotgun-based library preparation and consequently reduce the burden of sample preparation.

In contrast to in-solution hybridization of simple oligonucleotide probes, MIPs have the added advantage of being able to selectively enrich for unknown mutations and sequences. Indeed, they have become popularized in high-throughput studies that target not only SNPs but also copy number variations where conventional in-solution hybridization enrichment would not be applicable [9–15] (Fig. 7.2).

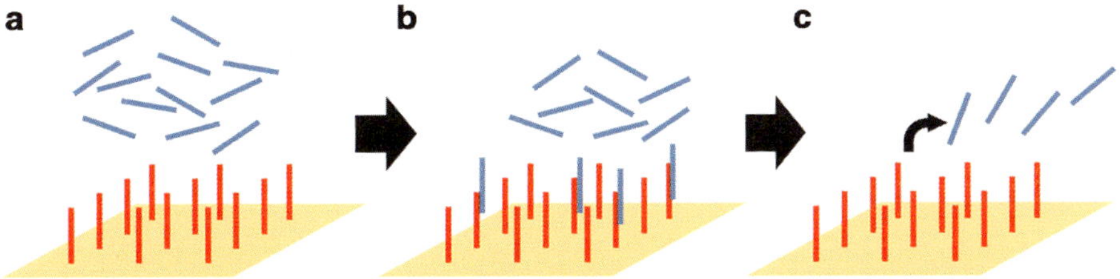

Fig. 7.3 Microarray-based target capture. (**a**) A microarray with probes (*red*) corresponding to regions of interest is exposed to a pool of fragments (*blue*). (**b**) Target fragments hybridize to the probes on the microarray, while the rest remain in solution. (**c**) The unbound fragments are then washed, and the hybridized fragments are then eluted for downstream sequencing. The microarray can then be reused

2.3 *Solid-Phase Capture on Microarrays*

Similar to the in-solution approach, hybridization on solid supports has been used to enrich for target regions of interest. In general, oligonucleotide pools are synthesized in a microarray format. A nucleic acid sample of interest is then hybridized against it, after which it is washed and the successfully hybridized fragments are eluted for downstream sequencing [16–18]. Whereas oligonucleotide pools for in-solution hybridizations are essentially used as consumable reagents, microarray-based hybridization platforms are reusable and are amenable to a variety of microarray-based cross-validation strategies. Oligonucleotide microarrays, being a mature technology, are also relatively inexpensive compared to other targeted sequencing strategies; therefore, the approach lends itself to higher flexibility towards iterative optimization.

The throughput of microarray-based capture, however, is dependent on the spatial resolution of the oligonucleotide probe array. The number of target regions that can be captured is consequently limited, although current commercial products by Roche that have 2.1 million unique probes may be sufficient for many targeted sequencing applications. In contrast, in-solution hybridization strategies do not suffer from any kind of spatial resolution and can scale up to virtually unlimited numbers of unique targeting probes (Fig. 7.3).

2.4 *Microfluidic Enrichment*

Microintegrated technologies offer an attractive path for performing high-throughput chemical reactions with minimum labor and quantity of reagents. One particular approach is the high-throughput encapsulation of reactions that results in a high number of physically segregated reactions. In effect, each molecule that may potentially participate in a desired chemical reaction is segregated from one another; in the application of targeted sequencing, each target nucleic acid is separated from

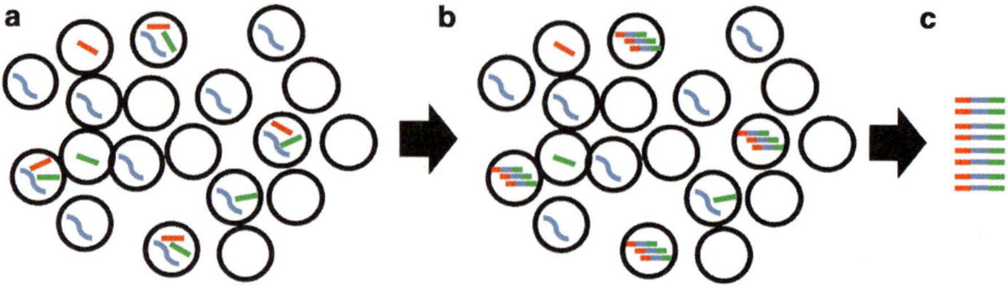

Fig. 7.4 Microfluidic target enrichment. (**a**) Nucleic acid fragments (*blue*) and targeting PCR primers (*red* and *green*) are encapsulated into independent PCR reactions, such as in an emulsion. (**b**) PCR amplification of the emulsion enriches for the fragment of interest. (**c**) The emulsion can then be broken, resulting in a pool of nucleic acids where the targets are highly enriched

one another allowing for single-molecule sensitivity. These microcompartmentalized reactions often enjoy higher kinetic reaction rates as well as reduced side reactions that may occur.

In comparison to conventional in-solution hybridization strategies, microfluidic target capture offers higher specificity due to physical encapsulation. Furthermore, at high dilutions of genomic DNA, individual fragments can be reliably targeted and then be amplified to quantities suitable for sequencing. As each individual compartment is physically separated from one another, biases in amplification are thus reduced. Sample usage can be consequently reduced; in fact, with microfluidic target enrichment strategies, pre-amplification of libraries would be unnecessary.

One recent implementation for nucleic acid enrichment has been demonstrated, where individual hybridization reactions are microencapsulated in microfluidic droplets [19]. In addition to the hybridization reaction itself, each individual library fragment that is successfully targeted is then amplified by PCR inside the droplets without any further reagent handling. Raindance Technologies, one of the few companies that provide microdroplet-based molecular biology solutions, currently offer screening panels for the enrichment of cancer-related genes. However, the emulsion-based strategy found in microdroplets complicates downstream sample processing for sequencing. Accordingly, other encapsulating methods have been developed that utilize nanowells embossed in a solid substrate [20, 21] and simply downstream workflows (Fig. 7.4).

2.5 ChIP-Seq

All targeting strategies are based on nucleic acid hybridization, which requires some knowledge of the regions of interest. However, there are often cases where the nucleic acid targets are unknown. For example, the regulatory mapping of gene interactions can be done in a genome-wide transcriptomics and proteomics approach, but it is difficult for conventional targeted sequencing strategies to

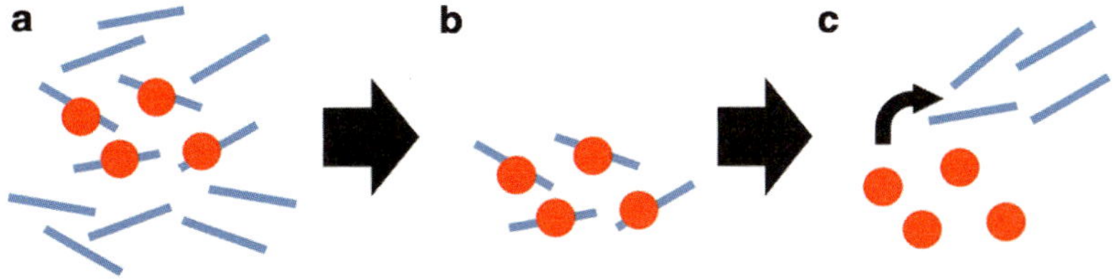

Fig. 7.5 Target capture by ChIP-Seq. (**a**) DNA-binding proteins (*red*) are bound to their associated nucleic acid fragments (*blue*) in solution. (**b**) The proteins can be selectively extracted by chromatin immunoprecipitation and the unbound nucleic acids washed away. The nucleic acids associated with these proteins can be eluted for downstream sequencing

provide more insightful information. In addition, the epigenetic modification of histones and nucleosome positioning has been known to regulate gene expression; however, the extent of this effect cannot be directly interrogated through nucleic acid probes.

ChIP-Seq is a sequencing strategy that combines immunoprecipitation enrichment with high-throughput sequencing in order to determine possible interactions between proteins and DNA [22]. It is a targeting strategy in that it specifically searches for protein-DNA interactions and enriches based on a target protein, but is also a tool for broadly discovering gene targets at high resolution. In this technique, antibodies are used to capture proteins of interest that have potential DNA-binding activity or specifically modified nucleosomes. After extraction, nucleic acids can be extracted out for subsequent sequencing (Fig. 7.5).

2.6 *Oligonucleotide-Selective Sequencing*

While there may exist many types of DNA capture technologies, many of them involve labor-intensive reagent handling and require extensive optimization. Automated and integrated techniques therefore become attractive for mass use as opposed to small-scale studies.

Oligonucleotide-selective sequencing (OS-Seq) is a targeted sequencing strategy for the Illumina platform that brings the hybridization reaction into the sequencing chamber itself [23]. Probes similar to those used for in-solution hybridization approaches with the addition of a 5′ overhang complementary to one of the two immobilized primer lawns are used. First, one of the two types of primers on the immobilized primer lawn is directly modified to contain an extra primer sequence complementary to sample regions of interest. A single-adapter library of the sample is then added, and a second extension step creates paired-end fragments that are compatible for cluster generation and sequencing.

In comparison to in-solution hybridization strategies, this technique significantly reduces the required hybridization time as well as reducing the amount of sample processing steps. Microarray-based hybridization strategies are similar to OS-Seq as both techniques hybridize against a solid surface; however, as OS-Seq directly pools all the hybridization probes together and generates the paired-end fragments inside an Illumina flow cell, the densities of probes can be much higher. In contrast, the low spatial resolution found in microarrays consequently leads to large required surface areas as the number of probes increases.

In-solution and microarray-based hybridization strategies suffer from the difficulty in tracing a particular sequenced fragment to its capture probe. Similar to MIP-based targeting strategies, OS-Seq exploits the use of polymerases in the targeting reaction such that the successfully captured nucleic acid is directly copied onto the probe itself. This in turn facilitates the alignment pipeline as each sequencing read is accordingly anchored to a specific genome location as defined by the capture probe.

In contrast to conventional oligonucleotide hybridization, a concurrent primer extension step by a DNA polymerase allows for thermodynamic stabilization of an annealed strand. Furthermore, the specific generation of paired-end fragments between the on-target DNA fragment and the capture primer prevents the secondary capture and consequent spurious sequencing of off-target DNA. Therefore, not only does this technique reduce the time and labor cost required for DNA capture, it also improves sensitivity and accuracy. In general, as next-generation sequencing technologies mature, we expect technologically minded research groups to follow suit by directly performing custom chemical reactions inside the sequencing device itself.

3 Applications of Targeted Next-Generation Sequencing to Cancer Research

The use of targeted next-generation sequencing, as discussed earlier, is justified as we face a growing number of technical limitations: (1) when limited amounts of nucleic acids from samples are available; (2) when high coverage of clinically actionable alleles from many individuals are required, as compared to moderate/low coverage of whole genomes from few individuals; (3) when resources such as reagents, budget, and bioinformatics infrastructure are limited; and (4) when researchers are ethically obliged to minimize the exposure of genetic information by only focusing on intended targets. The development of targeted sequencing technologies consequently enables researchers to address all of these issues simultaneously.

Targeted sequencing has gained widespread usage in cancer research. Broadly, the relevant subsets of the human genome associated with cancers are as follows: (1) all protein-coding DNA sequences, (2) an interesting gene or set of clinically actionable genes such as oncogenic genes or tumor suppressor genes, (3) a known region of a chromosome associated with cancer phenotypes, (4) regions or motifs of regulatory DNA, and (5) DNA sequences corresponding to clinically relevant microRNA. In addition, the research community has identified nonhuman sequences such as bacterial or viral sequences in cancer samples. We will briefly review the recent application of next-generation sequencing to each of these research interests.

3.1 Targeting Coding DNAs

Protein-coding DNAs of the all known genes are comprised of ~180,000 exons and represent 1–2 % of the genome. Mutations in coding DNAs can lead to direct alterations in the downstream protein structure and consequently affect their function. Therefore, sequencing of coding DNAs is an efficient way of identifying the genetic cause or association of not only tumor development but also other genetic diseases [24]. Depending on the specific research interest, all protein-coding regions can be subject to sequencing and are referred to as "exome sequencing." Coding DNAs from a smaller subset can also be directly targeted and sequenced. This is particularly useful for the detection of mutations in canonical genetic pathways related to oncogenesis.

3.2 Whole-Exome Sequencing

The merits of whole-exome sequencing have been thoroughly discussed in other reviews (see [25] for an in-depth discussion). Briefly, the primary advantage of exome sequencing lies in the direct correlation between mutations and their expressed phenotypes. Many mutations in long noncoding DNA regions are challenging to sequence, and as such statistically relevant conclusions would be difficult to reach. Furthermore, the total cost of exome sequencing (e.g., costs in both reagents and bioinformatics infrastructure) has dropped dramatically such that it is now within the budget for many cancer-related research projects.

Fundamentally, targeted sequencing and consequently whole-exome sequencing suffers from the fact that the researcher must know the targets a priori—the discovery of clinically relevant mutations is limited to the scope of the experiment. Whole-exome sequencing strategies attempt to address this bias in experimental design by targeting all known coding regions. However, multiple revisions from the Human Genome Project have over time changed the genomic regions currently accepted to be coding; therefore, any nucleotides outside of these specific targeted regions will also not be represented in any whole-exome sequencing studies. Furthermore, whole-exome sequencing is limited by a variety of other technical difficulties, including detecting structural variations and detecting major genetic variations in noncoding regulatory regions (e.g., transcription factor binding sites and enhancer regions).

Despite these technical issues with whole-exome sequencing, it has emerged as the most popular tool for discovering clinically relevant mutations in cancer research. To the best of our knowledge through the NCBI Pubmed database, 17 different tissue types in 4 cancer classes—hematopoietic, carcinomas, neuroectodermal, and sarcoma—have been surveyed by whole-exome sequencing in 71 studies up to October 2012 (Table 7.1). These studies sequenced 2,696 cancer samples mostly from primary patients as well as cell lines. Melanoma and lung carcinoma

Table 7.1 Summary of whole-exome sequencing on cancer samples

Class	Tissue type	# Study	# Samples
Hematopoietic	T and B cells	17	403
	Plasma cells	1	22
Carcinomas	Pancreas	5	28
	Ovary	2	217
	Head and neck	2	106
	Kidney	3	12
	Skin	9	207
	Gastric	3	38
	Colorectal	2	226
	Prostate	6	164
	Lung	4	280
	Breast	9	791
	Hepatobiliary	4	80
	Endometrium	1	13
Neuroectodermal	CNS	1	92
	CNS (oligodendrocytes)	1	16
Sarcoma	Blood vessels	1	1
Sum		71	2,696

have relatively higher mutation rates, while pancreatic and renal carcinomas have a relatively lower mutation rate (Fig. 7.6) [26]. Most of these studies have been done in the last 2 years—2011 and 2012—while few studies were published before 2010. As the cost of whole-exome sequencing continues to decrease, we expect the number of such studies to accelerate.

In the case of hematopoietic cancers, three types of cancer were sequenced [27–44]: leukemia, lymphoma, and myeloma (Table 7.2). Interestingly, leukemia is the cancer type that has been mostly sequenced. Two types of neuroectodermal cancer have been sequenced [45, 46], while one type was sequenced in sarcomas [47] (Table 7.3). Twelve tissue types with 38 subtypes have been sequenced in carcinomas, which comprise most cancers [48–98] (Table 7.4). As clearly evident in Tables 7.2 and 7.3, the platform of choice for the whole-exome capture of DNA is offered by Agilent and SureSelect.

3.3 Targeted Sequencing of Specific Genes

The early discovery of canonical oncogenes and gene families (*SRC*, *RAS*, *MYC*, *RAF*) and tumor suppressor genes (*TP53 and RB1*) allowed researchers for the first time to correlate mutations with tumor phenotypes. Most targeted therapeutics have been developed based on these genes. Historically, cancer researchers focused on these cancer genes due to the technical limitations of PCR and sequencing technology before the development of next-generation sequencing technology. Several initial studies systematically surveyed the mutations on *BRAF* [99], *PIK3CA* [100], and *EGFR* [101] in

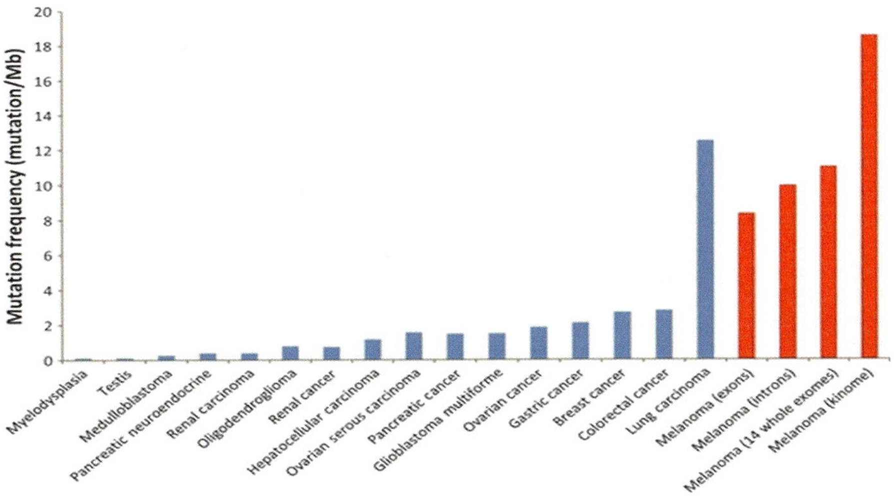

Fig. 7.6 The somatic mutation rate of different types of cancers. Reproduced from [26] with permission

cancer samples. As large-scale cancer genome projects started in 2006, two seminal studies sequenced the sets of genes: (1) 13,023 genes in 11 breast and 11 colorectal cancers [102] and (2) 518 protein kinase genes in 210 different human cancers [103]. Eventually, several studies started to use next-generation sequencing to survey several genes that are associated with clinical decision [104–110] (Table 7.5).

3.4 Miscellaneous Targeting: Noncoding DNA Regions and Nonhuman DNA

The ENCyclopedia of DNA Elements (ENCODE) project discovered that many noncoding DNA regions are actively involved in the regulation of gene expression [111–113]. Several studies showed that microRNAs play a critical role in cancer development (for an excellent review, see [114]). Two studies adopted next-generation sequencing technology to detect microRNA systemically [115, 116]. In addition, Chmielecki et al. tried to detect the conserved GXGXXG kinase motif associated with tyrosine kinase (TK) fusions by using 454 sequencing platform [117] (Table 7.6).

Parkin's study showed that about 20 % of cancer incidences were associated with infection globally in 2002 [118]. Therefore, we can expect to find the DNA sequences of infectious agents such as viral or bacterial sequences in cancer genomes. Several studies have identified viral sequences in the transcriptome of human samples [119–123]. Recent studies have adopted next-generation sequencing technology to detect viral sequences [124–127] (Table 7.7).

Table 7.2 Whole-exome sequencing of hematopoietic cancer

Type	Sample	#Samples	Sequencing platform	Enrichment	Key findings	Study	Reference
Leukemia	ALL	4	Illumina GAII	Roche SeqCap	14 genuine somatic mutations	Lilljebjörn et al. Leukemia 2012	[35]
	AML	6	Illumina GAII/HiSeq	Roche SeqCap	Internal tandem duplication in FLT3; early mutations on NPM1, TET2, and SMCIA	Jan et al. Sci Transl Med. 2012	[32]
	AML-M5	9	Illumina GAIIX	NimbleGen exome array	Recurrent mutation on DNMT3A	Dickinson et al. Blood 2011	[27]
	APL	3	Illumina GAIIX	Agilent SureSelect	Recurrent mutation on FLT3	Grief et al. Leukemia 2011	[30]
	CLL	105	Illumina GAIIX	Agilent SureSelect	Recurrent mutation on SF3B1	Quesade et al. Nat Genet 2011	[37]
		88	Illumina GAII	Own method	Recurrent mutation on TP53, ATM, MYD88, NOTCH2, SF3B1, ZMYM3, MAPK1, FBXW7, and DDX3X	Wang et al. N Engl J Med 2011	[43]
	CN-AML	5	Illumina GAIIX	Agilent SureSelect	Recurrent mutations on ZF1	Greif et al. Blood 2012	[29]
		1	Illumina GAII	Agilent SureSelect	Recurrent mutation on BCOR	Grossmann et al. Blood 2011	[31]
	HCL	1	Illumina GAII	Agilent SureSelect	BRAF V600E	Tiacci et al. N Engl J Med. 2011	[41]
	MCL	1	Illumina GAII	Roche SeqCap	KIT D816V		[40]
	MDS	29	Illumina GAII/HiSeq	Agilent SureSelect	Frequent mutations on RNA splicing machinery; U2AF35, ZRSR2, SRSF2, and SF3B1	Yoshida et al. Nature 2011	[44]
	T-LGL	76	Illumina GAII	Agilent SureSelect	Recurrent mutations on STAT3	Koskela et al. N Engl J Med. 2012	[34]

Lymphoma	SMZL	8	Illumina HiSeq	Agilent SureSelect	Mutations in NOTCH2	Rossi et al. J Exp Med 2012	[38]
	PCNSL	4	Illumina GAIIX	Agilent SureSelect	Recurrent mutations on MYD88, TBL1XR1	Gonzalez-Aguilar et al. Clin Cancer Res 2012	[28]
	NKTCL	4	Illumina GAIIx	Own method	Mutations on JAK3	Koo et al. Cancer Discov 2012	[33]
	DLBCL	55	Illumina HiSeq	Own method	Mutations on KRAS, BRAF, and NOTCH1	Lohr et al. Proc Natl Acad Sci USA 2012	[36]
	NLPHL	A family of four cousins	Illumina GAII	Agilent SureSelect	NPAT mutations as a candidate risk factor	Saarinen et al. Blood 2011	[39]
Myeloma	Myeloma	22	Illumina GAIIx	Agilent SureSelect	Recurrent RAS pathway mutations	Walker et al. Blood 2012	[42]

AML acute myeloid leukemia, *CN-AML* cytogenetically normal acute myeloid leukemia, *AML-M5* M5 subtype of acute myeloid leukemia, *APL* acute promyelocytic leukemia, *CLL* chronic myelocytic leukemia, *CN-AML* cytogenetically normal acute myeloid leukemia, *HCL* hairy-cell leukemia, *MCL* mast-cell leukemia, *MDS* myelodysplasia (predisposition to AML), *T-LGL* T-cell large granular lymphocytic leukemia, *SMZL* splenic marginal zone lymphoma, *PCNSL* primary central nervous system lymphoma, *NKTCL* natural killer/T-cell lymphoma, *DLBCL* diffuse large B-cell lymphoma, *NLPHL* nodular lymphocyte predominant Hodgkin lymphoma

Table 7.3 Whole-exome sequencing of neuroectodermal and sarcomas

Tissue	Sample	#Samples	Sequencing platform	Enrichment	Key findings	Study	Reference
Neuroectodermal	Medulloblastomas	92	Illumina HiSeq	RNA baits solution hybridization	DDX3X as a component of pathogenic β-catenin signalling	Pugh et al. Nature 2012	[46]
	Oligodendrogliomas	16	Illumina HiSeq	Agilent SureSelect	Concurrent CIC mutations, IDH mutations	Yip et al. J Pathol. 2012	[45]
Sarcoma	Kaposi sarcoma	1	Illumina GAIIx	SureSelect Human All Exon Kit	Homozygous splice-site mutation in STIM1	Byun et al. J Exp Med. 2010	[47]

Table 7.4 Whole-exome sequencing of carcinomas

Tissue	Sample	#Samples	Sequencing platform	Enrichment	Key findings	Study	Reference
Pancreas	IPMN	1	SOLD	Agilent SureSelect	Recurrent mutation GNAS	Furukawa et al. Sci Rep. 2011	[55]
	PDAC	1	Illumina GAII	NimbleGen SeqCap	12 genes with higher allele frequencies of functional mutations	Zhou et al. Caner Biol Ther. 2011	[96]
		3	Illumina GAIIx	Agilent SureSelect	TM as a predisposition gene	Roberts et al. Cancer Discov. 2012	[82]
	PDAC-derived cell lines	15	Illumina GAIIx	Agilent SureSelect	MLH1 hemizygous deletion	Wang et al. Genome Res. 2012	[82]
	SCA, IPMN, MCN, SPNs	8	Illumina GAII/HiSeq	Agilent SureSelect	Recurrent mutation on VHL, RNF43, CTNNB1	Wu et al. Proc Natl Acad Sci USA. 2011	[84]
Gastric	Metastatic gastric adenocarcinoma	1	Illumina HiSeq	Agilent SureSelect	Somatic protein-damaging variants (SCEL, SMC4, KBTBD10, PANK1, DSG3, SALL3, PCBP3, ITGAX, LAMA 1, GPER)	Chang et al. Pediatr Blodd Cancer 2012	[53]
	Gastric adenocarcinomas	15	Illumina GAII/HiSeq	Agilent SureSelect	Mutations on TP53, PIK3CA and ARID1A, FA T4, ARID1A, MLL3 and MLL	Zang et al. Nat Genet. 2012	[87]
	Gastric cancer	22	Illumina GAIIx	Agilent SureSelect	ARID1A as better prognosis indicator	Wang et al. Nat Genet 2011	[81]
Colorectal	Colon or rectum adenocarcinoma	224	Illumina HiSeq/ SOLiD	NimbleGen Exome array	Frequent mutations on APC, TP53, SMAD4, PIK3C, KRASA, RID1A, SOX9, and FAM123B	Cancer Genome Atlas Network Nature 2012	[97]
	G3 grade colon adenocarcinomas	2	454/SOLiD	Agilent SureSelect/ NimbleGen Exome array	Mutations on BMPR1A	Timmermann et al. PLoS One 2010	[77]

(continued)

Table 7.4 (continued)

Tissue	Sample	#Samples	Sequencing platform	Enrichment	Key findings	Study	Reference
Head and neck	HNSCC	32	Illumina GAIIx/ HiSeq	Agilent SureSelect	Frequent mutations on NOTCH1, FBXW7, CDKN2A, PIK3CA, and HRAS	Agrawal et al. Science 2011	[49]
		74	Illumina GAIIx/ HiSeq	Own method	Frequent mutations on NOTCH1, IRF6, and TP63, TP53, CDKN2A, PTEN, PIK3CA, and HRAS	Stransky et al. Science 2011	[75]
Kidney	Metastatic renal-cell carcinoma	4	Illumina GAIIx/ HiSeeq	Agilent SureSelect/ NimbleGen SeqCap	Mutational intratum or heterogenity on SETD2, PTEN, and KDM5C	Gerlinger et al. N Engl J Med. 2012	[57]
	ccRCC	1	Illumina HiSeq	Agilent SureSelect	Mutations on USP6, LRRK2, SRGAP3, AHNAK, SRGAP3, NIPBL, UBE4A, and SH3GL1	Xu et al. Cell 2012	[86]
		7	Illumina GAIIx	Agilent SureSelect	Frequent mutations on PBRM1	Varela et al. Nature 2011	[80]
Prostate	LNCaP cell line	1	Illumina GAIIx	Agilent SureSelect	1,802 non-synonymous SNVs, 218 small insertions and deletions	Spans et al. Prostate 2012	[73]
	Metastatic castrate-resistant prostate cancer	17	Illumina HiSeq	Agilent SureSelect	AR gain and PTEN loss	Lonigro et al. Neoplasia 2011	[64]
	Prostate tumor tissue	1	SOLiD	Agilent SureSelect	FFPE material is a reliable source of material for whole-exome sequencing	Menon et al. Int J Mol Sci. 2012	[65]
		120	Illumina HiSeq	Agilent SureSlect	Recurrent mutations on Sp	Barbieri et al. Nat Genet. 2012	[50]
	Metastatic prostate cancer xenografts	2	Illumina HiSeq	Agilent SureSelect	Mutations on NRAS, TP53, AURKA, FAS, and MYH11, HRAS	Roychowdhury et al. Sci Transl Med. 2011	[70]

	Prostate cancer xenografts	23	Illumina GAIIx/HiSeq	NimbleGen SeqCap	Mutations on TP53, DLK2,GPC6, and SDF4	Kumar et al. Proc Natl Acad Sci USA 2011	[62]
Ovary	HGSC	3	Illumina HiSeq	Agilent SureSelect	Mutations on genes associated with cell cycle checkpoint function and Golgi vesicle trafficking	Casterllarin et al. J Pathol 2012	[52]
	Ovarian carcinoma	214	Illumina	Own method	Mutations on TP53 and NF1, CDK12	Carter et al. Nat Biotechnol 2012	[51]
Skin	Acral melanoma cell lines	6	Illumina GAIIx	Agilent SureSelect	Mutations on BRAF, NRAS, KIT, PTEN and TP53, BRCA1, and BRCA2	Furney et al. Pigment Cell Melanoma Res 2012	[54]
	Highly metastatic uveal melanomas of the eye	2	Illumina GAIIx	Roche SeqCap	Loss of BAP1	Harbour et al. Science 2010	[59]
	Melanoma cell lines	7	Illumina GAIIx	Roche SeqCap	Recurrent mutations on MAP2K1and MAP2K2	Nikolaev et al. Nat Genet. 2012	[66]
	Melanomas	147	Illumina GAIIx/HiSeq	Roche SeqCap	Recurrent mutations on PPP6C, RAC1	Krauthammer et al. Nat Genet. 2012	[95]
		20	Illumina GAIIx/HiSeq	Agilent SureSelect	BRAF V600E	Shi et al. Nat Commun 2012	[72]
	Metastases melanoma	2	Illumina GAIIx	Agilent SureSelect	96% of the single-nucleotide variants are shared between the two metastases	Gartner et al. BMC Genomics 2012	[56]
	Metastatic melanoma cell lines	8	Illumina GAII/HiSeq	Roche SeqCap	Recurrent loss of heterozygosity of MAP3K5 and MAP3K9	Stark et al. Nat Genet. 2012	[74]
		14	Illumina GAII	Agilent SureSelect	Recurrent mutations on GRIN2A, TRRAP	Wei et al. Nat Genet 2011	[83]
	Primary acral melanoma	1	Illumina GAIIx	Agilent SureSelect	Nonsense mutation in ERCC5	Turajlic et al. Genome Res. 2012	[79]

(continued)

Table 7.4 (continued)

Tissue	Sample	#Samples	Sequencing platform	Enrichment	Key findings	Study	Reference
Lung	Lung adenocarcinoma	182	Illumina HiSeq	Agilent SureSelect	Recurrent mutations on U2AF1, RBM10 and ARID1A	Imielinski et al. Cell 2012	[61]
	NSCLC	14	Illumina HiSeq	Agilent SureSelect	MXRA5 as cancer gene	Xiong et al. Carcinogenesis 2012	[85]
		31	Illumina GAIIx	Agilent SureSelect	CSMD3 as the second most frequently mutated gene	Liu et al. Carcinogenesis 2012	[63]
	Small-cell lung cancer	53	Illumina HiSeq	Agilent SureSelect	22 significantly mutated genes	Rudin et al. Nat Genet. 2012	[71]
Breast	Breast cancer cell line	2	454	NimbleGen exome array	loss of heterozygosity and allele-specific expression on BRCA1, MSH3 and SETX	Zhao et al. Genome Biol 2010	[88]
	Breast cancer families	15	Illumina GAIIx/HiSeq	NimbleGen SeqCap	Deleterious mutations on FANCC and BLM	Thompson et al. PLoS Genet. 2012	[76]
		4	SOLiD	NimbleGen SeqCap	FAN1 as a breast cancer susceptibility gene	Park et al. Breast Cancer Res Treat. 2011	[68]
	Breast carcinoma	115	Illumina GAII	Own method	Recurrent mutations in the CBFB and deletions of RUNX1	Banerji et al. Nature 2012	[92]
		31	Illumina	Roche SeqCap	Eighteen significantly mutated genes	Ellis et al. Nature 2012	[91]
		1	Illumina GAII	NimbleGen exome array	33 nonsense mutations	He et al. Am J Cancer Res 2011	[90]
	Familial breast cancer	50	Illumina GAIIx	Agilent SureSelect	The median of 10 variants in individuals	Snape et al. Breast Cancer Res Treat. 2012	[93]
		66	SOLiD	NimbleGen SeqCap	XRCC2 as a breast cancer susceptibility gene	Park et al. Am J Hum Genet. 2012	[89]
	Invasive breast adenocarcinoma	507	Illumina HiSeq	Agilent SureSelect	TP53, PIK3CA, and GATA3 occurred at >10% incidence across all breast cancers	The Cancer Genome Atlas Network Nature 2012	[98]

Hepatobiliary	A primary hepatitis C virus-positive hepatocellular carcinoma	1	Illumina GAIIx	Agilent SureSelect	63 validated non-synonymous substitutions revealed enrichment of phosphoproteins	Totoki et al. Nat Genet. 2011	[78]
	Hepatocellular carcinoma	24	Illumina HiSeq	Agilent SureSelect	Recurrent alterations in ARID1A, RPS6KA3, NFE2L2, and IRF2	Guichard et al. Nat Genet. 2012	[58]
		47	SOLiD	Roche SeqCap	Functional framework of a comprehensive program for genomic analysis	Harring et al. World J Surg. 2011	[60]
	Opisthorchis viverrini-related cholangiocarcinoma	8	Illumina GAIIx/HiSeq	Agilent SureSelect	Recurrent mutations on MLL3, ROBO2, RNF43, PEG3, and GNAS	Ong et al. Nat Genet. 2021	[67]
Uterine	Endometrial cancer	13	SOLiD	Agilent SureSelect	12 potential driver cancer genes	Liang et al. Genome Res. 2012	[48]

PDAC pancreatic ductal adenocarcinoma, *IPMN* intraductal papillary mucinous neoplasm, *HNSCC* head and neck squamous cell carcinoma, *FFPE* formalin fixed paraffin embedded, *HGSC* high-grade serous carcinoma, *NSCLC* non-small cell lung carcinomas

Table 7.5 Targeted sequencing of specific genes

Targets	Sample	Sequencing platform	Study	Reference
137 actionable cancer genes	Formalin-fixed, paraffin-embedded (FFPE) tumor samples	Illumina HiSeq	Wagle et al. Cancer Discov. 2012	[110]
97 genes that are known to be involved in oncogenesis of T-ALL	T-ALL cell lines, primary T-ALL patient	454	Kalender Atak et al. PLoS One 2012	[107]
541 genes and miRNAs from 3p, UTX, JARID1C, TP53, and ISW	Primary ccRCC tumor tissue	Illumina HiSeq	Duns et al.	[106]
BRCA1	Coriell samples	HeliScope	Thompson et al.	[109]
280-kb region of the NF1	Neurofibromatosis type 1	454	Chou et al.	[105]
182 cancer-associated genes and 37 introns of 14 commonly rearranged genes	Advanced prostate cancer	Illumina HiSeq	Beltran et al. Eur Urol. 2012	[104]
BRCA1 and BRCA2 genomic regions	Familial breast cancer patients	Illumina GAII	Ozcelik et al. J Mol Diagn. 2012	[108]

Table 7.6 Surveying of microRNAs by next-generation sequencing

Targets	Sample	Sequencing platform	Study	Reference
microRNA	Ovarian cancer	454	Wyman et al. PLoS One 2009	[115]
microRNA	Human embryonic stem cells	Illumina	Morin et al. Genome Res. 2008	[116]
Upstream of a conserved GXGXXG kinase motif	Thyroid cancer cells, leukemia cells	454	Chmielecki et al. Nucleic Acids Res. 2010	[117]

Table 7.7 Identification of viral sequences by next-generation sequencing

Targets	Sample	Sequencing platform	Study	Reference
Merkel cell polyomavirus	Merkel cell carcinoma	454	Feng et al. Science 2008	[126]
Merkel cell polyomavirus (MCPyV) genome	Merkel cell carcinoma	Illumina GAIIx	Duncavage et al. J Mol Diagn. 2011	[125]
Human polyomaviruses	Merkel cell carcinoma	Illumina HiSeq	Sauvage et al. Emerg Infect Dis. 2011	[127]
Human papillomavirus (HPV)	Patients with HPV infection	454	Barzon et al. J Clin Virol. 2011	[124]

4 Conclusions

Targeted resequencing approaches integrate the potential to capture, partition, index, or target genomic DNA subsets with the power of the next-generation sequencing technologies. This snapshot resequencing offers cost-effective and feasible options for biomedical applications that require, from a statistical perspective, high-throughput deep resequencing of large gene sets from a large number of cancer samples. Although resequencing entire cancer genomes provides valuable information by systematically cataloguing mutations, reducing the complexity of the sequencing libraries and focusing the sequencing efforts towards genomic regions with highest clinical and biological interest provide a myriad of alternative approaches to sleuth the details of the cancer genomes. Furthermore, as emerging technological developments in DNA capture will allow researchers to conveniently extract and sequence genes of interest, targeted resequencing will soon become a primary research and analysis method. In the near future, targeted resequencing studies of large clinical sample cohorts will be accessible and, undoubtedly, advance our understanding of the molecular genetics and genomics of a variety of human cancers. Hopefully, that new insight and deeper knowledge will revolutionize the treatment of these grave diseases.

Acknowledgements Support came from National Institutes of Health Grants P01 HG000205 (B.T.L. and H.P.J.), R21 CA140089 (H.P.J.), and NIH U01 CS151920 (H.L. and H.P.J). Additional support came from Doris Duke Clinical Foundation Clinical Scientist Development Award (H.P.J) and the Howard Hughes Medical Foundation Early Career Grant (H.P.J.).

References

1. Teer JK, Mullikin JC. Exome sequencing: the sweet spot before whole genomes. Hum Mol Genet. 2010;19(R2):R145–51.
2. Turner EH, Ng SB, Nickerson DA, Shendure J. Methods for genomic partitioning. Annu Rev Genomics Hum Genet. 2009;10:263–84.
3. Mamanova L, Coffey AJ, Scott CE, Kozarewa I, Turner EH, Kumar A, et al. Target-enrichment strategies for next-generation sequencing. Nat Methods. 2010;7(2):111–8.
4. Metzker ML. Sequencing technologies – the next generation. Nat Rev Genet. 2009;11(1):31–46.
5. Gnirke A, Melnikov A, Maguire J, Rogov P, LeProust EM, Brockman W, et al. Solution hybrid selection with ultra-long oligonucleotides for massively parallel targeted sequencing. Nat Biotechnol. 2009;27(2):182–9.
6. Briggs AW, Good JM, Green RE, Krause J, Maricic T, Stenzel U, et al. Targeted retrieval and analysis of five Neanderthal mtDNA genomes. Science. 2009;325(5938):318–21.
7. Noonan JP, Coop G, Kudaravalli S, Smith D, Krause J, Alessi J, et al. Sequencing and analysis of Neanderthal genomic DNA. Science. 2006;314(5802):1113–8.
8. Horn S. Target enrichment via DNA hybridization capture. Methods Mol Biol. 2012;840:177–88.
9. Antson DO, Isaksson A, Landegren U, Nilsson M. PCR-generated padlock probes detect single nucleotide variation in genomic DNA. Nucleic Acids Res. 2000;28(12):E58.
10. Faruqi AF, Hosono S, Driscoll MD, Dean FB, Alsmadi O, Bandaru R, et al. High-throughput genotyping of single nucleotide polymorphisms with rolling circle amplification. BMC Genomics. 2001;2(1):4.

11. Lizardi PM, Huang X, Zhu Z, Bray-Ward P, Thomas DC, Ward DC. Mutation detection and single-molecule counting using isothermal rolling-circle amplification. Nat Genet. 1998;19(3):225–32.
12. Hardenbol P, Baner J, Jain M, Nilsson M, Namsaraev EA, Karlin-Neumann GA, et al. Multiplexed genotyping with sequence-tagged molecular inversion probes. Nat Biotechnol. 2003;21(6):673–8.
13. Hardenbol P, Yu F, Belmont J, Mackenzie J, Bruckner C, Brundage T, et al. Highly multiplexed molecular inversion probe genotyping: over 10,000 targeted SNPs genotyped in a single tube assay. Genome Res. 2005;15(2):269–75.
14. Turner EH, Lee C, Ng SB, Nickerson DA, Shendure J. Massively parallel exon capture and library-free resequencing across 16 genomes. Nat Methods. 2009;6(5):315–6.
15. Porreca GJ, Zhang K, Li JB, Xie B, Austin D, Vassallo SL, et al. Multiplex amplification of large sets of human exons. Nat Methods. 2007;4(11):931–6.
16. Okou DT, Steinberg KM, Middle C, Cutler DJ, Albert TJ, Zwick ME. Microarray-based genomic selection for high-throughput resequencing. Nat Methods. 2007;4(11):907–9.
17. Hodges E, Rooks M, Xuan Z, Bhattacharjee A, Gordon DB, Brizuela L, et al. Hybrid selection of discrete genomic intervals on custom-designed microarrays for massively parallel sequencing. Nat Protoc. 2009;4(6):960–74.
18. Albert TJ, Molla MN, Muzny DM, Nazareth L, Wheeler D, Song X, et al. Direct selection of human genomic loci by microarray hybridization. Nat Methods. 2007;4(11):903–5.
19. Tewhey R, Warner JB, Nakano M, Libby B, Medkova M, David PH, et al. Microdroplet-based PCR enrichment for large-scale targeted sequencing. Nat Biotechnol. 2009;27(11):1025–31.
20. Bau S, Schracke N, Kränzle M, Wu H, Stähler PF, Hoheisel JD, et al. Targeted next-generation sequencing by specific capture of multiple genomic loci using low-volume microfluidic DNA arrays. Anal Bioanal Chem. 2009;393(1):171–5.
21. Wei CW, Cheng JY, Huang CT, Yen MH, Young TH. Using a microfluidic device for 1 microl DNA microarray hybridization in 500 s. Nucleic Acids Res. 2005;33(8):e78.
22. Park PJ. ChIP|[ndash]|seq: advantages and challenges of a maturing technology. Nat Rev Genet. 2009;10(10):669–80.
23. Myllykangas S, Buenrostro JD, Natsoulis G, Bell JM, Ji HP. Efficient targeted resequencing of human germline and cancer genomes by oligonucleotide-selective sequencing. Nat Biotechnol. 2011;29:1024–7.
24. Ng SB, Buckingham KJ, Lee C, Bigham AW, Tabor HK, Dent KM, et al. Exome sequencing identifies the cause of a mendelian disorder. Nat Genet. 2010;42(1):30–5.
25. Biesecker LG, Shianna KV, Mullikin JC. Exome sequencing: the expert view. Genome Biol. 2011;12(9):128.
26. Walia V, Mu EW, Lin JC, Samuels Y. Delving into somatic variation in sporadic melanoma. Pigment Cell Melanoma Res. 2012;25(2):155–70.
27. Dickinson RE, Griffin H, Bigley V, Reynard LN, Hussain R, Haniffa M, et al. Exome sequencing identifies GATA-2 mutation as the cause of dendritic cell, monocyte, B and NK lymphoid deficiency. Blood. 2011;118(10):2656–8.
28. Gonzalez-Aguilar A, Idbaih A, Boisselier B, Habbita N, Rossetto M, Laurenge A, et al. Recurrent mutations of MYD88 and TBL1XR1 in primary central nervous system lymphomas. Clin Cancer Res. 2012;18(19):5203–11.
29. Greif PA, Dufour A, Konstandin NP, Ksienzyk B, Zellmeier E, Tizazu B, et al. GATA2 zinc finger 1 mutations associated with biallelic CEBPA mutations define a unique genetic entity of acute myeloid leukemia. Blood. 2012;120(2):395–403.
30. Greif PA, Yaghmaie M, Konstandin NP, Ksienzyk B, Alimoghaddam K, Ghavamzadeh A, et al. Somatic mutations in acute promyelocytic leukemia (APL) identified by exome sequencing. Leukemia. 2011;25(9):1519–22.
31. Grossmann V, Tiacci E, Holmes AB, Kohlmann A, Martelli MP, Kern W, et al. Whole-exome sequencing identifies somatic mutations of BCOR in acute myeloid leukemia with normal karyotype. Blood. 2011;118(23):6153–63.

32. Jan M, Snyder TM, Corces-Zimmerman MR, Vyas P, Weissman IL, Quake SR, et al. Clonal evolution of preleukemic hematopoietic stem cells precedes human acute myeloid leukemia. Sci Transl Med. 2012;4(149):149ra18.
33. Koo GC, Tan SY, Tang T, Poon SL, Allen GE, Tan L, et al. Janus kinase 3-activating mutations identified in natural killer/T-cell lymphoma. Cancer Discov. 2012;2(7):591–7.
34. Koskela HLM, Eldfors S, Ellonen P, van Adrichem AJ, Kuusanmäki H, Andersson EI, et al. Somatic STAT3 mutations in large granular lymphocytic leukemia. N Eng J Med. 2012;366(20):1905–13.
35. Lilljebjorn H, Rissler M, Lassen C, Heldrup J, Behrendtz M, Mitelman F, et al. Whole-exome sequencing of pediatric acute lymphoblastic leukemia. Leukemia. 2012;26(7):1602–7.
36. Lohr JG, Stojanov P, Lawrence MS, Auclair D, Chapuy B, Sougnez C, et al. Discovery and prioritization of somatic mutations in diffuse large B-cell lymphoma (DLBCL) by whole-exome sequencing. Proc Natl Acad Sci U S A. 2012;109(10):3879–84.
37. Quesada V, Conde L, Villamor N, Ordonez GR, Jares P, Bassaganyas L, et al. Exome sequencing identifies recurrent mutations of the splicing factor SF3B1 gene in chronic lymphocytic leukemia. Nat Genet. 2011;44(1):47–52.
38. Rossi D, Trifonov V, Fangazio M, Bruscaggin A, Rasi S, Spina V, et al. The coding genome of splenic marginal zone lymphoma: activation of NOTCH2 and other pathways regulating marginal zone development. J Exp Med. 2012;209(9):1537–51.
39. Saarinen S, Aavikko M, Aittomäki K, Launonen V, Lehtonen R, Franssila K, et al. Exome sequencing reveals germline NPAT mutation as a candidate risk factor for Hodgkin lymphoma. Blood. 2011;118(3):493–8.
40. Spector MS, Iossifov I, Kritharis A, He C, Kolitz JE, Lowe SW, et al. Mast-cell leukemia exome sequencing reveals a mutation in the IgE mast-cell receptor β chain and KIT V654A. Leukemia. 2012;26(6):1422–5.
41. Tiacci E, Trifonov V, Schiavoni G, Holmes A, Kern W, Martelli MP, et al. BRAF mutations in hairy-cell leukemia. N Eng J Med. 2011;364(24):2305–15.
42. Walker BA, Wardell CP, Melchor L, Hulkki S, Potter NE, Johnson DC, et al. Intraclonal heterogeneity and distinct molecular mechanisms characterize the development of t(4;14) and t(11;14) myeloma. Blood. 2012;120(5):1077–86.
43. Wang L, Lawrence MS, Wan Y, Stojanov P, Sougnez C, Stevenson K, et al. SF3B1 and other novel cancer genes in chronic lymphocytic leukemia. N Eng J Med. 2011;365(26):2497–506.
44. Yoshida K, Sanada M, Shiraishi Y, Nowak D, Nagata Y, Yamamoto R, et al. Frequent pathway mutations of splicing machinery in myelodysplasia. Nature. 2011;478(7367):64–9.
45. Yip S, Butterfield YS, Morozova O, Chittaranjan S, Blough MD, An J, et al. Concurrent CIC mutations, IDH mutations, and 1p/19q loss distinguish oligodendrogliomas from other cancers. J Pathol. 2012;226(1):7–16.
46. Pugh TJ, Weeraratne SD, Archer TC, Pomeranz Krummel DA, Auclair D, Bochicchio J, et al. Medulloblastoma exome sequencing uncovers subtype-specific somatic mutations. Nature. 2012;488(7409):106–10.
47. Byun M, Abhyankar A, Lelarge V, Plancoulaine S, Palanduz A, Telhan L, et al. Whole-exome sequencing-based discovery of STIM1 deficiency in a child with fatal classic Kaposi sarcoma. J Exp Med. 2010;207(11):2307–12.
48. Liang H, Cheung LW, Li J, Ju Z, Yu S, Stemke-Hale K, et al. Whole-exome sequencing combined with functional genomics reveals novel candidate driver cancer genes in endometrial cancer. Genome Res. 2012;22(11):2120–9.
49. Agrawal N, Frederick MJ, Pickering CR, Bettegowda C, Chang K, Li RJ, et al. Exome sequencing of head and neck squamous cell carcinoma reveals inactivating mutations in NOTCH1. Science. 2011;333(6046):1154–7.
50. Barbieri CE, Baca SC, Lawrence MS, Demichelis F, Blattner M, Theurillat J-P, et al. Exome sequencing identifies recurrent SPOP, FOXA1 and MED12 mutations in prostate cancer. Nat Genet. 2012;44(6):685–9.
51. Carter SL, Cibulskis K, Helman E, McKenna A, Shen H, Zack T, et al. Absolute quantification of somatic DNA alterations in human cancer. Nat Biotechnol. 2012;30(5):413–21.

52. Castellarin M, Milne K, Zeng T, Tse K, Mayo M, Zhao Y, et al. Clonal evolution of high-grade serous ovarian carcinoma from primary to recurrent disease. J Pathol. 2012;229:515–24.
53. Chang VY, Federman N, Martinez-Agosto J, Tatishchev SF, Nelson SF. Whole exome sequencing of pediatric gastric adenocarcinoma reveals an atypical presentation of Li-Fraumeni syndrome. Pediatr Blood Cancer. 2012;60:570–4.
54. Furney SJ, Turajlic S, Fenwick K, Lambros MB, MacKay A, Ricken G, et al. Genomic characterisation of acral melanoma cell lines. Pigment Cell Melanoma Res. 2012;25(4):488–92.
55. Furukawa T, Kuboki Y, Tanji E, Yoshida S, Hatori T, Yamamoto M, et al. Whole-exome sequencing uncovers frequent GNAS mutations in intraductal papillary mucinous neoplasms of the pancreas. Sci Rep. 2011;1:161.
56. Gartner J, Davis S, Wei X, Lin JC, Teer JK, Rosenberg SA, et al. Comparative exome sequencing of metastatic lesions provides insights into the mutational progression of melanoma. BMC Genomics. 2012;13(1):505.
57. Gerlinger M, Rowan AJ, Horswell S, Larkin J, Endesfelder D, Gronroos E, et al. Intratumor heterogeneity and branched evolution revealed by multiregion sequencing. N Eng J Med. 2012;366(10):883–92.
58. Guichard C, Amaddeo G, Imbeaud S, Ladeiro Y, Pelletier L, Maad IB, et al. Integrated analysis of somatic mutations and focal copy-number changes identifies key genes and pathways in hepatocellular carcinoma. Nat Genet. 2012;44(6):694–8.
59. Harbour JW, Onken MD, Roberson EDO, Duan S, Cao L, Worley LA, et al. Frequent mutation of BAP1 in metastasizing uveal melanomas. Science. 2010;330(6009):1410–3.
60. Harring TR, Guiteau JJ, Nguyen NTT, Cotton RT, Gingras M-C, Wheeler DA, et al. Building a comprehensive genomic program for hepatocellular carcinoma. World J Surg. 2011;35(8):1746–50.
61. Imielinski M, Berger AH, Hammerman PS, Hernandez B, Pugh TJ, Hodis E, et al. Mapping the hallmarks of lung adenocarcinoma with massively parallel sequencing. Cell. 2012;150(6):1107–20.
62. Kumar A, White TA, MacKenzie AP, Clegg N, Lee C, Dumpit RF, et al. Exome sequencing identifies a spectrum of mutation frequencies in advanced and lethal prostate cancers. Proc Natl Acad Sci U S A. 2011;108(41):17087–92.
63. Liu P, Morrison C, Wang L, Xiong D, Vedell P, Cui P, et al. Identification of somatic mutations in non-small cell lung carcinomas using whole-exome sequencing. Carcinogenesis. 2012;33(7):1270–6.
64. Lonigro RJ, Grasso CS, Robinson DR, Jing X, Wu Y-M, Cao X, et al. Detection of somatic copy number alterations in cancer using targeted exome capture sequencing. Neoplasia. 2011;13(11):1019–25.
65. Menon R, Deng M, Boehm D, Braun M, Fend F, Boehm D, et al. Exome enrichment and SOLiD sequencing of formalin fixed paraffin embedded (FFPE) prostate cancer tissue. Int J Mol Sci. 2012;13(7):8933–42.
66. Nikolaev SI, Rimoldi D, Iseli C, Valsesia A, Robyr D, Gehrig C, et al. Exome sequencing identifies recurrent somatic MAP2K1 and MAP2K2 mutations in melanoma. Nat Genet. 2012;44(2):133–9.
67. Ong CK, Subimerb C, Pairojkul C, Wongkham S, Cutcutache I, Yu W, et al. Exome sequencing of liver fluke-associated cholangiocarcinoma. Nat Genet. 2012;44(6):690–3.
68. Park DJ, Odefrey FA, Hammet F, Giles GG, Baglietto L, Abcfs, et al. FAN1 variants identified in multiple-case early-onset breast cancer families via exome sequencing: no evidence for association with risk for breast cancer. Breast Cancer Res Treat. 2011;130(3):1043–9.
69. Roberts NJ, Jiao Y, Yu J, Kopelovich L, Petersen GM, Bondy ML, et al. ATM mutations in patients with hereditary pancreatic cancer. Cancer Discov. 2012;2(1):41–6.
70. Roychowdhury S, Iyer MK, Robinson DR, Lonigro RJ, Wu Y-M, Cao X, et al. Personalized oncology through integrative high-throughput sequencing: a pilot study. Sci Transl Med. 2011;3(111):111ra21.

71. Rudin CM, Durinck S, Stawiski EW, Poirier JT, Modrusan Z, Shames DS, et al. Comprehensive genomic analysis identifies SOX2 as a frequently amplified gene in small-cell lung cancer. Nat Genet. 2012;44(10):1111–6.
72. Shi H, Moriceau G, Kong X, Lee M-K, Lee H, Koya RC, et al. Melanoma whole-exome sequencing identifies (V600E)B-RAF amplification-mediated acquired B-RAF inhibitor resistance. Nat Commun. 2012;3:724.
73. Spans L, Atak ZK, Van Nieuwerburgh F, Deforce D, Lerut E, Aerts S, et al. Variations in the exome of the LNCaP prostate cancer cell line. Prostate. 2012;72(12):1317–27.
74. Stark MS, Woods SL, Gartside MG, Bonazzi VF, Dutton-Regester K, Aoude LG, et al. Frequent somatic mutations in MAP3K5 and MAP3K9 in metastatic melanoma identified by exome sequencing. Nat Genet. 2012;44(2):165–9.
75. Stransky N, Egloff AM, Tward AD, Kostic AD, Cibulskis K, Sivachenko A, et al. The mutational landscape of head and neck squamous cell carcinoma. Science. 2011;333(6046):1157–60.
76. Thompson ER, Doyle MA, Ryland GL, Rowley SM, Choong DYH, Tothill RW, et al. Exome sequencing identifies rare deleterious mutations in DNA repair genes FANCC and BLM as potential breast cancer susceptibility alleles. PLoS Genet. 2012;8(9):e1002894.
77. Timmermann B, Kerick M, Roehr C, Fischer A, Isau M, Boerno ST, et al. Somatic mutation profiles of MSI and MSS colorectal cancer identified by whole exome next generation sequencing and bioinformatics analysis. PLoS One. 2010;5(12):e15661.
78. Totoki Y, Tatsuno K, Yamamoto S, Arai Y, Hosoda F, Ishikawa S, et al. High-resolution characterization of a hepatocellular carcinoma genome. Nat Genet. 2011;43(5):464–9.
79. Turajlic S, Furney SJ, Lambros MB, Mitsopoulos C, Kozarewa I, Geyer FC, et al. Whole genome sequencing of matched primary and metastatic acral melanomas. Genome Res. 2012;22(2):196–207.
80. Varela I, Tarpey P, Raine K, Huang D, Ong CK, Stephens P, et al. Exome sequencing identifies frequent mutation of the SWI/SNF complex gene PBRM1 in renal carcinoma. Nature. 2011;469(7331):539–42.
81. Wang K, Kan J, Yuen ST, Shi ST, Chu KM, Law S, et al. Exome sequencing identifies frequent mutation of ARID1A in molecular subtypes of gastric cancer. Nat Genet. 2011;43(12):1219–23.
82. Wang L, Tsutsumi S, Kawaguchi T, Nagasaki K, Tatsuno K, Yamamoto S, et al. Whole-exome sequencing of human pancreatic cancers and characterization of genomic instability caused by MLH1 haploinsufficiency and complete deficiency. Genome Res. 2012;22(2):208–19.
83. Wei X, Walia V, Lin JC, Teer JK, Prickett TD, Gartner J, et al. Exome sequencing identifies GRIN2A as frequently mutated in melanoma. Nat Genet. 2011;43(5):442–6.
84. Wu J, Jiao Y, Dal Molin M, Maitra A, de Wilde RF, Wood LD, et al. Whole-exome sequencing of neoplastic cysts of the pancreas reveals recurrent mutations in components of ubiquitin-dependent pathways. Proc Natl Acad Sci. 2011;108(52):21188–93.
85. Xiong D, Li G, Li K, Xu Q, Pan Z, Ding F, et al. Exome sequencing identifies MXRA5 as a novel cancer gene frequently mutated in non-small cell lung carcinoma from Chinese patients. Carcinogenesis. 2012;33(9):1797–805.
86. Xu X, Hou Y, Yin X, Bao L, Tang A, Song L, et al. Single-cell exome sequencing reveals single-nucleotide mutation characteristics of a kidney tumor. Cell. 2012;148(5):886–95.
87. Zang ZJ, Cutcutache I, Poon SL, Zhang SL, McPherson JR, Tao J, et al. Exome sequencing of gastric adenocarcinoma identifies recurrent somatic mutations in cell adhesion and chromatin remodeling genes. Nat Genet. 2012;44(5):570–4.
88. Zhao Q, Kirkness E, Caballero O, Galante P, Parmigiani R, Edsall L, et al. Systematic detection of putative tumor suppressor genes through the combined use of exome and transcriptome sequencing. Genome Biol. 2010;11(11):R114.
89. Park DJ, Lesueur F, Nguyen-Dumont T, Pertesi M, Odefrey F, Hammet F, et al. Rare mutations in XRCC2 increase the risk of breast cancer. Am J Hum Genet. 2012;90(4):734–9.

90. He M-L, Chen Y, Chen Q, He Y, Zhao J, Wang J, et al. Multiple gene dysfunctions lead to high cancer-susceptibility: evidences from a whole-exome sequencing study. Am J Cancer Res. 2011;1(4):562–73.
91. Ellis MJ, Ding L, Shen D, Luo J, Suman VJ, Wallis JW, et al. Whole-genome analysis informs breast cancer response to aromatase inhibition. Nature. 2012;486(7403):353–60.
92. Banerji S, Cibulskis K, Rangel-Escareno C, Brown KK, Carter SL, Frederick AM, et al. Sequence analysis of mutations and translocations across breast cancer subtypes. Nature. 2012;486(7403):405–9.
93. Snape K, Ruark E, Tarpey P, Renwick A, Turnbull C, Seal S, et al. Predisposition gene identification in common cancers by exome sequencing: insights from familial breast cancer. Breast Cancer Res Treat. 2012;134(1):429–33.
94. Liang H, Cheung LWT, Li J, Ju Z, Yu S, Stemke-Hale K, et al. Whole-exome sequencing combined with functional genomics reveals novel candidate driver cancer genes in endometrial cancer. Genome Res. 2012;22(11):2120–9.
95. Krauthammer M, Kong Y, Ha BH, Evans P, Bacchiocchi A, McCusker JP, et al. Exome sequencing identifies recurrent somatic RAC1 mutations in melanoma. Nat Genet. 2012;44(9):1006–14.
96. Zhou B, Irwanto A, Guo YM, Bei JX, Wu Q, Chen G, et al. Exome sequencing and digital PCR analyses reveal novel mutated genes related to the metastasis of pancreatic ductal adenocarcinoma. Cancer Biol Ther. 2012;13(10):871–9.
97. Network CGA. Comprehensive molecular characterization of human colon and rectal cancer. Nature. 2012;487(7407):330–7.
98. Network CGA. Comprehensive molecular portraits of human breast tumours. Nature. 2012;490(7418):61–70.
99. Davies H, Bignell GR, Cox C, Stephens P, Edkins S, Clegg S, et al. Mutations of the BRAF gene in human cancer. Nature. 2002;417(6892):949–54.
100. Samuels Y, Wang Z, Bardelli A, Silliman N, Ptak J, Szabo S, et al. High frequency of mutations of the PIK3CA gene in human cancers. Science. 2004;304(5670):554.
101. Lynch TJ, Bell DW, Sordella R, Gurubhagavatula S, Okimoto RA, Brannigan BW, et al. Activating mutations in the epidermal growth factor receptor underlying responsiveness of non–small-cell lung cancer to gefitinib. N Eng J Med. 2004;350(21):2129–39.
102. Sjöblom T, Jones S, Wood LD, Parsons DW, Lin J, Barber TD, et al. The consensus coding sequences of human breast and colorectal cancers. Science. 2006;314(5797):268–74.
103. Greenman C, Stephens P, Smith R, Dalgliesh GL, Hunter C, Bignell G, et al. Patterns of somatic mutation in human cancer genomes. Nature. 2007;446(7132):153–8.
104. Beltran H, Yelensky R, Frampton GM, Park K, Downing SR, Macdonald TY, et al. Targeted next-generation sequencing of advanced prostate cancer identifies potential therapeutic targets and disease heterogeneity. Eur Urol. 2012;63:920–6.
105. Chou L-S, Liu CSJ, Boese B, Zhang X, Mao R. DNA sequence capture and enrichment by microarray followed by next-generation sequencing for targeted resequencing: neurofibromatosis type 1 gene as a model. Clin Chem. 2010;56(1):62–72.
106. Duns G, Hofstra RMW, Sietzema JG, Hollema H, van Duivenbode I, Kuik A, et al. Targeted exome sequencing in clear cell renal cell carcinoma tumors suggests aberrant chromatin regulation as a crucial step in ccRCC development. Hum Mutat. 2012;33(7):1059–62.
107. Kalender Atak Z, De Keersmaecker K, Gianfelici V, Geerdens E, Vandepoel R, Pauwels D, et al. High accuracy mutation detection in leukemia on a selected panel of cancer genes. PLoS One. 2012;7(6):e38463.
108. Ozcelik H, Shi X, Chang MC, Tram E, Vlasschaert M, Di Nicola N, et al. Long-range PCR and next-generation sequencing of BRCA1 and BRCA2 in breast cancer. J Mol Diagn. 2012;14(5):467–75.
109. Thompson JF, Reifenberger JG, Giladi E, Kerouac K, Gill J, Hansen E, et al. Single-step capture and sequencing of natural DNA for detection of BRCA1 mutations. Genome Res. 2012;22(2):340–5.

110. Wagle N, Berger MF, Davis MJ, Blumenstiel B, Defelice M, Pochanard P, et al. High-throughput detection of actionable genomic alterations in clinical tumor samples by targeted, massively parallel sequencing. Cancer Discov. 2012;2(1):82–93.
111. Birney E, Stamatoyannopoulos JA, Dutta A, Guigó R, Gingeras TR, Margulies EH, et al. Identification and analysis of functional elements in 1% of the human genome by the ENCODE pilot project. Nature. 2007;447(7146):799–816.
112. Dunham I, Kundaje A, Aldred SF, Collins PJ, Davis CA, Doyle F, et al. An integrated encyclopedia of DNA elements in the human genome. Nature. 2012;489(7414):57–74.
113. Material SO, Web S, Press H, York N, Nw A. The ENCODE (ENCyclopedia Of DNA Elements) project. Science. 2004;306(5696):636–40.
114. Lee YS, Dutta A. MicroRNAs in cancer. Ann Rev Pathol. 2009;4(1):199–227.
115. Wyman SK, Parkin RK, Mitchell PS, Fritz BR, O'Briant K, Godwin AK, et al. Repertoire of microRNAs in epithelial ovarian cancer as determined by next generation sequencing of small RNA cDNA libraries. PLoS One. 2009;4(4):e5311.
116. Morin RD, O'Connor MD, Griffith M, Kuchenbauer F, Delaney A, Prabhu A-L, et al. Application of massively parallel sequencing to microRNA profiling and discovery in human embryonic stem cells. Genome Res. 2008;18(4):610–21.
117. Chmielecki J, Peifer M, Jia P, Socci ND, Hutchinson K, Viale A, et al. Targeted next-generation sequencing of DNA regions proximal to a conserved GXGXXG signaling motif enables systematic discovery of tyrosine kinase fusions in cancer. Nucleic Acids Res. 2010;38(20):6985–96.
118. Parkin DM. The global health burden of infection-associated cancers in the year 2002. Int J Cancer. 2006;118(12):3030–44.
119. Feng H, Taylor JL, Benos PV, Newton R, Waddell K, Lucas SB, et al. Human transcriptome subtraction by using short sequence tags to search for tumor viruses in conjunctival carcinoma. J Virol. 2007;81(20):11332–40.
120. MacConaill L, Meyerson M. Adding pathogens by genomic subtraction. Nat Genet. 2008;40(4):380–2.
121. Palacios G, Druce J, Du L, Tran T, Birch C, Briese T, et al. A new arenavirus in a cluster of fatal transplant-associated diseases. N Eng J Med. 2008;358(10):991–8.
122. Weber G, Shendure J, Tanenbaum DM, Church GM, Meyerson M. Identification of foreign gene sequences by transcript filtering against the human genome. Nat Genet. 2002;30(2):141–2.
123. Xu Y, Stange-Thomann N, Weber G, Bo R, Dodge S, David RG, et al. Pathogen discovery from human tissue by sequence-based computational subtraction. Genomics. 2003;81(3):329–35.
124. Barzon L, Militello V, Lavezzo E, Franchin E, Peta E, Squarzon L, et al. Human papillomavirus genotyping by 454 next generation sequencing technology. J Clin Virol. 2011;52(2):93–7.
125. Duncavage EJ, Magrini V, Becker N, Armstrong JR, Demeter RT, Wylie T, et al. Hybrid capture and next-generation sequencing identify viral integration sites from formalin-fixed, paraffin-embedded tissue. J Mol Diagn. 2011;13(3):325–33.
126. Feng H, Shuda M, Chang Y, Moore PS. Clonal integration of a polyomavirus in human Merkel cell carcinoma. Science. 2008;319(5866):1096–100.
127. Sauvage V, Foulongne V, Cheval J, Ar Gouilh M, Pariente K, Dereure O, et al. Human polyomavirus related to African green monkey lymphotropic polyomavirus. Emerg Infect Dis. 2011;17(8):1364–70.

Chapter 8
Comprehensive Genomic Alterations in Common Cancer Cell Lines Revealed by Exome Sequencing

Han Chang, Donald G. Jackson, Paul S. Kayne, Petra B. Ross-Macdonald, Rolf-Peter Ryseck, and Nathan O. Siemers

Abstract It is well established that genomic alterations play an essential role in oncogenesis, disease progression, and response of tumors to therapeutic intervention. The advances of next-generation sequencing technologies (NGS) provide unprecedented capabilities to scan genomes for changes such as mutations, deletions, and alterations of chromosomal copy number. However, the cost of full-genome sequencing still prevents the routine application of NGS in many areas. Capturing and sequencing the coding exons of genes (the "exome") can be a cost-effective approach for identifying changes that result in alteration of protein sequences. We applied an exome sequencing technology (Roche NimbleGen capture paired with 454 sequencing) to identify sequence variation and mutations in eight commonly used cancer cell lines from a variety of tissue origins (A2780, A549, Colo205, GTL16, NCI-H661, MDA-MB468, PC3, and RD). We showed that this technology can accurately identify sequence variation, providing ~95 % concordance with Affymetrix SNP Array 6.0 performed on the same cell lines. Furthermore, we detected 19 of the 21 mutations reported in Sanger COSMIC database for these cell lines. We identified an average of 2,779 potential novel sequence variations/mutations per cell line, of which 1,904 were non-synonymous. Many non-synonymous changes were identified in kinases and known cancer-related genes. In addition we confirmed that the read depth of exome sequence data can be used to estimate high-level gene amplifications and identify homologous deletions. In summary, we demonstrate that exome sequencing can be a reliable and cost-effective way for identifying alterations in cancer genomes, and we have generated a comprehensive catalogue of genomic alterations in coding regions of eight cancer cell lines. These findings could provide important insights into cancer pathways and mechanisms of resistance to anticancer therapies.

H. Chang (✉) • D.G. Jackson • P.S. Kayne
• P.B. Ross-Macdonald • R.-P. Ryseck • N.O. Siemers
Research and Development, Bristol-Myers Squibb Company,
P.O. Box 5400, HPW3A-0.07, Princeton, NJ 08543, USA
e-mail: Han.Chang@bms.com

W. Wu and H. Choudhry (eds.), *Next Generation Sequencing in Cancer Research: Volume 1: Decoding the Cancer Genome*, DOI 10.1007/978-1-4614-7645-0_8,

Keywords Exome sequencing • Cancer • Mutation • Genome • Cell line • Copy number

1 Introduction

All cancer cells have somatic mutations in their genomes, such as single-nucleotide mutations, insertions, deletions, and copy-number gain or loss. Genomic lesions in cancer cells disrupt normal functions and pathways such as proliferation and apoptosis and are essential for tumor genesis, growth, and metastasis. In addition, each tumor carries a unique combination of mutations in its genome, leading to heterogeneity in cancer prognosis and responses to therapeutic intervention. Our limited understanding of the more common mutations has already affected therapeutic regimens. For example, treatment with small-molecule inhibitors of the epidermal growth factor receptor (EGFR) has been shown to primarily benefit lung cancer patients that carry certain somatic mutations in their EGFR gene [1, 2]. Similarly, certain antibody therapies directed against EGFR only show efficacy in the subset of colorectal cancer patients with a wild-type KRAS gene [3, 4]. Deep systematic characterization of somatic mutations in cancer genomes promises to be a powerful tool for both understanding cancer pathways and developing targeted therapeutics.

Over the last two decades, focused studies on candidate genes have led to the identification of mutations occurring with high frequency in crucial cancer pathway genes such TP53, KRAS, and PTEN [5]. In recent years, the coding regions of breast, lung, colon, and brain tumor genomes have been analyzed using capillary-based sequencing technologies. These efforts have led to the identification of causative mutations in previously unsuspected genes such as IDH1, highlighting the power and importance of unbiased, genomic-scale mutation discovery [6–8]. However, large-scale capillary-based sequencing technologies are time consuming and expensive and thus not feasible for wider use.

Next-generation sequencing (NGS) technologies have increased the throughput and decreased the cost of DNA sequencing by several orders of magnitude. A number of studies have applied NGS technologies to sequence cancer genomes, as summarized in recent reviews [9, 10]. However, even with decreasing cost of NGS, sequencing the whole genome is still not cost-effective for many potentially valuable applications.

One alternative to whole-genome methods is exome sequencing, which captures and sequences only coding exons in the genome. Exome sequencing methods can deliver sequencing information for much of the functionally relevant genome at increased coverage and reduced cost. Recent studies have successfully applied exome sequencing to identify causal mutations of Mendelian diseases [11, 12]. Large cancer genome initiatives such as The Cancer Genome Atlas project also include exome sequencing as part of their strategy to characterize cancer genomes [7].

Protein kinases are the most ubiquitous family of signaling molecules in human cells and play essential roles in regulating most cellular functions [13]. Since the protein kinase family is one of the most frequently mutated gene families in cancers [5], it has been subjected to several focused genomic sequencing studies. Bardelli et al. conducted the first systematic screen of mutations in the receptor tyrosine kinase subfamily of protein kinases, in colorectal cancer samples [14]. Since then, studies in primary tissues and cell lines have identified many mutations in protein kinases across multiple tumor types [15–17]. The interest in mutations of kinases has continued with recent genome-wide mutation discovery studies [6–8].

Cell line models of human cancer have played a critical role in our understanding of cancer disease pathways, identification and validation of cancer target genes, and our ability to screen potential anticancer drugs. These cell lines carry genomic mutations inherited from their source tumor cells, although additional mutations can be acquired during the course of cell line development and passage. In general, comparisons between cell lines reveal substantial heterogeneity in genomic mutations and reflect cancer pathways similar to those found in primary tumors. For example, comparison of a panel of breast cancer cell lines with a collection of primary breast samples showed that gene expression and copy-number profiles in cell lines mirror those found in the primary tumors [18]. Similarly, genomic mutations reported in the COSMIC database for cell lines have a similar spectrum to those in primary tumors [19]. As additional large-scale tumor genome sequencing results become available, there is a growing need for corresponding cell models to determine how novel variants affect protein function. Comprehensive characterization of genomic alterations in cancer cell lines will advance our understanding of cancer biology and could also provide a basis for choosing relevant cell line models to study a particular aspect of cancer disease biology or to screen for antagonists of certain cancer pathways.

To evaluate NGS technologies and to characterize genomic mutations in cancer cell lines, we have analyzed data from the Roche NimbleGen exome capturing array and Roche 454 NGS technologies, applied to eight commonly used cell lines representing several major cancer types [20]. We demonstrate that exome sequencing can be a reliable and cost-effective way for identifying genomic alterations in cancer genome and generated a comprehensive catalogue of genomic alterations in coding regions of eight cancer cell lines.

2 Materials and Methods

2.1 DNA Preparation

A2780, A549, Colo205, GTL16, NCI-H661, MDA-MB468, PC3, and RD cell lines were originally obtained from ATCC. Cell lines were grown in RPMI 1640 (Gibco) with 10 % heat-inactivated fetal bovine serum (FBS; CellGro) with the exception of RD (additional 25 mM HEPES) and A549 (Ham's F12 (Gibco), with 10 % FBS). Genomic DNA (10 μg) was prepared by QIAamp DNA Mini Kit (Qiagen) using manufacturer's protocols and provided to the Roche 454 Sequencing Center.

2.2 Exome Capture and Next-Generation Sequencing

Exome capture and next-generation sequencing were performed by Roche NimbleGen and Roche 454 Life Science according to manufacturer's protocols. Genomic DNA was captured on the NimbleGen Capture Human Exome 2.1M Array, which has 197,218 total regions (capture regions) covering about 175,278 exons and miRNA regions (target regions, large target region may consist of several capture regions). For each cell line, captured DNA was sequenced with two runs of the 454 GS FLX Titanium Sequencing technology.

2.3 Array-Based Genotyping and Copy-Number Analysis

Two aliquots of 250 ng genomic DNA per sample were digested by restriction enzymes NspI and StyI, respectively. The resulted products were ligated to the corresponding adaptors and PCR amplified. The labeled PCR products were hybridized to the Affymetrix Genome-Wide Human SNP Array 6.0 according to the manufacturer's recommendations. The Birdseed algorithm [21] implemented in Affymetrix Power Tools (APT) Software Package (version 1.10.0) was used for genotype determination. For copy-number analysis, the Cel files were processed using the aroma.affymetrix package [22] for the R-project. Segmentation of normalized raw copy-number data was performed with the CBS algorithm [23] implemented in the aroma.affymetrix package.

2.4 Bioinformatics Analysis

The Human genome NCBI36/hg18 reference assembly (http://www.ncbi.nlm.nih.gov/genome/guide/human/release_notes.html#b36) was used as the framework for all analyses. Sequence data processing, mapping to the human genome, and initial calls of variation from the reference sequence were performed by Roche 454 Life Science using GS Reference Mapper software (Roche Inc.). To qualify as a variant from the reference genome sequence, there must be at least two independent reads that (1) show the difference, (2) have at least five bases on both sides of the difference, and (3) have few other isolated sequence differences in the read. Variants identified as "high confidence" were subject to a more stringent filter, requiring at least three independent reads with the variant comprising at least 40 % of all independent reads covering the allele genomic position. To identify non-synonymous variants, the impact of each variant on translated protein sequence was assessed by mapping its genomic coordinates back to genes in RefSeq collection [24] release 37 and identifying changes in codon specificity.

We calculated the theoretical rate of detection at heterozygous positions as a function of different read depth as follows: N sequencing reads covering a heterozygous position could be considered as random sampling of the two alleles repeated N times, thus should follow the binomial distribution. Assuming that allele A is reported in the human reference genome and allele B is the variant allele, we require at least two sequencing reads with the B allele for declaring the detection of allele B. The probability of detecting both A and B alleles at a heterozygous position can be calculated as follows: $PAB = 1 - P1 - P2$. P1 is the probability of finding 0 or 1 read with the A allele in N sequencing reads according to the binomial distribution, which would lead to a genotype call of AA. P2 is the probability of finding N reads with the B allele in N sequencing reads according to the binomial distribution, which will lead to a genotype call of BB.

3 Results

3.1 *Exome Capture and Sequencing Results*

Exome capture and 454 sequencing technologies were applied to DNA samples from eight cancer cell lines (A2780, A549, COLO205, GTL16, NCI-H661, MDA-MB468, PC3, and RD, as described in Sect. 2). The results of initial data processing are summarized in Table 8.1. For each cell line, about 1.9 million sequencing reads (688 million bases; 98.5 % of total sequencing reads) could be successfully mapped to the human genome NCBI36/hg18 reference assembly (http://www.ncbi.nlm.nih.gov). The average read length across all cell lines is 364 bases, consistent with the long read length reported for the 454 sequencing technology. On average, 89.5 % of the circa 180,000 exons on the NimbleGen 2.1M human exome array (target regions) were covered with at least one sequencing read, and the average sequencing read depth for all cell lines is 7.3 in target regions. The exome capture and sequencing results are within the normal range of performance specified by the manufacturer and are comparable with published results using the same technology [25].

We detected on average 14,340 sequence variants (differences from the human reference genome) per cell line. The majority of these differences are known polymorphisms in normal human population (i.e., recorded in NCBI dbSNP database, build 130). On average 2,779 variants per cell line are not found in the dbSNP database and therefore represent novel sequence variations and/or somatic mutations. On average, 1,904 of the 2,779 novel variants are non-synonymous, i.e., they alter codon specificity. These variants are more likely to change protein functions and impact cellular phenotypes. Since germline-matched normal cells are not available for cancer cell lines used in this study, detected sequence variants could be either due to somatic mutations in cancer genome or due to germline variations present in cancer patients. Recent publications reported results of two large-scale surveys of germline variations in noncancer population, the 1000 genome project (1000g) [26], and the NHLBI GO Exome Sequencing Project (esp6500) [27]. We downloaded

Table 8.1 Exome capture and sequencing results

Cell line	A2780	A549	COLO205	GTL16	NCI-H661	MDA-MB468	PC3	RD	Average
Cancer type	Ovary	Lung	Colon	Stomach	Lung	Breast	Prostate	Soft tissue	
Number of mapped reads (% total reads)	2,112,926 (98.29 %)	1,906,737 (98.87 %)	1,707,216 (98.39 %)	1,932,251 (98.42 %)	1,774,006 (98.55 %)	1,843,735 (98.36 %)	1,789,248 (98.42 %)	2,150,836 (98.76 %)	1,902,119 (98.5 %)
Number of mapped bases (% total bases)	751 Mb (99.14 %)	730 Mb (99.53 %)	613 Mb (99.11 %)	715 Mb (98.42 %)	641 Mb (99.41 %)	665 Mb (99.3 %)	657 Mb (99.4 %)	735 Mb (98.76 %)	688 Mb (99.1 %)
Average read length	355	382	360	368	360	359	365	366	364
Target regions coverage	87.6 %	90.8 %	90.8 %	89.4 %	91.7 %	85.6 %	87.1 %	92.6 %	89.5 %
Average read depth (target regions)	7.7	8.1	6.7	7.6	6.7	6.6	6.7	8.5	7.3
Total variant detected (target regions)	16,036	14,283	13,768	14,296	13,966	14,931	12,701	14,741	14,340
Novel variant detected (target regions)	3,563	2,769	2,075	3,111	2,759	3,021	2,150	2,786	2,779
Novel non-synonymous variant (target regions)/after remove new SNP	2,243/1,658	1,977/1,774	1,463/1,248	2,121/1,808	1,974/1,792	1,949/1,470	1,538/1,359	1,967/1,745	1,904/1,607

data from 1000g and esp6500 projects and compared to our results. As shown in the last row in Table 8.1, significant portions of detected novel variants are likely due to germline variations.

3.2 Concordance with Genotyping Results

As another means to assess the accuracy of exome sequencing, we compared the data with genotyping results across the eight cell lines (Table 8.2). The Affymetrix Genome-Wide Human SNP Array 6.0 is designed to detect genotype information for about one million known SNP positions. It can therefore provide independent verification of variations observed in the exome sequence data. For each cell line, we identified SNP Array 6.0 positions with successful genotype calls that were also covered by at least two unique exome sequencing reads. The overlap yielded between 26,407 and 29,650 SNP positions (depending on cell line) for further analysis. Overall, there was an average of 91 % concordance between genotype calls from SNP Array 6.0/Birdseed and those determined by exome sequencing. In the RD cell line, for example, 26,154 (91.5 %) out of 28,594 SNP positions have the same genotype call (i.e., AA, AB, or BB) by SNP Array 6.0 and by exome sequencing (Table 8.2).

It is expected that the accuracy of genotype detection by sequencing will be influenced both by sequencing read depth and by heterozygosity at a given genomic location. We calculated concordance of genotype calls at difference sequencing read depth and separately for homozygous or heterozygous SNPs. As shown in Fig. 8.1, concordance is high for homozygous SNPs (average 97 %) regardless of sequencing read depth. Concordance for heterozygous alleles is lower, but increases with sequence read depth, starting with 31 % concordance at a read depth of 3 and reaching >90 % at a read depth of 10 or higher. In theory, sequencing DNA fragments from a region that contains a heterozygous SNP is a process of random sampling. At lower sequencing depth, there is a higher chance of missing one of the two alleles. We calculated the theoretical rate of detecting both alleles by sequencing at different read depths, assuming no error in sequencing (Fig. 8.1, dashed line). At low read depths, our experimental observations are close to the theoretical rate, indicating that low concordance at low read depths is likely due to the random sampling process rather than poor quality of sequence data.

3.3 Comparison of Exome Sequencing to the COSMIC Database of Cancer Mutations

The protein-coding exons and immediate flanking intron sequences of 61 common cancer genes have previously been systematically determined in about 800 cell lines by the Wellcome Trust Sanger Institute, using capillary-based sequencing [19].

Table 8.2 Comparison of genotype calls by SNP 6 chip and exome sequencing (Reproduced from Chang et al. 2011 with permission from PLOS ONE)

	All SNP			Homozygous SNP			Heterozygous SNP		
Cell line	All SNP	Same genotype call	Percent	Homozygous SNP	Same genotype call	Percent	Heterozygous SNP	Same genotype call	Percent
A2780	29,193	25,493	87 %	22,079	21,559	98 %	7,114	3,934	55 %
A549	26,407	24,065	91 %	22,627	22,188	98 %	3,780	1,877	50 %
Colo205	27,638	24,780	90 %	23,526	22,797	97 %	4,112	1,983	48 %
GTL16	29,650	27,474	93 %	26,447	25,691	97 %	3,203	1,783	56 %
NCI-H661	29,117	26,901	92 %	26,737	25,756	96 %	2,380	1,145	48 %
MDA-MB468	29,360	27,088	92 %	25,914	25,221	97 %	3,446	1,867	54 %
PC3	27,914	25,505	91 %	24,421	23,729	97 %	3,493	1,776	51 %
RD	28,594	26,154	91 %	24,616	23,870	97 %	3,978	2,284	57 %
Average	28,484	25,933	91 %	24,546	23,851	97 %	3,938	2,081	52 %

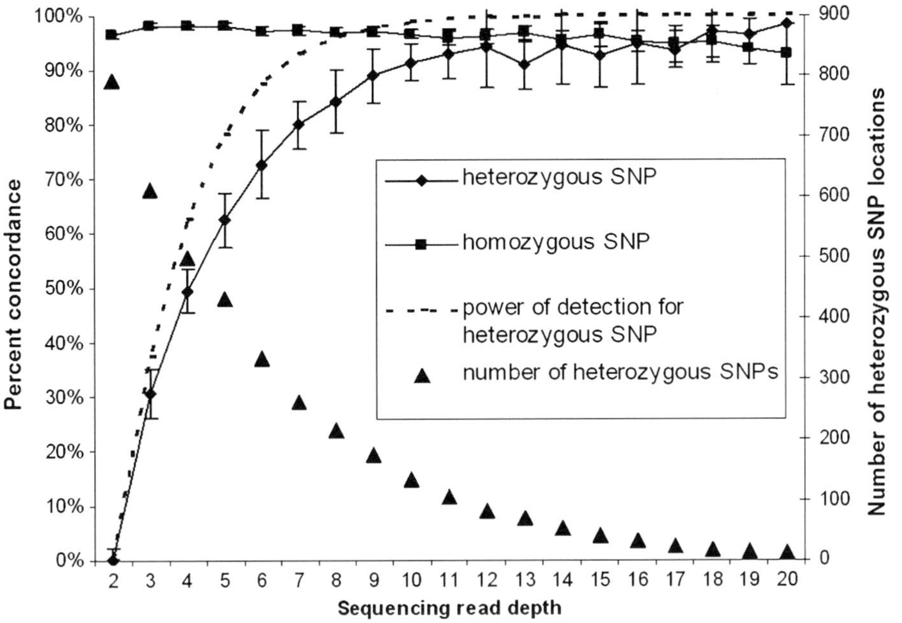

Fig. 8.1 Sequencing depth and concordance between genotype. The graph displays a plot of average concordance of genotype calls obtained from the Affymetrix SNP Array 6.0 and from exome sequencing, as a function of sequencing read depths. *Square markers* indicate concordance at homozygous positions; *diamond markers* indicate concordance at heterozygous positions. The *dashed line* shows the theoretical rate of detecting heterozygous positions by sequencing (as described in Sect. 2). *Triangle markers* display average number of heterozygous SNP locations per cell line as a function of sequencing read depths (*Y*-axis on the *right*) (Modified from Chang et al. 2011 with permission from PLOS ONE)

Of the eight cell lines in this study, all except one (GTL16) have been screened in that project. We compared somatic mutation information from the Sanger COSMIC database with our exome sequencing results for the seven cell lines. As shown in Table 8.3, exome sequencing rediscovered most of the 21 mutations reported in the COSMIC database, including point mutations and small insertion/deletions. The two missing cases are due to lack of sequence coverage in the locus of interest: the documented STK11 mutation in A549 is not measurable due to lack of STK11 gene coverage in the NimbleGen 2.1M human exome arrays, and the TP53 gene is covered by the NimbleGen array but lacks sufficient reads in the PC3 line to verify in this study (there are sufficient reads for the TP53 gene in other lines, as in Table 8.3).

Large homozygous deletions, such as the known deletions of the CDKN2A gene in A549 and SMAD4 in Colo205 cells, cannot be directly observed with exome sequencing. But a deletion of gene regions can be inferred where the read depth is zero for several consecutive exons (see next section for detailed discussion). All five genomic deletions reported in the COSMIC database are identifiable from exome

Table 8.3 Comparison of exome sequencing results with mutations reported in the COSMIC database (Reproduced from Chang et al. 2011 with permission from PLOS ONE)

Cell line	Gene	COSMIC report	Exome sequencing (reference allele reads to variant allele reads)	Notes
A2780	PTEN	KGR128-130 del (Hom)	KGR 128-130 del (0:3)	
A549	KRAS	G12S (Hom)	G12S (0:5)	
A549	SMARCA4	Q729fs (23 bp del) (Hom)	Q729fs (23 bp del) (0:4)	
A549	CDKN2A	Large region deletion (Hom)	Large region deletion (Hom)	Zero read depth in 14 consecutive regions
A549	STK11	Q37* (Hom)	Not in exome capture array	Gene not in exome capture array
Colo205	BRAF	V600E (Het)	V600E (1:2)	
Colo205	TP53	26 bp del (Hom)	26 bp del (0:4)	
Colo205	SMAD4	904 bp del (Hom)	Large region deletion (Hom)	Zero read depth in four consecutive regions
Colo205	APC	T1556fs (insertion A) (Hom)	T1556fs (insertion A) (6:5)	Within a stretch of 6 A
NCI-H661	TP53	R158L (Hom)	R158L (5:2)	
NCI-H661	TP53	S215I (Het)	S215I (2:4)	
NCI-H661	SMARCA4	L1161fs (deletion G) (Hom)	L1161fs (deletion G) (0:6)	Within a stretch of 6 G
NCI-H661	CDKN2A	chr9_21960900 G->T (Hom)	chr9_21960900 G->T (0:7)	Splicing site
MDA-MB468	PTEN	chr10_89680827 G->T (Hom)	chr10_89680827 G->T (0:9)	Splicing site
MDA-MB468	TP53	R273H (Hom)	R273H (0:4)	
MDA-MB468	RB1	Large region deletion (Hom)	Large region deletion (Hom)	Zero read depth in 35 consecutive regions
MDA-MB468	SMAD4	Large region deletion (Hom)	Large region deletion (Hom)	Zero read depth in 16 consecutive regions
PC3	PTEN	Large deletion (Hom)	Large region deletion (Hom)	Zero read depth in 34 consecutive regions
PC3	TP53	K139fs (Hom)	No sequencing read	No sequencing read
RD	NRAS	Q61H (Hom)	Q61H (5:10)	
RD	TP53	R248H (Hom)	R248H (Hom)	

Q37* mutation means that this is a Glutamine (Q) to stop codon (*) mutation

sequencing results (Table 8.3). For example, in the A549 cell line, we observed 14 consecutive regions around CDKN2A gene with a read depth of zero. In the Colo205 cell line, a documented 904-base deletion in the SMAD4 gene manifests as four consecutive target regions with a read depth of zero.

3.4 Detecting Gene Amplification and Deletion

Deletions or amplifications of chromosomal segments are common alterations in cancer genomes. In principle, the sequencing read depth in a region should be proportional to its copy number. However, the relatively modest read depth of the current study could give undue weight to random variations in read depth. Variability in read depth could also arise from technical aspects of the exome sequencing process. For example, the exome capturing array could vary in efficiencies for different exon regions due to diverse sequence composition. To assess the possibility of estimating copy-number information from our exome sequencing data, we compared average sequence read depths with copy-number data estimated from SNP6 platform. As shown in Fig. 8.2, there is a positive correlation between sequence read depth and copy number, with Pearson correlation coefficient of 0.41. The variation in read depth makes it challenging to accurately detect low-level copy-number changes. On the other hand, we find that accurate detection of high-level gene amplifications and homozygous deletions is possible.

Homozygous deletion of the SMAD4 gene region has been reported in the MDA-MB468 cell line (Sanger COSMIC database) and is thus illustrative for comparing deletion detection methods. The sequencing read depths of exon regions in SMAD4 gene and surrounding area were determined for MDA-MB468 and plotted according to their chromosomal location (Fig. 8.3a). Sixteen consecutive exon regions on chromosome 18 have a read depth of zero in the data for MDA-MB468. The genomic locations of the 16 exon regions are from 46.75 to 46.86 MB, which spans the SMAD4 gene. For comparison, we performed copy-number analysis of the Affymetrix SNP Array 6.0 data as described in Sect. 2. For MDA-MB468, this analysis indicated a homozygous deletion of genomic region 46.76–46.86 Mb on chromosome 18 (Fig. 8.3b), in good agreement with results from read depth analysis.

A read depth of zero could result from technical issues, such as probe design in the NimbleGen 2.1M array. In fact, we identified 2,513 exon regions that have a read depth of zero for all eight cell lines. However, since the median read depth across all eight cell lines is greater than zero for all of the 16 exon regions (Fig. 8.3a), it is unlikely that the observed depth of zero in the MDA-MB468 cell line is due to a systematic failure of exome capture. Random variation in read depth is another reason for lack of sequencing coverage. In the MDA-MB468 cell line, there are 17,161 exon regions with a read depth of zero (from 194,706 total regions, excluding the 2,513 regions mentioned above). It is highly unlikely that 16 consecutive exon regions around SMAD4 gene would have a read depth of zero due to random variation ($p = 1.3e{-17}$, calculated from the binomial distribution).

We were also able to reidentify previously documented gene amplification events using the read depth data. For example, amplification of EGFR1 in the MDA-MB468 cell line has been documented by fluorescence in situ hybridization and by quantitative PCR [28]. We observed that the 53 exon regions around the EGFR gene on chromosome 7 have very high read depths in the MDA-MB468 data (Fig. 8.4a; the exons between 55.58 and 55.73 Mb have an average read depth of 107). Our

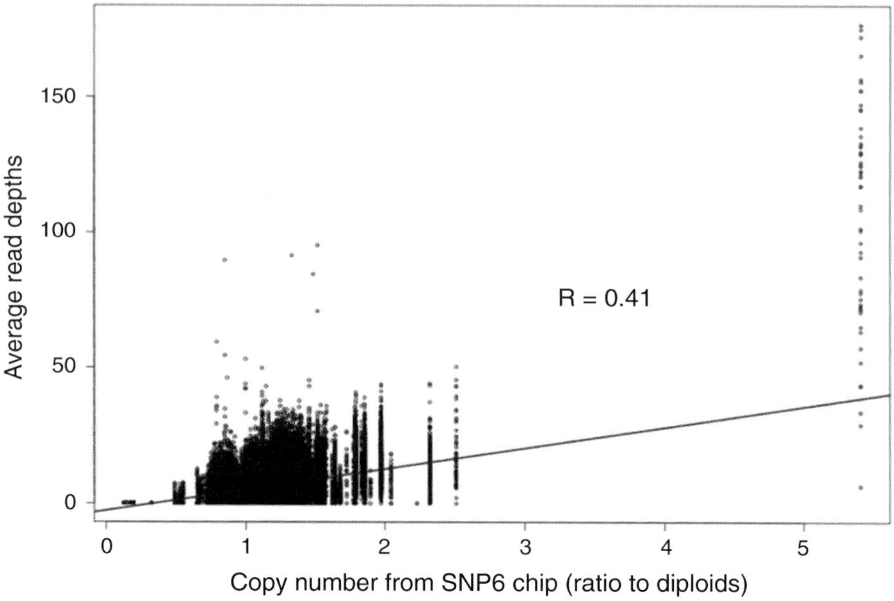

Fig. 8.2 Comparison of sequencing read depth with copy-number data in MDA-MB468 cell line. Average sequencing read depths in capture regions were plotted against copy-number data estimated from Affymetrix SNP 6.0 data as described in Sect. 2. The *line* shows the linear regression line. The Pearson correlation coefficiency ($r = 0.41$) of sequencing read depth and copy-number data is printed on the figure (Modified from Chang et al. 2011 with permission from PLOS ONE)

copy-number analysis of the Affymetrix SNP Array 6.0 data also indicated that the EGFR gene region is highly amplified in the MDA-MB468 line (Fig. 8.4b, genomic region 55.48–55.81 Mb).

3.5 Novel Non-synonymous Variants in Protein Kinases

Since mutations in protein kinases have important roles in cancer biology, we chose to examine the sequence data for protein kinases and focus on non-synonymous variations, which produce amino acid substitutions that may have functional consequences. As noted above, exome sequencing revealed circa 2,000 novel non-synonymous variants in each of the eight cell lines. After applying a stringent filter (as described in Sect. 2), between 199 and 479 genes have novel non-synonymous variants, depending on the cell line. The NimbleGen 2.1M capture array used in this study included exons for 440 of the 518 protein kinases in the human genome [13]. In each cell line, an average of 122 non-synonymous variations was detected in kinase genes. After removing likely germline variants (found in dbSNP) and applying a stringent filter described above, each cell line

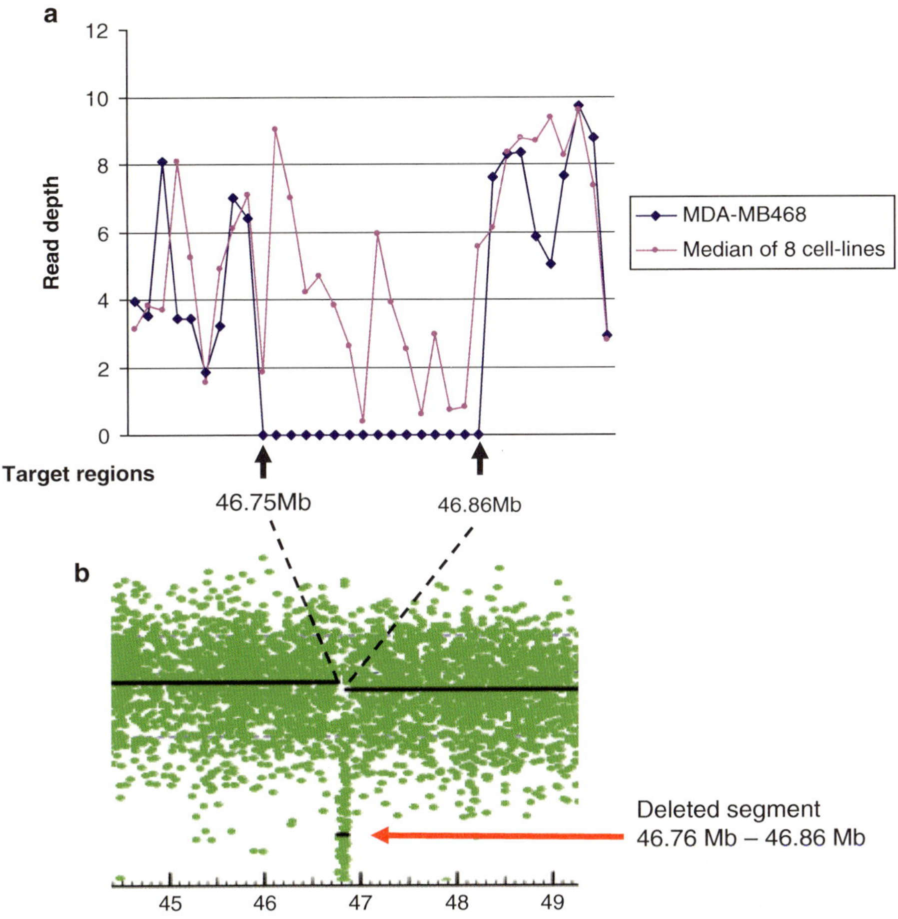

Fig. 8.3 Sequencing read depth around the SMAD4 gene on chromosome 18 in the MDA-MB468 cell line. (**a**) Plots of read depth data on consecutive exons around the SMAD4 gene region on chromosomal 18. The *blue line* shows sequencing read depth data for MDA-MB468, and the *pink line* shows the median sequencing read depth of all eight cell lines. (**b**) Copy-number data from Affymetrix SNP6 chip data around the SMAD4 gene region on chromosomal 18. The *black line* shows the segmented copy-number data (log_2 ratio to normal samples) generated by the aroma. affymetrix package in R as described in Sect. 2 (Reproduced from Chang et al. 2011 with permission from PLOS ONE)

has an average of eight kinases with non-synonymous variations. These sequence variations in protein kinases are listed in Table 8.4. Most of these sequence variations are not reported in the COSMIC database or reported in the literature, but several have independent confirmation. For example, we identified EGFR variant A1048V in the GTL16 gastric cell line. The same variant in EGFR has been reported in the MKN45 gastric cell line [29], which is the parental cell line

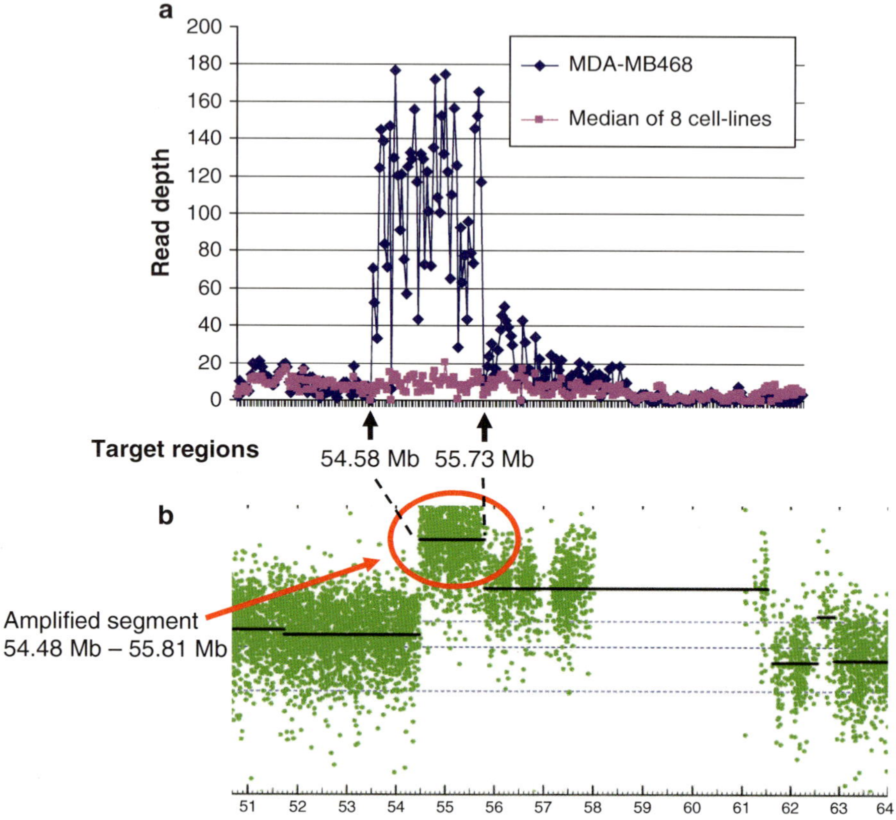

Fig. 8.4 Sequencing read depth and amplification around the EGFR gene on chromosome 7 in the MDA-MB468 cell line. (**a**) Plots of read depth data on consecutive exons around the EGFR gene region on chromosomal 7. The *blue line* shows sequencing read depth data for MDA-MB468, and the *pink line* shows the median sequencing read depth of all eight cell lines. (**b**) Copy-number data from Affymetrix SNP6 chip data around the EGFR gene region on chromosomal 7. The *black line* shows the segmented copy-number data (log_2 ratio to normal samples) generated by the aroma. affymetrix package in R as described in Sect. 2 (Reproduced from Chang et al. 2011 with permission from PLOS ONE)

of GTL16 [30]. A second example is the R796S variant of the insulin receptor gene (INSR) in the RD cell line (Table 8.4). We had previously identified this variant in the RD cell line using capillary sequencing technology (data not shown). Sequence variants were further compared to recently published germline polymorphisms from 1000g and esp6500 projects. Sequence variants that match data from 1000g or esp6500 projects were labeled with "(g)" in Table 8.4, since they are likely to be germline variations.

Table 8.4 High-confidence[a] non-synonymous variants in protein kinase genes in each of eight cell lines

A2780		A549		Colo205		GTL16	
ALPK2	721fs	ALPK2	G286C(g)	ADRBK1	443fs	EGFR	A1048V
EPHA2	D232G	BRD3	K508-	EPHA2	R315Q(g)	MAP3K4	395fs
FLT3	Q771P(g)	HIPK3	D713G	EPHA8	L559F(g)	PSKH2	E227G
HIPK3	G586R(g)	MKNK1	406fs	MTOR	P1193L	STK31	S160T
HSPB8	E179Q(g)	PAK6	302fs	LRRK2	G1213S		
LATS2	D1013N(g)	SGK196	169fs	NEK9	D84N(g)		
MAP3K5	M375V	ZAP70	I342T(g)	NTRK2	A203T(g)		
MYO3A	248fs			RNASEL	G59S(g)		
TRPM6	C943G			TAOK2	A867V		
				TTN	S597C		
H661		MB468		PC3		RD	
ATR	2428fs	ADCK2	R390C(g)	BCR	316fs	INSR	R796S(g)
CAMKV	A475D	CAMK4	R338C	DAPK2	G223R(g)	MYLK3	747fs
ICK	KPSP368 del	CDKL2	T471I	PRKCA	398fs	PHKG1	M220T
IRAK3	426fs	CSNK1G1	260fs	TRRAP	R696H(g)	PTK2B	D424Y(g)
JAK3	E698K	DDR2	V824A(g)	TYRO3	E489K(g)	STK33	S504SGS
LRRK2	1638fs; Q1648P	GAK	L1218R(g)	ULK1	T452S(g)		
NTRK2	K308Q	MAP4K3	A163G				
PAK6	Y471C	RIPK3	R422[a](g)				
ROS1	A1443S						
SIK1	615fs						
STK33	S469[a]						
TEC	P587L						
TRIO	R1593W						
TSSK4	T45S						
ZAK	G23C						

Variants that match 1000genome or ESP6500 genome SNV are labeled with (g)

[a]At least three reads show the variation and allele frequency for the variant ≥40 %

4 Discussion

Analysis of data from eight diverse cancer cell lines shows that Roche NimbleGen and 454 exome sequencing technologies can be successfully applied to identify variations in gene-coding regions. From sequencing data with an average of 7.3-fold coverage, variants from the NCBI36 reference genome were identified in about 8 % (14,340 regions) of all target regions on the exome capture array. While the majority of these variants could be confirmed in dbSNP database, on average, 0.16 % (2,779) of total target regions carry a novel variant.

A comparison of SNP genotype calls from exome sequencing with data generated on the Affymetrix Genome-Wide Human SNP Array 6.0 showed that there is high concordance between the two technology platforms. The concordance is 97 %

for homozygous sites and ranges from 30 to >90 % at heterozygous positions, with accuracy dependent on sequencing read depth. Our analysis of the relationship between read depth and power of detection suggested that a minimum of tenfold read depth is required for reliably detecting both alleles at heterozygous sites. These results provide guidance in planning future genome sequencing projects.

For the seven examined cell lines that are also present in the COSMIC database, we show that 19 of 21 known mutations can be rediscovered by exome sequencing. Two previously described mutations were missing due to lack of sequence coverage. In one case this was due to incomplete coverage of the human exome in the NimbleGen 2.1M capture array. Recent versions of exome capture arrays have much better coverage of human exome.

By successful reidentification of the EGFR amplification and the SMAD4 homozygous deletion in the MDA-MB468 cell line, we demonstrate that copy-number alterations can be inferred from the sequencing read depth data. However, because of the stochastic nature of sequencing read depth and likely unevenness in the exome capturing process, in general it is not possible to reliably estimate copy-number information from our data. Applying the technology to more samples would help improve our ability to estimate and correct for systematic biases in the platform, and increasing the depth of sequencing reads would reduce the variance due to random fluctuation in read number.

To bring context to the genomic variation identified in this study, we chose to focus on protein kinases as an illustrative class. In this work, we identified with high confidence at least four novel variant protein kinases in each cell line. Most of the novel sequence variations in protein kinases identified in this study have not previously been reported and probably reflect the high diversity of genomic alteration in cancer. Our results expand the knowledge of sequence variations in protein kinases and other potential cancer-related genes. These novel variants could be either germline SNPs not yet reported in the dbSNP database or somatic mutations in these cancerous cells. Comparison with data from two large-scale human genome sequencing projects (1000g and esp6500) suggested that significant portions of novel variants are likely to be low-frequency germline variations. Because of the abundance of rare genetic variants in normal population [27], distinguishing somatic mutations from unknown germline variations will remain a challenge for cancer cell lines (or any cancer sample without matched normal cells) in the future.

In conclusion, we showed that exome sequencing can be a reliable and cost-effective approach to identify genomic alterations in cancer cell lines and suggest ways to further improve exome sequencing technologies for applications in cancer genomics. A comprehensive catalogue of genomic alterations in the coding regions of eight cancer cell lines was generated, which should contribute not only to our knowledge of these models in particular but also to our understanding of cancer genomics and cancer biology in general.

Acknowledgments We thank Charles Tilford and Jansen Lim for their help in providing genomic coordinates for genes in the human genome. We also thank Roche NimbleGen and Roche 454 Life Science for performing exome capture, 454 GS FLX Titanium Sequencing runs, and initial data processing.

References

1. Paez JG, Janne PA, Lee JC, Tracy S, Greulich H, Gabriel S, et al. EGFR mutations in lung cancer: correlation with clinical response to gefitinib therapy. Science. 2004;304(5676):1497–500.
2. Lynch TJ, Bell DW, Sordella R, Gurubhagavatula S, Okimoto RA, Brannigan BW, et al. Activating mutations in the epidermal growth factor receptor underlying responsiveness of non-small-cell lung cancer to gefitinib. N Engl J Med. 2004;350(21):2129–39.
3. Khambata-Ford S, Garrett CR, Meropol NJ, Basik M, Harbison CT, Wu S, et al. Expression of epiregulin and amphiregulin and K-ras mutation status predict disease control in metastatic colorectal cancer patients treated with cetuximab. J Clin Oncol. 2007;25(22):3230–7.
4. Lievre A, Bachet J-B, Le Corre D, Boige V, Landi B, Emile J-F, et al. KRAS mutation status is predictive of response to cetuximab therapy in colorectal cancer. Cancer Res. 2006;66(8):3992–5.
5. Futreal PA, Coin L, Marshall M, Down T, Hubbard T, Wooster R, et al. A census of human cancer genes. Nat Rev Cancer. 2004;4(3):177–83.
6. Jones S, Zhang X, Parsons DW, Lin JC-H, Leary RJ, Angenendt P, et al. Core signaling pathways in human pancreatic cancers revealed by global genomic analyses. Science. 2008;321(5897):1801–6.
7. Parsons DW, Jones S, Zhang X, Lin JC-H, Leary RJ, Angenendt P, et al. An integrated genomic analysis of human glioblastoma multiforme. Science. 2008;321(5897):1807–12.
8. Wood LD, Parsons DW, Jones S, Lin J, Sjöblom T, Leary RJ, et al. The genomic landscapes of human breast and colorectal cancers. Science. 2007;318(5853):1108–13.
9. Meyerson M, Gabriel S, Getz G. Advances in understanding cancer genomes through second-generation sequencing. Nat Rev Genet. 2010;11(10):685–96.
10. Robison K. Application of second-generation sequencing to cancer genomics. Brief Bioinform. 2010;11(5):524–34.
11. Choi M, Scholl UI, Ji W, Liu T, Tikhonova IR, Zumbo P, et al. Genetic diagnosis by whole exome capture and massively parallel DNA sequencing. Proc Natl Acad Sci. 2009;106(45):19096–101.
12. Ng SB, Buckingham KJ, Lee C, Bigham AW, Tabor HK, Dent KM, et al. Exome sequencing identifies the cause of a mendelian disorder. Nat Genet. 2010;42(1):30–5.
13. Manning G, Whyte DB, Martinez R, Hunter T, Sudarsanam S. The protein kinase complement of the human genome. Science. 2002;298(5600):1912–34.
14. Bardelli A, Parsons DW, Silliman N, Ptak J, Szabo S, Saha S, et al. Mutational analysis of the tyrosine kinome in colorectal cancers. Science. 2003;300(5621):949.
15. Greenman C, Stephens P, Smith R, Dalgliesh GL, Hunter C, Bignell G, et al. Patterns of somatic mutation in human cancer genomes. Nature. 2007;446(7132):153–8.
16. Stephens P, Edkins S, Davies H, Greenman C, Cox C, Hunter C, et al. A screen of the complete protein kinase gene family identifies diverse patterns of somatic mutations in human breast cancer. Nat Genet. 2005;37(6):590–2.
17. Davies H, Hunter C, Smith R, Stephens P, Greenman C, Bignell G, et al. Somatic mutations of the protein kinase gene family in human lung cancer. Cancer Res. 2005;65(17):7591–5.
18. Neve RM, Chin K, Fridlyand J, Yeh J, Baehner FL, Fevr T, et al. A collection of breast cancer cell lines for the study of functionally distinct cancer subtypes. Cancer Cell. 2006;10(6):515–27.
19. Forbes SA, Tang G, Bindal N, Bamford S, Dawson E, Cole C, et al. COSMIC (the Catalogue of Somatic Mutations in Cancer): a resource to investigate acquired mutations in human cancer. Nucleic Acids Res. 2010;38(Database issue):D652–7.
20. Chang H, Jackson DG, Kayne PS, Ross-Macdonald PB, Ryseck RP, Siemers NO. Exome sequencing reveals comprehensive genomic alterations across eight cancer cell lines. PLoS One. 2011;6(6):e21097 [Research Support, Non-U.S. Gov't].
21. Korn JM, Kuruvilla FG, McCarroll SA, Wysoker A, Nemesh J, Cawley S, et al. Integrated genotype calling and association analysis of SNPs, common copy number polymorphisms and rare CNVs. Nat Genet. 2008;40(10):1253–60.

22. Bengtsson H, Wirapati P, Speed TP. A single-array preprocessing method for estimating full-resolution raw copy numbers from all Affymetrix genotyping arrays including GenomeWideSNP 5 & 6. Bioinformatics. 2009;25(17):2149–56.
23. Olshen AB, Venkatraman ES, Lucito R, Wigler M. Circular binary segmentation for the analysis of array-based DNA copy number data. Biostatistics. 2004;5(4):557–72.
24. Pruitt KD, Tatusova T, Maglott DR. NCBI reference sequences (RefSeq): a curated non-redundant sequence database of genomes, transcripts and proteins. Nucleic Acids Res. 2007;35(Database issue):D61–5.
25. Hedges DJ, Burges D, Powell E, Almonte C, Huang J, Young S, et al. Exome sequencing of a multigenerational human pedigree. PLoS One. 2009;4(12):e8232.
26. Abecasis GR, Auton A, Brooks LD, DePristo MA, Durbin RM, Handsaker RE, et al. An integrated map of genetic variation from 1,092 human genomes. Nature. 2012;491(7422):56–65 [Research Support, N.I.H., Extramural Research Support, Non-U.S. Gov't].
27. Fu W, O'Connor TD, Jun G, Kang HM, Abecasis G, Leal SM, et al. Analysis of 6,515 exomes reveals the recent origin of most human protein-coding variants. Nature. 2013;493(7431):216–20. doi:10.1038/nature11690.
28. Agelopoulos K, Greve B, Schmidt H, Pospisil H, Kurtz S, Bartkowiak K, et al. Selective regain of egfr gene copies in CD44+/CD24-/low breast cancer cellular model MDA-MB-468. BMC Cancer. 2010;10:78.
29. Kimura T, Maesawa C, Ikeda K, Wakabayashi G, Masuda T. Mutations of the epidermal growth factor receptor gene in gastrointestinal tract tumor cell lines. Oncol Rep. 2006;15(5):1205–10.
30. Rege-Cambrin G, Scaravaglio P, Carozzi F, Giordano S, Ponzetto C, Comoglio PM, et al. Karyotypic analysis of gastric carcinoma cell lines carrying an amplified c-met oncogene. Cancer Genet Cytogenet. 1992;64(2):170–3.

Chapter 9
Application of Next-Generation Sequencing in RNA Biomarker Discovery in Cancer Research

Stephen P. Fink and Kishore Guda

Abstract The advent of next-generation sequencing (NGS) technology has opened up a plethora of possibilities in cancer research by allowing for an unprecedented characterization of the cancer genome. The sensitivity, broad dynamic range, speed, and reduced cost per sample make the NGS technology a highly attractive platform in biomedical research compared to other sequencing and expression profiling techniques. NGS is being currently employed in several malignancies for both quantitative and qualitative profiling of nucleic acids and has already uncovered novel genetic determinants that play an important role during tumor development. In particular, the use of NGS technology for profiling the transcriptome from tumor tissues and body fluids has led to the identification of novel molecular targets that could potentially be translated in the clinic as diagnostic, prognostic, and therapeutic biomarkers. Currently, efforts are also being undertaken in the clinic to characterize an individual's cancer genome for guiding evidence-based molecular therapies tailored for individual patients. In this chapter, we review recent advances in the use of NGS technology for RNA-based biomarker studies in cancer and its potential implications in the overall management of the disease.

Keywords miRNA • *lnc*RNA • mRNA • Cancer • miR-31 • miR-375 • Sequencing • Transcriptome • Biomarker • Barrett's • Esophagus • Diagnostic • Prognostic • Expression • Fusion

---

S.P. Fink, Ph.D.
Division of Hematology and Oncology, Case Western Reserve University Comprehensive Cancer Center, School of Medicine, 2103, Cornell Rd, WRB 3101, Cleveland, OH 44106, USA
e-mail: sxf14@case.edu

K. Guda, D.V.M., Ph.D. (✉)
Division of General Medical Sciences (Oncology), Case Western Reserve University Comprehensive Cancer Center, School of Medicine, 2103, Cornell Rd, WRB 3143, Cleveland, OH 44106, USA
e-mail: gkishore@yahoo.com

W. Wu and H. Choudhry (eds.), *Next Generation Sequencing in Cancer Research: Volume 1: Decoding the Cancer Genome*, DOI 10.1007/978-1-4614-7645-0_9,

Cancer is a worldwide health problem that is responsible for many disease-related deaths each year. Many cancers are detected at an advanced stage where treatment options are limited leading to poor survival probability. It is therefore critical to identify risk factors associated with tumor development and to develop effective screening strategies for early cancer detection when the disease is most curable. Understanding the molecular mechanisms involved in the initiation and progression of tumors helps not only in identifying the causative factors associated with tumor development but also in identifying potential biomarkers and novel therapeutic targets for respective malignancies. In particular, given that genome-wide expression changes in RNAs are widespread in almost all cancer types, high-throughput technologies such as microarrays have enabled the assessment of RNA expression levels comprehensively in patient samples and are both time- and cost-effective. The resulting patterns of RNA expression or "signatures" can be of potential clinical value as diagnostic, prognostic, and therapeutic biomarkers, a possibility that is being rigorously tested in various malignancies. Some of these RNA expression signatures are already being utilized in the clinic as prognostic markers in combination with other clinical and pathological features to guide treatment strategies for specific disease contexts. For example, MammaPrint and Onco*type* DX are two commercial gene expression-based molecular assays widely utilized by clinical oncologists to assess the prognosis and benefits of chemotherapy in a subgroup of breast cancer patients [1]. Recently, an 18-gene expression signature panel termed as ColoPrint was shown to improve the overall prognostic accuracy in patients with stage II and III colon cancers and further classifies patients with stage II cancers who do not require chemotherapy [2]. Furthermore, RNA expression signatures have also been helpful in identifying tissue origin of cancers of unknown primary [3], which in turn would immensely aid in guiding diagnostic and therapeutic decisions in the clinic. In addition to protein-coding RNAs, there is growing evidence highlighting the significance of the noncoding RNAs, including microRNAs (miRNA/miRs) and long noncoding RNAs (*lnc*RNAs), as important players during tumor development. miRNAs and *lnc*RNAs are a relatively new class of RNA molecules that have been shown to have oncogenic and tumor suppressor potential [4, 5]. In particular, miRNAs are attractive as candidate biomarkers owing to their extreme stability in body fluids and therefore will be ideal as noninvasive diagnostic and prognostic markers [6].

Although earlier studies have utilized traditional array-based techniques for profiling coding and noncoding RNAs in clinical samples, more recently there has been a preferential shift towards the NGS platform for characterizing the cancer transcriptome. Besides being an agnostic platform, NGS offers unique advantages over microarrays, including digital quantitation of RNA expression, having a broad dynamic range, and detection of low-abundant and novel transcripts. More importantly, NGS allows for the detection of qualitative changes in the transcriptome, including detection of cancer-specific splicing and RNA-editing events, identification of gene fusions, mutational changes, and allelic imbalances in the gene expression, all of which could be important in driving tumorigenesis besides being potential diagnostic and prognostic biomarkers. The following sections highlight findings from recent studies that clearly demonstrate the power of NGS-based approaches in the development of potential noncoding and coding RNA biomarkers in cancer research.

1 MicroRNA (miRNA/miR) Biomarkers

As shown in Table 9.1, a very active area of RNA biomarker discovery using NGS involves the analysis of miRNA expression profiles to find cancer-specific biomarkers [7–23]. MiRs are attractive candidates for biomarker development because they are expressed in a variety of tissues and biological fluids such as whole blood, serum, plasma, and urine and are frequently deregulated in cancer [24–27]. Furthermore, the ability of miRs to circulate stably in biological fluids makes them optimal for the development of noninvasive biomarker assays [24, 28]. Finally, the finding that there is a high level of correlation in miR expression levels between matched frozen and formalin-fixed paraffin-embedded tissues (FFPE) opens up a large repository of archival material for biomarker discovery studies [29].

1.1 miRs as Diagnostic Markers

The ability to detect and remove a cancer in its early stages before it metastasizes allows for the highest chance of patient survival when treated with traditional therapies. For example, colon cancer survival rates are the highest (>90 %) when tumors are detected early (stage I) and removed [30]. Thus, finding biomarkers that are expressed in the early stages of disease is of significant importance for early detection and improved patient survival. Because cancer is a heterogenic disease, it is difficult to find useful diagnostic markers of early disease that are common in the majority of cancers of a specific type. miRs offer the potential as early diagnostic markers due to their deregulation in cancer. As such, several groups have used NGS technology to screen large numbers of samples for novel as well as known miRs to discover potential miR-based diagnostic markers of early disease [7, 10, 14–19, 21–23]. In the majority of these studies, after this initial discovery phase, candidate miRs were then further validated by an independent method usually by qRT-PCR.

To date, many of the miR-based diagnostic papers involve screens in breast cancer patient samples for either serum- or tissue-based biomarkers [7, 10, 18, 19]. In a study looking for serum-based miR diagnostic biomarkers, Hu et al. found four miRs (miR-16, miR-25, miR-222, and miR-324-3p) significantly overexpressed in breast cancer patients compared to controls in a discovery set of 48 pooled cases versus 48 pooled controls that was further validated in an additional 152 individual cases versus controls by qRT-PCR [10]. Using this four-miR panel gave an area under the receiver operating characteristic curve (AUC) value of 0.954 with a sensitivity of 91.7 % and a specificity of 89.6 % in the discovery set and an AUC value

Table 9.1 Summary of potential RNA biomarkers found by next-generation sequencing

RNA type	Cancer type	Sample type	NGS platform	RNA found	Potential utility	Reference
miRNA	Astrocytoma	Serum	Illumina	miR-15b*, miR-23a, miR-133a, miR-150*, miR-197, miR-497, miR-548b-5p	Diagnostic Predictive	Yang [21]
miRNA	Breast	Serum	Illumina	miR-16, miR-25, miR-222, miR-324-3p	Diagnostic	Hu [10]
miRNA	Breast	Serum	Illumina GAIIx	miR-122, miR-375	Prognostic	Wu [20]
miRNA	Breast	Serum	Life Tech SOLiDv2.0	miR-29a, miR-21	Diagnostic	Wu [18]
miRNA	Breast	Tissue/serum	Life Tech SOLiDv2.0	miR-222	Diagnostic	Wu [19]
miRNA	Breast	Tissue	Illumina GAIIx	Cluster-miR-423, Cluster-miR-184, miR-19a	Diagnostic Prognostic	Farazi [7]
miRNA	Cervical	Tissue	Illumina 1G GA	67 miRs including miR-143, miR-205, miR-944, and miR-200 family	Diagnostic	Witten [17]
miRNA	Esophageal[a]	Serum	Illumina	miR-10a, miR-22, miR-100, miR-148b, miR-223, miR-133a, miR-127-3p	Diagnostic	Zhang [23]
miRNA	Esophageal[b]	Tissue	Illumina GAII	miR-31, miR-375	Prognostic	Leidner [12]
miRNA	Gastric	Serum	Illumina	miR-1, miR-20a, miR-27a, miR-34, miR-423-5p	Diagnostic	Liu [16]
miRNA	Gastric	Tissue	Life Tech SOLiDv3.0	miR-548d-3p, miR-20b, miR-135b, miR-140-3p, miR-93, miR-19a	Prognostic	Kim [11]
miRNA	Liver	Serum	Illumina	miR-1, miR-25, miR-92a, miR-206, miR-375, let-7f	Diagnostic	Li [14]
miRNA	Liver	Tissue	Illumina	miR-99a	Prognostic	Li [13]
miRNA	Liver	Tissue	Illumina	miR-199a/b-3p	Prognostic	Hou [8]
miRNA	Lung[c]	Serum	Illumina	miR-486, miR-30d, miR-1, miR-499	Prognostic	Hu [9]
miRNA	Pancreatic	Serum	Illumina	miR-20a, miR-21, miR-24, miR-25, miR-99a, miR-185, miR-191	Diagnostic Prognostic	Liu [15]
miRNA	Thyroid[d]	Serum	Illumina	Let-7e, miR-151-5p, miR-222	Diagnostic Predictive	Yu [22]

lncRNA	Breast	Tissue	Illumina GAIIx	Novel RNA Peak 13741	Diagnostic	Brunner [37]
mRNA[e]	Colon	Tissue	Illumina HiSeq 2000	EIF3E-RSPO2, PTPRK-RSPO3	Diagnostic Prognostic	Seshagiri [43]
mRNA	Lung	Tissue	Illumina GAIIx	RAC1b isoform	Predictive	Liu [39]
mRNA	Ovarian	Tissue	Illumina GAII	IGFBP-4	Diagnostic	Mosig [40]
mRNA[e]	Prostate	Tissue	Illumina GAII	KLK2-ETV1, FKBP5-ERG, FKBP5-TMPRSS2, ALG5-PIGU, TNPO1-IKBKB, CDKN1A-CD9, MIER2-RSRC2	Diagnostic Prognostic	Pflueger [42]
mRNA[e]	Prostate	Tissue	Illumina GAII	TMEM79-SMG5	Diagnostic	Kannan [38]
mRNA[e]	Prostate	Tissue	Illumina	SLC45A3-ELK4, MSMB-NCOA4, AZGP1-GJC3, HDAC8-CITED1, ADCK4-NUMBL, DUS4L-BCAP29	Diagnostic Prognostic	Nacu [41]

[a]Esophageal squamous cell carcinoma
[b]Esophageal adenocarcinoma
[c]Non-small-cell lung cancer
[d]Papillary thyroid cancer
[e]Indicates fusion transcripts

of 0.928 with a sensitivity of 92.1 % and a specificity of 93.4 % in the validation set. In addition to finding a serum miR diagnostic panel, Hu et al. also performed an analysis to find candidate endogenous control miRs that were stable across ten samples, with each sample being a pool of either healthy control samples or samples from a particular cancer type. They found that miR-191 and miR-484 were stable across all ten pooled samples which were further validated by qRT-PCR. In two studies by the group Wu et al. looking for serum and/or tissue miRs for breast cancer detection, significant upregulation of miR-29a and miR-21 was found in 20 breast cancer patient samples versus 20 control serums [18]. In a larger follow-up study of 20 cases versus control tissues and 50 cases versus control serums, neither miR-29a nor miR-21 repeated as potential diagnostic biomarkers. Instead, miR-222 was shown to be significantly increased in breast cancer cases and demonstrated a moderate discriminatory value with an AUC of 0.67 [19]. Finally, in a breast cancer tissue study analyzing miR expression levels in 17 HER2-postive noninvasive and 151 invasive breast cancers along with 11 normal breast tissue samples, Farazi et al. found upregulation of miR-19a as a potential diagnostic marker of triple negative breast cancer [7].

In addition to breast cancer, NGS has been used in single studies of a variety cancer types to screen for serum miRs as useful diagnostic markers for cancers of the brain (astrocytoma) [21], esophagus (squamous cell) [23], stomach [16], liver [14], pancreas [15], and thyroid (papillary) cancer [22]. In an NGS screen of 44 pooled serums from patients with astrocytoma versus 43 pooled normal serums, Yang et al. identified seven serum miRs (miR-15b, miR-23a, miR-133a, miR-150, miR-197, miR-497, and miR-548b-5p) that were decreased in astrocytoma. Subsequent training set (33 cancers and 33 controls) and validation set (45 cancers and 47 controls) studies using qRT-PCR demonstrated that the 7-miR panel was significantly decreased in grade II–IV astrocytoma patients as compared to controls and demonstrated an AUC value of 0.972 with a sensitivity of 88 % and a specificity of 97.9 % [21]. Furthermore, a subpanel of five of these miRs demonstrated the potential for discriminating malignant astrocytomas from both astrogliosis and benign astrocytoma. In esophageal squamous cell carcinoma (ESCC), Zhang et al. used NGS to screen miR concentrations in pooled samples of 85 nonmetastatic patients, 55 metastatic patients, and 40 healthy controls to identify a panel of seven serum miRs (miR-10a, miR-22, miR-100, miR-148b, miR-223, miR-133a, and miR-127-3p) that were upregulated in cancer samples versus controls [23]. They further went on to validate these miRs in 149 ESCC patients and 100 controls by qRT-PCR and found that this panel was able to distinguish stage I/II ESCC patients from controls, thus indicating its potential as a biomarker for ESCC detection at its early stages [23]. In an NGS initial screen of pooled serums from 20 gastric cancer cases and 20 controls, Liu et al. discovered a panel of five serum miRs (miR-1, miR-20, miR-27a, miR-34, and miR-423-5p) that were upregulated in gastric cancer, which was further confirmed in a training set (22 cases versus controls) as well as in a validation set (142 cases versus 147 controls) by qRT-PCR [16]. Using this panel as a biomarker for gastric cancer detection resulted in an AUC of 0.879 and 0.831 for the two separate sets of serum samples, and the sensitivity and specificity of gastric cancer detection were 80 % and 81 %, respectively, for all samples combined. In a similar study by the same group looking for biomarkers of pancreatic cancer, Liu et al. screened a pool of 25 cases and 25 controls by NGS and found seven miRs (miR-20a, miR-21, miR-24, miR-25, miR-99a, miR-185, and miR-191) that were upregulated in cancer and were further validated by qRT-PCR in both training set (25 cases versus controls) and validation set (95 cases versus 81 controls) of individual samples [15]. Using this panel as a biomarker for pancreatic cancer detection resulted in an AUC values of 0.992 and 0.985 for the training and validation sets, respectively, for discriminating between pancreatic cancer and controls, with a validation set sensitivity and specificity of 94 % and 93 %, respectively. Furthermore, this profile was able to accurately discriminate cancer patients from those with chronic pancreatitis (AUC of 0.993) [15]. For liver cancer, Li et al. performed NGS on a discovery set of 30 pooled cases versus 30 pooled controls and identified six serum miRs (miR-1, miR-25, miR-92a, miR-206, miR-375, and let-7f) that were subsequently validated by qRT-PCR as being significantly upregulated in a training set of 30 cases and controls as well as in a validation set of 55 cases and 50 controls [14]. Using miR-25, miR-375, and let-7f as biomarkers gave an AUC value of 0.997 with a sensitivity of 97.9 % and a specificity of 99.1 %. In an analysis of papillary thyroid carcinoma (PTC) serums, Yu et al. used NGS to screen pooled samples of ten cases versus ten benign papillary nodule cases and ten controls and found three serum miRs (let-7e, miR-151-5p, and miR-222) that were subsequently validated by qRT-PCR as significantly upregulated in a validation set of 106 PTC cases versus 95 benign thyroid nodule patients and 44 controls [22]. Using this panel of markers to distinguish PTC versus benign nodules or control groups resulted in an AUC of 0.917 with 87.8 % sensitivity and 88.4 % specificity for the benign nodule group comparison and an AUC of 0.897 with 86.8 % sensitivity and 79.5 % specificity for the control group comparison [22].

Finally, in a tissue-based study using NGS to screen cervical cancer tissues for diagnostic miRs, Witten et al. [17] examined differences in miR expression between 29 matched N/T pairs using a FDR cutoff of <0.0001 and found 67 miRs that were deregulated in tumor as compared to normal tissue. Two miRs, miR-205 and miR-944, were significantly increased in expression in cervical cancer, while miR-143 expression was lost in cervical cancer, a finding similar to other cancer types, and may have a suppressor role [17]. miR-205 has been found to be upregulated in a variety of cancer types and may serve as both a potential diagnostic and a therapeutic target, while mir-944 appears to be unique to cervical cancer and, thus, may be a potential diagnostic marker.

1.2 miRs as Prognostic Markers

The ability to determine patients that are at high risk for developing cancer is important for ascertaining the type and frequency of screening intervention for patients at risk. Furthermore, accurate prognostic markers are important in the prediction of disease outcome of patients and can potentially direct the type of therapy given to a patient with disease. Despite this knowledge, there is a lack of prognostic biomarkers that meet these needs, with the best prognostic factors for most solid tumors still being simple clinical features such as tumor grade, whether there is spread to lymph nodes, and number of metastatic sites, if any [31]. Using NGS methodology, several groups have now begun to look for miRs that could potentially serve as clinically useful prognostic markers [7–9, 11–13, 15, 20] (Table 9.1).

Finding useful prognostic markers of cancer risk can facilitate better decision-making as to who should undergo increased early cancer detection screenings, thus improving cost-effectiveness by reducing the number of unnecessary screenings for patients who are not at risk. For example, our laboratory is focused on the development of prognostic biomarkers to follow esophageal adenocarcinoma (EAC) disease progression. Barrett's esophagus is the only known precursor of EAC, yet the rate of neoplastic progression is only 1 per 200 patient-years, thus making it difficult to identify the small number of patients who will progress to EAC [12, 32, 33]. Endoscopy and histology-based assessment for high-grade dysplasia in Barrett's patients have shown only a minimal impact and is highly prone to interobserver variability besides not being cost-effective. Therefore, identification of novel biomarkers that reliably aid in risk stratification in Barrett's esophagus would allow for the development of more rational surveillance strategies aimed at early cancer detection. Recently, we published a study in which we used an Illumina-based platform to perform RNA-Seq on a test set of nine paired EAC/normal esophageal squamous (nSQ) samples to identify miRs associated with disease progression and thus of potential use as prognostic markers [12]. We identified 26 miRs that were highly deregulated (≥4-fold variation) in at least five of the nine EACs as compared to the paired nSQs, which we then proceeded to individually validate in a separate, follow-up set of laser-microdissected biopsy pairs of Barrett's metaplasia (BM)/nSQ ($n=15$) and high-grade dysplasia (HGD)/nSQ ($n=14$) by qRT-PCR. Of the 26 candidate miRs for disease progression, 23 miRs were found to be deregulated at the BM stage, thus demonstrating that the majority of the deregulated miRs in EACs are metaplasia derived and therefore unsuitable as markers of disease progression. Nonetheless, we found two miRs (miR-31 and miR-31*) that had frequent downregulation only in HGD and EAC cases, suggesting association with transition from BM to HGD, and we found one miR (miR-375) having marked downregulation exclusively in EACs, suggesting its association with progression to invasive carcinoma. Follow-up studies are now ongoing testing the clinical utility of these miRs using a cohort of Barrett's progressors and nonprogressors. Additionally, since our study was the first to look for tissue-based biomarker miRs in EAC, the importance of finding the proper "housekeeping" miRs that are moderately expressed and have minimal variability between matched EAC/nSQ samples was paramount in order to effectively normalize the miR expression counts from the NGS data as well as for normalizing qRT-PCR results. The use of NGS in our study was critical in facilitating the identification of such endogenous controls. As such, we found miR-103, miR-191, and miR-374a that met the above criteria and proceeded to use miR-103 and miR-191 for normalizing miRs in our study. Interestingly, miR-103 and miR-191 have been shown previously to be highly consistent in their expression across a wide variety of normal and tumor tissues by Peltier and Latham [34], a finding that was confirmed for miR-191 by Hu et al. in their search for normalization miRs in their breast cancer studies [10].

Finding useful prognostic markers to determine the disease outcome of patients is not only important for predicting survival but can also be relevant for selecting the choice of initial treatment a patient receives. Using NGS for the initial discovery phase, several groups have been attempting to find serum-based miRs for use as noninvasive prognostic markers of patient outcome. In a breast cancer study using a training cohort of 42 stage II–III breast cancer patients who received neoadjuvant chemotherapy (NCT), Wu et al. found two circulating miRs whose upregulation (miR-375) or downregulation (miR-122) predicted better clinical outcome as determined by complete response to NCT and absence of relapse [20]. In a validation set of 26 stage II–III cases, though, only higher circulating levels of miR-122 levels could predict metastasis relapse with a sensitivity of 80 % and specificity of 100 % [20]. In a study by the group of Hu et al. looking for miRs that could predict survival of non-small-cell lung cancer patients, four miRs (miR-486, miR-30d, miR-1, and miR-499) were found to be significantly different in expression between 30 long-term survivors versus 30 short-term survivors [9]. Subsequent validation studies using qRT-PCR, in which an additional 243 patients were divided into training and testing sets, found that the panel of four miRs was significantly associated with, and was an independent predictor of, overall survival. In addition to looking for diagnostic biomarkers of pancreatic cancer, as reviewed above, Liu et al. also screened for prognostic biomarkers and found that high miR-21 serum levels were significantly associated with poorer overall survival independent of other clinical/pathological factors, a finding that confirmed an earlier, independent study in pancreatic cancer using in situ hybridization [15].

Analysis of tumor tissues by NGS has also been performed by several groups in the search for miRs of prognostic utility. To identify miRs that play a role in gastric cancer development, Kim et al. analyzed 24 tumor and 6 normal tissue samples by RNA-Seq and looked for miRs with altered expression in cancer versus normal tissue and had detectable repression on the expression of their target genes [11]. The group was able to identify six miRs (miR-548d-3p, miR-20p, miR-135b, miR-140-3p, miR-93, and miR-19a) that were differentially expressed in gastric cancer versus normal tissue and had significant anticorrelation with their putative targets. As such, these miRs have the potential for the prognosis of gastric cancer but warrant further follow-up studies [11]. In liver cancer, the downregulation of miR-99a [13] and miR-199a/b-3p [8] in liver tumor tissues has been associated with poor patient outcome. Low miR-199a/b-3p expression was an independent predictor for reduced tumor-free survival as well as overall survival [8], while decreased miR-99a levels in patients were an independent predictor of poorer disease-free survival [13]. In both studies, RNA-Seq of a screen set of tissues samples resulted in these candidate prognostic markers, which were then validated by qRT-PCR in subsequent larger training and validation sets. In addition to looking for diagnostic biomarkers of breast cancer, as reviewed earlier, Farazi et al. also screened for prognostic biomarkers and found that increased expression of miR-423 and decreased expression of miR-184 were associated with patients that went on to develop metastasis ($P=0.013$ and $P=0.041$, respectively), with miR-423 being an independent predictor of outcome [7].

1.3 miRs as Predictive Markers

Determining patients that will respond to a particular therapy and those who will not is important in helping clinicians tailor treatments that will provide maximal therapeutic benefit with minimal toxicity. In order to accomplish this, there is a need for biomarkers that can monitor or predict response outcome to treatment. Two studies have been published using NGS technology to begin looking for miRs that can predict patient outcome to treatment [21, 22] (Table 9.1). In astrocytomas, Yang et al. found that their panel of seven miRs used for the detection of grade II–IV astrocytomas may also have utility as a predictor of treatment response, with the finding that expression of the seven miRs increased in the serum of all 14 patients after surgery to remove their tumors [21]. Similarly for PTC, Yu et al. found that at least two of the three miRs in their diagnostic panel (miR-151-5p and miR-222) may be useful predictors of treatment response with postoperative serum levels decreasing significantly in nine of nine patients after surgery to healthy control levels, while let-7e decreased in seven of nine patients [22].

2 Long Noncoding RNA (*lnc*RNA) Biomarkers

Recently studies have begun to also look at a new class of noncoding RNAs as potential biomarkers, those of long noncoding RNAs (*lnc*RNAs). As the name suggests, *lnc*RNAs are nonprotein-coding RNAs that are generally considered greater than 200 nucleotides in length. Though the majority of the predicted *lnc*RNAs have not been well characterized, an increasing number of studies are demonstrating that *lnc*RNAs may play an important role in gene regulation through chromatin modification, transcription, and posttranscriptional regulation [35]. Because studies are beginning to demonstrate that *lnc*RNAs tend to have greater tissue-specific expression that protein coding mRNAs [36] and that deregulation of *lnc*RNAs may play an important role in tumor biology [5], *lnc*RNAs have the strong potential for being clinically useful biomarkers. Currently, though, there is only one published paper utilizing NGS to look for potential *lnc*RNA biomarkers in cancer by Brunner et al. [37] (Table 9.1). In this paper, a panel of 64 tumor samples comprising of 17 diagnostic subtypes of adenocarcinomas, squamous cell carcinomas, and sarcomas were examined for *lnc*RNA expression differences between normal and cancerous tissues as well as between various cancer types [37]. Though the work by Brunner et al. is in its early stages of biomarker development, their results found several hundred known *lnc*RNAs that had variability in transcriptional level between the tumor types, thus having the potential as biomarker candidates. Additionally, they found 1,071 novel intergenic transcribed regions suggestive of *lnc*RNAs that also showed similar patterns of variability between tumor types. Follow-up in situ hybridization experiments of one such novel transcript that was a potential breast cancer-specific biomarker resulted in the finding that the transcript's expression was significantly correlated with low tumor grade, estrogen receptor expression (ER+), and progesterone receptor expression (PR+), suggesting that it could be a biomarker for distinguishing between ER+/PR+ and ER − breast cancer subtypes [37]. The utility of NGS to analyze large numbers of samples and to find novel *lnc*RNAs quickly will enable the field of *lnc*RNA biomarker discovery to expand rapidly and will quite likely yield intriguing results.

3 Coding RNA Biomarkers

3.1 mRNAs as Diagnostic Markers

Use of NGS to interrogate the transcriptome enables the discovery of cancer-specific changes such as aberrantly expressed genes, mRNA splicing alterations, and novel structural variations, like gene fusions, that might be missed by other analysis methods [38–43] (Table 9.1). Several papers have now been published looking at altered gene expression or gene fusions/transcription-induced chimeras (TICs) for utility as possible diagnostic biomarkers. For example, in a study of 22 ovarian cancer samples looking for strong expression of genes encoding secreted proteins for possible use as serum diagnostic biomarkers, Mosig et al. found the gene IGFBP-4 present in the top 7.5 % of all expressed genes in all 22 tumor samples [40]. As a result of this initial screen, they subsequently tested for IGFBP-4 protein levels in the serum of 21 cases and 4 controls in a discovery set and in 78 cases and 82 controls in a validation set and demonstrated that serum IGFBP-4 protein levels were significantly increased in patients with disease versus controls [40].

After the initial finding in prostate cancer of a TMPRSS2 fusion to the ETS transcription factor family members ETV1 and ERG [44], a number of subsequent papers have been published using RNA-Seq looking for further, novel gene fusions in solid tumors in prostate cancer [38, 41, 42] and colon cancer [43]. In prostate cancer, a variety of intriguing gene fusions have been found by Pflueger et al. (KLK2-ETV1, FKBP5-ERG, FKBP5-TMPRSS2, ALG5-PIGU, TNPO1-IKBKB, CDKN1A-CD9, MIER2-RSRC2) [42], Nacu et al. (SLC45A3-ELK4, MSMB-NCOA4, AZGP1-GJC3, HDAC8-CITED1, ADCK4-NUMBL, DUS4L-BCAP29) [41], and Kannan et al. (TMEM79-SMG5) [38], which offer the potential of being diagnostic or prognostic biomarkers due to prostate cancer-specific expression and/or putative functions as drivers of tumor progression. In a similar study to find fusion transcripts in colon cancer by Seshagiri et al., the gene fusions of R-spondin family members RSPO2 and RSPO3 (EIF3E-RSPO2 and PTPRK-RSPO3) were found in approximately 10 % of colon cancers and may act as drivers of colon tumorigenesis [43]. Though follow-up studies need to be performed on the above-mentioned gene fusion/TIC discoveries, it is not unreasonable to speculate that some of these fusion products may have potential diagnostic and/or prognostic value and results from future studies will be of considerable interest.

3.2 mRNAs as Predictive Markers

There is a growing interest in the utility of cancer-specific splice isoforms as biomarkers in the clinic [45]. In a recent paper using RNA-Seq to analyze the transcriptome of three lung cancer tissue samples with matched normal, Liu et al. found a tenfold higher expression of an isoform of RAC1 (RAC1b isoform) in tumor versus the corresponding normal samples, a finding corroborated by others studying colon and breast cancer, and suggest preferential upregulation of RAC1b in cancer [39]. Since RAC1b results in a RAC1 isoform that is predominately in a GTP-bound active state that can influence RAS-MEK-ERK signaling activity, Liu et al. treated cell lines with a MAP2K (MEK) inhibitor, PD-0325901, to determine if high RAC1b-expressing lines were sensitive to the drug. Indeed, they found that cells lines sensitive to PD-0325901 had significantly higher expression of RAC1b ($P=0.019$), indicating that the RAC1b isoform may serve as a predictive biomarker for response to PD-0325901 [39].

4 Conclusions

While the above studies clearly demonstrate the utility of NGS- and RNA-based biomarkers in cancer diagnostics, there however is significant room for improvement, including development and standardization of robust algorithms for bioinformatics analysis to reduce technical- and platform-based artifacts and development of reliable post hoc analysis and filtering tools for biological interpretation of the massive NGS data. Furthermore, rather than limiting the focus on either DNA or RNA markers, future studies can include a comprehensive characterization of the genome, transcriptome, and methylome from individual cancers, thereby increasing the chances of identifying evidence-based biomarkers besides providing significant insights in to the molecular mechanisms of tumorigenesis. Following the initial discovery of potential tumor biomarkers using NGS, efforts should also be focused on developing easier and robust techniques for routine assaying of such biomarkers in a wide array of clinical specimens including body fluids and frozen and formalin-fixed samples for respective malignancies. This in turn would be highly cost-effective and will provide a standardized platform for assessing these biomarkers in the clinic for routine diagnostic and prognostic purposes. Nonetheless, the advent of NGS technology has already revolutionized the field of cancer research and biomarker discovery with more significant advances in the field to be expected in the next decade.

References

1. Kim C, Paik S. Gene-expression-based prognostic assays for breast cancer. Nat Rev Clin Oncol. 2010;7:340–7.
2. Salazar R, Roepman P, Capella G, Moreno V, Simon I, Dreezen C, et al. Gene expression signature to improve prognosis prediction of stage II and III colorectal cancer. J Clin Oncol. 2011;29:17–24.
3. Tothill RW, Kowalczyk A, Rischin D, Bousioutas A, Haviv I, van Laar RK, et al. An expression-based site of origin diagnostic method designed for clinical application to cancer of unknown origin. Cancer Res. 2005;65:4031–40.
4. Calin GA, Croce CM. MicroRNA signatures in human cancers. Nat Rev Cancer. 2006;6:857–66.
5. Prensner JR, Chinnaiyan AM. The emergence of lncRNAs in cancer biology. Cancer Discov. 2011;1:391–407.
6. Hoshino I, Matsubara H. MicroRNAs in cancer diagnosis and therapy: from bench to bedside. Surg Today. 2012;43:467–78.
7. Farazi TA, Horlings HM, ten Hoeve JJ, Mihailovic A, Halfwerk H, Morozov P, et al. MicroRNA sequence and expression analysis in breast tumors by deep sequencing. Cancer Res. 2011;71:4443–53.
8. Hou J, Lin L, Zhou W, Wang Z, Ding G, Dong Q, et al. Identification of miRNomes in human liver and hepatocellular carcinoma reveals miR-199a/b-3p as therapeutic target for hepatocellular carcinoma. Cancer Cell. 2011;19:232–43.
9. Hu Z, Chen X, Zhao Y, Tian T, Jin G, Shu Y, et al. Serum MicroRNA signatures identified in a genome-wide serum MicroRNA expression profiling predict survival of non-small-cell lung cancer. J Clin Oncol. 2010;28:1721–6.
10. Hu Z, Dong J, Wang LE, Ma H, Liu J, Zhao Y, et al. Serum microRNA profiling and breast cancer risk: the use of miR-484/191 as endogenous controls. Carcinogenesis. 2012;33: 828–34.
11. Kim YH, Liang H, Liu X, Lee J-S, Cho JY, Cheong J-H, et al. AMPKα modulation in cancer progression: multilayer integrative analysis of the whole transcriptome in Asian gastric cancer. Cancer Res. 2012;72:2512–21.
12. Leidner RS, Ravi L, Leahy P, Chen Y, Bednarchik B, Streppel M, et al. The microRNAs, MiR-31 and MiR-375, as candidate markers in Barrett's esophageal carcinogenesis. Genes Chromosomes Cancer. 2012;51:473–9.
13. Li D, Liu X, Lin L, Hou J, Li N, Wang C, et al. MicroRNA-99a inhibits hepatocellular carcinoma growth and correlates with prognosis of patients with hepatocellular carcinoma. J Biol Chem. 2011;286:36677–85.
14. Li LM, Hu ZB, Zhou ZX, Chen X, Liu FY, Zhang JF, et al. Serum microRNA profiles serve as novel biomarkers for HBV infection and diagnosis of HBV-positive hepatocarcinoma. Cancer Res. 2010;70:9798–807.
15. Liu R, Chen X, Du Y, Yao W, Shen L, Wang C, et al. Serum microRNA expression profile as a biomarker in the diagnosis and prognosis of pancreatic cancer. Clin Chem. 2012;58:610–8.
16. Liu R, Zhang C, Hu Z, Li G, Wang C, Yang C, et al. A five-microRNA signature identified from genome-wide serum microRNA expression profiling serves as a fingerprint for gastric cancer diagnosis. Eur J Cancer. 2011;47:784–91.
17. Witten D, Tibshirani R, Gu SG, Fire A, Lui WO. Ultra-high throughput sequencing-based small RNA discovery and discrete statistical biomarker analysis in a collection of cervical tumours and matched controls. BMC Biol. 2010;8:58.
18. Wu Q, Lu Z, Li H, Lu J, Guo L, Ge Q. Next-generation sequencing of MicroRNAs for breast cancer detection. J Biomed Biotechnol. 2011;2011:597145.
19. Wu Q, Wang C, Lu Z, Guo L, Ge Q. Analysis of serum genome-wide microRNAs for breast cancer detection. Clin Chim Acta. 2012;413:1058–65.
20. Wu X, Somlo G, Yu Y, Palomares MR, Li AX, Zhou W, et al. De novo sequencing of circulating miRNAs identifies novel markers predicting clinical outcome of locally advanced breast cancer. J Transl Med. 2012;10:42.

21. Yang C, Wang C, Chen X, Chen S, Zhang Y, Zhi F, et al. Identification of seven serum microRNAs from a genome-wide serum microRNA expression profile as potential noninvasive biomarkers for malignant astrocytomas. Int J Cancer. 2013;132:116–27.
22. Yu S, Liu Y, Wang J, Guo Z, Zhang Q, Yu F, et al. Circulating microRNA profiles as potential biomarkers for diagnosis of papillary thyroid carcinoma. J Clin Endocrinol Metab. 2012; 97:2084–92.
23. Zhang C, Wang C, Chen X, Yang C, Li K, Wang J, et al. Expression profile of microRNAs in serum: a fingerprint for esophageal squamous cell carcinoma. Clin Chem. 2010;56:1871–9.
24. Koshiol J, Wang E, Zhao Y, Marincola F, Landi MT. Strengths and limitations of laboratory procedures for microRNA detection. Cancer Epidemiol Biomarkers Prev. 2010;19:907–11.
25. Li J, Smyth P, Flavin R, Cahill S, Denning K, Aherne S, et al. Comparison of miRNA expression patterns using total RNA extracted from matched samples of formalin-fixed paraffin-embedded (FFPE) cells and snap frozen cells. BMC Biotechnol. 2007;7:36.
26. Xi Y, Nakajima G, Gavin E, Morris CG, Kudo K, Hayashi K, et al. Systematic analysis of microRNA expression of RNA extracted from fresh frozen and formalin-fixed paraffin-embedded samples. RNA. 2007;13:1668–74.
27. Zhang B, Farwell MA. microRNAs: a new emerging class of players for disease diagnostics and gene therapy. Journal of Cellular and. Mol Med. 2008;12:3–21.
28. Chen X, Ba Y, Ma L, Cai X, Yin Y, Wang K, et al. Characterization of microRNAs in serum: a novel class of biomarkers for diagnosis of cancer and other diseases. Cell Res. 2008;18: 997–1006.
29. Zhang X, Chen J, Radcliffe T, Lebrun DP, Tron VA, Feilotter H. An array-based analysis of microRNA expression comparing matched frozen and formalin-fixed paraffin-embedded human tissue samples. J Mol Diagn. 2008;10:513–9.
30. Markowitz SD, Dawson DM, Willis J, Willson JK. Focus on colon cancer. Cancer Cell. 2002;1:233–6.
31. Ludwig JA, Weinstein JN. Biomarkers in cancer staging, prognosis and treatment selection. Nat Rev Cancer. 2005;5:845–56.
32. Prasad GA, Bansal A, Sharma P, Wang KK. Predictors of progression in Barrett's esophagus: current knowledge and future directions. Am J Gastroenterol. 2010;105:1490–502.
33. Reid BJ, Li X, Galipeau PC, Vaughan TL. Barrett's oesophagus and oesophageal adenocarcinoma: time for a new synthesis. Nat Rev Cancer. 2010;10:87–101.
34. Peltier HJ, Latham GJ. Normalization of microRNA expression levels in quantitative RT-PCR assays: identification of suitable reference RNA targets in normal and cancerous human solid tissues. RNA. 2008;14:844–52.
35. Mercer TR, Dinger ME, Mattick JS. Long non-coding RNAs: insights into functions. Nat Rev Genet. 2009;10:155–9.
36. Cabili MN, Trapnell C, Goff L, Koziol M, Tazon-Vega B, Regev A, et al. Integrative annotation of human large intergenic noncoding RNAs reveals global properties and specific subclasses. Genes Dev. 2011;25:1915–27.
37. Brunner AL, Beck AH, Edris B, Sweeney RT, Zhu SX, Li R, et al. Transcriptional profiling of lncRNAs and novel transcribed regions across a diverse panel of archived human cancers. Genome Biol. 2012;13:R75.
38. Kannan K, Wang L, Wang J, Ittmann MM, Li W, Yen L. Recurrent chimeric RNAs enriched in human prostate cancer identified by deep sequencing. Proc Natl Acad Sci. 2011;108:9172–7.
39. Liu J, Lee W, Jiang Z, Chen Z, Jhunjhunwala S, Haverty PM, et al. Genome and transcriptome sequencing of lung cancers reveal diverse mutational and splicing events. Genome Res. 2012;22:2315–27.
40. Mosig RA, Lobl M, Senturk E, Shah H, Cohen S, Chudin E, et al. IGFBP-4 tumor and serum levels are increased across all stages of epithelial ovarian cancer. J Ovarian Res. 2012;5:3.
41. Nacu S, Yuan W, Kan Z, Bhatt D, Rivers CS, Stinson J, et al. Deep RNA sequencing analysis of readthrough gene fusions in human prostate adenocarcinoma and reference samples. BMC Med Genomics. 2011;4:11.

42. Pflueger D, Terry S, Sboner A, Habegger L, Esgueva R, Lin PC, et al. Discovery of non-ETS gene fusions in human prostate cancer using next-generation RNA sequencing. Genome Res. 2011;21:56–67.
43. Seshagiri S, Stawiski EW, Durinck S, Modrusan Z, Storm EE, Conboy CB, et al. Recurrent R-spondin fusions in colon cancer. Nature. 2012;488:660–4.
44. Tomlins SA, Rhodes DR, Perner S, Dhanasekaran SM, Mehra R, Sun X-W, et al. Recurrent Fusion of TMPRSS2 and ETS Transcription Factor Genes in Prostate Cancer. Science. 2005;310:644–8.
45. Pajares MJ, Ezponda T, Catena R, Calvo A, Pio R, Montuenga LM. Alternative splicing: an emerging topic in molecular and clinical oncology. Lancet Oncol. 2007;8:349–57.

Chapter 10
Understanding the Role of Long Noncoding RNAs in the Cancer Genome

Wei Wu and Jennifer A. Chan

Abstract The secrets of long noncoding RNAs, like other noncoding RNAs that are often referred to as the "dark matter" of the genome, are gradually coming to light. Knowledge from the Encyclopedia of DNA Elements project revealed that greater than 70 % of the genome is actively transcribed and thousands of long noncoding RNA (lncRNA) transcripts, which are greater than 200 base pairs without protein-coding potential, have been annotated. Increasing evidence shows that lncRNAs play crucial roles in development, stem cell pluripotency, differentiation, cell-cycle control, and diseases including cancer. The molecular actions of lncRNAs are involved in regulation of gene expression, interaction with chromatin-modifying complexes, and DNA–RNA–protein hybrid. In this chapter, we introduce the biogenesis of lncRNAs, their functionality and mechanisms of action, and strategies for discovery of lncRNAs and highlight the functional role of lncRNAs in cancer development. We briefly discuss our own work with lncRNAs. Our aim is to pinpoint the complex and diverse functions of these enigmatic molecules in cancer biology and expand our understanding of the cancer initiatome.

Keywords Long noncoding RNA • LncRNA • Cancer biology • RNA-seq • Cancer Initiatome

W. Wu (✉)
Department of Pathology & Laboratory Medicine,
University of Calgary, Calgary, AB, Canada
e-mail: wuwei@ucalgary.ca

J.A. Chan (✉)
Department of Pathology & Laboratory Medicine,
University of Calgary, Calgary, AB, Canada

Department of Clinical Neurosciences, University of Calgary,
Calgary, AB, Canada

Department of Oncology, University of Calgary, Calgary, AB, Canada
e-mail: jawchan@ucalgary.ca

W. Wu and H. Choudhry (eds.), *Next Generation Sequencing in Cancer Research: Volume 1: Decoding the Cancer Genome*, DOI 10.1007/978-1-4614-7645-0_10,

1 Introduction

The human genome project revealed approximately 21,000 protein-coding genes, but this represents less than 2 % of the total genome sequence. Subsequently, ENCODE (encyclopedia of DNA element project) presented evidence that >70 % of the genome is actively transcribed [1]. Once thought to be the "dark matter" of the genome, non-protein-coding RNAs (ncRNAs) have emerged as an integral component of the mammalian transcriptome. NcRNAs can be grouped into two major classes according to an arbitrary length cutoff: the small ncRNAs such as microRNA, which are less than 200 nucleotides (nt), and the long noncoding RNAs (lncRNAs), which range from 200 nt to greater than 100 kilobases (kb). Readers may also see other names such as large RNAs, marcroRNAs, and long intergenic ncRNAs (lincRNAs). NcRNA research has focused primarily on the small subclass of microRNAs and their regulatory roles in gene expression, and we have recently reviewed this work [2]. Here, we turn our attention to lncRNAs and review their evolutionary complexity and the growing evidence for their diverse mechanisms of action and functional roles in basic molecular and cellular biology and in cancer biology.

The existence of lncRNAs has been known since the 1980s, which is earlier than the finding of microRNAs in 1993. The lncRNAs were not fully appreciated until the introduction of large-scale sequencing of full-length cDNA libraries which allowed an unbiased assessment of a larger proportion of the mammalian transcriptome which does not code proteins [3]. The imprinting of H19 was the first lncRNA studied in the mouse and human genome [4], followed quickly by the X-inactive-specific transcript (Xist) lncRNA genes of the X chromosome [5]. These lncRNAs were discovered using traditional gene-mapping approaches and were only realized to be nonprotein-coding RNAs after their RNA sequences were analyzed. Due to the functional characterization of a handful of lncRNAs, the scientific community is revisiting lncRNA biology and starting a new wave of genome-wide screening to identify and characterize lncRNAs using similar strategies that were used to study protein-coding genes and microRNAs. To date, there are about 23,000 annotated transcripts of lncRNA, but less than 1 % has been characterized [6]. LncRNAs are exquisitely regulated during development and in response to diverse experimental stimuli [7]. LncRNAs are being shown to play critical roles in a variety of cellular processes, including differentiation, development, and tumorigenesis [8]. In this chapter, we will discuss the biogenesis of lncRNA and its nomenclature, the functional mechanisms of lncRNAs in gene regulation, and the strategies for identifying lncRNAs, particularly the RNA-seq procedure, and we will highlight the role of lncRNAs in cancer biology.

2 Biogenesis of lncRNAs

LncRNAs are endogenous cellular RNA transcripts of more than 200 nt and up to 100 kb in length, and they lack an open reading frame of significant length (less than 100 amino acids) [9]. The current estimated number of individual lncRNAs ranges from 7,000 to 23,000 in the human genome [10], and they are expected to surpass the estimate protein-coding genes in their abundance. LncRNAs are transcribed at any region in the genome by RNA polymerase II/III, they lack a significant open reading frame, and while the majority of the identified lncRNAs are polyadenylated [6], an increasing number of nonpolyadenylated transcripts are being reported [11]. LncRNA may contain few exons and long introns, and some undergo alternative splicing [12]. LncRNAs may develop secondary structures, even tertiary structures, forming specific functional domains that interact with DNA or RNA or proteins [13]. LncRNAs are distributed on all 22 autosomes and the sex chromosomes, and they tend not to be correlation with protein-coding genes and microRNAs [14].

Based on the genomic locations of the lncRNAs, they can be further classified into diverse subtypes illustrated in Fig. 10.1a: intergenic lncRNAs (transcribed in the space between genes); intronic lncRNAs (transcribed between exons of genes); antisense lncRNAs that are lncRNAs that initiate inside or 3′ of a protein-coding gene, are transcribed in the opposite direction of this gene, and overlap at least one coding exon of the gene; and overlapping lncRNAs that are more complex as they overlap both intron and exon of a coding gene. Each of these types of lncRNAs may also be in the sense or in the antisense direction. Over 50 % of lncRNAs are intergenic lncRNA, followed by 19 % intronic lncRNA [6] (Fig. 10.1b). LncRNAs are generally less expressed than protein-coding genes and display more tissue-specific and cell-specific expression patterns [14, 15]. Moreover, lncRNAs seem to be particularly enriched in the nucleus relative to the cytoplasm, when compared to protein-coding genes [16]. A study with human cell lines suggests that ~30 % of lncRNAs are found exclusively in the nucleus, ~15 % are found exclusively in the cytoplasm, while ~50 % show both nuclear and cytoplasmic localization [16]. Within the nucleus, they are particularly enriched in the chromatin fraction [6]. The turnover of lncRNAs varies from several minutes to greater than 16 h. In general, intronic lncRNAs are less stable than intergenic and antisense lncRNAs, whereas spliced lncRNAs are more stable than unspliced transcripts. In addition, cytoplasmic lncRNAs are more stable than nuclear lncRNAs, as exemplified by the extremely labile lncRNA *NEAT1*, involved in nuclear paraspeckle assembly (a subcellular compartment in the interchromatin space within a nucleus) [17]. The dynamic nature of lncRNA turnover emphasizes the complexity of regulating RNA metabolism, sometimes elicited by other RNAs that are, themselves, subject to posttranscriptional control.

3 Functions and Mechanisms of lncRNAs

In the postgenomic era, deciphering the 3 billion DNA code (6 billion nt in the diploid organism) is expanding our understanding of the regulatory regions in the genome, the epigenetic regulation of gene expression, and the interactions between these components [1]. In the genome space, lncRNAs exhibit a variety of cellular functions with *cis*-(on neighboring genes) or *trans*-(on distantly located genes)

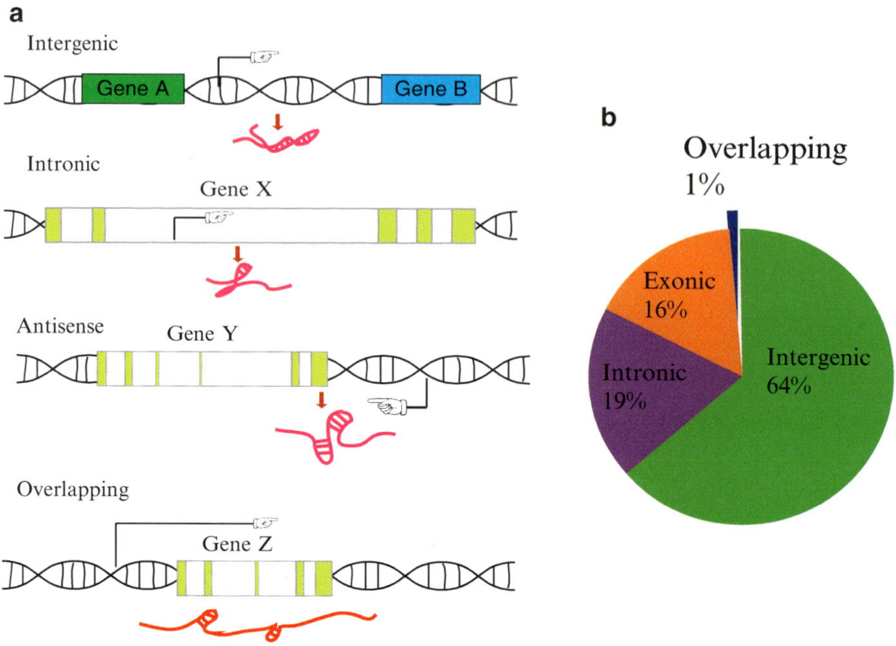

Fig. 10.1 Schematic of long noncoding RNA (lncRNA) category and distribution in the genome. (**a**) The region of lncRNA subtypes: intergenic lncRNAs between the two genes, intronic lncRNAs in an intron of a gene, and antisense lncRNAs where RNA molecules that are transcribed from antisense strands and overlap in part with well-defined spliced sense messenger RNAs. Overlapping lncRNAs where transcript variants of protein-coding RNAs overlap with known annotated genes at same genomic strands. *Light yellow bars* represent exons in genes; *white bars* represent introns in genes; *red hairpin* structures represent lncRNA molecules. (**b**) Estimated percentage of each subtype of lncRNAs in the genome. Intergenic lncRNAs are the majority, followed by intronic lncRNA, and the least is overlapping lncRNA [6]

regulatory behavior [18]. This genomic regulation by lncRNAs is unraveling more complexity than we previously imagined, and what we know about functions of lncRNAs is now the tip of an iceberg. However, it is becoming obvious that lncRNAs play an important role in regulating gene expression at different levels including chromatin modification and both transcriptional and posttranscriptional processing.

3.1 Functions of lncRNAs

Like other components in the genome, lncRNAs exhibit broad physiological functions such as in development [19], transcription, splicing [20], cellular differentiation [21, 22], reprogramming of stem cells [23], and diseases including cancer [24]. The earliest discovered and well-studied lncRNAs, Xist (for X-inactive-specific transcript) and H19, are involved in two related, fundamental epigenetic processes, namely, dosage compensation and genomic imprinting, where certain genes are expressed in a parent-of-origin-specific manner. Xist plays an essential role in X chromosome inactivation. During female development, Xist RNA is expressed from the inactive X chromosome and suppresses the function of the same X chromosome in order to make the dosage of gene expression the same as in males with a single X chromosome. The Xist allele on the active X chromosome is repressed by its antisense partner ncRNA, TSIX [25]. The imprinting-associated lncRNAs of H19 and KCNQ1OT1 are expressed on the maternal and paternal human chromosome 11, respectively, and they maintain silencing of the IGF2 and KCNQ1 genes on those chromosomes [4, 26]. More recent study revealed that lncRNA CRNDE is highly expressed in pluripotent stem cells and decreased during differentiation; furthermore, CRNDE is expressed in the fetal brain but is only selectively expressed in compartments of the adult normal brain [15]. This implies that lncRNAs may command a much larger sequence space than proteins, and therefore, they can achieve very precise spatiotemporal control of development.

3.2 *Mechanisms of lncRNAs*

Emerging data indicate that the broad range of lncRNA functions are governed by gene regulatory circuits in transcriptional and epigenetic mechanisms: the recruitment of transcription factors or chromatin-modifying complexes to specific nuclear and genomic sites, alternative splicing and other posttranscriptional RNA modifications through the assembly of nuclear domains containing RNA-processing factors, nuclear–cytoplasmic shuttling, and translational control. LncRNAs can also act as precursors for small ncRNAs, such as small nucleolar RNAs (snoRNAs) and microRNAs [18].

3.2.1 Chromatin Modification

There is evidence that lncRNAs bind chromatin-modifying enzyme complexes to execute gene expression regulation [27–29]. For example, numerous lncRNAs are pulled down by RNA immunoprecipitation (RIP) of PRC2 (polycomb repressive complex 2) and other chromatin-modifying partners [29]. PRC2 comprises of histone H3 lysine 27 (H3K27) methylase EZH2, SUZ12, and EED1, and the PRC2 complex may directly bind to lncRNA stem-loop structures via EZH2 [30]. Besides PRC2, lncRNAs may also interact with LSD1/CoREST/REST protein complexes. The lncRNA HOTAIR (HOX antisense intergenic RNA) is expressed from an intergenic region of the HOXC cluster in chromosome 12 and is necessary for PRC2 occupancy, H3K27me3, and silencing of the HOXD locus in chromosome 2 [31]. Analysis of HOTAIR revealed that a 5′ end-domain binds PRC2 (H3K27me3) and a 3′ end-domain binds an LSD1 (H3K4me2 demethylase) containing complex [28].

This leads to removal of the active histone modification H3K4me2 while methylating H3K27 in the direction of a repressive mode. Therefore, lncRNAs act as scaffolding to bring functional proteins to specific genomic regions, resulting in a change of chromatin architecture.

3.2.2 Posttranscriptional Regulation of Gene Expression

Several lncRNAs have been reported to regulate gene expression at posttranscriptional levels such as mRNA processing, mRNA stability, and mRNA translation.

lncRNA and mRNA Splicing

The nuclear lncRNAs MALAT1 and NEAT1 are genomic neighbors on Chr11q13, and they are thought to contribute to gene expression by regulating mRNA splicing, editing, and export [20]. MALAT1 interacts with serine/arginine (SR) protein and directs the phosphorylated SR splicing factor to targeted pre-mRNA in nuclear speckle domain (known as interchromatin granule clusters).

lncRNAs and mRNA Stability

Similar to microRNAs that form duplexes with mRNA for gene regulation, lncRNAs have been shown to interact with mRNA sequences either in untranslated regions or in coding regions. However, differing from microRNA, lncRNAs bearing ancestral Alu repeats are able to interact with protein-coding mRNA with imperfect hybrids, followed by recruitment of an RNA degradation protein STAU1 to promote the decay of a specific mRNA [32]. Indirectly, lncRNAs, like transcribed pseudogenes, serve as decoys or "molecular sponges" for miRNA that target protein-coding mRNAs [33]. For example, PTENP1 is a pseudogene of PTEN tumor suppressor gene, and it competes for miRNA binding sites with PTEN; this releases the inhibition from miRNA and allows biological function of PTEN.

lncRNAs and mRNA Translation

Cytoplasmic lncRNAs could regulate the translational machinery. P53-induced lincRNA-p21 was recently shown to co-distribute with ribosomes and to repress beta-catenin and JunB protein synthesis via the interaction of the lincRNA-p21-mRNA-translation repressor protein complex (RcK and Fmrp proteins) [34].

When taken all together, the mechanisms of action of the lncRNAs can be categorized broadly as regulatory elements that interact with substrate proteins to be a decoy, or to tether, guide, or signal and modify chromatin architecture, or to modify the expression of coding genes at various levels of control (see Fig. 10.2). And, at

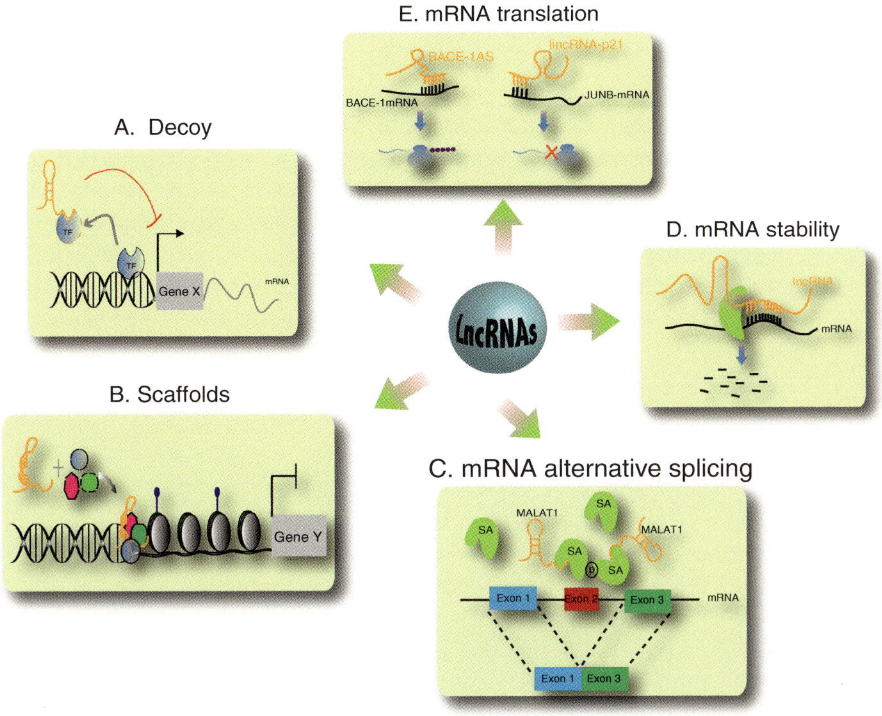

Fig. 10.2 The mechanisms of long noncoding RNAs (lncRNAs). LncRNAs function as regulatory elements in the genome to either "tether," guide, or scaffold the proteins with their functional motif to modify chromatin architecture or modulate coding gene expression in various levels. (**a**) A lncRNA binds a transcription factor (e.g., glucocorticoid receptors) and pulls the transcription factor away from the promoter of a gene and therefore turns off the gene expression; (**b**) lncRNAs bind chromatin-modifying enzymes (e.g., EZH2, LSD1) and change chromatin architecture that results in inhibition of gene expression; (**c–e**) some lncRNAs regulate gene expression through posttranscriptional modification from mRNA alternative splicing, mRNA stability to mRNA translation

the same time, there can be either a cis-regulation or trans-regulation of gene expression. Therefore, mechanisms of action of the lncRNAs are, to say the least, dramatically and extremely heterogeneous.

4 Strategies to Identify lncRNAs

Several lncRNAs have been identified and characterized in the past 20 years, but genome-wide identification of lncRNAs has only recently become possible with the advent of high-throughput or massively parallel sequencing technology. Here, we briefly describe four widely used approaches to purposely uncover potential functional lncRNAs.

4.1 Cloning Methods

In the early 1990s, cloning methods helped to find numerous functional genes and a few novel long noncoding RNAs. For example, Ji and Diederichs et al. established a cDNA library from subtractive hybridization between primary and metastatic lung cancer and subcloned each differentially displayed transcripts into vectors. Then, they sequenced each clone and finally identified MALAT1 as a long noncoding RNA associated with NSCLC [35]. This conventional molecular technique could be used in most experimental conditions, but it is tedious and labor intensive and it has been gradually replaced by more high-throughput approaches described below.

4.2 Microarray and Tiling Array

Regular microarray-based approaches are the most commonly used genome-wide screening methods to identify target genes, microRNAs [36, 37]. With annotation of lncRNAs, the lncRNA microarrays are increasingly applied to profile the lineage- or tissue-specific or disease-related lncRNA expressions [38, 39]. The limitation of regular lncRNA microarray is that the detection of lncRNAs depends entirely on known-lncRNA probes on the microarray; therefore, they are inherently incapable of identifying novel lncRNAs and splicing variants. Unlike regular microarrays, DNA tiling arrays contain oligonucleotide probes covering an entire length of a defined genomic DNA segmentation. Variable mapping resolutions are obtainable by adjusting the amount of sequence overlap between probes or the amount of known base pairs between probe sequences as well as by adjusting the probe length. This unique design is capable of "fishing out" novel lncRNAs without prior knowledge of their precise locations within the particular region [40, 41]. In order to understand the role of the region of human HOX genes in development, Rinn et al. [31] synthesized 400,000 probes of 50 bases in length with each probe overlapping the next one by 45 bases to cover all four human HOX gene clusters. This configuration allowed for the identification of hybridized DNA sequence at 5-bp resolution. HOTAIR was identified with this particular HOX tiling array, and then using the same HOX tiling array, HOTAIR was found to be associated with breast cancer metastasis [42].

4.3 RNA-seq

With the advent of massively parallel sequencing technology, high-throughput RNA sequencing is increasingly employed in the discovery of long noncoding RNAs in human genome and disease-associated studies.

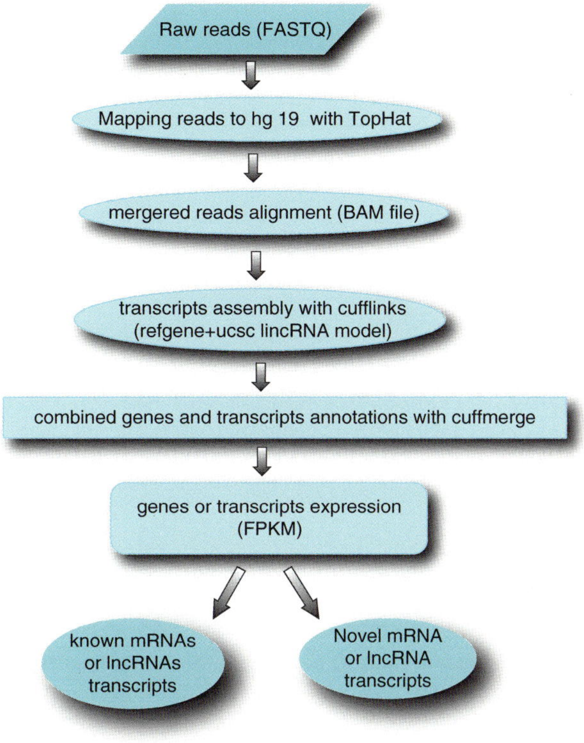

Fig. 10.3 RNA-seq analysis to identify long-noncoding RNAs with Galaxy platform (https://main.g2.bx.psu.edu/). Short DNA sequencing reads with fastq format are mapped to latest human reference genome hg19 with TopHat software; this generates files with BAM format which is binary version of sequence alignment and map format. Under the gene and transcript model of refgene and UCSC lincRNA database, the mapped BAM files are analyzed with Cufflink and Cuffmerge software to generate genes or transcripts expression level measured with FPKM (fragments per kilobase of transcripts per million mapped reads)

Depending on the way the RNA library is established, several modified methods of high-throughput RNA-seq are used, such as SAGEs (serial analysis of gene expression) [14], 3-SEQ (3′-end sequencing for expression quantification) [43], and whole transcriptome RNA-seq. RNA-seq has two clear advantages: (1) genome-wide scale at single-nucleotide resolution and (2) ability to search for novel lncRNA and their variants. The major drawbacks are high cost and the need for special data analysis. In general, RNA-seq is begun with either total RNA or polyadenylated RNAs. Polyadenylated RNA is frequently used for RNA-seq to avoid the presence of large abundant rRNAs and tRNAs, but this will compromise the prevalence of nonpolyadenylated lncRNAs. To solve that problem, selectively removing rRNA or tRNA from total RNA could preserve the nonpolyadenylated lncRNA.

We are participating in a collaborative project, referred to Medulloblastoma Advanced Genomics International Consortium (MAGIC), and we are focusing on discovery of lncRNAs in pediatric medulloblastoma tissues. With pilot RNA-seq data, we used the RNA-seq analysis pipeline with the available Galaxy platform to identify a noncoding RNA called PVT1 (for *p*lasmacytoma *v*ariant *t*ranslocation) that is significantly more highly expressed in a subgroup of medulloblastoma.

Figure 10.3 illustrates our workflow to analyze the RNA-seq to discover known or novel lncRNA transcripts.

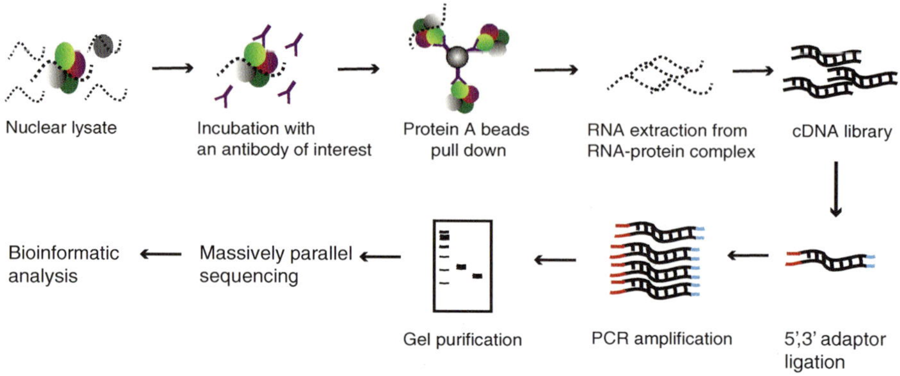

Fig. 10.4 Schematic of RNA immunoprecipitation sequencing assay (RIP-seq). RIP-seq is a combination of an antibody-based immunoprecipitation technique and high-throughput DNA sequence technology to map RNA–protein interaction in vivo and allows to identify all protein-of-interest-bound transcripts including long noncoding RNAs

4.4 RIP-seq/CLIP-seq

To study the interaction between lncRNA transcripts and proteins, RNA immunoprecipitation (RIP) could be applied, followed by RNA-seq procedure (RIP-seq) to reveal the genome-wide lncRNA transcripts binding particular proteins [30]. Figure 10.4 illustrates the workflow of RIP-seq. Alternatively, cross-link immunoprecipitation (CLIP) can also be used to find the lncRNA transcript interacting with proteins.

5 LncRNAs and Cancer

The cancer transcriptome is more complex than previously believed. In addition to protein-coding genes and microRNAs, dysregulation of lncRNAs is becoming a pervasive component in the gene regulatory network for cancer development and progression [44–47].

The cancer-related lncRNA, MALAT1, was identified during screening early non-small cell lung cancer (NSCLC) with or without metastasis via a subtractive hybridization approach [35]. High MALAT1 expression could be predictive of poor prognosis and a worse survival time in early stage NSCLC. The well-studied lncRNA, HOTAIR (HOX gene antisense transcript in Chr12 of HOXC locus), was uncovered in a panel of breast cancer with metastasis utilizing HOX tiling array. Subsequently, highly expressed HOTAIR has been found in several solid tumors (colorectal, pancreatic, laryngeal squamous, and gastrointestinal stromal sarcoma) and is associated with poor prognosis and less survival time [38, 44, 45, 48]. Another lncRNA, taurine upregulated gene 1 (TUG1), is overexpressed in bladder urothelial carcinoma comparing to paired normal urothelia tissue [49].

Prensner et al. [50] employed RNA-seq on a cohort of 102 prostate tissues and cell lines and discovered 121 prostate cancer-associated unannotated noncoding RNA transcripts (PCATs) with cancer-specific expression patterns. Among these, the authors characterized PCAT-1 as a novel prostate-specific regulator of cell proliferation and target of the polycomb repressive complex 2 (PRC2). Higher PCAT-1 and PRC2 expression stratified patient tissues into molecular subtypes distinguished by expression signatures of PCAT-1-repressed target genes.

Antisense lncRNA, ANRIL (antisense lncRNA of the *INK4* locus), spans an estimated region of 30–40 kb on Chr 9p21, and its expression correlates with *INK4a* epigenetic silencing [51]. Altered ANRIL activity might result in dysregulated silencing of the *INK4b/ARF/INK4a* locus, contributing to cancer initiation. These lncRNAs are functionally considered as "Onco-lncRNAs."

Tumor suppression is a cellular defense mechanism that prevents the neoplastic transformation of normal cells, and the tumor suppressor p53 plays a central role in tumor suppression. Not only are protein-coding genes activated by p53 to achieve the tumor suppressor effects of p53 through cell-cycle arrest or apoptosis, but lncRNAs are also involved in this p53-mediated regulatory network. Several p53-dependent lncRNAs such as lncRNA-p21 have been identified in response to DNA damage [47]. LncRNA-p21 is a ~3 kb transcript, located in the proximity of the cell-cycle regulator gene, *Cdkn1a*, and it was found to be a direct target of p53 in response to DNA damage. LincRNA-p21 serves as a repressor in p53-dependent transcriptional responses, and the transcriptional repression by lincRNA-p21 is mediated through the physical association with hnRNP-K, a protein that binds the promoters of genes involved in p53 signaling [47]. Nevertheless, it is unclear how lncRNA-p21 expression changes in human cancer tissues.

MEG3 lncRNA is a maternally expressed imprinted lncRNA on Chr14q32, and it is believed to be a tumor suppressor lncRNA [52]. This is evident with the following observations: (a) loss of MEG3 lncRNA was detected in most of human cancer cell lines, (b) overexpression MEG3 lncRNA inhibits cell proliferation, and (c) MEG3 has been linked to activate p53 signaling pathway through promoting p53 binding to target gene promoters [53]. Hypermethylation of MEG3 lncRNA promoter seems to be the main mechanism of its expression inhibition [54]. PVT1 is a large (>300 kb) locus located adjacent to the c-Myc locus on human chromosome 8q24, and it produces a wide variety of spliced noncoding RNAs as well as a cluster of six annotated microRNAs [55]. PVT1was found to fuse with the Myc gene in a subtype of pediatric brain tumors based on whole genome sequencing. This Myc–PVT1 fusion may contribute to the genesis of medulloblastoma [56]. Conversely, p53-dependent induction of PVT1 may cause cell death [57]. Therefore, the dual effect of PVT1 in the genome seems to depend upon the cellular information content, and the detailed functionality of PVT1 remains a mystery.

It is now clear that the expression of lncRNAs changes dynamically during different stages of cancer development and the course of treatment. When HeLa cells or breast cancer MCF-7 cells were treated with DNA-damage reagents (either

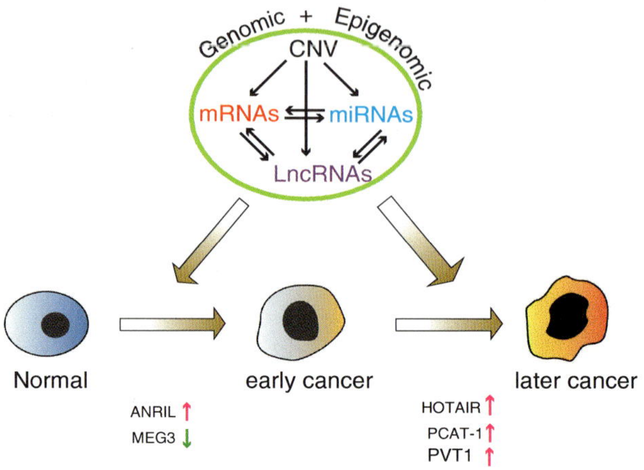

Fig. 10.5 The role of long noncoding RNAs in cancer development. Cancer is a genetic disorder, resulting from the accumulated genetic and epigenetic changes in the genome. With the decoding of cancer cell genomes, there are interactions between copy number variations, aberrant gene expression, and small noncoding RNAs (i.e., microRNAs) that contribute to pathogenesis of cancer. Long noncoding RNAs are the challenging new players in the processing of information that leads to malignant transformation

chemo- or radiotherapy), HOTAIR and MALAT1 expressions are decreased, while expressions of lncRNA-p21, GAS5, MEG3, ANRIL, and ncRNA-CCND1 are upregulated and some others, such as TUG1, UCA1, and PANDA, are not affected [58]. The mechanisms of such changes remain to be further investigated.

Taken together, the lncRNAs are new players in the cancer genome, and whole transcriptional profiling of lncRNAs (the "lncRNAome") in all cancer types will deepen our understanding of how lncRNAs interact with well-known protein-coding genes and microRNAs. LncRNAs are clearly involved in the information-processing networks in the transformation of normal cells to specific malignant states (see Fig. 10.5). We suggest that lncRNAs may be new targets to consider for reengineering genomically encoded regulatory networks and for novel cancer treatments that reprogram cell fates.

6 Summary

The central dogma of molecular biology relegates RNAs to the role of "messengers" of genetic information, with proteins as the end products that perform key roles as regulators and effectors of biological processes. Our understanding of the role of RNA in information processing is rapidly changing to better fit its abundance in the genome. There are numerous noncoding transcripts in addition to those housekeeping transcripts (ribosomal RNAs and transfer RNAs) that participate in

Table 10.1 Selected list of aberrant expression of lncRNA in cancers

Type of cancer	Associated lncRNA	Methods used	Functionality and mechanism	References
Breast cancer	HOTAIR on Chr.12q13.13	HOX tiling array	Promote metastasis through reprogramming chromatin	[42]
	GAS5	Functional expression cloning	Promote growth arrest and apoptosis	[59]
Colorectal cancer	HOTAIR	RT-PCR and gene expression array	Promote invasion and metastasis via increasing SUZ2, EZH2, and H3K27me3	[38]
Prostate cancer	PRNCR1 on Chr.8q24	Mapping and re-sequencing	Activation of androgen receptors	[60]
	PCAT-1 on Chr.8q24	RNA-seq	Inhibition of target gene expression	[50]
Lung cancer	MALAT1 on Chr.11q13	Subtractive hybridization	Regulator of metastatic gene expression and promote migration	[35, 61]
Pancreatic cancer	HOTAIR overexpression	Q-RT-PCR	Binding EZH2 to modify chromatin architecture	[45]
Laryngeal squamous cell carcinoma	HOTAIR overexpression	Q-RT-PCR	Binding EZH2 to modify chromatin architecture	[44]
Gastrointestinal stromal tumor	HOTAIR overexpression	Q-RT-PCR	Binding EZH2 to modify chromatin architecture	[48]
Medulloblastoma	PVT1	RNA-seq	Fusion with c-Myc gene	[56]

the processes of protein production within cells, and they include the large number of small RNAs such as microRNAs, piwiRNAs, and snoRNAs. In addition to these relatively small transcripts, there are considerably longer noncoding RNA transcripts that play important roles in genetic information processing within the genome, and we have reviewed our current understanding of lncRNA biology. We presented experimental approaches and strategies for systematic characterization of lncRNA activities, with applications to known gene regulatory networks and diseases. The list of lncRNAs associated with cancer is increasing (Table 10.1); however, relatively little is known about the precise functions of most cancer-associated lncRNAs. Nevertheless, what is currently known suggests that lncRNA transcripts function in many different ways to process information within cells and promote cancer development and progression from normal to multiple malignant cell states.

If the cancer initiatome is the collection of all causes of malignant transformation, it now extends well beyond the protein-coding genome and it goes deep into the noncoding genome. A more thorough and deeper understanding of the structure of the cancer initiatome will show us how information flows through this genomically encoded system. We believe this understanding will suggest novel targets for reengineering the cancer genome and offer new approaches to cancer treatment.

Acknowledgements J.A.C. is an Alberta Innovates Health Solutions (AIHS) Clinical Investigator. J.A.C. and W.W. are supported by funds from the Kids Cancer Care Foundation of Alberta, Genome Canada, the Clark H. Smith Brain Tumor Centre, and the Ross Family Fund and Family of Kathleen Lorette. We are grateful to Dr. Fred Biddle for the stimulating discussion about cancer initiatome and for critical reading of the manuscript.

References

1. Dunham I, Kundaje A, Aldred SF, Collins PJ, Davis CA, Doyle F, et al. An integrated encyclopedia of DNA elements in the human genome. Nature. 2012;489(7414):57–74. Epub 2012/09/08.
2. Wu W, Sun M, Zou GM, Chen J. MicroRNA and cancer: current status and prospective. Int J Cancer. 2007;120(5):953–60. Epub 2006/12/14.
3. Okazaki Y, Furuno M, Kasukawa T, Adachi J, Bono H, Kondo S, et al. Analysis of the mouse transcriptome based on functional annotation of 60,770 full-length cDNAs. Nature. 2002;420(6915):563–73. Epub 2002/12/06.
4. Brannan CI, Dees EC, Ingram RS, Tilghman SM. The product of the H19 gene may function as an RNA. Mol Cell Biol. 1990;10(1):28–36. Epub 1990/01/01.
5. Brown CJ, Ballabio A, Rupert JL, Lafreniere RG, Grompe M, Tonlorenzi R, et al. A gene from the region of the human X inactivation centre is expressed exclusively from the inactive X chromosome. Nature. 1991;349(6304):38–44. Epub 1991/01/03.
6. Derrien T, Johnson R, Bussotti G, Tanzer A, Djebali S, Tilgner H, et al. The GENCODE v7 catalog of human long noncoding RNAs: analysis of their gene structure, evolution, and expression. Genome Res. 2012;22(9):1775–89. Epub 2012/09/08.
7. Guttman M, Amit I, Garber M, French C, Lin MF, Feldser D, et al. Chromatin signature reveals over a thousand highly conserved large non-coding RNAs in mammals. Nature. 2009;458(7235):223–7. Epub 2009/02/03.
8. Calin GA, Liu CG, Ferracin M, Hyslop T, Spizzo R, Sevignani C, et al. Ultraconserved regions encoding ncRNAs are altered in human leukemias and carcinomas. Cancer Cell. 2007;12(3):215–29. Epub 2007/09/06.
9. Chen LL, Carmichael GG. Long noncoding RNAs in mammalian cells: what, where, and why? Wiley Interdiscip Rev RNA. 2010;1(1):2–21. Epub 2010/07/01.
10. Lipovich L, Johnson R, Lin CY. MacroRNA underdogs in a microRNA world: evolutionary, regulatory, and biomedical significance of mammalian long non-protein-coding RNA. Biochim Biophys Acta. 2010;1799(9):597–615. Epub 2010/10/19.
11. Cheng J, Kapranov P, Drenkow J, Dike S, Brubaker S, Patel S, et al. Transcriptional maps of 10 human chromosomes at 5-nucleotide resolution. Science. 2005;308(5725):1149–54. Epub 2005/03/26.
12. Cabili MN, Trapnell C, Goff L, Koziol M, Tazon-Vega B, Regev A, et al. Integrative annotation of human large intergenic noncoding RNAs reveals global properties and specific subclasses. Genes Dev. 2011;25(18):1915–27. Epub 2011/09/06.
13. Novikova IV, Hennelly SP, Sanbonmatsu KY. Sizing up long non-coding RNAs: do lncRNAs have secondary and tertiary structure? Bioarchitecture. 2012;2(6):189–99. Epub 2012/12/06.
14. Gibb EA, Vucic EA, Enfield KS, Stewart GL, Lonergan KM, Kennett JY, et al. Human cancer long non-coding RNA transcriptomes. PLoS One. 2011;6(10):e25915. Epub 2011/10/13.

15. Ellis BC, Molloy PL, Graham LD. CRNDE: a long non-coding RNA involved in CanceR, Neurobiology, and DEvelopment. Front Genet. 2012;3:270. Epub 2012/12/12.
16. Kapranov P, Cheng J, Dike S, Nix DA, Duttagupta R, Willingham AT, et al. RNA maps reveal new RNA classes and a possible function for pervasive transcription. Science. 2007;316(5830):1484–8. Epub 2007/05/19.
17. Clark MB, Johnston RL, Inostroza-Ponta M, Fox AH, Fortini E, Moscato P, et al. Genome-wide analysis of long noncoding RNA stability. Genome Res. 2012;22(5):885–98. Epub 2012/03/13.
18. Rinn JL, Chang HY. Genome regulation by long noncoding RNAs. Annu Rev Biochem. 2012;81:145–66. Epub 2012/06/06.
19. Mercer TR, Qureshi IA, Gokhan S, Dinger ME, Li G, Mattick JS, et al. Long noncoding RNAs in neuronal-glial fate specification and oligodendrocyte lineage maturation. BMC Neurosci. 2010;11:14. Epub 2010/02/09.
20. Tripathi V, Ellis JD, Shen Z, Song DY, Pan Q, Watt AT, et al. The nuclear-retained noncoding RNA MALAT1 regulates alternative splicing by modulating SR splicing factor phosphorylation. Mol Cell. 2010;39(6):925–38. Epub 2010/08/28.
21. Kretz M, Webster DE, Flockhart RJ, Lee CS, Zehnder A, Lopez-Pajares V, et al. Suppression of progenitor differentiation requires the long noncoding RNA ANCR. Genes Dev. 2012;26(4):338–43. Epub 2012/02/04.
22. Kretz M, Siprashvili Z, Chu C, Webster DE, Zehnder A, Qu K, et al. Control of somatic tissue differentiation by the long non-coding RNA TINCR. Nature. 2013;493:231–5. Epub 2012/12/04.
23. Loewer S, Cabili MN, Guttman M, Loh YH, Thomas K, Park IH, et al. Large intergenic non-coding RNA-RoR modulates reprogramming of human induced pluripotent stem cells. Nat Genet. 2010;42(12):1113–7. Epub 2010/11/09.
24. Wapinski O, Chang HY. Long noncoding RNAs and human disease. Trends Cell Biol. 2011;21(6):354–61. Epub 2011/05/10.
25. Lee JT. Lessons from X-chromosome inactivation: long ncRNA as guides and tethers to the epigenome. Genes Dev. 2009;23(16):1831–42. Epub 2009/08/18.
26. Lee MP, DeBaun MR, Mitsuya K, Galonek HL, Brandenburg S, Oshimura M, et al. Loss of imprinting of a paternally expressed transcript, with antisense orientation to KVLQT1, occurs frequently in Beckwith-Wiedemann syndrome and is independent of insulin-like growth factor II imprinting. Proc Natl Acad Sci U S A. 1999;96(9):5203–8. Epub 1999/04/29.
27. Zhao J, Sun BK, Erwin JA, Song JJ, Lee JT. Polycomb proteins targeted by a short repeat RNA to the mouse X chromosome. Science. 2008;322(5902):750–6. Epub 2008/11/01.
28. Tsai MC, Manor O, Wan Y, Mosammaparast N, Wang JK, Lan F, et al. Long noncoding RNA as modular scaffold of histone modification complexes. Science. 2010;329(5992):689–93. Epub 2010/07/10.
29. Khalil AM, Guttman M, Huarte M, Garber M, Raj A, Rivea Morales D, et al. Many human large intergenic noncoding RNAs associate with chromatin-modifying complexes and affect gene expression. Proc Natl Acad Sci U S A. 2009;106(28):11667–72. Epub 2009/07/03.
30. Zhao J, Ohsumi TK, Kung JT, Ogawa Y, Grau DJ, Sarma K, et al. Genome-wide identification of polycomb-associated RNAs by RIP-seq. Mol Cell. 2010;40(6):939–53. Epub 2010/12/22.
31. Rinn JL, Kertesz M, Wang JK, Squazzo SL, Xu X, Brugmann SA, et al. Functional demarcation of active and silent chromatin domains in human HOX loci by noncoding RNAs. Cell. 2007;129(7):1311–23. Epub 2007/07/03.
32. Gong C, Maquat LE. lncRNAs transactivate STAU1-mediated mRNA decay by duplexing with 3′ UTRs via Alu elements. Nature. 2011;470(7333):284–8. Epub 2011/02/11.
33. Poliseno L, Salmena L, Zhang J, Carver B, Haveman WJ, Pandolfi PP. A coding-independent function of gene and pseudogene mRNAs regulates tumour biology. Nature. 2010;465(7301):1033–8. Epub 2010/06/26.
34. Yoon JH, Abdelmohsen K, Srikantan S, Yang X, Martindale JL, De S, et al. LincRNA-p21 suppresses target mRNA translation. Mol Cell. 2012;47(4):648–55. Epub 2012/07/31.

35. Ji P, Diederichs S, Wang W, Boing S, Metzger R, Schneider PM, et al. MALAT-1, a novel noncoding RNA, and thymosin beta4 predict metastasis and survival in early-stage non-small cell lung cancer. Oncogene. 2003;22(39):8031–41. Epub 2003/09/13.
36. Lu J, Getz G, Miska EA, Alvarez-Saavedra E, Lamb J, Peck D, et al. MicroRNA expression profiles classify human cancers. Nature. 2005;435(7043):834–8.
37. O'Donnell KA, Wentzel EA, Zeller KI, Dang CV, Mendell JT. c-Myc-regulated microRNAs modulate E2F1 expression. Nature. 2005;435(7043):839–43.
38. Kogo R, Shimamura T, Mimori K, Kawahara K, Imoto S, Sudo T, et al. Long noncoding RNA HOTAIR regulates polycomb-dependent chromatin modification and is associated with poor prognosis in colorectal cancers. Cancer Res. 2011;71(20):6320–6. Epub 2011/08/25.
39. Ng SY, Johnson R, Stanton LW. Human long non-coding RNAs promote pluripotency and neuronal differentiation by association with chromatin modifiers and transcription factors. EMBO J. 2012;31(3):522–33. Epub 2011/12/24.
40. Kapranov P, Cawley SE, Drenkow J, Bekiranov S, Strausberg RL, Fodor SP, et al. Large-scale transcriptional activity in chromosomes 21 and 22. Science. 2002;296(5569):916–9. Epub 2002/05/04.
41. Rinn JL, Euskirchen G, Bertone P, Martone R, Luscombe NM, Hartman S, et al. The transcriptional activity of human Chromosome 22. Genes Dev. 2003;17(4):529–40. Epub 2003/02/26.
42. Gupta RA, Shah N, Wang KC, Kim J, Horlings HM, Wong DJ, et al. Long non-coding RNA HOTAIR reprograms chromatin state to promote cancer metastasis. Nature. 2010;464(7291):1071–6. Epub 2010/04/16.
43. Brunner AL, Beck AH, Edris B, Sweeney RT, Zhu SX, Li R, et al. Transcriptional profiling of lncRNAs and novel transcribed regions across a diverse panel of archived human cancers. Genome Biol. 2012;13(8):R75. Epub 2012/08/30.
44. Li D, Feng J, Wu T, Wang Y, Sun Y, Ren J, et al. Long intergenic noncoding RNA HOTAIR is overexpressed and regulates PTEN methylation in laryngeal squamous cell carcinoma. Am J Pathol. 2013;182:64–70. Epub 2012/11/13.
45. Kim K, Jutooru I, Chadalapaka G, Johnson G, Frank J, Burghardt R, et al. HOTAIR is a negative prognostic factor and exhibits pro-oncogenic activity in pancreatic cancer. Oncogene. 2013;32:1616–25. Epub 2012/05/23.
46. Tsai MC, Spitale RC, Chang HY. Long intergenic noncoding RNAs: new links in cancer progression. Cancer Res. 2011;71(1):3–7. Epub 2011/01/05.
47. Huarte M, Guttman M, Feldser D, Garber M, Koziol MJ, Kenzelmann-Broz D, et al. A large intergenic noncoding RNA induced by p53 mediates global gene repression in the p53 response. Cell. 2010;142(3):409–19. Epub 2010/08/03.
48. Niinuma T, Suzuki H, Nojima M, Nosho K, Yamamoto H, Takamaru H, et al. Upregulation of miR-196a and HOTAIR drive malignant character in gastrointestinal stromal tumors. Cancer Res. 2012;72(5):1126–36. Epub 2012/01/20.
49. Han Y, Liu Y, Gui Y, Cai Z. Long intergenic non-coding RNA TUG1 is overexpressed in urothelial carcinoma of the bladder. J Surg Oncol. 2013;107:555–9. Epub 2012/09/11.
50. Prensner JR, Iyer MK, Balbin OA, Dhanasekaran SM, Cao Q, Brenner JC, et al. Transcriptome sequencing across a prostate cancer cohort identifies PCAT-1, an unannotated lincRNA implicated in disease progression. Nat Biotechnol. 2011;29(8):742–9. Epub 2011/08/02.
51. Yap KL, Li S, Munoz-Cabello AM, Raguz S, Zeng L, Mujtaba S, et al. Molecular interplay of the noncoding RNA ANRIL and methylated histone H3 lysine 27 by polycomb CBX7 in transcriptional silencing of INK4a. Mol Cell. 2010;38(5):662–74. Epub 2010/06/15.
52. Zhou Y, Zhang X, Klibanski A. MEG3 noncoding RNA: a tumor suppressor. J Mol Endocrinol. 2012;48(3):R45–53. Epub 2012/03/07.
53. Zhou Y, Zhong Y, Wang Y, Zhang X, Batista DL, Gejman R, et al. Activation of p53 by MEG3 non-coding RNA. J Biol Chem. 2007;282(34):24731–42. Epub 2007/06/16.
54. Benetatos L, Hatzimichael E, Dasoula A, Dranitsaris G, Tsiara S, Syrrou M, et al. CpG methylation analysis of the MEG3 and SNRPN imprinted genes in acute myeloid leukemia and myelodysplastic syndromes. Leuk Res. 2010;34(2):148–53. Epub 2009/07/15.

55. Beck-Engeser GB, Lum AM, Huppi K, Caplen NJ, Wang BB, Wabl M. Pvt1-encoded microRNAs in oncogenesis. Retrovirology. 2008;5:4. Epub 2008/01/16.
56. Northcott PA, Shih DJ, Peacock J, Garzia L, Morrissy AS, Zichner T, et al. Subgroup-specific structural variation across 1,000 medulloblastoma genomes. Nature. 2012;488(7409):49–56. Epub 2012/07/27.
57. Barsotti AM, Beckerman R, Laptenko O, Huppi K, Caplen NJ, Prives C. p53-Dependent induction of PVT1 and miR-1204. J Biol Chem. 2012;287(4):2509–19. Epub 2011/11/24.
58. Ozgur E, Mert U, Isin M, Okutan M, Dalay N, Gezer U. Differential expression of long non-coding RNAs during genotoxic stress-induced apoptosis in HeLa and MCF-7 cells. Clin Exp Med. 2012;13(2):119–26. Epub 2012/04/11.
59. Mourtada-Maarabouni M, Pickard MR, Hedge VL, Farzaneh F, Williams GT. GAS5, a non-protein-coding RNA, controls apoptosis and is downregulated in breast cancer. Oncogene. 2009;28(2):195–208. Epub 2008/10/07.
60. Chung S, Nakagawa H, Uemura M, Piao L, Ashikawa K, Hosono N, et al. Association of a novel long non-coding RNA in 8q24 with prostate cancer susceptibility. Cancer Sci. 2011;102(1):245–52. Epub 2010/09/30.
61. Gutschner T, Hammerle M, Eissmann M, Hsu J, Kim Y, Hung G, et al. The noncoding RNA MALAT1 is a critical regulator of the metastasis phenotype of lung cancer cells. Cancer Res. 2013;73:1180–9. Epub 2012/12/18.

Chapter 11
Identification of piRNAs in HeLa Cells by Massive Parallel Sequencing

Yilu Lu and Yongxin Ma

Abstract Piwi proteins and Piwi-interacting RNAs (piRNAs) have been implicated in transposon control in germ line from Drosophila to mammals. To examine the profile of small RNA transcriptome and explore the potential roles of Human Piwi-like 2 gene (alias HILI) and its associated piRNAs in human cancer cells, small RNA libraries prepared from HILI-overexpressed, HILI-knockdown, and control HeLa cells were respectively sequenced using Solexa, a next-generation massive parallel sequencing technology. A set of piRNAs and other repeat-associated small RNAs were observed in HeLa cells. By using in situ hybridization, piR-49322 was localized in the nucleolus and around the periphery of nuclear membrane in HeLa cells. Following the overexpression of HILI, the retrotransposon element LINE1 was significantly repressed, while LINE1-associated small RNAs decreased in abundance. The present study demonstrated that HILI along with piRNAs plays a role in LINE1 suppression in HeLa cancer cell line.

Keywords HILI • piRNA • Retrotransposon • HeLa • Solexa • Sequencing

1 Introduction

The Argonaute gene family, encoding basic proteins that contain both PAZ and Piwi domains and complex with distinct classes of small RNAs, takes part in an RNA silencing machinery using small RNA molecules as guides to identify homologous sequences and can induce histone and DNA methylation, deletion of DNA sequences, mRNA breakdown, and inhibition of translation [1, 2].

Y. Lu • Y. Ma (✉)
Division of Morbid Genomics, Department of Medical genetics, State Key Laboratory of Biotherapy, West China Hospital, Sichuan University, Chengdu 610041, China
e-mail: mayongxin@gmail.com

W. Wu and H. Choudhry (eds.), *Next Generation Sequencing in Cancer Research: Volume 1: Decoding the Cancer Genome*, DOI 10.1007/978-1-4614-7645-0_11,

Based on the phylogenetic analysis, known Argonaute proteins can be classified into two distinct subfamilies: the ubiquitous Ago clade and the predominantly germ line-specific Piwi clade. Recent discoveries showed that Piwi subfamily proteins can be found in all animals tested so far and play essential roles in stem-cell self-renewal, gametogenesis, and RNA silencing in diverse organisms [3–8]. Mutations of three Piwi homologs in mice (miwi, mili, and miwi2) respectively, cause arrestment of spermatogenesis and male sterility [9–11].

In contrast to their Ago-family cousins, the Piwi proteins do not associate with siRNAs or miRNAs. Only recently a separate class of 24- to 30-nt RNA molecules termed Piwi-interacting RNAs (piRNAs) was identified associating with Piwi proteins in germ line cells [12–15]. The biogenesis, characteristics, and functions of piRNAs have been extensively studied in *Caenorhabditis elegans*, Drosophila, zebrafish, and mice. It is now clear that piRNAs can be divided into two subclasses: repeat-associated interfering RNA (rasiRNA) and non-repetitive piRNA, which may be involved in different biogenesis mechanism and physiological functions [6, 7, 12, 14–24].

The Piwil2 gene, alias HILI in human, is a member of Piwi subfamily and required for spermatogonial stem-cell self-renewal [25, 26]. The loss of Piwil2 in zebrafish results in an increased level of transposon transcript and the inability of germ cells differentiation [27]. And in mice, Piwil2 knockout leads to meiotic defects in germ line stem cell and correlated with increased transposon activity [10]. In humans, the ectopic expression of Piwil2 is highly correlated with cancers, suggesting that Piwil2 may disturb cell division, inhibit apoptosis, and play a role as dose-dependent oncogenic fate determinants [28, 29]. Evidences indicated that Piwil2 can play a role in the development of precancer stem cells as well as directly regulate apoptosis related pathways in human cancers [30–33]. Recent study also showed that Piwil2 is a key factor in regulating chromatin modifications in response to DNA damage and therefore involves in cisplatin resistance in ovarian cancer cells [34, 35].

The relationship between Piwil2 and cancers suggested the potential roles of piRNAs in cancer cells. However, little has known about the expression and functions of piRNAs in human cancer cells. Here we discuss a newly developed massive parallel sequencing technology, Solexa, employed to investigate the expression profile of piRNAs in human cancer cells.

2 Materials and Methods

2.1 Cell Culture

HeLa cells were maintained in PRMI1640 (Gibco, USA) containing 10 % heat-inactivated FBS, 100 U/ml penicillin, and 100 μg/ml streptomycin and cultured at 37 °C in a humidified atmosphere of 5 % CO_2. Cells were placed in a freshly prepared 25 cm^2 culture flask and transfected with a Lipofectamine 2000 solution (Invitrogen, USA). The expression vector of HILI was constructed by obtaining complete CDS of human Piwil2 gene with a RT-PCR method and cloning it into a pcDNA3.1+ plasmid. The siRNA of HILI was synthesized and purchased from GenePharma Co. Ltd. (Shanghai, China). The transfection was carried out according to the manufacturer's protocol, and stable overexpression transfectants were selected by cultured in medium containing 1 mg/ml of G418 (Solarbio, China) for 1 month. The cell line was designated as HILI+, while the normal HeLa cell line was designated as NC- and siRNA-transfected cells as HILI−.

2.2 *Small RNA Libraries Preparation and Sequencing*

Total RNA was extracted from the prepared cells with TRIzol reagent (Invitrogen, USA) according to manufacturer's protocol. The quality of RNA samples was evaluated by a 2100 Bioanalyzer (Agilent, USA) and then size-fractionated on a 15 % PAGE gel to collect the 10–30-nt fraction. The 5′ and 3′ RNA adaptors were ligated to the RNA pool, following with a RT-PCR to produce sequencing libraries. PCR products were purified and sequenced using Solexa 1G Genome Analyzer (Beijing Genomics Institute, China).

After filtering out low-quality reads, trimming the adaptor and acceptor sequences, and cleaning up contaminants formed by ligation, clean reads were grouped and converted into unique sequences with associated counts of the individual reads. Those small RNA tags were mapped to genome by a SOAP program [36] to analyze their expression and distribution on the genome. Then small RNA tags were aligned with the sequences of noncoding RNAs (rRNA, tRNA, snRNA, snoRNA) available in Rfam (http://www.sanger.ac.uk/software/Rfam) and the GenBank noncoding RNA database (http://www.ncbi.nlm.nih.gov/) to get rid of matched tags from unannotated tags. In addition, all sequences were searched for miRNA precursor/mature miRNA in miRBase14.0 (http://www.mirbase.org/) and piRNA in NCBI database with screening program developed by Beijing Genomics Institute.

2.3 *In Situ Hybridization*

The DIG-labeled LNA (locked nucleic acid) detection probe was purchased (Exiqon, Denmark) to localize piR-49322 (GenBank accession number: DQ581210.1) in cultured HeLa cells. The cells were fixed for 10 min at room temperature in PBS solution containing 4 % formaldehyde and then permeabilized with 70 % ethanol at 4 °C overnight. After rehydrated in 2×SSC containing 50 % formamide for 5 min at room temperature, the cells are hybridized at 37 °C for 1 h in 40 μl 2×SSC of a mixture containing 10 % dextran sulfate, 2 mM vanadyl-ribonucleoside complex, 0.02 % RNAse-free BSA, 50 % formamide, 40 μg *E. coli* tRNA, and 30 ng of probe.

Then the cells were washed in 0.2×SSC at 60 °C for 1 h and incubated with 1:2,000 diluted anti-DIG-alkaline phosphatase antibody at 4 °C overnight. The method above referenced manufacturer's protocol and Singer's protocol of ISH of mammalian cells (http://www.singerlab.org/protocols). As negative control, cells were pretreated with RNAse (100 μg/ml) for 1 h before fixation.

2.4 Real-Time RT-PCR

Reverse transcription reactions were performed in a final volume of 20 μl containing 2 μl purified total RNA, 1×RT Buffer (Fermentas, USA), 0.25 mM each of dNTPs, 10 U/μl RevertAid M-MuLV Reverse Transcriptase (Fermentas, USA), 1 U/μl RNase inhibitor (Fermentas, USA), and 0.2 μg random primers. The reactions were incubated in Dyad Peltier Thermo Cycler (BioRad, USA) at 37 °C for 10 min, 42 °C for 60 min, 70 °C for 10 min, and then held in 4 °C.

Following reverse transcription, 2 μl equivalents of cDNAs were used as a template for each PCR reaction in a final volume of 30 μl containing 1×PCR Buffer, 2.5 mM $MgCl_2$, 0.3 mM each of dNTPs, 1.5 U PCR Polymerase (TaKaRa, Japan), 300 nM TaqMan Probe, and 300 nM each of primers.

Quantitative PCRs were performed in an iCycler IQ real-time PCR Detection System (BioRad, USA), with a first denaturation step at 94 °C for 10 min, followed by 45 cycles comprising denaturation at 94 °C for 20s, annealing at 58 °C for 30s, and extension at 60 °C for 40s. To normalize for differences in the amount of total RNA in each RT-PCR, the expression of β-actin gene was performed as an internal control as well as positive control. Negative controls were also established and all experiments were run in triplicate. All synthetic oligonucleotides were purchased from (Invitrogen, USA).

3 Results

3.1 Overall Analysis of Small RNAs Among the Libraries

By using Solexa, a newly developed high-throughput sequencing technology producing highly accurate, reproducible, and quantitative readouts of small RNAs, three libraries of small RNAs from normal (NC), HILI-overexpressed (HILI+), and HILI-knockdown (HILI−) human HeLa cells were established and sequenced. After filtering out low-quality reads, trimming the adaptor and acceptor sequences, and cleaning up contaminants formed by ligation, 6,704,180 clean reads out of 9,435,055 in total (77.39 %) were obtained from NC library, while 8,850,724 (86.68 %) from HILI+ and 10,410,335 (91.85 %) from HILI−. For analysis, raw reads were grouped and converted into unique sequences with associated counts of the individual reads,

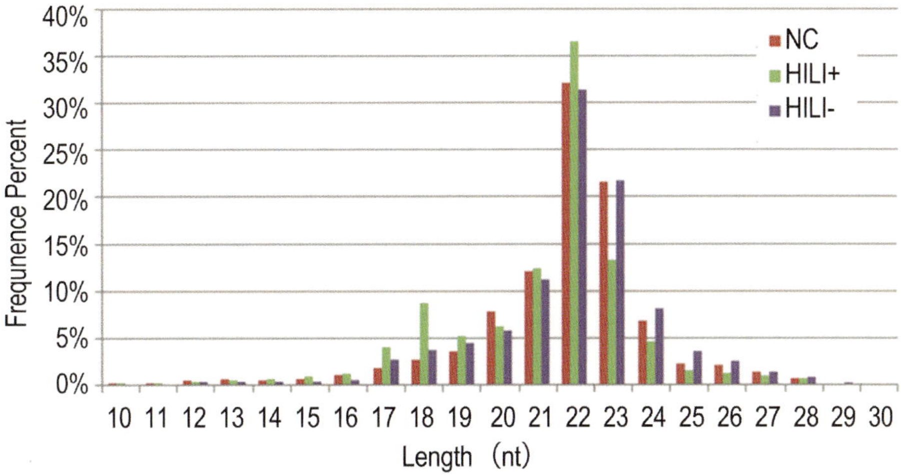

Fig. 11.1 Overview of small RNA libraries. The histogram shows length distribution of small RNAs (18–30 nt) in three libraries. The figure is adapted from Lu et al. [47] with permission

generating 403,212 (NC), 464,987 (HILI+), and 473,365 (HILI–) identical unique sequences, respectively.

The length distribution of high-quality reads showed approximately consistency among the three RNA libraries. In NC library, 73.88 % of clean reads were at a range of 20–23 nt, which was considered as a canonical length of miRNA, while 68.57 % and 70.26 % of reads met this length in HILI+ and HILI– libraries, respectively. Interestingly, a significant 18 nt peak including 837,685 of 9,649,654 reads (8.68 %) was observed in HILI+, but not HILI– (3.68 %) and NC (2.80 %) libraries (Fig. 11.1).

3.2 Small RNA Alignment to Genome

The clean reads from three libraries were used to map the human genome (GRCh37) using the SOAP program and then localized on each chromosome. 78.68 % of clean reads (5,274,732 out of 6,704,180) in NC, 79.79 % in HILI+ and 84.35 % in HILI– were successfully mapped to genome and the distribution of small RNAs across different chromosomes had a generally similarity among three libraries. However, some significant difference may suggest potential targets of Piwi-piRNA silencing system or region of chromatin modification regulated by Piwil2. For example, a great number of sequences (39.29 %) in HILI+ were mapped on the plus strand of chromosome 9, while only 13.18 % in NC and 18.17 % in HILI– were located on the same region. Another significant difference was on the minus strand of chromosome X, as 25.61 % in HILI+ but only 5.42 % and 10.24 %, respectively, in NC and HILI– libraries (Table 11.1).

Table 11.1 The chromosome distribution of small RNAs mapped to human genome in each libraries

	NC (%)		HILI+ (%)		HILI− (%)	
Chr.[a]	+	−	+	−	+	−
1	3.99	4.76	7.89	7.98	3.69	3.95
2	2.24	3.21	2.47	2.14	4.73	2.16
3	1.36	9.41	1.51	7.52	1.04	5.59
4	0.59	0.66	0.72	0.63	0.64	0.78
5	6.95	1.87	6.34	2.24	3.24	1.86
6	4.36	5.45	7.93	5.59	4.10	7.78
7	1.81	8.79	1.62	6.14	4.13	4.95
8	0.83	3.96	0.89	3.75	0.81	5.40
9	13.18	1.74	39.29	1.35	18.17	1.65
10	0.56	1.06	0.43	1.01	0.51	3.89
11	1.02	12.66	1.86	15.82	0.99	10.41
12	4.25	2.18	2.99	1.48	4.30	1.60
13	1.29	0.70	0.95	1.34	1.06	0.87
14	1.20	0.90	1.44	0.86	1.02	1.00
15	0.71	0.89	0.77	0.58	0.68	0.74
16	3.36	2.01	6.04	4.95	2.28	1.45
17	13.47	10.37	6.20	5.49	12.24	5.73
18	0.54	0.48	0.67	0.40	0.40	0.52
19	3.18	3.25	1.68	2.27	1.64	2.70
20	1.09	0.56	0.88	0.33	1.16	0.57
21	2.12	1.16	1.45	0.35	1.23	0.80
22	19.76	0.37	19.35	0.23	15.40	0.25
X	2.13	5.42	1.82	25.61	4.75	10.24
Y	0.40	0.11	0.55	0.09	0.36	0.11

[a]The plus (+) and minus (−) strands are counted relatively

By annotating the small RNA tags with known RNA sequences from database such as GenBank and Rfam, matched tags were divided into different categories of small RNAs. The majority fraction of clean reads (57.33 %, 58.86 %, and 49.49 % in each library, respectively) was identified as miRNA sequences. Another two significant fractions of the clean reads were derived from unannotated sequences (18.30 %, 17.22 %, and 12.29 %, respectively) and putative degradation products of rRNAs, tRNAs, small nuclear RNAs, and other noncoding RNAs (22.31 %, 20.88 %, 35.88 %, respectively). Read counts and genomic distribution categories were summarized in Tables 11.2, 11.3, and 11.4.

3.3 *piRNAs in HeLa Cells*

To investigate whether piRNAs, which previously reported to express only in germ cells, also exist in cancer cell lines such as HeLa cells, we compared our datasets to

Table 11.2 Identified small RNA species in NC library

Category	Unique sRNA	%	Total sRNA	%
Total	403,212	100	6,704,180	100
rRNA	60,377	14.97	949,400	14.16
scRNA	1,695	0.42	21,595	0.32
snRNA	4,118	1.02	79,121	1.18
snoRNA	2,530	0.63	42,321	0.63
srpRNA	96	0.02	222	0.00
tRNA	22,850	5.67	403,677	6.02
Exon_antisense	2,009	0.50	2,906	0.04
Exon_sense	49,066	12.17	73,321	1.09
Intron_antisense	5,724	1.42	9,416	0.14
Intron_sense	13,496	3.35	33,526	0.50
miRNA	5,017	1.24	3,843,528	57.33
piRNA	433	0.11	10,778	0.16
Repeat	18,425	4.57	142,453	2.12
LINE	4,764	25.86[a]	70,638	49.59[a]
SINE	5,254	28.52[a]	14,753	10.36[a]
LTR	3,094	16.79[a]	6,114	4.29[a]
unann	221,346	54.90	1,226,919	18.30

[a]Percentages are given with respect to repeat-derived small RNAs

Table 11.3 Identified small RNA species in HILI+ library

Category	Unique sRNA	%	Total sRNA	%
Total	464,987	100	8,850,724	100
rRNA	41,172	8.85	763,343	8.62
scRNA	1,300	0.28	24,878	0.28
snRNA	4,271	0.92	202,273	2.29
snoRNA	2,101	0.45	34,263	0.39
srpRNA	117	0.03	697	0.01
tRNA	19,801	4.26	822,665	9.29
Exon_antisense	2,154	0.46	2,907	0.03
Exon_sense	113,183	24.34	180,164	2.04
Intron_antisense	7,848	1.69	11,865	0.13
Intron_sense	13,829	2.97	24,702	0.28
miRNA	4,611	0.99	5,225,012	59.03
piRNA	460	0.10	11,406	0.13
Repeat	25,859	5.56	138,238	1.56
LINE	9,520	36.82[a]	59,448	43.00[a]
SINE	4,669	18.06[a]	8,485	6.14[a]
LTR	5,019	19.41[a]	7,420	5.37[a]
unann	232,100	49.92	1,524,279	17.22

[a]Percentages are given with respect to repeat-derived small RNAs

non-repetitive piRNAs from NCBI database. Among 6,704,180 clean reads screened in NC library, 433 unique sequences with 10,778 counts were successfully aligned to known human piRNAs from NCBI database, while 460 unique sequences in HILI+ and 525 in HILI–, respectively.

Table 11.4 Identified small RNA species in HILI– library

Category	Unique sRNA	%	Total sRNA	%
Total	473,365	100	10,410,335	100
rRNA	59,758	12.62	2,825,796	27.14
scRNA	1,510	0.32	20,044	0.19
snRNA	5,451	1.15	161,157	1.55
snoRNA	2,664	0.56	32,828	0.32
srpRNA	162	0.03	666	0.01
tRNA	22,681	4.79	694,163	6.67
Exon_antisense	3,704	0.78	5,626	0.05
Exon_sense	114,955	24.28	140,980	1.35
Intron_antisense	11,309	2.39	14,557	0.14
Intron_sense	22,811	4.82	42,580	0.41
miRNA	5,153	1.09	5,177,074	49.73
piRNA	525	0.11	16,867	0.16
Repeat	30,251	6.39	162,256	1.56
LINE	8,907	29.44[a]	64,519	39.76[a]
SINE	7,511	24.81[a]	22,173	13.67[a]
LTR	5,619	18.57[a]	8,173	5.04[a]
unann	197,343	41.69	1,279,039	12.29

[a]Percentages are given with respect to repeat-derived small RNAs

By normalizing and comparing the expression of piRNAs in two libraries, Fig. 11.2a showed 151 piRNAs (78 up-expressed and 73 down-expressed) which were significantly changed (fold changes > 2) corresponding to the library NC and HILI+ and 140 piRNAs (79 up-expressed and 61 down-expressed) significantly changed corresponding to NC and HILI–. Compared with the two transfected cell libraries, a number of piRNAs showed no significant changes were excluded as the influence of external simulation, and we obtained 46 piRNAs that are upregulated by HILI and50 piRNAs that are downregulated.

Analysis also showed that the population of piRNAs from HILI+ library had a lower nucleotide bias for Uridine at position 1 (21.15–26.45 %), while piRNAs from HILI– library exhibited higher 5′ U-bias at position 1 (31.14 %). However, the bias for Adenine at position 10 showed little significant among the piRNA populations (Fig. 11.2b).

Considering that in Drosophila and mice, Piwi-associated RNAs are primarily derived from transposons and other repeated sequence elements, leading to their alternative designation as repeat-associated small-interfering RNAs (rasiRNAs), which are thought as a subset of piRNAs [6, 20, 21, 24, 37, 38], we aligned small RNA tags previously mapped to genome to repeat-associated RNAs, providing a pool of potential human piRNAs associated with transposable elements. In mammalian genome, transposable elements are mainly composed of three major classes: long terminal repeats (LTR), long interspersed nuclear elements (LINE), and short interspersed nuclear elements (SINE). Our dataset revealed 3,094 LTR-derived,

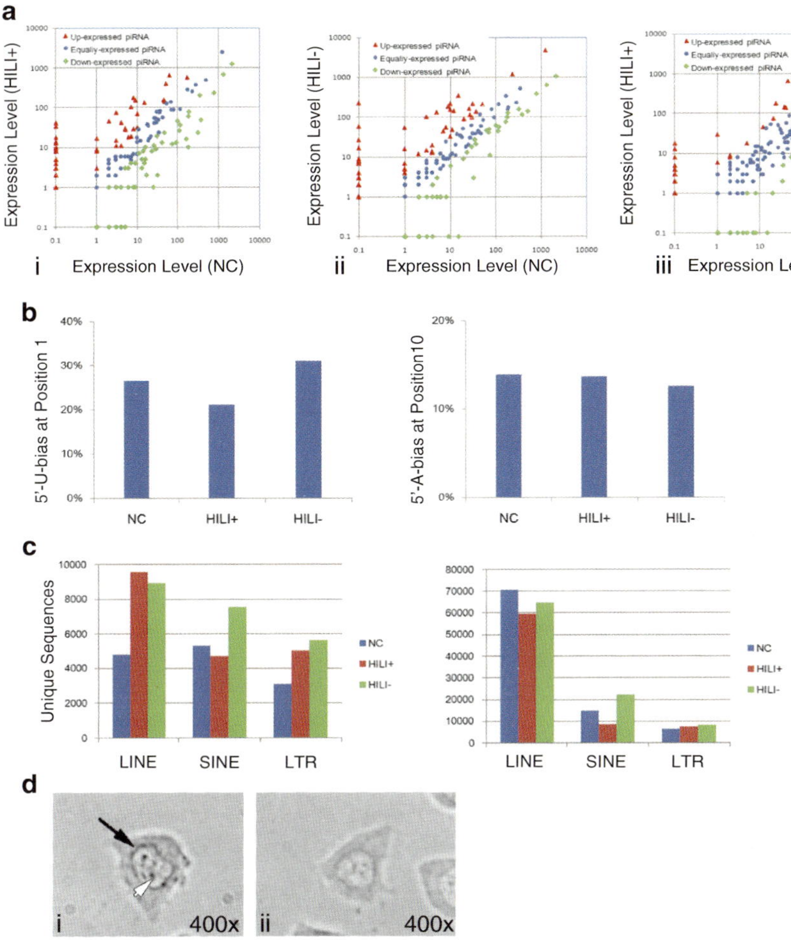

Fig. 11.2 piRNAs were analyzed in HeLa cells. (**a**) The scatted diagrams showed changes in expression of these piRNAs. Significantly upregulated (*red*, folder>2), significantly downregulated (*green*, folder>2), and no significant changes (*blue*). piRNAs were marked according to counting number among libraries. (i) *X*-axis NC, *Y*-axis HILI+; (ii) *X*-axis NC, *Y*-axis HILI−; (iii) *X*-axis HILI−, *Y*-axis HILI+. (**b**) The nucleotide bias showed by populations of piRNAs from each library. *Left*, 5′ Uridine bias at position 1; *Right*, 5′ Adenine-bias at position 10. (**c**) Small RNAs derived from three major classes of transposable elements: LINE, SINE, and LTR. Numbers of matched unique tags (*Left*) and total reads (*right*) were shown in histograms. (**d**) piR-49322 was subcellular localized using in situ hybridization. Signals of DIG-labeled probe (*brown*) were detected concentrating around the periphery of nuclear membrane (*black arrow*) and scatting in the nucleolus (*white arrow*). (i) Cultured HeLa cells; (ii) negative control treated with RNase. The figure is adapted from Lu et al. [47] with permission

4,764 LINE-derived, and 5,254 SINE-derived small RNAs in NC. In HILI+ library, there were 5,019 LTR-derived, 9,520 LINE-derived, and 4,669 SINE-derived small RNAs, while 5,619 LTR-, 8,907 LINE-, and 7,511 SINE-derived in HILI– library (Fig. 11.2c).

To localize the expression of piRNAs in HeLa cells, an in situ hybridization assay was performed using digoxin-labeled oligonucleotide probe. As shown in Fig. 11.2d, signal of piRNA piR-49322 was detected concentrating around the periphery of nuclear membrane and scatting in the nucleolus.

3.4 HILI Repressed LINE1 and LINE1-Associated piRNA in HeLa Cells

Our analysis revealed that in NC library, 14,910 reads (4,937 to antisense and 9,973 to sense) representing 2,689 unique sequences (1,230 to antisense and 1,459 to sense) were matched to LINE1, a retrotransposon element composing approximately 18 % of the mass of human genome. In HILI+ library, unique sequences matched to LINE1 increased in species (1,775 to antisense and 4,125 to sense), while the total reads decreased in abundance (4,181 to antisense and 8,829 to sense). And in HILI– library, 17,955 reads (7,582 to antisense and 10,373 to sense) representing 4,865 unique sequences (1,747 to antisense and 3,118 to sense) were matched to LINE1 (Fig. 11.3a).

The length distribution of LINE1-derived small RNAs also changed following the expression of HILI. The fraction of 23 to 30 nt in length decreased from 58.9 % (NC) to 26.2 % (HILI+), while fraction of 18 to 20 nt increased from 20.4 % (NC) to 59.1 % (HILI+). And the corresponding fractions from HILI– library accounted for 45.1 % and 31.7 %, respectively (Fig. 11.3b).

To further analyze whether HILI regulates the expression of LINE1, a real-time RT-PCR was employed and LINE1 relative expression level was examined. The results showed that there was a significant decrease in LINE1 mRNA expression following the overexpression of HILI, while the silence of HILI slightly decreased LINE1 transcription (Fig. 11.3c).

4 Discussion

Traditionally, piRNAs were considered as a class of small RNA associated with Piwi family proteins and predominately localized in germ line or gonads [6]. However, the increasingly explored roles of Piwi proteins in somatic tissues provide new insights of potential functions of piRNAs. Lee (2011) reported a set of piRNAs detected in mouse hippocampus, consistent with the fact that mouse Piwil1 expressed predominately in testis but also in kidney and brain [39]. In human, Piwi family members have been reported expressing in most tumors and classified into the category of cancer-testis antigens. And Piwil2, which is most widely studied in

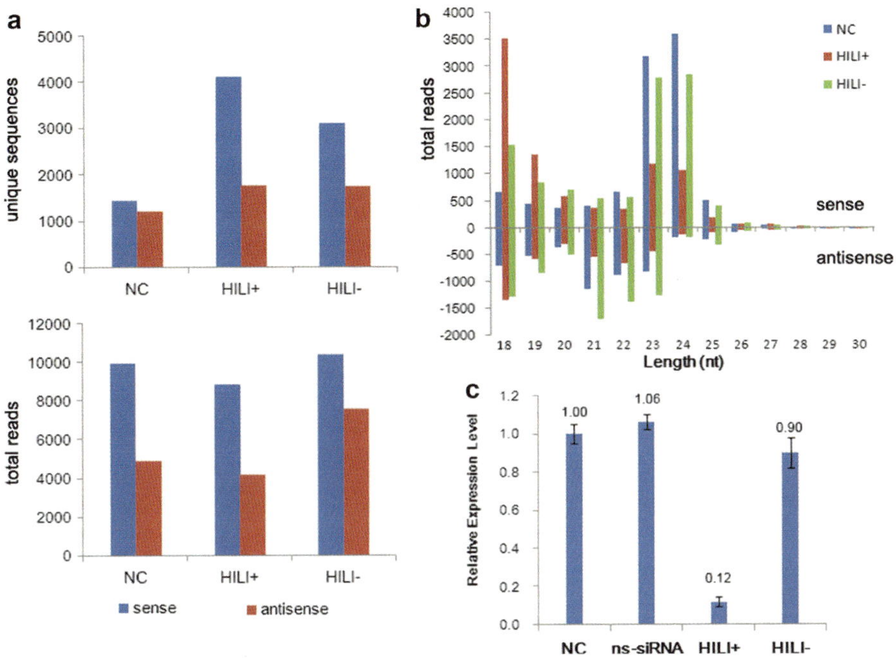

Fig. 11.3 HILI repressed LINE1 retrotransposon in HeLa cells. (**a**) Histograms showed the number of LINE1-derived small RNAs. *Upper*, number of unique sequences mapped to LINE1; *Lower*, number of total reads. (**b**) Length distribution of LINE1-derived small RNAs (18–30 nt) in three libraries. *Upper*, small RNAs matched to sense strand of LINE1; *Lower*, small RNAs matched to antisense strand. (**c**) The expression of LINE1 mRNA was evaluated using real-time RT-PCR method. ns-siRNA, nonspecific control siRNA-transfected HeLa cells. The figure is adapted from Lu et al. [47] with permission

human cancers, plays important roles in regulating signaling pathways and epigenetic modifications in cancer cells [29–31, 33–35, 40, 41].

By employing the newly developed sequencing technology Solexa, we screened the transcriptome of small RNAs in normal and HILI expression-modulated HeLa cells. The generated dataset revealed the existence of piRNAs and other repeat-associated small RNAs in HeLa cells and that piRNAs in cancer can be both derived from transposable elements and non-repetitive sequences.

Further research by using in situ hybridization showed that piR-49322 expressed in nucleolus and cytoplasm, especially concentrated around the periphery of nuclear membrane. This result was consistent with previous report that mature piRNAs are generated in a germ line-specific perinuclear structure called nuage and then imported into the nucleus in the form of Piwi-piRNA complexes [24]. However, in contrast to the enormous population of known piRNAs in male germ cells, the population of piRNAs revealed in HeLa cells is dramatically limited.

Research on *C. elegans* suggested that Piwi is not essential for piRNA biogenesis, but necessary to stabilize piRNAs and facilitate their accumulation, and the expression level of piRNAs is dramatically reduced in Piwi mutants [23]. Whereas in Drosophila, evidence indicated that a Piwi-dependent mechanism called ping-pong cycle amplification exists during piRNA biogenesis [20]. It is also demonstrated that Piwil2, alias zili in zebrafish and mili in mice, plays an essential role in this mechanism in both species. In this amplification cycle, sense/primary piRNAs with a 5′ Uridine-bias at position 1 are bound to Piwil2 and take part in a complex that generates antisense/secondary piRNAs with Adenine bias at position 10 from target transcripts [6, 16]. Thus, a knockout of mili would significantly block piRNA production in mouse germ line [15]. Our sequencing result revealed a fraction of piRNAs whose expression can be regulated by HILI in dose response in human cancer cell line HeLa. Interestingly, the populations of piRNAs from three libraries exhibited difference in 5′ Uridine-bias at position 1 but not in Adenine-bias at position 10 (Fig. 11.2b), suggesting that HILI expression may regulate the abundance of primary rather than secondary piRNAs in HeLa cells. However, the biogenesis and functional of piRNAs in cancer cells remained unknown yet.

Piwi family proteins had been well studied for their ability of transposable elements silencing in Drosophila and mouse germ line. Retrotransposition-competent LINE1 (long interspersed nuclear element 1) is a class of highly abundant retrotransposons comprising ~17 % of the human genome. As LINE1 activity can disrupt the integrity of the genome, an RNAi silencing mechanism is crucially required especially in the periods of gametogenesis and early development, when DNA methylation-dependent silencing is not activated [2, 11, 42–44]. In human cancer cells, LINE1 can be activated by loss of genomic methylation and induce p53-dependent apoptosis [45]. In our study, it was observed that in HeLa cells over-expressed HILI significantly repressed the expression of LINE1 mRNA, while also decreased LINE1-derived small RNA in abundance (Fig. 11.3). Aravin (2008) indicated that in mouse germ line, the initiating input for piRNA production was the transcription of active elements that generates sense piRNAs, rather than specialized piRNA clusters generating antisense piRNAs [16]. Thus, increases in LINE piRNAs might be linked to derepression of these elements and the increased capacity of transposon mRNAs for entry into the piRNA pathway in germ line. And this might be supported by the significant change of length distribution of LINE1-derived small RNAs in our study. In NC library, the LINE1-derived small RNAs were predominantly composed of small RNAs above 23 nt in length, whereas in HILI+ library 18- to 20-nt-sized RNAs which were more likely degradation product of LINE1 transcripts composed the larger fraction of LINE1-derived small RNAs (Fig. 11.3b). Interestingly, compared to the significant reduction in HILI+ cells, the LINE1 expression level was unexpectedly a little lower in HILI-knockdown cells than NC. As reported by Aravin (2008), miwi2, another Piwi family member, delocalized from nucleus to cytoplasm in mili-deficient germ cells, suggesting that the compensation of other Piwi genes may reduce the effect of Piwil2 knockdown [16]. On the other hand, Ye (2010) reported a group of Piwil2-like proteins generated by spliced mRNA variants in human testis and cancer cells [46]. These variants, predominately expressing in HeLa cells, may have different functions contrast to exogenous full-length Piwil2, even though they can be silenced by transfected siRNA as well. Accordingly, additional studies are necessary to elucidate the role of HILI and piRNAs in human cancers.

In the present study, we have sequenced small RNAs in normal and HILI-regulated HeLa cells. Our data revealed the expression of piRNAs and other repeat-associated small RNAs in HeLa cell line and demonstrated that human HILI repressed retrotransposon LINE1 and decreased the expression of LINE1-derived small RNAs in HeLa cells. These investigations will provide insights into roles of HILI and piRNAs in cancer cells.

References

1. Zamore PD, Haley B. Ribo-gnome: the big world of small RNAs. Science. 2005;309(5740): 1519–24.
2. Aravin AA, Hannon GJ, Brennecke J. The Piwi-piRNA pathway provides an adaptive defense in the transposon arms race. Science. 2007;318(5851):761–4.
3. Cox DN, Chao A, Baker J, Chang L, Qiao D, Lin H. A novel class of evolutionarily conserved genes defined by piwi are essential for stem cell self-renewal. Genes Dev. 1998;12(23): 3715–27.
4. Cox DN, Chao A, Lin H. Piwi encodes a nucleoplasmic factor whose activity modulates the number and division rate of germline stem cells. Development. 2000;127(3):503–14.
5. Carmell MA, Xuan Z, Zhang MQ, Hannon GJ. The Argonaute family: tentacles that reach into RNAi, developmental control, stem cell maintenance, and tumorigenesis. Genes Dev. 2002;16(21):2733–42.
6. Houwing S, Kamminga LM, Berezikov E, Cronembold D, Girard A, van den Elst H, et al. A role for piwi and piRNAs in germ cell maintenance and transposon silencing in zebrafish. Cell. 2007;129:69–83.
7. Seto AG, Kingston RE, Lau NC. The coming of age for Piwi proteins. Mol Cell. 2007;26(5): 603–9.
8. Farazi TA, Juranek SA, Tuschl T. The growing catalog of small RNAs and their association with distinct Argonaute/Piwi family members. Development. 2008;135(7):1201–14.
9. Deng W, Lin H. Miwi, a murine homolog of piwi, encodes a cytoplasmic protein essential for spermatogenesis. Dev Cell. 2002;2(6):819–30.
10. Kuramochi-Miyagawa S, Kimura T, Ijiri TW, Isobe T, Asada N, Fujita Y, et al. Mili, a mammalian member of piwi family gene, is essential for spermatogenesis. Development. 2004;131(4):839–49.
11. Carmell MA, Girard A, van de Kant HJ, Bourc'his D, Bestor TH, de Rooij DG, et al. MIWI2 is essential for spermatogenesis and repression of transposons in the mouse male germline. Dev Cell. 2007;12:503–14.
12. Aravin A, Gaidatzis D, Pfeffer S, Lagos-Quintana M, Landgraf P, Iovino N, et al. A novel class of small RNAs bind to MILI protein in mouse testes. Nature. 2006;442(7099):203–7.
13. Lau NC, Seto AG, Kim J, Kuramochi-Miyagawa S, Nakano T, Bartel DP, et al. Characterization of the piRNA complex from rat testes. Science. 2006;313(5785):363–7.
14. Girard A, Sachidanandam R, Hannon GJ, Carmell MA. A germline-specific class of small RNAs binds mammalian Piwi proteins. Nature. 2006;442(13):199–202.
15. Grivna ST, Beyret E, Wang Z, Lin H. A novel class of small RNAs in mouse spermatogenic cells. Genes Dev. 2006;20:1709–14.
16. Aravin AA, Sachidanandam R, Bourc'his D, Schaefer C, Pezic D, Toth KF, et al. A piRNA pathway primed by individual transposons is linked to de novo DNA methylation in mice. Mol Cell. 2008;31(6):785–99.

17. Ruby JG, Jan C, Player C, Axtell MJ, Lee W, Nusbaum C, et al. Large-scale sequencing reveals 21U-RNAs and additional microRNAs and endogenous siRNAs in *C. elegans*. Cell. 2006;127(6):1193–207.
18. Vagin VV, Sigova A, Li C, Seitz H, Gvozdev V, Zamore PD. A distinct small RNA pathway silences selfish genetic elements in the germline. Science. 2006;313(5785):320–4.
19. Watanabe T, Takeda A, Tsukiyama T, Mise K, Okuno T, Sasaki H, et al. Identification and characterization of two novel classes of small RNAs in the mouse germline: retrotransposon-derived siRNAs in oocytes and germline small RNAs in testes. Genes Dev. 2006;20(13):1732–43.
20. Brennecke J, Aravin AA, Stark A, Dus M, Kellis M, Sachidanandam R, et al. Discrete small RNA-generating loci as master regulators of transposon activity in Drosophila. Cell. 2007;128(6):1089–903.
21. Gunawardane LS, Saito K, Nishida KM, Miyoshi K, Kawamura Y, Nagami T, et al. A slicer-mediated mechanism for repeat-associated siRNA 5′ end formation in Drosophila. Science. 2007;315(5828):1587–90.
22. Batista PJ, Ruby G, Claycomb JM, Chiang R, Fahlgren N, Kasschau KD, et al. PRG-1 and 21U-RNAs interact to form the piRNA complex required for fertility in *C. elegans*. Mol Cell. 2008;31(1):67–78.
23. Das PP, Bagijn MP, Goldstein LD, Woolford JR, Lehrbach NJ, Sapetschnig A, et al. Piwi and piRNAs act upstream of an endogenous siRNA pathway to suppress Tc3 transposon mobility in the Caenorhabditis elegans germline. Mol Cell. 2008;31(1):79–90.
24. Klattenhoff C, Theurkauf W. Biogenesis and germline functions of piRNAs. Development. 2008;135(1):3–9.
25. Lee JH, Engel W, Nayernia K. Stem cell protein Piwil2 modulates expression of murine spermatogonial stem cell expressed genes. Mol Reprod Dev. 2006;73:173–9.
26. Wang J, Saxe JP, Takashi T, Chuma S, Lin H. Mili interacts with tudor domain-containing protein 1 in regulating spermatogenesis. Curr Biol. 2009;19(8):640–4.
27. Houwing S, Berezikov E, Ketting RF. Zili is required for germ cell differentiation and meiosis in zebrafish. EMBO J. 2008;27:2702–11.
28. Sasaki T, Shiohama A, Minoshima S, Shimizu N. Identification of eight members of the Argonaute family in the human genome. Genomics. 2003;82(3):323–30.
29. Lee JH, Schütte D, Wulf G, Füzesi L, Radzun H-J, Schweyer S, et al. Stem-cell protein Piwil2 is widely expressed in tumors and inhibits apoptosis through activation of Stat3/Bcl-XL pathway. Hum Mol Genet. 2006;15(2):201–11.
30. Chen L, Shen R, Ye Y, Pu X-A, Liu X, Duan W, et al. Precancerous stem cells have the potential for both benign and malignant differentiation. PLoS One. 2007;2(3):e293.
31. He G, Chen L, Ye Y, Xiao Y, Hua K, Jarjoura D, et al. Piwil2 expressed in various stages of cervical neoplasia is a potential complementary marker for p16INK4a. Am J Transl Res. 2010;2(2):156–69.
32. Lee JH, Jung C, Javadian-Elyaderani P, Schweyer S, Schütte D, Shoukier M, et al. Pathways of proliferation and antiapoptosis driven in breast cancer stem cells by stem cell protein Piwil2. Cancer Res. 2010;70(11):4569–79.
33. Lu Y, Zhang K, Li C, Yao Y, Tao D, Liu Y, et al. Piwil2 Suppresses P53 by inducing phosphorylation of signal transducer and activator of transcription 3 in tumor cells. PLoS One. 2012;7(1):e30999.
34. Wang Q-E, Han C, Milum K, Wani AA. Stem cell protein Piwil2 modulates chromatin modifications upon cisplatin treatment. Mutat Res. 2011;708(1–2):59–68.
35. Yin D-T, Wang Q, Chen L, Liu M-Y, Han C, Yan Q, et al. Germline stem cell gene PIWIL2 mediates DNA repair through relaxation of chromatin. PLoS One. 2011;6(11):e27154.
36. Li R, Li Y, Kristiansen K, Wang J. SOAP: short oligonucleotide alignment program. Bioinformatics. 2008;24(5):713–4.
37. Aravin AA, Lagos-Quintana M, Yalcin A, Zavolan M, Marks D, Snyder B, et al. The small RNA profile during drosophila melanogaster development. Dev Cell. 2003;5(2):337–50.
38. Saito K, Nishida KM, Mori T, Kawamura Y, Miyoshi K, Nagami T, et al. Specific association of piwi with rasiRNAs derived from retrotransposon and heterochromatic regions in the drosophila genome. Genes Dev. 2006;20(16):2214–22.

39. Lee EJ, Banerjee S, Zhou H, Jammalamadaka A, Arcila M, Manjunath BS, et al. Identification of piRNAs in the central nervous system. RNA. 2011;17(6):1090–9.
40. Qiao D, Zeeman A-M, Deng W, Looijenga LHJ, Lin H. Molecular characterization of hiwi, a human member of the piwi gene family whose overexpression is correlated to seminomas. Oncogene. 2002;21(25):3988–99.
41. Liu X, Sun Y, Guo J, Ma H, Li J, Dong B, et al. Expression of hiwi gene in human gastric cancer was associated with proliferation of cancer cells. Int J Cancer. 2006;118(8):1922–9.
42. Kalmykova AI, Klenov MS, Gvozdev VA. Argonaute protein PIWI controls mobilization of retrotransposons in the Drosophila male germline. Nucleic Acids Res. 2005;33(6):2052–9.
43. Soifer HS, Rossi JJ. Small interfering RNAs to the rescue: blocking L1 retrotransposition. Nat Struct Mol Biol. 2006;13(9):758–9.
44. Kuramochi-Miyagawa S, Watanabe T, Gotoh K, Totoki Y, Toyoda A, Ikawa M, et al. DNA methylation of retrotransposon genes is regulated by Piwi family members MILI and MIWI2 in murine fetal testes. Genes Dev. 2008;22(7):908–17.
45. Haoudi A, Semmes OJ, Mason JM, Cannon RE. Retrotransposition-competent human LINE-1 induces apoptosis in cancer cells with intact p53. J Biomed Biotechnol. 2004;2004(4):185–94.
46. Ye Y, Yin D-T, Chen L, Zhou Q, Shen R, He G, et al. Identification of Piwil2-like (PL2L) proteins that promote tumorigenesis. PLoS One. 2010;5(10):e13406.
47. Lu Y, Li C, Zhang K, Sun H, Tao D, Liu Y, et al. Identification of piRNAs in Hela cells by massive parallel sequencing. BMB Rep. 2010;43(9):635–41.

Chapter 12
MicroRNA Expression in Breast Cancer Revealed by Deep Sequencing Technology

Thalia A. Farazi, Carl S. Leonhardt, and Thomas Tuschl

Abstract MicroRNAs (miRNAs) regulate many genes critical for tumorigenesis; thus, studying their expression in breast cancer could enhance our understanding of tumorigenesis pathways and/or identify prognostic and diagnostic markers. Although studies in various malignancies point to miRNA dysregulation in cancer, many studies show conflicting results, partly because they use different methodologies. Deep sequencing technology has been used to identify differentially expressed miRNAs in tissue biopsy specimens as well as in circulation from breast cancer patients, also enabling identification of nucleotide variations and novel miRNAs. In a large sequencing-based study of breast tissue specimens, the differentiation of tumor subtypes and prediction of metastasis based on miRNA levels were statistically possible but not driven by dysregulation of abundant miRNAs, implicating fewer miRNAs involved in tumorigenic processes than previously suggested. None of the highly abundant miRNAs showed nucleotide variation within their mature sequence, in agreement with studies of other malignancies. Analysis of circulating miRNAs also suggests differentiation of tumor subtypes and miRNAs with prognostic potential, but clinical application requires standardized methods for sample processing and larger validation studies.

Keywords microRNA • Breast cancer • Deep sequencing • Nucleotide variation • Differential expression • Prognosis • Diagnosis • Clustering • 3′ UTR • Argonaute

---

T.A. Farazi • C.S. Leonhardt • T. Tuschl (✉)
Laboratory of RNA Molecular Biology, Howard Hughes Medical Institute,
The Rockefeller University, 1230 York Avenue, Box 186, New York, NY 10065, USA
e-mail: tfarazi@rockefeller.edu; carl.s.leonhardt@gmail.com; ttuschl@rockefeller.edu

W. Wu and H. Choudhry (eds.), *Next Generation Sequencing in Cancer Research: Volume 1: Decoding the Cancer Genome*, DOI 10.1007/978-1-4614-7645-0_12,

1 Introduction

1.1 miRNA Overview

Mature miRNAs are single-stranded RNA molecules 20–23 nucleotides (nt) in length. They control gene expression in cellular processes that mediate tumorigenesis by generally reducing stability of target mRNAs [1–4]. These processes include cell cycle regulation, differentiation, apoptosis, invasion, and migration. miRNA recognition of its mRNA targets is mostly achieved through specific base-pairing interactions between the 5′ end of the miRNA (referred to as the "seed" region) and sites within coding and untranslated regions (UTRs) of mRNAs [3]. Target sites in the 3′ UTR lead to more effective mRNA destabilization than targets in coding regions. Since miRNAs frequently target hundreds of mRNAs, miRNA regulatory pathways are complex [5]. miRNA-mediated regulatory functions are concentration dependent: regulation by miRNAs requires a minimum number of miRNA copies per cell for meaningful biological regulation of the copies of hundreds of mRNA targets. Thus, determination of miRNA abundance is an important parameter that can easily be obtained from sequencing-based profiling. miRNA pathways could involve feedback loops, coordinated regulation of multiple members of cellular pathways, or regulation of critical factors orchestrating cellular responses. In turn, miRNAs themselves are under tight regulation by distinct cellular factors, RNA-binding proteins, and transcription pathways [6, 7].

The mature miRNA molecules are expressed from primary transcripts (pri-miRNAs) that contain one or more ~70 nt hairpin miRNA precursors (pre-miRNA) and either have their own promoters or share promoters with coding genes (when located within gene intronic regions, referred to as intronic miRNAs). miRNA biogenesis is reviewed in [1, 8], where many primary references are listed. miRNAs are excised from pri-miRNAs in a multistep process. First, these hairpin precursor structures are recognized in the nucleus by DGCR8, a double-stranded RNA-binding protein (dsRBP) and RNASEN (otherwise known as RNase III Drosha), and excised to yield pre-miRNAs. Subsequently, these molecules are transported to the cytoplasm by XPO5 (exportin 5), where they are processed by DICER1 (Dicer) in complex with the dsRBPs TARBP2 (TRBP) and/or PRKRA to yield an RNA duplex processing intermediate, comprising mature miRNA and miRNA* sequences. Most miRNA profiling methods, including sequencing-based approaches, are designed to identify the processed mature miRNA and miRNA* sequences. Sequencing-based approaches can in fact evaluate the prototypical processing mechanism by RNASEN and DICER1 by aligning sequence reads corresponding to the well-defined regions of mature miRNA and miRNA* to the miRNA precursor, confirming the predicted hairpin structure of the precursor pre-miRNA.

Some miRNAs employ distinct biogenesis mechanisms, bypassing the general miRNA processing order: some miRNAs can mature independently of DGCR8 and RNASEN, such as *mir-320* or *mir-484*, while other miRNAs are DICER1 independent, such as *mir-451*. DGCR8- and RNASEN-independent miRNAs include mirtrons and tailed mirtrons, which release their pre-miRNA by splicing and exonuclease trimming (mirtrons are miRNAs located in intronic regions of their mRNA host genes and are processed through splicing instead of Drosha). miR-451 is generated from an unusual hairpin structure that is processed by Argonaute2/EIF2C2 (Ago2) instead of DICER1 (see below).

After processing, the mature miRNA is loaded into the Ago proteins that are at the core of the miRNA-containing ribonucleoprotein complex (miRNP), sometimes also referred to as miRNA-induced silencing complex (miRISC). The miRNA* is usually released from the miRNP and degraded, resulting in lower levels than the mature miRNA sequence. The miRNPs contain a member of the Ago family (1–4), which binds the miRNA and mediates target mRNA recognition. Several other RBPs have been implicated in miRNA biogenesis, including LIN28 suggested to play a role in neuroblastoma tumorigenesis [9]. After transcription, miRNAs may be modified by different enzymes: deaminase activity results in miRNA editing, whereas terminal uridyl transferases cause pre-miRNA uridylation. These modifications potentially affect the sequence, amount, and ratio of miRNA and miRNA*. Sequencing-based miRNA profiling can capture these modifications and explore any potential differential regulation in normal and tumor samples.

1.2 Tissue miRNAs as Diagnostic and Prognostic Markers

miRNA dysregulation in tissues could be used as a diagnostic and/or prognostic tool, even if the particular miRNAs do not serve any regulatory function. miRNAs demonstrated their potential as diagnostic tumor markers early on when their profiles were shown to correlate with the embryonic origin of different tumors; miRNAs define tumors of unknown origin otherwise indistinguishable by histology and previously assigned based solely on clinical information [10]. miRNA expression patterns have been linked to clinical outcomes, as suggested by their ability to modulate tumor behavior, such as tumor progression and metastasis. Expression of *let-7* is downregulated in non-small cell lung cancer patients [11] and is associated with poor prognosis [12, 13]. Similarly in CLL a miRNA signature was identified to be associated with prognosis [14].

Alternatively, miRNA dysregulation could drive tumorigenesis via the roles miRNAs can adopt as tumor suppressors or oncogenes. miRNAs that are up- or downregulated in malignancies are respectively referred to as oncogenic or tumor-suppressor miRNAs, sometimes even if there is no evidence for their causative role in tumorigenesis. Functional studies performed in cancer cell lines or mouse models of various malignancies, through overexpression or knockdown of miRNAs, have supported a role for some of these miRNAs in tumorigenesis [1]. Examples include (1) overexpression of tumor suppressor miRNAs, such as *let-7g,* reduced tumor burden in a K-RAS murine lung cancer model; (2) overexpression of the oncogenic *mir-17~92* cluster led to a lymphoproliferative disorder, and an even higher-level expression of the cluster in MYC-driven B-cell lymphomas dramatically increased tumorigenicity; and (3) overexpression of the oncogenic *mir-21*, frequently highly expressed in many different malignancies, resulted in a pre-B malignant lymphoid-like phenotype.

Metastasis-related miRNAs have been identified in various malignancies, mainly from cell line and xenograft experiments [15]. The regulatory roles of these miRNAs have not always been validated in large clinical studies, pointing to difficulties resulting from differences in methodologies for acquisition of miRNA profiles and the tumor biological characteristics. For example, challenges arise from attempting to reconcile cell line, xenograft model, and patient materials, due to (1) the tissue heterogeneity which makes it hard to dissect contributions from subpopulations of cells that may be important for tumor characteristics, such as metastasis, (2) the heterogeneous nature of the malignancy, and (3) timing of clinical specimen acquisition, usually collected at time of diagnosis, by which time a tumor is already well established and cannot unravel early changes that may be critical in tumor initiation or later changes important in metastasis.

1.3 Circulating miRNAs as Biomarkers

Detection of circulating biomarkers, which are minimally invasive, can help address some of the challenging issues in treating breast cancer, such as (1) the ability to detect tumors early and therefore increase the chances of cure and (2) predict the response of patients to treatment. Mammography—the current "gold standard" in breast cancer diagnostics—has been reported to have a false-positive rate of 8–10 % [16]. Additionally, ionizing radiation remains a concern for use of this diagnostic modality, particularly in younger women. Furthermore, the value of reported protein-based biomarkers, such as carcinoembryonic antigen (CEA) and carbohydrate antigen (CA) 15-3, for disease management is limited. These markers usually exhibit low sensitivity and do not effectively discriminate between disease-free patients and patients with metastatic relapse [16].

An ideal biomarker would be sensitive enough to detect early stages of tumor growth, relapse, and progression during treatment and classify patients into risk-adapted therapeutic groups. Interestingly, in a recent paper, the investigators estimated that peptide tumor biomarkers, using current detection methodologies, would require a tumor to grow for over 8 years and reach a ~1 cm diameter before being detectable [17]. The discovery that circulating miRNAs are stable in blood, plasma, and serum makes them promising biomarkers [18, 19]. Serum has roughly the same composition as plasma, with the main difference being the absence of several clotting factors. Due to a higher level of miRNAs in total blood compared to plasma or serum, total blood has also been proposed to constitute a suitable specimen [20]. When using total blood, the composition of miRNAs in immune cells and circulating cancer cells would be the studied biomarker. Using RT-PCR for a subset of miRNAs (miR-15b, miR-16, miR-19b, miR-24), Mitchell et al. [18] suggested that miRNA levels in plasma and serum strongly correlate. Several groups have published reports about disease types associated with altered levels of circulating miRNAs [21–24], employing different methodologies.

Despite these emerging advances and links to dysregulated circulating miRNA levels in disease, there has not yet been a comprehensive and quantitative analysis of the miRNA landscape in peripheral circulation. Moreover, the absolute contribution of a tumor to the circulating miRNA pool has not been assessed. Furthermore, it remains unclear how miRNAs enter circulation. Some simple explanations include cell death, cell lysis by hemodynamic turbulence in vessels, and sample handling and processing. Alternatively, miRNAs could be incorporated in microparticles released by different cell types [25–27]. The exosome, derived from intracellular late endosomes/multivesicular bodies (MVBs), represents one such microparticle [28, 29]. In fact MVBs have been linked to RISC functioning and activity [30, 31]. "Secreted" miRNAs could possibly reflect differences in nuclease degradation in MVBs or serum nuclease stability. A recent study reported cell type-specific release and retention of miRNAs using the breast cancer cell lines MDA-MB-231, MCF-7, SK-BR-3, and BT-20 and the noncancerous mammary epithelial cell line MCF-10A [28]. Interestingly, these miRNAs did not represent the most abundant cellular miRNA in the respective cell type [28]. These observations of "secreted" miRNAs led to the speculation that selectively released miRNAs may have important hormone-like functions in cell-cell communication in both a tumorigenic and non-tumorigenic environment.

A recent deep sequencing-based study from our lab addressed some of these issues. We measured plasma miRNAs from healthy individuals using deep sequencing and observed that the miRNA landscape did not directly reflect the contribution of endothelial and blood cells and suggested a degree of preference in stability, or "secretability," of circulating miRNAs (Williams Z, manuscript accepted). We also studied miRNAs from the placenta, considering this a model of tumor mass, and compared their levels in the mother and newborn. Our data demonstrated that 1 mg of placenta generated only 0.03 copies of a placenta-specific miRNA per ml of blood and, using the currently available technology, a tumor would have to grow to 0.3 g before a miRNA expressed to a similar degree and specificity could be identified. This study suggested that dramatically improved methods for miRNA detection are required before miRNAs can serve as early, reliable, and useful biomarkers. Finally, our data argued against the notion that miRNAs may exert hormone-like effects on target tissues [32]. The plasma concentration of total miRNA, in the 100 fM range, was below the levels of any known hormone. Steroid hormones such as progesterone and estrogen are present at nM concentrations, while trace peptide hormones such as parathyroid hormone and ACTH are still present in pM amounts. Moreover, while hormones bind receptors that amplify their signals, miRNAs, as discussed earlier, act at a stoichiometric ratio with their mRNA targets and are thought to require >1,000 copies in order to exert measureable cellular activity.

2 miRNAs and Breast Cancer

2.1 Breast Cancer Overview

Breast cancer is the most common non-cutaneous tumor in women, with an estimated 226,870 new cases of invasive disease (plus 63,000 cases of in situ disease) and 39,510 deaths in 2012 [33]. In terms of our clinical approach, this heterogeneous disease is categorized into three basic main subtypes, usually defined by immunohistochemistry (IHC): estrogen receptor (ER) positive, HER2-positive (overexpressing *ERBB2*), and triple negative (TNBC), not expressing ER, progesterone receptor (PR) or overexpressing *ERBB2*. The ER-positive subtype is both the most common subtype and the most diverse, making it hard to predict patient outcome. Different mRNA-based tests have been used to provide supplementary information to predict patient outcome for the ER-positive subtype [34]. The HER2-positive subtype (also known as *ERBB2* amplified subtype) is distinguished by its response to targeted therapies utilizing antibodies against HER2. TNBCs, sometimes also referred to as basal-like carcinomas, based on their mRNA profiles [35], are seen more frequently in patients with germline BRCA1 mutations [36] or patients of African descent [37], and TNBCs usually receive nonspecific chemotherapy treatment. Despite our success in the treatment of breast cancer, there are still a number of unanswered questions, such as how to detect the disease early in its progression, predict patient outcome, and appropriately tailor treatment of metastatic disease. Given the critical roles miRNAs play in various cellular processes and their implication in tumorigenesis, breast cancer research quickly incorporated the relatively new and fast-growing field of miRNAs.

2.1.1 miRNA Dysregulation in Breast Cancer Tissues

miRNA dysregulation in breast tumor tissues was first described in 2005 [38]. Since then, over 800 studies have been published on this topic including cell line, animal model, and patient-based studies. Some of these studies address the potential of miRNAs as diagnostic markers for breast tumor subtypes, and others address the potential of miRNAs as prognostic markers for patient outcome. Most of these studies were conducted using microarrays or real-time PCR (RT-PCR) and thus limited to a subset of miRNAs. Many of these studies have been performed in cell lines and animal models and are only supported by small patient cohorts.

2.1.2 miRNAs as Breast Cancer Diagnostic and Prognostic Markers

Various studies have reported differentiation of already established tumor subtypes by miRNAs (e.g., [39]). Given the heterogeneity of breast cancer, small patient cohorts make it impossible to correlate their results to clinical relevance. Moreover, the differences in miRNA levels in patients that developed metastasis involved lowly abundant miRNAs, which is challenging to translate into prognostic markers given the detection limits of currently available experimental methods. For example, even though overexpression/knockdown of several miRNAs, including miR-126, miR-335, miR-10b, and miR-31, was shown to play a role in metastasis [40–42], correlation between these miRNAs and metastasis has not been identified in large clinical studies. Most of these studies based on cell line and animal models were validated with smaller ($n \sim 20$) patient sample collections. In larger patient studies, only miR-210 was shown to be inversely correlated with time to metastasis, disease-free, and overall survival [43–46].

2.1.3 Coordinated miRNA and mRNA Target Levels as Diagnostic and Prognostic Markers

Recent studies have used bioinformatic approaches to correlate miRNA and mRNA levels in breast tumors to further validate and confirm the role of miRNA dysregulation in breast cancer, each with different results. The first large patient study ($n=93$, including various molecular subtypes), using same sample miRNA/mRNA levels in breast cancer [39], identified a number of differentially expressed miRNAs between tumor subtypes. However, the study detected only a few instances of down- or upregulation of computationally predicted target mRNAs (TargetScan) consistent with changes in miRNA expression. Another study focused on 82 ER-positive and HER2-negative patients with variable lymph node involvement and identified anti-correlated miRNA-mRNA pairs in the node-negative and node-positive group (let-7i, miR-143, miR-16, miR-196a, miR-26a, miR-27a, miR-375, miR-503, miR-519b-3p (negative), and miR-361-5p (positive)) [47]. Enerly et al. studied 101 primary tumor specimens, mainly including luminal A and basal-like breast tumors, and observed correlation between some miRNAs found to be differentially expressed in tumor subtypes and their predicted targets, revealing regulated pathways (e.g., miR-17~92 cluster members, differentially expressed in basal-like tumors, were correlated to cell cycle genes) [48]. Hannafon et al. investigated miRNA/mRNA expression in 16 ductal carcinoma in situ (DCIS) specimens, identifying differentially expressed miRNAs [49]. They highlighted negative correlations for a set of TargetScan computationally predicted targets of the differentially expressed miR-125b, miR-182, and miR-183 and these miRNAs. A study by van de Auwera et al. assessed miRNA expression in inflammatory and noninflammatory breast cancer ($n=70$). Expression levels for miR-29a, miR-30b, miR-342-3p, and miR-520a-5p were able to differentiate between the two types of breast cancer and were anticorrelated with their predicted targets [50]. Finally, a study using 207 tumors consisting of a mix of breast tumor subtypes identified that for miR-210, miR-218, and miR-27b, the coordinated dysregulation of miRNA/mRNA levels conferred a significantly worse prognosis [45]. Moreover, they evaluated these levels in other patient cohorts (with already published mRNA profiles) and confirmed their prognostic potential.

3 Summary

Despite all the studies described above, miRNA dysregulation in breast cancer has not yet been translated into an improved classification of tumor subtypes. Studies of miRNA-mRNA target identification in breast cancer have also not been consistent with identifying the most biologically relevant mRNA targets. Sequencing methodology offers the promise of clarifying some of our unanswered questions and allows us to place our understanding of miRNA dysregulation in the context of miRNA expression, modification, mutation, and regulatory potential. We conducted the first large miRNA sequencing study in breast cancer tissues to address some of the limitations of the earlier studies and summarize our findings in this book chapter. The results of an even larger miRNA sequencing study in breast cancer from the Cancer Genome Atlas (TCGA) were just published after completion of this manuscript [51] and suggest further heterogeneity within tumor subtypes as well as similarity to ovarian cancer. The TCGA dataset is an information-rich database that can further clarify the importance of miRNAs as diagnostic, and, in the future, as prognostic markers (patient follow-up not long enough at this point in time to allow evaluation of prognosis). Finally, we also summarize the results of the largest so far published study of circulating miRNAs in breast cancer patients conducted to evaluate the potential of miRNAs as biomarkers.

4 miRNA Expression Determined by Deep Sequencing Technology

4.1 Methodologies for miRNA Isolation and Detection

4.1.1 RNA Isolation

The RNA input material initially used for miRNA profiling studies comprised total RNA recovered from high-quality preserved fresh-frozen samples. However, recently protocols have been developed to obtain reproducible and comparable profiles using formalin-fixed paraffin-embedded tissues (FFPE), making these archived tumor collections accessible for study [52–54]. Moreover, methods for isolating miRNAs in body fluids, such as serum or plasma, have now allowed profiling of miRNAs in circulation. Separating red blood cells prior to sample storage avoids profiling of red blood cell-associated miRNAs. Different methods have been applied for isolation of RNA suitable for miRNA profiling. Some profiling methods use an input of total RNA, given that they involve later steps that select for the relevant miRNA size, whereas other methods use an input of size-fractionated RNA.

4.1.2 miRNA Detection by Microarray and RT-PCR Methodologies

The main methods currently used for miRNA profiling are microarray and quantitative real-time RT-PCR-based approaches [55, 56]. While they can be high throughput, these approaches are limited to the study of previously identified miRNAs deposited at miRBase (www.mirbase.org). Another caveat of using RT-PCR or microarray methodology for miRNA profiling is that the concentration of most circulating miRNAs is very low and the sensitivity of current technologies remains a challenge ([25]; (Williams Z, manuscript accepted)).

Microarrays generally provide fold changes in miRNA expression between samples, with members of miRNA sequence families prone to cross-hybridization. Due to the short length of miRNAs, use of short probes often makes it complex to achieve similar melting temperature (T_m) values for the hundreds of probes tested [57]. Differences of more than 20 °C are sometimes encountered, affecting hybridization and thus assay specificity and sensitivity [57]. A comparison of six commercially available microarrays showed that some miRNAs are only reported dysregulated in one array, suggesting false-positives, which are hard to detect and evaluate [57]. More recently, calibration cocktails of synthetic miRNAs were used in array experiments to derive absolute abundance of miRNAs [58].

RT-PCR methods are lower throughput and, as mentioned earlier, require normalization; if external miRNA standards are used for quantification, the most abundant miRNA, which may vary in length due to 3′ end heterogeneity, should be used as a calibration standard [59]. Other RT-PCR normalization methods have also been recently proposed that utilize the overall miRNA expression to avoid some of these caveats [60].

4.1.3 miRNA Detection by Sequencing Methodologies

Sequencing methods, besides their obvious potential to identify novel miRNAs, RNA editing, potential single-nucleotide polymorphisms (SNPs), and mutation events, estimate miRNA abundance based on the frequency of sequence reads. Moreover, sequencing-based approaches allow reevaluation of the data obtained as our knowledge of miRNA annotation evolves. As discussed earlier, miRNA abundance is of particular importance in understanding the contribution of miRNAs to cellular processes. Every miRNA potentially regulates hundreds of transcripts, and it is those abundantly expressed miRNAs that have been shown to control gene expression in a quantifiable manner. Deep sequencing of a limited set of tissues is sufficient to detect virtually every annotated miRNA species, although most are detected at low frequency. Typically, a small subset of miRNAs (30–50 in mammalian tissues) represents 90 % of all sequence reads annotated as miRNAs, and it is this small subset that confers the majority of miRNA regulation within the cell.

Given the dramatic increase in sequencing power, barcoding samples can allow multiple specimens to be processed at the same time, reducing the cost and effort of profiling, thus enabling profiling of large specimen collections. We describe below the barcoded miRNA deep sequencing approach we applied to a breast tumor sample collection [61]. Ligation biases exist in sequencing methods, which may affect comparable representation of certain miRNAs obscuring absolute expression levels, but are irrelevant when monitoring fold changes between samples. To evaluate the ligation biases in our barcoded miRNA deep sequencing approach, we conducted a study with a synthetic pool of 776 miRNA sequences [62]. This study showed that, overall, these biases did not prevent identification of miRNAs and allowed estimation of these biases for the small subset of sequences affected.

4.1.4 miRNA Profile Differential Expression Analysis

Finally, choosing the appropriate statistical analysis tools to evaluate differential expression depends on the methodology used to obtain the profiles, ranging from established Significance Analysis of Microarrays (SAM) for evaluation of microarray data to newly developed techniques for sequencing data, such as edgeR and DESeq [63].

4.2 Barcoded miRNA Deep Sequencing Approach

Our barcoded miRNA profiling approach is detailed in two recent methodology articles [63, 64]. The experimental process (Fig. 12.1) of small RNA cDNA library generation consists of small RNA isolation, ligation of barcoded 3′ adapters to up to 20 individual samples, pooling of samples, ligation of a 5′ adapter, reverse transcription and PCR to generate a cDNA library, and Illumina sequencing. In contrast to a number of RNA cDNA library generation protocols for RNA-seq applications, ligation of 3′ and 5′ adapters of different sequence to the small RNA preserves the orientation of the RNA insert and allows for the determination of its origin from the sense or antisense strand of the genome after cDNA sequencing. As discussed above, ligation biases have been quantified and can be used to better elucidate abundance of dysregulated miRNAs. Adding a cocktail of synthetic RNAs allows for absolute quantitation of miRNA levels per μg of total input RNA. Downstream bioinformatic analysis starts with the separation of sequence reads according to the barcode sequence and is followed by mapping and annotation of the extracted sequence reads.

5 Determination of miRNA Expression in Breast Cancer Tissues by Deep Sequencing

5.1 Summary of Studies Performed to Date

Our deep sequencing breast cancer study included 11 normal breast tissues, 17 non-invasive, 151 invasive breast carcinomas, and 6 cell lines (49,479,978 miRNA reads) [61]. The invasive carcinomas were comprised mostly of invasive ductal carcinomas (IDC), with a mixture of TNBC, HER2-, or ER-positive tumors.

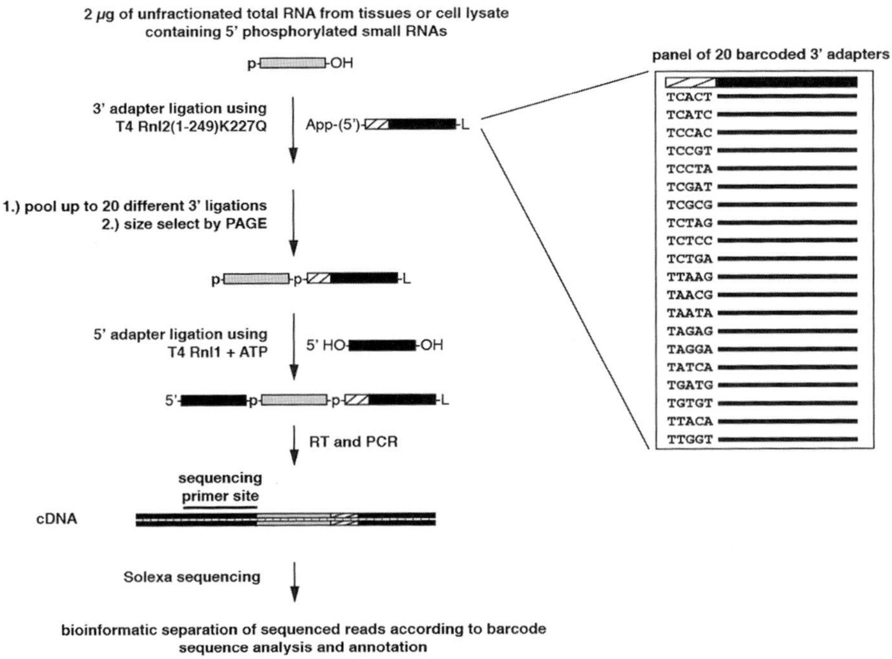

Fig. 12.1 Schematic representation of barcoded small RNA cDNA library preparation. The barcode represents a 5 nt unique sequence at the 5′ end of the 3′ adapter oligodeoxynucleotides. Illumina HiSeq sequencing at its smallest scale yields approximately 100 million sequence reads per sequencing lane of which more than 70 % contain recognizable barcode sequences, resulting in more than 3 million sequence reads per sample when using the full set of 20 barcoded adapters. L, 3′ aminohexyl blocking group that prevents adapter circularization. Figure adapted from [66]

The TCGA has conducted the largest survey of miRNA profiling by Illumina sequencing in ~ 700 breast specimens, which was just published (minimum of 750,000 reads per sample) [51]. Smaller studies have explored the same and additional questions using sequencing technologies, suggesting identification of novel miRNAs. A study by Tu et al. used another pair-barcode sequencing approach for SOLID sequencing (Applied Biosystems Inc.) to analyze 32 independent, pair-barcoded, miRNA libraries (26 tumor and 6 normal) by the combination of 4 unique forward barcodes and 8 unique reverse barcodes within one sequencing run [65]. They validated their barcoded method and identified the top 10 most abundant miRNAs in all their samples. Then, they interrogated miRNAs known to be dysregulated in breast cancer, confirming miR-155 and miR-31 and identifying other dysregulated miRNAs. Another study analyzed ten FFPE tissues (five invasive micropapillary carcinomas and five IDC of no special types) using Illumina sequencing [66]. They identified differentially expressed miRNAs between these two groups of tumors and miRNA sequence features, including RNA editing (5′ variability, deaminase activity, and 3′ untemplated addition).

Finally, further smaller sequencing-based studies identified novel miRNAs from breast cancer cell lines [67] and patient tissues [68]. Persson et al. sequenced matched tumor, normal and tumor-adjacent breast tissue from five patients, and reported 361 new miRNA precursors, approximately half of which were incorporated in Ago2 complexes in MCF7 cells [68]. The abundance of most of these novel miRNAs is very low, questioning their biological significance but also suggesting more stringent criteria for miRNA annotation and validation to uncover miRNAs that may have important roles in tumorigenesis.

5.2 *Characteristics of miRNA Expression Profiles*

5.2.1 miRNA Abundance

We quantified miRNA levels by adding a known amount of synthetic RNAs to the total RNA subjected to deep sequencing [61]. Specifically, we added a cocktail of an equimolar amount of 10 synthetic 22 nt 5′ phosphorylated RNAs distinct from human sequences per μg of total RNA. Synthetic RNAs on average represented 18 % of the total reads. We found that miRNAs were expressed in similar amounts in normal breast compared to cell lines, in situ and invasive tumors, even though previous reports suggested that miRNAs were less abundant in tumor compared with normal breast samples [10, 69, 70]. Consistent with these results, examination of mRNA levels in the same samples for miRNA pathway components or other factors implicated in miRNA biogenesis did not suggest globally differential miRNA processing in normal breast and carcinomas.

miRNA profiles of a sample can be presented as the relative percentage of miRNA read frequencies (rf) by dividing miRNA read counts by total miRNA reads per library. Defining the copy numbers of miRNAs per cell helps to better appreciate their functional potential, as discussed earlier. On the basis of calculations in MCF-7 cells, each tumor cell contained 145,000 miRNA molecules, illustrating that miRNAs expressed at 1 % of the total miRNA content in each cell (relative rf) would represent 1,500 copies per cell. Quantitative Western blot analysis for Ago2 protein, the main component of the miRNA effector complexes, in MCF-7 showed the presence of approximately 42,000 copies per cell. Assuming similar abundance for the often co-expressed Ago members, the number of effector complexes may match the miRNA copy number. Recent results suggested deviation in miRNA:Ago ratios in a cervical cancer cell line, with miRNAs expressed in excess of Agos (13-fold) and miRNAs bound to mRNAs in a sevenfold excess relative to Agos, implying the existence of miRNA-mRNA duplexes not stoichiometrically bound by Agos [71]. The existence of Ago-free miRNA-mRNA duplexes that can be functionally bound by Agos suggested that miRNA-mediated repression can be catalytic.

5.2.2 Clustering by Individual miRNAs, miRNA Genomic Clusters, and Sequence Families

miRNA profiles can be condensed either by assigning individual miRNA or miRNA* reads to their originating miRNA genomic clusters or to sequence families (denoted cluster-mir and sf-miR, respectively) and listing number of cluster/family members in parenthesis. The genomic cluster profiles represent promoter-controlled miRNA expression units, whereas the sequence families are most informative for characterization of seed sequence-dependent miRNA target regulation (reviewed in [1]); either approach reduces the complexity of the data.

We conducted unsupervised hierarchical clustering for 179 clinical samples using individual miRNAs, precursor clusters, and sequence families expressed at 85 % of the total miRNA reads in at least 1 sample, limiting our analysis to higher-abundance miRNAs (Fig. 12.2 shows clustering for individual miRNAs). Normal breast samples clustered together, close to a small group of ER- and HER2-positive tumors characterized by lower expression of sf-miR-21(1) and higher expression of sf-miR-22(1) as compared to the remainder of tumor samples. Some DCIS samples also clustered together, breaking up groups of invasive tumor samples and suggesting that DCIS samples accumulate changes in miRNA expression early in the tumorigenic process. Invasive tumor samples positive for 1 or more IHC marker clustered together, whereas TNBCs emerged as several groups distinct from the other tumors. Samples did not cluster according to other pathological and clinical characteristics (overall survival and future metastatic status included in Fig. 12.2).

5.2.3 Comparison of Clustering by miRNAs vs. mRNAs

We compared clustering of 179 samples based on miRNAs (using 98 % of miRNAs expressed in at least 1 sample) and clustering of 161 of these samples according to their mRNA profiles (using the same number of genes as miRNAs, 221, selecting genes with most variance) (Fig. 12.3). mRNA profiles better separated HER2-positive samples, suggesting that the HER2 pathway is not related to miRNA expression changes. miRNA clustering was correlated with TNBC. Three TNBC groups emerged by clustering miRNA profiles, one of which included mostly special histologic types; these groups did not show distinct patient characteristics or outcome.

The TCGA study identified seven miRNA subtypes by consensus nonnegative matrix factorization clustering using an abundance matrix containing the 25 % most variable miRNAs (306 miRNAs) [51]. These subtypes correlated with mRNA subtypes, ER, PR, and HER2 clinical status. Of note, two miRNA groups showed high overlap with the basal-like mRNA subtype and contained many TP53 mutations. The remaining miRNA groups comprised a mixture of luminal A, luminal B, and HER2 enriched, with little correlation with molecular subtypes (based on mRNA profiles). With the exception of TP53, which showed a strong positive correlation, and PIK3CA and GATA3, which showed negative associations with groups 4 and 5, respectively, there was little correlation with mutation status and miRNA subtype.

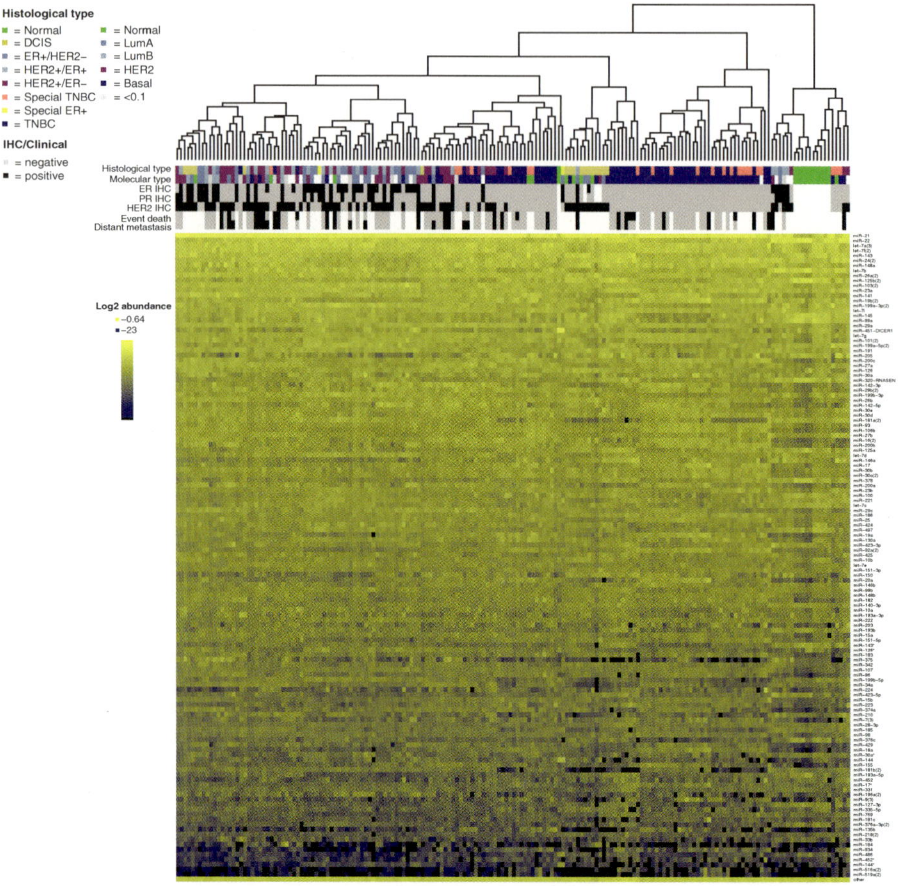

Fig. 12.2 Clustering of breast cancer specimens by individual miRNAs. Unsupervised hierarchical clustering with complete linkage and Spearman correlation for patient samples conducted using the miRNA sequence families making up 85 % of the sequence reads within at least 1 sample. Color histogram represents $\log_2$ miRNA abundance (relative read frequency (rf)). Figure adapted from [1]

5.2.4 Novel miRNAs

The rapid acquisition of deep sequencing small RNA cDNA profiles from many human tissues led to a rapid expansion of entries in miRBase, currently in its 19th version. When we mapped reads obtained in our lab from over 1,000 diverse human diseased and normal tissues (561,200,705 reads) against miRBase 16 entries, we noticed that some read alignments did not correspond to prototypical miRNA biogenesis patterns and were more likely erroneously annotated as miRNA genes [61]. Curation of miRNAs was performed by analyzing their expression levels and cistronic expression pattern, the mapping pattern of reads to existing miRNA regions,

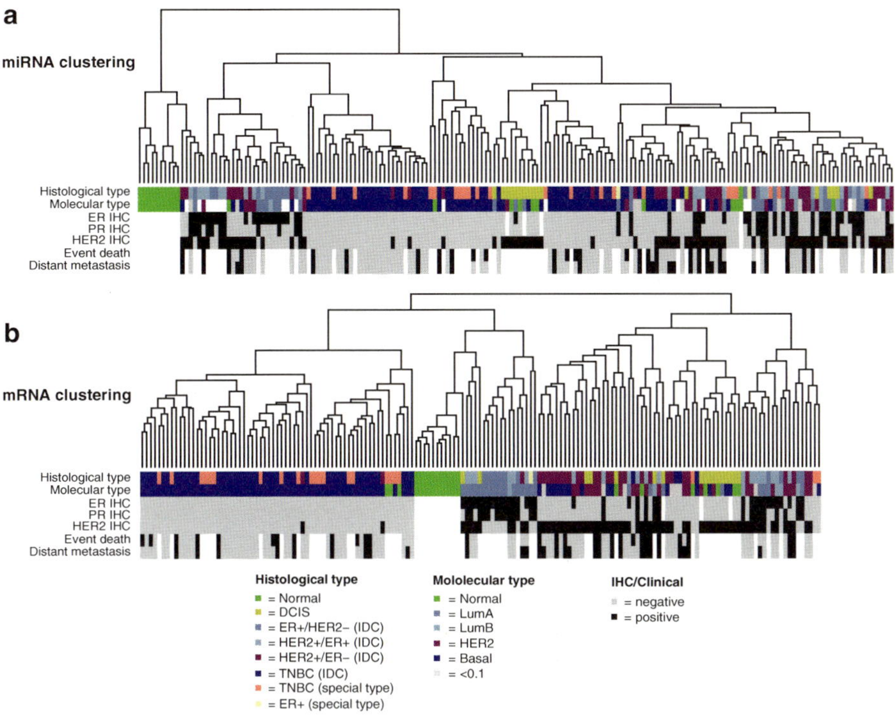

Fig. 12.3 Comparison of miRNA and mRNA clustering. Comparison of clustering using miRNA sequence families (characterizing seed sequence-dependent miRNA target regulation) and mRNA profiles. (**a**) Unsupervised hierarchical clustering conducted for miRNA sequence families, using the top 98 % expressed families within at least 1 sample (221 families; Spearman correlation). (**b**) Unsupervised hierarchical clustering of 161 samples with available mRNA profiles, using the 221 genes with the highest variance (Pearson correlation). Figure adapted from [1]

as well as secondary structure prediction for the expected fold-back structure of precursor molecules, described in [63]. Of the 1,045 miRBase 16 annotated human precursor sequences, 488 failed one or more of our criteria (expression, mapping pattern, and RNA fold), with 282 having little or no expression evidence. We did not identify any novel miRNAs with the same stringent criteria we use to curate miRBase.

5.3 *Clinical Applications of miRNA Expression Profiles*

5.3.1 Tissue Heterogeneity

When considering molecular diagnostics based on RNA isolation from tumor tissue (e.g., RT-PCR or sequencing), tissue heterogeneity can skew results by introducing

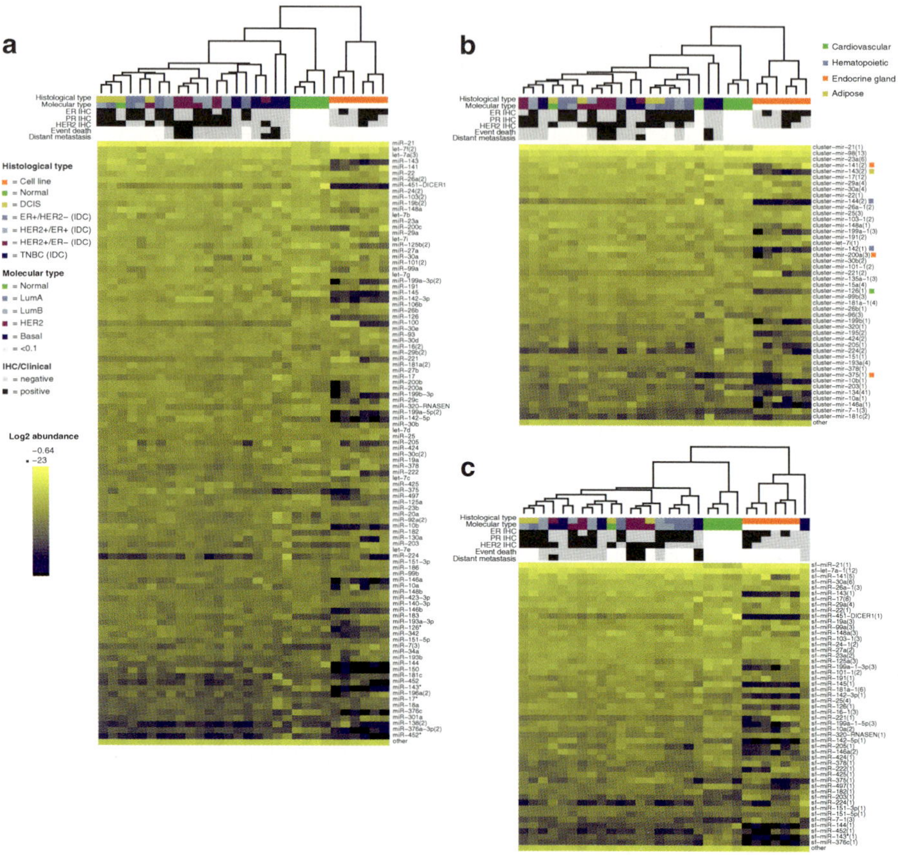

Fig. 12.4 Tissue heterogeneity as seen through comparison of human tissue specimens and cell lines. Unsupervised hierarchical clustering with complete linkage and Spearman correlation depicting (**a**) the 85 % top expressed mature miRNAs, (**b**) miRNA genomic clusters, and (**c**) sequence families. Every clustering includes four samples from each ER/HER2 IHC category, normal breast, and 6 cell lines. Color histogram represents $\log_2$ miRNA abundance (relative read frequency (rf)). Examples of lineage-specific miRNA genomic clusters are noted. Figure adapted from [1]

multiple cell types, including tumor-adjacent normal tissue. To understand miRNA expression in the context of tissue heterogeneity, we compared the 6 ductal cell lines to human specimens (Fig. 12.4). We identified miRNAs present at similar levels in cell lines and tumors that may be involved in tumorigenic processes as well as miRNAs absent from cell lines that are likely expressed by other cell types, such as cluster-mir-126(1), cardiovascular system; cluster-mir-143(2), adipose tissue; and cluster-mir-144(2) and cluster-mir-142(1) hematopoietic system.

5.3.2 Differential Expression Analysis Focusing on Tumorigenesis, Invasion, and Tumor Subtypes

We used the edgeR package to identify individual mature miRNAs, miRNA genomic clusters, and miRNA sequence families differentially represented between normal, DCIS, and invasive tumor samples. The edgeR algorithm compensates for the difference in sequence reads between samples, as well as the different number of samples within the compared categories. Similarly to the clustering analysis, we only focused on miRNAs that were among the most highly expressed.

Tumorigenesis: To identify miRNAs that were dysregulated during tumorigenesis, we compared normal samples to HER2-positive IDC and DCIS samples. Interestingly, most miRNA changes in IDC were already apparent in DCIS samples. sf-miR-21(1), sf-miR-142-3p(1), and sf-miR-142-5p(1) had higher abundance in IDC compared to normal breast (Fig. 12.5a). Sf-miR-22(1), sf-miR-99a(3), sf-miR-125a(3), sf-miR-99a(3), sf-miR-145(1), sf-miR-451(1), sf-miR-378(1), sf-miR-497(1), and sf-miR-320(1) were examples of abundant sf in normal breast but reduced in IDC. As discussed earlier, miRNAs present both in ductal cell lines and tumors likely represent tumor-downregulated miRNAs (e.g., sf-miR-125a(3) and sf-miR-22(1)). miRNAs absent in ductal cell lines likely reflect differences in tissue composition (e.g., adipose-specific sf-miR-145(1) is likely related to the presence of adipose tissue in the biopsies).

Tumor Invasion: To identify miRNAs altered in tumor invasion, we compared HER2-positive DCIS and IDC samples. Sf-miR-142-3p(1) members were overrepresented in HER2-positive IDC compared with DCIS. Given that sf-miR-142-3p(1) is hematopoietic lineage specific, this overrepresentation most likely reflects a change in tissue composition.

Tumor Subtypes: To differentiate among breast tumor subtypes, we determined whether miRNA levels correlated with IHC or special histologic characteristics (Fig. 12.5b). Differences in miRNA representation between IHC subtypes involved less abundant miRNAs. TNBC showed the largest number of miRNA level changes. Sf-miR-19a(3), expressed in ductal cell lines, sf-miR-205(1), and sf-miR-146a(2) were higher expressed in TNBC compared with ER- and/or HER2-positive tumors. Sf-miR-451(1) was less abundant in TNBC. Special histologic TNBCs (atypical medullary, metaplastic, adenoid cystic, and apocrine carcinomas) were differentiated from IDC TNBCs by less abundant sf-miR-224(1), absent from some ductal cell lines.

5.3.3 Nucleotide Variation Analysis

miRNAs have been proposed to act as tumor suppressors or oncogenes, yet after implementing an algorithm for identification of nucleotide variations relative to the reference human genome, focusing on variations that could represent SNPs or somatic mutations, we did not obtain evidence for somatic mutations in miRNAs.

For our analysis, we required 10 or more varied sequence reads covering a given position per sample, with 25 % or above variation frequency. On the basis of analysis

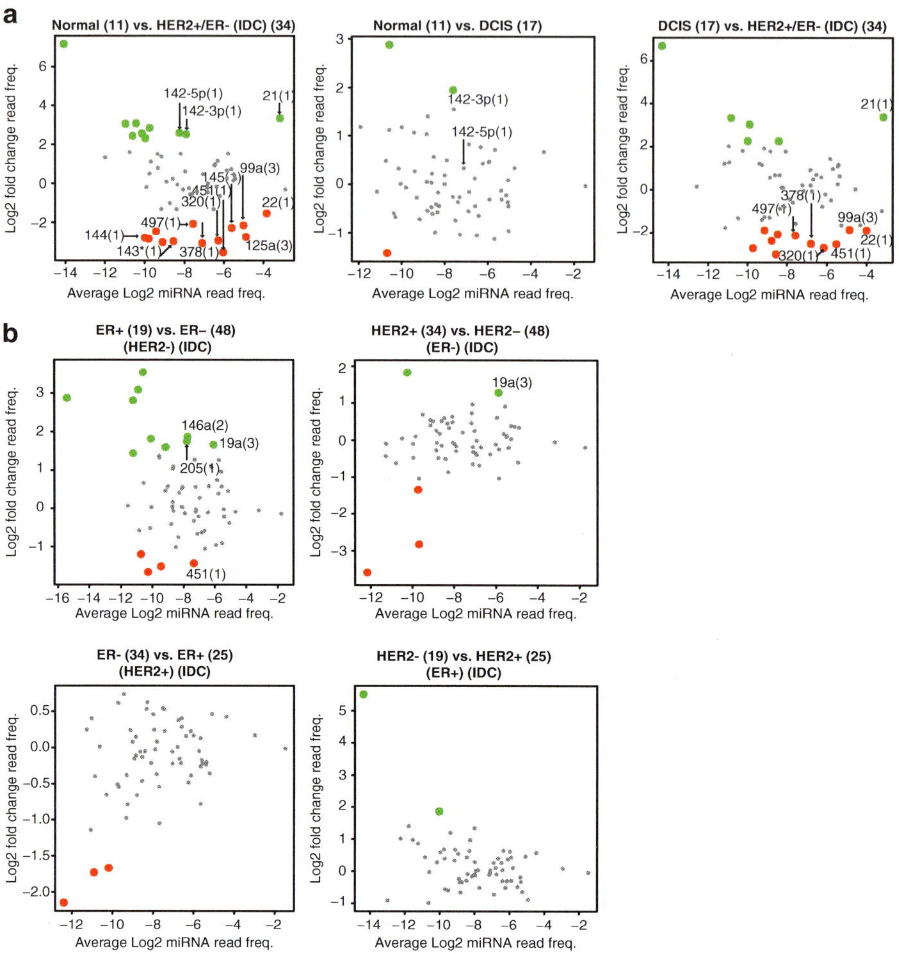

Fig. 12.5 miRNA sequence families (sf) differentially represented in tumor stages and tumor IHC types. Results of edgeR comparison analysis between groups of samples. Results plotted as $\log_2$ of the fold change between normal and/or tumor categories as a function of the $\log_2$ of the average miRNA abundance in the 2 categories compared. Colored dots represent miRNAs that are significantly differentially expressed ($P \leq 0.001$). miRNAs in *green* are overexpressed, whereas miRNAs in *red* are underexpressed in the second category of samples within each comparison. Abundant miRNAs are labeled in *black*, whereas low abundant miRNAs are labeled in *gray*. Figure adapted from [1]

of deep sequencing data from the pool of 770 synthetic miRNAs [62], 10 % or higher variation frequency guaranteed that 98 % of the identified variations were not random events due to sequencing errors, but were likely due to the ambiguity of mapping miRNAs similar in sequence (if allowing for mapping errors).

We identified 144 distinct nucleotide variations located within 117 mature and star miRNA sequences (excluding the 3′ most terminal residues of all sequence reads from somatic mutational analysis if it was altered, because it frequently contained 3′ untemplated nucleotide addition). There were 109 variations in the last

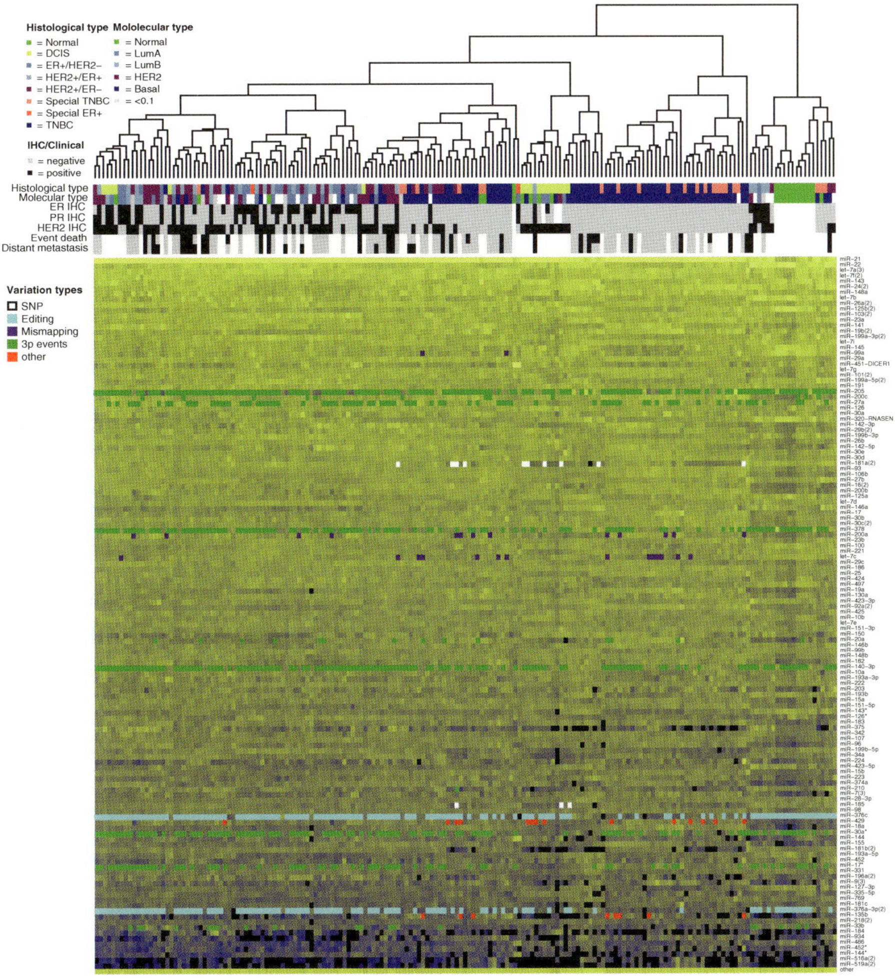

Fig. 12.6 Examples of nucleotide variations in the mature or star miRNA sequences marked in the dendrogram/heatmap of the top 85 % expressed mature miRNAs. Color histogram represents $\log_2$ miRNA abundance (relative read frequency (rf)). *Color codes* represent different types of nucleotide variations, suggested by the distribution of the frequency of the varied nucleotide among samples (editing or SNP), or sequence similarity to other miRNAs (mismapping from more abundant miRNA). Figure adapted from [1]

two positions of the predominant mature sequence read, likely representing instances of untemplated 3′ terminal addition that were insufficiently repressed by our computational approach of not considering the 3′ most nucleotides (102 variations represented changes into A or U). None of the remaining 35 variations were detected in abundant miRNAs (>1 % rf). These sequence variations could represent RNA editing events (including deamination, polyuridylation, and polyadenylation), SNPs, or somatic mutations (Fig. 12.6). The most common of the 35 variations observed in the mature and star miRNA sequences were A to G, likely representing A to I RNA editing by dsRNA-specific adenosine deaminases. This observation was further supported by a well-represented unimodal distribution of the nucleotide variation frequency for edited miRNAs (e.g., miR-376a).

We detected two known SNPs (SNPdb version 131) in less abundant miR-196a-2* and miR-146a, which have been studied in the context of breast cancer risk (rs11614913 and rs2910164; see below). Epidemiological studies have studied breast cancer risk in the context of these two SNPs but are plagued with conflicting results in different ethnic cohorts and patients with different genetic backgrounds (i.e., BRCA-positive patients). For example, some studies have supported the association of SNP rs11614913 in miR-196a-2* with breast cancer susceptibility risk [72], while other studies have not confirmed it [73]. In another example, some studies have revealed an association between miR-146a rs2910164 SNP and tumor age-of-onset in BRCA1/BRCA2-negative familial breast and ovarian cancer patients [74, 75]. In our study, rs2910164 (C5G) was detected in 5 carcinomas, whereas rs11614913 (C18T) was detected in 39 samples, including 1 normal breast sample and 3 cell lines. Given the low number of normal samples included in our study, we were not able to draw statistically significant conclusions as to enrichment of a particular SNP within tumors compared to normal breast. We did identify 10 nucleotide variations that are candidate new SNPs, based on the trimodal distribution of their variation frequency. Two of these variations were identified in miRNAs present within 85 % of all miRNA reads in at least 1 sample: miR-181a (T19G) was observed in ten carcinomas and miR-185 (T16G) was observed in three carcinomas.

The TCGA study so far has been analyzed without allowing any mismatches in miRNA sequences during the annotation process [51], so it has not commented on the presence or absence of nucleotide variations. However, the data can be reanalyzed, in the context of the sequencing technology and data processing methodology used to also assess the frequency of miRNA variations.

5.3.4 miRNAs as Prognostic Markers

We did not identify any statistically significant prognostic miRNAs while evaluating 48 TNBC patients and 49 HER2-positive patients, for which clinical follow-up information was available (distant metastasis-free survival and overall survival; mean follow-up of 5.6 years).

Cluster-mir-423(1) members and cluster-mir-375(1), that are normally less abundant, were more abundant in patients that went on to develop metastasis, whereas cluster-mir-184(1) was less abundant (Fig. 12.7a). Kaplan–Meier analyses supported the findings for cluster-mir-423(2) in HER2-positive patients (Fig. 12.7b). Univariate and multivariate Cox regression indicated that cluster-mir-423(1) was an independent predictor of outcome in the presence of other clinical parameters (tumor size, grade and lymph node status). However, these findings should be

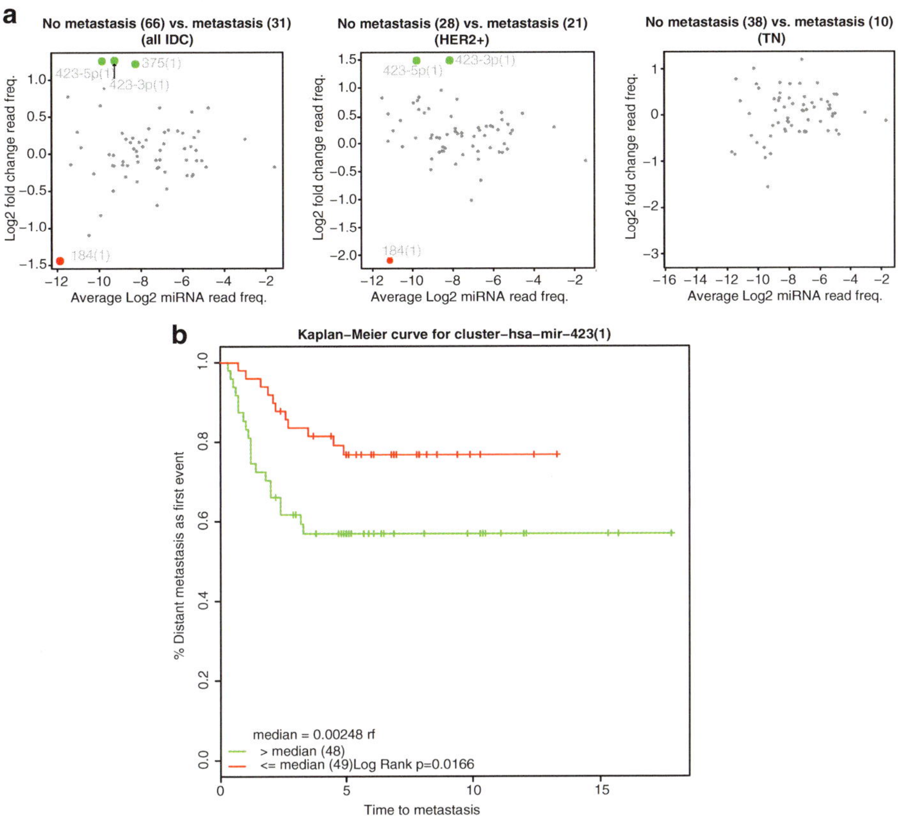

Fig. 12.7 Testing prognostic value of miRNAs in breast cancer. (**a**) miRNA sequence families (sf) that predict metastasis in IDC, also separately evaluated in HER2-positive and TNBC patients. edgeR analysis described in legend for Fig. 12.4. (**b**) Kaplan–Meier curves for cluster-mir-423(1). Figure adapted from [1]

interpreted with caution, as these *P* values did not correct for multiple testing. Interestingly, a genetic variant located in *mir-423* is associated with reduced breast cancer risk [76]. Our study showed higher levels of miR-210 in invasive compared with noninvasive carcinomas (discussed in an earlier section) but did not confirm a statistically significant prognostic role for this miRNA [61].

A follow-up study reanalyzed our results to identify prognostic markers [77]. The prognostic value of specific miRNAs depended on the statistical tools used, highlighting the need for larger validation studies to draw conclusive evidence. Volinia et al. confirmed that the miRNA profile established for the normal breast to ductal carcinoma in situ transition was largely maintained in the in situ to invasive ductal carcinoma transition. They suggested a 9-miRNA signature that differentiated invasive from in situ carcinoma, which mostly involved lower abundant miRNAs. Specifically, let-7d, miR-210, and miR-221 were down-regulated in the normal to in situ carcinoma transition and up-regulated in the in situ to invasive transition, thus featuring an expression reversal along the cancer progression path. Additionally, they identified miRNAs that were prognostic for overall survival and time to metastasis: miR-210, miR-21, miR-106b*, miR-197, and let-7i, with miR-210 as the only one also involved in the invasive transition. As mentioned above, the breast cancer TCGA study does not comment on prognosis, as data is still being collected and patient follow-up time is currently a rather short 17 months [51].

6 Determination of Circulating miRNAs in Breast Cancer Patients by Deep Sequencing

6.1 Circulating miRNAs in Breast Cancer Patients by Non-sequencing Based-Approaches

Several studies used non-sequencing-based techniques, either microarray or RT-PCR, to determine expression of circulating miRNAs in breast cancer patients. One of the first studies linking clinical and histopathological characteristics to circulating miRNA levels comes from Heneghan et al. (148 breast cancer patients, including different tumor subtypes, invasive and in situ disease, and 44 age-matched normal controls) [20]. This RT-PCR-based study used total blood and miR-16 as endogenous control, a miRNA frequently used as an endogenous control for circulating miRNAs [18, 78], and evaluated the levels of 7 miRNAs. miR-195 and let-7a were significantly more abundant in cancer patients compared to healthy controls, even though in breast tissues both miRNAs are more abundant in normal breast tissue compared to tumor tissue [61]. Following curative tumor resection, the levels of both miRNAs decreased to levels similar to healthy controls. ER-negative patients exhibited higher levels of miR-10b and miR-21; no other differences were found in regard to cancer type (in situ vs. invasive) or cancer subtype.

A study by Schooneveld et al. used TaqMan low-density arrays (TLDA) to survey circulating miRNAs in 84 breast cancer and 8 normal patients, identifying a different set of miRNAs (miR-215, miR-299-5p, miR-411, and miR-452) differentiating normal from cancer specimens [79]. These miRNAs are not abundant in normal and/or cancer breast tissues [61]. Interestingly, they noted a better differentiation in untreated patients with metastatic disease, suggesting that higher tumor burden enhances miRNA detection. In another study, literature searches and bioinformatic analysis were used to select a panel of 10 miRNAs and correlate their serum levels to clinicopathological data in 100 breast cancer patients and 20 controls [80]. Decreased levels of miR-92a and increased levels of miR-21 were associated with tumor size and positive lymph node status.

6.2 Circulating miRNAs in Breast Cancer Patients by Deep Sequencing

Studies using deep sequencing to profile miRNAs in breast cancer patients from plasma, serum, or total blood are limited. Wu et al. published the largest study profiling serum miRNAs by deep sequencing in breast cancer patients [78]. They used Illumina sequencing to profile the sera of 42 patients with pathologically confirmed stage II and stage III locally advanced and inflammatory breast cancer. An additional cohort of 26 stage II and III breast cancer patients was used for validation. In total, more than 800 miRNAs were detected in patient sera (in at least two patients). Three hundred seventy-three miRNAs (>50 reads in at least 10 % of the samples) were subjected to unsupervised hierarchical clustering, which separated ER-positive and ER-negative samples. Subsequent correlative analysis of miRNA level association with clinical parameters identified that miR-122 and miR-375 could differentiate between patients who later developed metastatic relapse and those who did not, with a follow-up of 2.5 years. Only miR-122 was confirmed in the validation cohort, showing 88 % sensitivity and 78 % specificity. Of note, miR-122 is a liver-specific miRNA, involved in liver metabolism [81]. Downregulation of miR-122 was reported in hepatocellular carcinoma tissues [82], whereas higher levels of miR-122 were reported in circulating hepatocellular carcinoma and liver injury samples [83]. miR-375 is a pancreatic islet-specific miRNA regulating insulin secretion [84]. Seven miRNAs were able to predict response to neoadjuvant therapy: miR-375, miR-184, miR-1299, miR-196a, miR-381, miR-410, and miR-1246 (these miRNAs are not abundant in normal or breast tissues [61]).

Another recent study employed Illumina sequencing and TDLA to evaluate circulating biomarkers in the serum of 2 pools (48 breast cancer patients and 48 female control cases) [85]. The two pools were subjected to sequencing and further candidates were then selected for validation using RT-PCR. 10 miRNAs were selected from sequencing based on abundance, fold change between normal and tumor samples, and unidirectional correlation in TLDA and Illumina sequencing. 4 miRNAs (miR-16, miR-25, miR-222, and miR-324-3p) were validated by RT-PCR, showing significant fold change in tumor versus normal samples, and separated these samples based on other statistical analyses [85]. miR-16 was used in a study described earlier as a normalizer, pointing to discrepancies in the putative miRNAs linked to breast cancer between the different studies. These discrepancies might be due to small study sizes, different RNA extraction and processing, different statistical analysis, or insufficient control for potential confounding factors, but discrepancies are also due to the abovementioned biases of each technology. Further studies are needed to investigate the use for circulating miRNAs as biomarkers in breast cancer patients. The ability to better detect lowly abundant miRNAs by deep sequencing compared to traditionally used microarrays and RT-PCR approaches might be particularly useful [78].

7 Identification of miRNA-mRNA Targets Using Deep Sequencing

Despite all of the recent advances in defining miRNA dysregulation in breast cancer, the miRNA-mRNA targets that mediate the miRNA's oncogenic and tumor suppressor functions have not conclusively been identified. Computational prediction methods, analysis of patient coordinated miRNA and mRNA datasets, and perturbation of miRNAs in cell culture experiments followed by recording of mRNA expression levels are among the methods that have been used to define miRNA-mRNA targets. Given the complexity of identifying high-confidence miRNA targets, sequencing technology can also be applied to target identification.

The currently available target prediction databases (reviewed in [3]) do not easily allow prioritization of the involvement of reported targets in certain phenotypes, thus necessitating the selection of a few targets from a list of hundreds for further study and validation based on a priori knowledge of potentially involved biological pathways. Since the prediction algorithms do not always produce identical target lists, use of multiple algorithms and comparison or intersection of their results narrows the list to higher-confidence targets. Targets are only relevant to a specific phenotype if they are expressed in the studied tissue, an issue not addressed by most computational prediction algorithms. Bioinformatic analysis of coordinated patient data also requires input from computational prediction approaches, thus also facing some similar biases.

Biochemical identification methods in cell lines and tissues are being established and further refine our understanding of miRNA-mRNA target binding recognition. These methods involve two approaches: overexpression or downregulation of studied miRNAs, followed by assessment of transcriptome-wide mRNA levels by mRNA microarray analysis or proteomic analysis. Recently, these biochemical approaches have been coupled to deep sequencing technology; after immunoprecipitation of miRNAs and mRNAs complexed with the main component of the miRNA effector complex, Ago, deep sequencing can not only identify mRNA targets but also localize their precise binding sites [86, 87]. These sequencing-based methods are summarized in recent methodology papers [88, 89].

8 Conclusion

Sequencing technology for miRNA profiling has become high throughput and cost-effective and yields comparable results to other technologies, with the added advantage that sequence information allows reanalysis of data in the future to keep up with our advancing knowledge of miRNA annotation and the addition of new members to the miRNA repertoire. In breast cancer, so far, miRNAs have not proven to be more powerful at differentiating tumor subtypes or contributing to prognosis and early detection. However, with advances in miRNA detection methodology, especially given their stability in body fluids compared to mRNAs, and bioinformatic technologies to allow definition of combination marker profiles, miRNAs offer an attractive addition to tumor detection and classification. Appreciating how the heterogeneous genetic makeup of each breast cancer patient may suggest individual treatment options for each patient, miRNAs join the ranks of multiple levels of genetic and molecular information that will make personalized medicine a reality.

References

1. Farazi TA, Spitzer JI, Morozov P, Tuschl T. miRNAs in human cancer. J Pathol. 2011;223(2):102–15. Epub 2010/12/03.
2. Ventura A, Jacks T. MicroRNAs and cancer: short RNAs go a long way. Cell. 2009;136(4): 586–91. Epub 2009/02/26.
3. Bartel DP. MicroRNAs: target recognition and regulatory functions. Cell. 2009;136(2):215–33. Epub 2009/01/27.
4. Volinia S, Galasso M, Costinean S, Tagliavini L, Gamberoni G, Drusco A, et al. Reprogramming of miRNA networks in cancer and leukemia. Genome Res. 2010;20(5):589–99. Epub 2010/05/05.
5. Inui M, Martello G, Piccolo S. MicroRNA control of signal transduction. Nat Rev Mol Cell Biol. 2010;11(4):252–63. Epub 2010/03/11.
6. Newman MA, Hammond SM. Emerging paradigms of regulated microRNA processing. Genes Dev. 2010;24(11):1086–92. Epub 2010/06/03.
7. Winter J, Jung S, Keller S, Gregory RI, Diederichs S. Many roads to maturity: microRNA biogenesis pathways and their regulation. Nat Cell Biol. 2009;11(3):228–34. Epub 2009/03/04.
8. Bartel DP. MicroRNAs: genomics, biogenesis, mechanism, and function. Cell. 2004;116(2):281–97. Epub 2004/01/28.
9. Molenaar JJ, Domingo-Fernandez R, Ebus ME, Lindner S, Koster J, Drabek K, et al. LIN28B induces neuroblastoma and enhances MYCN levels via let-7 suppression. Nat Genet. 2012;44(11):1199–206.
10. Lu J, Getz G, Miska EA, Alvarez-Saavedra E, Lamb J, Peck D, et al. MicroRNA expression profiles classify human cancers. Nature. 2005;435(7043):834–8. Epub 2005/06/10.
11. Johnson CD, Esquela-Kerscher A, Stefani G, Byrom M, Kelnar K, Ovcharenko D, et al. The let-7 microRNA represses cell proliferation pathways in human cells. Cancer Res. 2007;67(16):7713–22. Epub 2007/08/19.
12. Yanaihara N, Caplen N, Bowman E, Seike M, Kumamoto K, Yi M, et al. Unique microRNA molecular profiles in lung cancer diagnosis and prognosis. Cancer Cell. 2006;9(3):189–98. Epub 2006/03/15.
13. Takamizawa J, Konishi H, Yanagisawa K, Tomida S, Osada H, Endoh H, et al. Reduced expression of the let-7 microRNAs in human lung cancers in association with shortened postoperative survival. Cancer Res. 2004;64(11):3753–6. Epub 2004/06/03.
14. Calin GA, Ferracin M, Cimmino A, Di Leva G, Shimizu M, Wojcik SE, et al. A MicroRNA signature associated with prognosis and progression in chronic lymphocytic leukemia. N Engl J Med. 2005;353(17):1793–801. Epub 2005/10/28.
15. Hurst DR, Edmonds MD, Welch DR. Metastamir: the field of metastasis-regulatory microRNA is spreading. Cancer Res. 2009;69(19):7495–8. Epub 2009/09/24.
16. Zhao H, Shen J, Medico L, Wang D, Ambrosone CB, Liu S. A pilot study of circulating miRNAs as potential biomarkers of early stage breast cancer. PLoS One. 2010;5(10):e13735. Epub 2010/11/10.
17. Hori SS, Gambhir SS. Mathematical model identifies blood biomarker-based early cancer detection strategies and limitations. Sci Transl Med. 2011;3(109):109ra16.

18. Mitchell PS, Parkin RK, Kroh EM, Fritz BR, Wyman SK, Pogosova-Agadjanyan EL, et al. Circulating microRNAs as stable blood-based markers for cancer detection. Proc Natl Acad Sci USA. 2008;105(30):10513–8. Epub 2008/07/30.
19. Chen X, Ba Y, Ma L, Cai X, Yin Y, Wang K, et al. Characterization of microRNAs in serum: a novel class of biomarkers for diagnosis of cancer and other diseases. Cell Res. 2008;18(10):997–1006. Epub 2008/09/04.
20. Heneghan HM, Miller N, Lowery AJ, Sweeney KJ, Newell J, Kerin MJ. Circulating microRNAs as novel minimally invasive biomarkers for breast cancer. Ann Surg. 2010;251(3):499–505. Epub 2010/02/06.
21. Chen X, Hu Z, Wang W, Ba Y, Ma L, Zhang C, et al. Identification of ten serum microRNAs from a genome-wide serum microRNA expression profile as novel noninvasive biomarkers for nonsmall cell lung cancer diagnosis. International journal of cancer Journal international du cancer. 2012;130(7):1620–8. Epub 2011/05/11.
22. Yang C, Wang C, Chen X, Chen S, Zhang Y, Zhi F, et al. Identification of seven serum microRNAs from a genome-wide serum microRNA expression profile as potential noninvasive biomarkers for malignant astrocytomas. Int J Cancer. 2013;132(1):116–27.
23. Yu S, Liu Y, Wang J, Guo Z, Zhang Q, Yu F, et al. Circulating microRNA profiles as potential biomarkers for diagnosis of papillary thyroid carcinoma. J Clin Endocrinol Metabol. 2012;97(6):2084–92. Epub 2012/04/05.
24. Liu R, Zhang C, Hu Z, Li G, Wang C, Yang C, et al. A five-microRNA signature identified from genome-wide serum microRNA expression profiling serves as a fingerprint for gastric cancer diagnosis. Eur J Cancer. 2011;47(5):784–91. Epub 2010/11/30.
25. Engelhardt S. Small RNA, biomarkers come of age. J Am Coll Cardiol. 2012;60(4):300–3. Epub 2012/07/21.
26. Collino F, Deregibus MC, Bruno S, Sterpone L, Aghemo G, Viltono L, et al. Microvesicles derived from adult human bone marrow and tissue specific mesenchymal stem cells shuttle selected pattern of miRNAs. PLoS One. 2010;5(7):e11803. Epub 2010/07/30.
27. Chen TS, Lai RC, Lee MM, Choo AB, Lee CN, Lim SK. Mesenchymal stem cell secretes microparticles enriched in pre-microRNAs. Nucleic Acids Res. 2010;38(1):215–24. Epub 2009/10/24.
28. Pigati L, Yaddanapudi SC, Iyengar R, Kim DJ, Hearn SA, Danforth D, et al. Selective release of microRNA species from normal and malignant mammary epithelial cells. PLoS One. 2010;5(10):e13515. Epub 2010/10/27.
29. Valadi H, Ekstrom K, Bossios A, Sjostrand M, Lee JJ, Lotvall JO. Exosome-mediated transfer of mRNAs and microRNAs is a novel mechanism of genetic exchange between cells. Nat Cell Biol. 2007;9(6):654–9. Epub 2007/05/09.
30. Gibbings DJ, Ciaudo C, Erhardt M, Voinnet O. Multivesicular bodies associate with components of miRNA effector complexes and modulate miRNA activity. Nat Cell Biol. 2009;11(9):1143–9. Epub 2009/08/18.
31. Lee YS, Pressman S, Andress AP, Kim K, White JL, Cassidy JJ, et al. Silencing by small RNAs is linked to endosomal trafficking. Nat Cell Biol. 2009;11(9):1150–6. Epub 2009/08/18.
32. Cortez MA, Bueso-Ramos C, Ferdin J, Lopez-Berestein G, Sood AK, Calin GA. MicroRNAs in body fluids–the mix of hormones and biomarkers. Nat Rev Clin Oncol. 2011;8(8):467–77. Epub 2011/06/08.
33. American Cancer Society: Cancer Facts and Figure 2012. Atlanta, Ga: American Cancer Society, 2012 [database on the Internet]. 2012 [cited Last accessed September 24, 2012].
34. van 't Veer LJ, Dai H, van de Vijver MJ, He YD, Hart AA, Mao M, et al. Gene expression profiling predicts clinical outcome of breast cancer. Nature. 2002;415(6871):530–6.
35. Perou CM. Molecular stratification of triple-negative breast cancers. Oncologist. 2011;16 Suppl 1:61–70. Epub 2011/02/10.
36. Sorlie T, Tibshirani R, Parker J, Hastie T, Marron JS, Nobel A, et al. Repeated observation of breast tumor subtypes in independent gene expression data sets. Proc Natl Acad Sci USA. 2003;100(14):8418–23. Epub 2003/06/28.

37. Carey LA, Perou CM, Livasy CA, Dressler LG, Cowan D, Conway K, et al. Race, breast cancer subtypes, and survival in the Carolina Breast Cancer Study. JAMA. 2006;295(21):2492–502. Epub 2006/06/08.
38. Iorio MV, Ferracin M, Liu CG, Veronese A, Spizzo R, Sabbioni S, et al. MicroRNA gene expression deregulation in human breast cancer. Cancer Res. 2005;65(16):7065–70. Epub 2005/08/17.
39. Blenkiron C, Goldstein LD, Thorne NP, Spiteri I, Chin SF, Dunning MJ, et al. MicroRNA expression profiling of human breast cancer identifies new markers of tumor subtype. Genome Biol. 2007;8(10):R214. Epub 2007/10/10.
40. Tavazoie SF, Alarcon C, Oskarsson T, Padua D, Wang Q, Bos PD, et al. Endogenous human microRNAs that suppress breast cancer metastasis. Nature. 2008;451(7175):147–52. Epub 2008/01/11.
41. Ma L, Teruya-Feldstein J, Weinberg RA. Tumour invasion and metastasis initiated by microRNA-10b in breast cancer. Nature. 2007;449(7163):682–8. Epub 2007/09/28.
42. Valastyan S, Reinhardt F, Benaich N, Calogrias D, Szasz AM, Wang ZC, et al. A pleiotropically acting microRNA, miR-31, inhibits breast cancer metastasis. Cell. 2009;137(6):1032–46. Epub 2009/06/16.
43. Foekens JA, Sieuwerts AM, Smid M, Look MP, de Weerd V, Boersma AW, et al. Four miRNAs associated with aggressiveness of lymph node-negative, estrogen receptor-positive human breast cancer. Proc Natl Acad Sci USA. 2008;105(35):13021–6. Epub 2008/08/30.
44. Camps C, Buffa FM, Colella S, Moore J, Sotiriou C, Sheldon H, et al. hsa-miR-210 Is induced by hypoxia and is an independent prognostic factor in breast cancer. Clin Cancer Res. 2008;14(5):1340–8.
45. Buffa FM, Camps C, Winchester L, Snell CE, Gee HE, Sheldon H, et al. microrna-associated progression pathways and potential therapeutic targets identified by integrated mRNA and microRNA expression profiling in breast cancer. Cancer Res. 2011;71(17):5635–45.
46. Rothe F, Ignatiadis M, Chaboteaux C, Haibe-Kains B, Kheddoumi N, Majjaj S, et al. Global microRNA expression profiling identifies MiR-210 associated with tumor proliferation, invasion and poor clinical outcome in breast cancer. PLoS One. 2011;6(6):e20980. Epub 2011/07/09.
47. Smeets A, Daemen A, Vanden Bempt I, Gevaert O, Claes B, Wildiers H, et al. Prediction of lymph node involvement in breast cancer from primary tumor tissue using gene expression profiling and miRNAs. Breast Cancer Res Treat. 2011;129(3):767–76. Epub 2010/12/01.
48. Enerly E, Steinfeld I, Kleivi K, Leivonen SK, Aure MR, Russnes HG, et al. miRNA-mRNA integrated analysis reveals roles for miRNAs in primary breast tumors. PLoS One. 2011;6(2):e16915.
49. Hannafon BN, Sebastiani P, de las Morenas A, LU J, Rosenberg CL. Expression of microRNA and their gene targets are dysregulated in preinvasive breast cancer. Breast cancer Res. 2011;13(2):R24.
50. Van der Auwera I, Limame R, van Dam P, Vermeulen PB, Dirix LY, Van Laere SJ. Integrated miRNA and mRNA expression profiling of the inflammatory breast cancer subtype. Br J Cancer. 2010;103(4):532–41. Epub 2010/07/29.
51. TCGA. Comprehensive molecular portraits of human breast tumours. Nature. 2010;490(7418):61–70.
52. Lawrie CH, Soneji S, Marafioti T, Cooper CD, Palazzo S, Paterson JC, et al. MicroRNA expression distinguishes between germinal center B cell-like and activated B cell-like subtypes of diffuse large B cell lymphoma. Int J Cancer. 2007;121(5):1156–61. Epub 2007/05/10.
53. Weng L, Wu X, Gao H, Mu B, Li X, Wang JH, et al. MicroRNA profiling of clear cell renal cell carcinoma by whole-genome small RNA deep sequencing of paired frozen and formalin-fixed, paraffin-embedded tissue specimens. J Pathol. 2010;222(1):41–51. Epub 2010/07/02.
54. Xi Y, Nakajima G, Gavin E, Morris CG, Kudo K, Hayashi K, et al. Systematic analysis of microRNA expression of RNA extracted from fresh frozen and formalin-fixed paraffin-embedded samples. RNA. 2007;13(10):1668–74. Epub 2007/08/19.

55. Creighton CJ, Reid JG, Gunaratne PH. Expression profiling of microRNAs by deep sequencing. Brief Bioinform. 2009;10(5):490–7. Epub 2009/04/01.
56. Meyer SU, Pfaffl MW, Ulbrich SE. Normalization strategies for microRNA profiling experiments: a "normal" way to a hidden layer of complexity? Biotechnol Lett. 2010;32(12):1777–88. Epub 2010/08/13.
57. Git A, Dvinge H, Salmon-Divon M, Osborne M, Kutter C, Hadfield J, et al. Systematic comparison of microarray profiling, real-time PCR, and next-generation sequencing technologies for measuring differential microRNA expression. RNA. 2010;16(5):991–1006. Epub 2010/04/03.
58. Bissels U, Wild S, Tomiuk S, Hafner M, Scheel H, Mihailovic A, et al. Combined characterization of microRNA and mRNA profiles delineates early differentiation pathways of CD133+ and CD34+ hematopoietic stem and progenitor cells. Stem Cells. 2011;29(5):847–57. Epub 2011/03/12.
59. Fiedler SD, Carletti MZ, Christenson LK. Quantitative RT-PCR methods for mature microRNA expression analysis. Methods Mol Biol. 2010;630:49–64. Epub 2010/03/20.
60. Mestdagh P, Van Vlierberghe P, De Weer A, Muth D, Westermann F, Speleman F, et al. A novel and universal method for microRNA RT-qPCR data normalization. Genome Biol. 2009;10(6):R64. Epub 2009/06/18.
61. Farazi TA, Horlings HM, Ten Hoeve JJ, Mihailovic A, Halfwerk H, Morozov P, et al. MicroRNA sequence and expression analysis in breast tumors by deep sequencing. Cancer Res. 2011;71(13):4443–53. Epub 2011/05/19.
62. Hafner M, Renwick N, Brown M, Mihailovic A, Holoch D, Lin C, et al. RNA-ligase-dependent biases in miRNA representation in deep-sequenced small RNA cDNA libraries. RNA. 2011;17(9):1697–712. Epub 2011/07/22.
63. Farazi TA, Brown M, Morozov P, Ten Hoeve JJ, Ben-Dov IZ, Hovestadt V, et al. Bioinformatic analysis of barcoded cDNA libraries for small RNA profiling by next-generation sequencing. Methods. 2012;58(2):171–87.
64. Hafner M, Renwick N, Farazi TA, Mihailovic A, Pena JT, Tuschl T. Barcoded cDNA library preparation for small RNA profiling by next-generation sequencing. Methods. 2012;58(2):164–70.
65. Tu J, Ge Q, Wang S, Wang L, Sun B, Yang Q, et al. Pair-barcode high-throughput sequencing for large-scale multiplexed sample analysis. BMC Genomics. 2012;13:43. Epub 2012/01/27.
66. Li S, Yang C, Zhai L, Zhang W, Yu J, Gu F, et al. Deep sequencing reveals small RNA characterization of invasive micropapillary carcinomas of the breast. Breast Cancer Res Treat. 2012;136(1):77–87. Epub 2012/09/15.
67. Ryu S, Joshi N, McDonnell K, Woo J, Choi H, Gao D, et al. Discovery of novel human breast cancer microRNAs from deep sequencing data by analysis of pri-microRNA secondary structures. PLoS One. 2011;6(2):e16403. Epub 2011/02/25.
68. Persson H, Kvist A, Rego N, Staaf J, Vallon-Christersson J, Luts L, et al. Identification of new microRNAs in paired normal and tumor breast tissue suggests a dual role for the ERBB2/Her2 gene. Cancer Res. 2011;71(1):78–86. Epub 2011/01/05.
69. Kumar MS, Pester RE, Chen CY, Lane K, Chin C, Lu J, et al. Dicer1 functions as a haploinsufficient tumor suppressor. Genes Dev. 2009;23(23):2700–4. Epub 2009/11/12.
70. Kumar MS, Lu J, Mercer KL, Golub TR, Jacks T. Impaired microRNA processing enhances cellular transformation and tumorigenesis. Nat Genet. 2007;39(5):673–7. Epub 2007/04/03.
71. Janas MM, Wang B, Harris AS, Aguiar M, Shaffer JM, Subrahmanyam YV, et al. Alternative RISC assembly: Binding and repression of microRNA-mRNA duplexes by human Ago proteins. RNA. 2012;18(11):2041–55. Epub 2012/09/29.
72. Gao LB, Bai P, Pan XM, Jia J, Li LJ, Liang WB, et al. The association between two polymorphisms in pre-miRNAs and breast cancer risk: a meta-analysis. Breast Cancer Res Treat. 2011;125(2):571–4. Epub 2010/07/20.
73. Jedlinski DJ, Gabrovska PN, Weinstein SR, Smith RA, Griffiths LR. Single nucleotide polymorphism in hsa-mir-196a-2 and breast cancer risk: a case control study. Twin Res Hum Genet. 2011;14(5):417–21. Epub 2011/10/04.

74. Pastrello C, Polesel J, Della Puppa L, Viel A, Maestro R. Association between hsa-mir-146a genotype and tumor age-of-onset in BRCA1/BRCA2-negative familial breast and ovarian cancer patients. Carcinogenesis. 2010;31(12):2124–6. Epub 2010/09/03.
75. Tian T, Xu Y, Dai J, Wu J, Shen H, Hu Z. Functional polymorphisms in two pre-microRNAs and cancer risk: a meta-analysis. Int J Mol Epidemiol Genet. 2010;1(4):358–66. Epub 2010/01/01.
76. Smith RA, Jedlinski DJ, Gabrovska PN, Weinstein SR, Haupt L, Griffiths LR. A genetic variant located in miR-423 is associated with reduced breast cancer risk. Cancer Genomics Proteomics. 2012;9(3):115–8. Epub 2012/05/18.
77. Volinia S, Galasso M, Sana ME, Wise TF, Palatini J, Huebner K, et al. Breast cancer signatures for invasiveness and prognosis defined by deep sequencing of microRNA. Proc Natl Acad Sci USA. 2012;109(8):3024–9. Epub 2012/02/09.
78. Wu X, Somlo G, Yu Y, Palomares MR, Li AX, Zhou W, et al. De novo sequencing of circulating miRNAs identifies novel markers predicting clinical outcome of locally advanced breast cancer. J Transl Med. 2012;10:42. Epub 2012/03/10.
79. van Schooneveld E, Wouters MC, Van der Auwera I, Peeters DJ, Wildiers H, Van Dam PA, et al. Expression profiling of cancerous and normal breast tissues identifies microRNAs that are differentially expressed in serum from patients with (metastatic) breast cancer and healthy volunteers. Breast Cancer Res. 2012;14(1):R34. Epub 2012/02/23.
80. Si H, Sun X, Chen Y, Cao Y, Chen S, Wang H, et al. Circulating microRNA-92a and microRNA-21 as novel minimally invasive biomarkers for primary breast cancer. J Cancer Res Clin Oncol. 2013;139(2):223–9.
81. Krutzfeldt J, Rajewsky N, Braich R, Rajeev KG, Tuschl T, Manoharan M, et al. Silencing of microRNAs in vivo with "antagomirs". Nature. 2005;438(7068):685–9. Epub 2005/11/01.
82. Kutay H, Bai S, Datta J, Motiwala T, Pogribny I, Frankel W, et al. Downregulation of miR-122 in the rodent and human hepatocellular carcinomas. J Cell Biochem. 2006;99(3):671–8. Epub 2006/08/23.
83. Xu J, Wu C, Che X, Wang L, Yu D, Zhang T, et al. Circulating microRNAs, miR-21, miR-122, and miR-223, in patients with hepatocellular carcinoma or chronic hepatitis. Mol Carcinog. 2011;50(2):136–42. Epub 2011/01/14.
84. Poy MN, Eliasson L, Krutzfeldt J, Kuwajima S, Ma X, Macdonald PE, et al. A pancreatic islet-specific microRNA regulates insulin secretion. Nature. 2004;432(7014):226–30. Epub 2004/11/13.
85. Hu Z, Dong J, Wang LE, Ma H, Liu J, Zhao Y, et al. Serum microRNA profiling and breast cancer risk: the use of miR-484/191 as endogenous controls. Carcinogenesis. 2012;33(4):828–34. Epub 2012/02/03.
86. Hafner M, Landthaler M, Burger L, Khorshid M, Hausser J, Berninger P, et al. Transcriptome-wide identification of RNA-binding protein and microRNA target sites by PAR-CLIP. Cell. 2010;141(1):129–41. Epub 2010/04/08.
87. Chi SW, Zang JB, Mele A, Darnell RB. Argonaute HITS-CLIP decodes microRNA-mRNA interaction maps. Nature. 2009;460(7254):479–86. Epub 2009/06/19.
88. Hafner M, Lianoglou S, Tuschl T, Betel D. Genome-wide identification of miRNA targets by PAR-CLIP. Methods. 2012;58(2):94–105.
89. Jaskiewicz L, Bilen B, Hausser J, Zavolan M. Argonaute CLIP—a method to identify in vivo targets of miRNAs. Methods. 2012;58(2):106–12.

Chapter 13
RNA-Seq in Prostate Cancer Research

ShanCheng Ren, Min Qu, and Yinghao Sun

Abstract The emergence of next-generation sequencing (NGS) is providing novel insights into cancer genomes as part of large-scale efforts by the International Cancer Genome Consortium (ICGC) as well as individual genome centers. RNA-seq, as one of the applications of NGS, is wildly used in biological and medical researches for the discovery of novel genetic changes in the pathogenesis of the disease. Prostate cancer (PCa) still affects the life of men in the developed and developing countries with growing incidence and becomes the second leading cause in population in the United States. Current understanding of the prostate cancer genome is based on extensive RNA sequencing for novel gene fusions and the first whole genome sequencing effort with great surprise which refutes the conventional opinion that few recurrent genetic mutations are related with this disease. Future efforts in RNA-seq will help to determine the specific transcriptomic rearrangements, their association with other alterations, and the effect on PCa progression. These findings might help the biologists and clinicians have completely understanding a specific molecular diagnostic criteria and targeted therapies for prostate cancer.

Keywords Next-generation sequencing (NGS) • RNA-seq • Prostate cancer (PCa) • Diagnosis • Targeted therapies

1 Introduction

Over the past decades, the introduction of massively parallel sequencing platforms for NGS protocols dramatically changed the landscape of the genetic studies. The production of large numbers of low-cost reads makes the NGS useful for many

S. Ren • M. Qu • Y. Sun (✉)
Department of Urology, Shanghai Changhai Hospital, The Second Military Medical University, Shanghai, China
e-mail: sunyh@medmail.com.cn

W. Wu and H. Choudhry (eds.), *Next Generation Sequencing in Cancer Research: Volume 1: Decoding the Cancer Genome*, DOI 10.1007/978-1-4614-7645-0_13,

applications which include the discovery of variant targeted regions, whole genomes, specific transcriptomes of interest, and profiling of epigenetic marks and chromatin structure [1]. Among them, RNA-seq might be one of the most complex NGS applications. It covers the analysis of RNA expression, alternative splicing, allele-specific expression, and gene fusion with accuracy in the level of transcriptome in any species. As the powerful tool for the complexity of transcriptomes, RNA-seq is wildly used in the researches of biology, clinical medicine, and drug discovery.

Prostate cancer (PCa) continues to be an important world health problem and the second leading cause of cancer-related death in men in developed countries [2]. The incidence of prostate cancer is increasing worldwide but varies remarkably among races and countries [2]. The highest incidence is found in the Western countries such as the United States, whereas the lowest incidence is found in the East Asian countries such as China, which may partly be due to genetic differences among the different races. The growth of prostate cancer is originally androgen dependent and metastatic tumors are generally treated with androgen deprivation therapy (ADT), with or without antiandrogen supplementation [3, 4]. However, resistance to hormonal therapy occurs within 12–18 months, referred to as hormone-refractory or castration-resistant prostate cancer (CRPC) [5, 6]. Currently, <5 % of men with newly diagnosed prostate cancer have distant metastases at first presentation, compared with >20–25 % 20 years ago [7].

Considering the fact that the morbidity and mortality of prostate cancer is increasing in recent few years world widely, and a variety of complex forms of PCa result in different therapeutic regimens and prognosis, the etiology, the mechanism of CRPC with progressive stage, and the therapeutic targets for those with poor prognosis are urgently to be studied for improving the patients' survival. RNA-seq, as one of the applications of NGS, can solve the problems with high-throughput technique, which has been proved. With the help of RNA-seq, PCa will be much clearer to the researchers and clinicians, which is a glad tiding for the patients. Here, we present the introduction of RNA-seq including its principle and applications and the applications in PCa researches.

2 Principle of RNA-Seq

The principle of RNA-seq is based on high-throughput technology. In general, a population of RNA is converted to a library of cDNA fragments with adaptors attached to one or both ends. Each molecule is sequenced in a high-throughput manner to obtain from one end or both ends. The reads are differing from 30 to 400 bp, which is depended by sequencing technology used. There are already three sequencing systems designed by companies for RNA-seq, including Illumina IG, Applied Biosystems SOLiD, and Roche 454 Life Science. The results are either aligned to a reference genome or transcripts or assembled de novo without the genomic sequence to produce a genome-scale transcription map that consists of both the transcriptional structure and level of expression for each gene [8].

2.1 Sample Isolation and Library Preparation

The isolation of RNA samples is the first step in RNA-seq. The samples consist of both mRNA and non-mRNA, and a large amount of rRNA species (90–95 %) which might affect the results of sequencing as the whole transcriptomes in a specific period. Thus, depletion of abundant rRNA molecules is carried out by hybridization of rRNA with oligonucleotide probes which is followed by removal with streptavidin-coated magnetic beads [9].

The step of library preparation becomes crucial after the isolation of RNA samples. The protocols which were firstly developed for the analysis of genomic DNA have been effectively adapted to the double-strand (ds) cDNA sequencing for transcriptomes studies [10]. Usually, a double-strand cDNA library is prepared by (1) fragmented ds cDNA and (2) hydrolyzed or fragmented RNA. The fragmented pieces should be particularly cared from degradation in the procedure of library construction.

The next step of the sequencing is the clonal amplification of the cDNA fragmentation. Illumina, Roche 454, and SOLiD use clonally amplified templates. In particular, the last two platforms use emulsion PCR (emPCR) to prepare sequencing templates in a cell-free system as an innovative procedure by carrying cDNA library onto beads in separated single strands, which accomplishes the aim of one DNA molecule per bead. After the enrichment of beads by PCR, 200 million spatially separated template clusters are produced for NGS reaction.

2.2 Sequencing and Imaging

Different NGS platforms use different sequencing procedures. Illumina combines microarray and cyclic reversible termination (CRT), which captures cDNA fragments onto the surface of flow cell for bridge amplification. In contrast, SOLiD system is based on continuous ligation between oligonucleotides with fluorescence labeling instead of traditional PCR for amplification and sequencing as a unique feature. Another approach is Roche 454 system by pyrosequencing with a nonelectrophoretic bioluminescence method which is different from the other two sequencing approaches. The characteristic of the sequencing is that the signal of visible light is measured after enzymatic reaction by release of pyrophosphate [11].

2.3 Mapping and Analysis

After transferring the data from RNA sequencers to the software in the computers, the analysis of any RNA-seq data consists of mapping the sequence reads to a reference genome, or known transcribed sequences, or de novo assembling to produce a genome-scale transcriptional map. In most cases, the reference genome is available and the mapping can be carried out using either the whole genome or known transcribed sequences. However, in some cases, read length, error profile, amount of data, and data format affect the results of analysis. To avoid or lower the incidence of false results, most RNA-seq packages are built on top of optimized short read core mappers and multiple runs are performed as mapping strategy. To date, there are several major methods for the analysis of transcriptomes, including hybridization of gene chip (or microarray), serial analysis of gene expression (SAGE), and massively parallel signature sequencing (MPSS). Gene expression technologies have greatly matured over the past years, but it has become clear that hybridization-based approaches have obvious limitations in cross-species comparisons [12]. Furthermore, the short sequence reads of serial analysis of gene expression (SAGE) and related techniques are severely limited by the requirement of reliable annotated genomes [13].

3 The Advantage of RNA-Seq

After knowing the whole principle of RNA-seq described above, we have a general landscape of how it works more effectively for sequencing than Sanger sequencing. Although RNA-seq is still under active development, it provides several more useful advantages when compared with existing technologies. Firstly, it makes the identification, characterization, and quantification of new splice isoforms possible because of the ability to detect known transcripts by RNA-seq, which cannot be achieved by hybridization technologies [8]. Furthermore, it is clear that RNA-seq will quickly supersede microarray-based methods in the researching fields of differential splicing activity, antisense transcription, and discovery of novel regions of transcription. There are another two principle differences between sequencing and microarray approaches in the resolution of the output and in the dynamic range of changes in addition to the advantages described above [14].

Generally speaking, the advantage of RNA-seq can be concluded as follows (Table 13.1): (1) Digitized signals. RNA-seq can directly determinate the sequences of each transcriptome, elevate the accuracy of mononucleotide's resolution, and detect the differences between two single-base and different transcriptional expression in similar genes in the gene family or caused by alternative splicing. At the same time, there are no cross-reaction and background noise problems in RNA-seq. And it covers the dynamic range of changes with strong signals. (2) The whole genome analysis in any species. RNA-seq can directly analyze the whole transcriptomes in any species without knowing about the genetic information of the species in the absence of the design of specific probes. (3) A wider range for sequencing. RNA-seq can identify and sequence the whole transcriptomes with a much greater dynamic range while microarray has a relative smaller range because of lacking sensitivity of genes expressed either at low or very high levels.

Table 13.1 Advantage of RNA-seq compared with other transcriptomic methods

Technology	Tiling microarray	cDNA or EST sequencing	RNA-seq
Technology specifications			
Principle	Hybridization	Sanger sequencing	High-throughput sequencing
Resolution	From several to 100 bp	Single base	Single base
Throughput	High	Low	High
Reliance on genomic sequence	Yes	No	In some cases
Background noise	High	Low	low
Application			
Simultaneously map transcribed regions and gene expression	Yes	Limited for gene expression	Yes
Dynamic range to quantify gene expression level	Up to a few hundredfold	Not practical	>8,000 fold
Ability to distinguish different isoforms	Limited	Yes	Yes
Ability to distinguish allelic expression	Limited	Yes	Yes
Practical issues			
Required amount of RNA	High	High	Low
Cost for mapping transcriptomes of large genomes	High	High	Relatively low

Quote from: Wang Z, Gerstein M, Snyder M (2009) RNA-Seq: a revolutionary tool for transcriptomics. Nat Rev Genet 10: 57–63

4 The Application and Translating Medicine of RNA-Seq

RNA-seq can not only provide digital measure of gene expression but also reveal many other aspects of the transcriptome such as structure of transcriptome, gene fusion, previously unknown coding or noncoding transcripts, alternative splicing isoform composition, and nucleotide variations. All these aspects that detected by RNA-seq are possibly related with diseases. Many researchers are dedicated themselves into the discovery of the relationship between the genetic changes and some certain diseases (Table 13.2). It gives the possibility to detect even low-expressed genes, to accurately quantify their expression levels in each condition (pathology, drug treatment, different developmental stages), a more accurate estimate of sense/antisense transcription of genes, and also to analyze transcription starting sites (TSS) of genes.

4.1 Map of Transcriptome

For a long time, the understanding of transcriptome was greatly obtained by gene predictions and limited EST evidence, which made it partial and biased. RNA-seq

Table 13.2 RNA-seq in cancer researches

Cancer type	Analysis type	Results
Hodgkin lymphoma	PE WT	Identification of gene fusions, among which fusions CIITA-involving
Non-Hodgkin lymphoma	PE poly-A+	Detection of 109 genes with multiple somatic mutations, including those involved in histone modifications
MDS	FR small RNA	Discovery of novel miRNA differentially expressed in tumor
Breast cancer	FR poly-A+	Alternative splicing and alterations in gene expression (i.e., LOX, ATP5L, GALNT3, and MME) have been identified in modulated ERBB2 overexpressing mammary cells
	PE poly-A+	Identification of 3 known and 24 novel fusion transcripts (including VAPB-IKZF3)
	SE, PE poly-A+	Discovery of gene fusions in breast cancer transcriptomes with BRCA1 mutations, including novel in-frame WWC1-ADRBK2fusion in HCC3153 cell line and ADNP-C20orf132 in a primary tumor
	FR poly-A+	Investigation of EMT-associated alternative splicing events regulated by different classes of splicing factors (RBFOX, MBNL, CELF, hnRNP, or ESRP)
Prostate cancer	SE poly-A+	Detection of transcription-induced chimeras in prostate adenocarcinoma
	PE WT	Discovery and characterization of seven novel cancer-specific gene fusions (four involving non-ETS)
	PE poly-A+	Identification of 121 unannotated prostate cancer-associated ncRNA transcripts, including the characterization of PCAAT-1
	FR poly-A+	25 Previously undescribed alternative splicing events involving known exons, and high-quality single-nucleotide discrepancies, have been detected in prostate cancer cell line LNCaP
Melanoma	PE poly-A+	Identification of 11 novel gene fusions, 12 read-through transcripts, somatic mutations, and unannotated splice variants
	FR poly-A+	Somatic CNVs affecting gene expression and new potential genes and pathways involved in tumorigenesis have been identified in seven human metastatic melanoma cell lines
Ovarian cancer	PE poly-A+	Discovery of the first gene fusions in ovarian cancer through a novel computational method
Sarcoma	PE poly-A+	Detection of novel gene fusions in sarcoma through a novel computational method
	FR ribodepletion	Evidence of a closer relationship between gene expression levels and protein expression in a human osteosarcoma cell line
Hepatocellular carcinoma	SE WT	Characterization of HBV-related HCC transcriptome, including identification of exon-level expression changes and novel splicing variants

Quote from: Costa V, Aprile M, Esposito R, Ciccodicola A (2013) RNA-Seq and human complex diseases: recent accomplishments and future perspectives. Eur J Hum Genet 21: 134–142 can greatly enrich the aspect of genetic sequencing with the help of single basal resolution, including identification of 5′ and 3′ ends, UTRs, new transcribed regions, and mapping exon/intron boundaries.

Mortazavi et al. [15] detected 80 % and 85 % of 5′ end and 3′ end, respectively, in the genome of *S. cerevisiae* by RNA-seq sample (RNA-Seq), discovering >90 % mapped reads fell within the known exons with the other 10 % suggesting new and revised gene models which consisted of exons, 3′ untranscribed regions, candidate microRNA precursors, and additional promoters. In 2008, Nagalakshmi et al. [16] detected 80 % and 85 % of 5′ end and 3′ end, respectively, in the genome of *S. cerevisiae* by RNA-seq.

4.2 Gene Fusion

In recent years, gene fusions have gained significant recognition as biomarkers. They can assist diagnosis and treatment decisions based on the fact that they are seldom found in normal tissue and are detectable through RNA-seq. The detection of gene fusion mainly consists of four steps: filtering, clustering, inclusion of splice-junction information, and scoring. The filtering removes read-pairs if one or both pairs do not uniquely map to the genome, if the pairs align within the expected distance, or if any pair is contained in an artificial pileup. All pairs are considered for the next step after passing the filters and being mapped to transcripts from different loci. In the second step, gene fusion predictions are generated through the clustering of mate-pairs that map to proximal positions and have the same strand orientation. For each gene fusion prediction, single-end reads spanning (spliced-alignment) the transcripts are used to determine the exact breakpoint. In the last step, the gene fusion candidates are ranked by several scores [17]. Furthermore, genomic DNA-based approaches cannot identify fusion events due to nongenomic factors, such as trans-splicing [18] and read-through events between adjacent transcripts [19]. Paired-end RNA-seq can be particularly advantageous for fusion identification because of the increased physical coverage it offers. This approach has led to important biological findings in oncology [20, 21], offering potential targets for therapeutic modulation.

Zhou et al. [22] used RNA-seq to detect gene fusion events in a patient with a 6-year history of chronic myelocytic leukemia (CML), finding classic BCR-ABL1 gene fusion and a novel gene fusion between the ring finger protein 213 (RNF213) and solute carrier family 26 member 11 (SLC26A11). The results showed that in addition to classic BCR-ABL fusion, many other alternative gene fusions take part in the development of CML. In a recent study on B-cell lymphomas, MHC class II transactivator (CIITA) has been identified as a novel partner of various fusions transcripts, suggesting a possible novel intriguing genetic mechanism underlying the onset of lymphoid cancers [23].

The challenges faced in gene fusion detection generally include that it cannot detect fusion events that involve the exchange of the promoter of a gene with the coding sequence of another gene. Furthermore, in the procedure of cDNA library construction, RNA-seq data include chimeric cDNA artifacts which might affect the results that lead to false positives in gene fusion identification [24]. These difficulties may be partly alleviated when long-read RNA-sequencing technologies with sufficient throughput and sequencing performance become available [25].

4.3 Noncoding RNA Expression

Another important researching aspect of RNA-seq is ncRNA. It is reported that more than 93 % genes can be transcribed into RNA in human, in which 2 % will express protein for important function of cells, and the other 91 % are transcribed into noncoding RNA (ncRNA) for regulation in whole process in genetic expression , including long ncRNA and microRNA. Accumulating evidence has indicated that long noncoding RNAs (lncRNAs) or long intergenic noncoding RNAs (lincRNAs) can be regarded as oncogenes or tumor-suppressor genes under some certain circumstances [26]. Therefore, profiling noncoding RNAs in cancer transcriptome is a new method for finding potential biomarkers for cancer diagnosis and potential targets for cancer intervention.

A recent work [27] has demonstrated in a mouse model of Alzheimer's disease (AD), the abnormal expression of miR-34a induces the inverse expression of bcl2 and contributed to AD pathogenesis by RNA microarray and sequencing technology. In spite of the predicted role of PABPN1 in mRNA polyadenylation, little is known about the impact of PABPN1 deficiency on human gene expression. Beaulieu et al. [28] showed that the large majority of protein-coding genes expressed normal levels of mRNA in PABPN1-deficient cells, raising the question that PABPN1 might not be required for the bulk of mRNA expression. At the same time, they identified a class of PABPN1-sensitive long noncoding RNAs (lncRNAs), the majority of which accumulated in conditions of PABPN1 deficiency by using RNA-seq. A recent study [29] characterized the transcriptome of melanoma skin cancer cell lines. Their results indicated that BRAF-regulated lncRNA1 (BANCR) might reduce melanoma cell migration.

4.4 Alternative Splicing

Furthermore, as a form of genetic alteration, RNA splice events are detected directly by mapping splice-crossing sequence reads, which is superior to other methods including standard gene expression microarray or serial analysis of gene expression in the aspect of higher resolution measurement of gene expression levels, revealing unknown transcripts and splicing isoforms, and providing quantitative measurement of alternatively spliced isoforms. Alternative splices were prominent with 3,500 different genes expressing one or more alternate internal splices in 1.45×10^5 distinct splices observed [15].

The study of Twine et al. [30] has provided, for the first time, an extensive transcriptome analysis of postmortem frontal and temporal lobes of AD patients, discovering a differential expression of known causative genes and also of previously unannotated expressed regions, which may probably be due to alternative splicing and alternative promoter usage. The detection of alternative splicing events associated with epithelial–mesenchymal transition (EMT) in breast cancer samples by using RNA-seq suggested the classification of cancer cell lines into basal and luminal subtypes, based on their EMT-associated splicing pattern [31].

4.5 Somatic Mutations

In addition, RNA-seq provides an efficient approach for studying variations in genomic regions including exon regions of protein-coding genes and various types of noncoding genes. The availability of the entire chromatic sequence (GRCh37/hg19) has allowed researchers to easily identify disease-causing mutations in more than 2,850 genes responsible for a huge number of Mendelian disorders and to detect statistically significant associations between about 1,100 loci and more than 165 complex diseases and traits [32]. About 88 % of the genetic variants (single-nucleotide polymorphisms (SNPs)) relying within intronic or intergenic regions currently associated to complex diseases and traits by genome-wide association studies (GWAS) [33], which revealed that genetic changes are more in charge of the causes of diseases than the changes of protein.

Shah et al. [34] sequenced the transcriptomes of four ovarian adult-type granulosa cell tumors (GCT) and 11 non-GCT ovarian tumors. They analyzed RNA-seq data of point mutations and small deletions/insertions, finding novel variants which revealed that mutation in FOXL2 might be a key cancer-causing factor of adult-type GCTs. Moreover, RNA-seq discovered a valuable resource to identify new ERBB2-mediated events and private fusions in some BRCA1-mutated transcriptomes, novel potential biomarkers for diagnosis, and treatment in breast cancer [35, 36].

Reported evidences strongly suggest with the help of RNA-seq, scientists and clinicians will have an increasingly understanding of the role of alternative genes detection in cancer research for both diagnosis and prognosis and for the improvement of surgical and therapeutic interventions. It is believed that somatic chromosomal number variation (CNV) affecting gene expression and potential new candidate genes involved in tumorigenesis and other diseases will be detected to a greater extent. We would like to discuss the application of RNA-seq in PCa researching in order to find out meaningful results for the future practice in clinic.

5 RNA-Seq in Prostate Cancer

Prostate cancer (PCa) continues to be an important world health problem and the second leading cause of cancer-related death in men in developed countries. The incidence of prostate cancer is increasing worldwide but varies remarkably among races and countries [2]. Nowadays, although the disease is greatly popular, it progresses only in a small proportion of patients, and the majority of patients under active surveillance will eventually die of other causes [37]. However, advanced prostate cancer is vital to all patients for the fact that it is usually treatment resistant (also we call it castration-resistant prostate cancer, CRPC) and has a huge possibility of metastasizing from the prostate to other parts of the body, particularly the bones and lymph nodes, which eventually cause the metastatic organ dysfunction. The etiology, early diagnosis, and early treatment are the research hotspots in the field of prostate cancer, as a potential life-threatening disease in the world.

5.1 Etiology

Many factors, including genetics, diet, and environment, have been implicated in the development of prostate cancer. Especially the genetic factors play more significant roles in the pathogenesis. For decades, physicians and researchers have recognized that family history is related with prostate cancer, as a significant risk factor. There are significant geographic and ethnic disparities due to several factors including differences in rates of prostate screening, disparate environmental exposures, and genetic variation in the incidence of disease [38]. Studies of twins in Scandinavia suggest that 40 % of prostate cancer risk can be explained by inherited factors [39].

No single gene is responsible for prostate cancer. Many different genes have been implicated. Mutations in BRCA1 and BRCA2 which are important risk factors for ovarian cancer and breast cancer in women have also been implicated in prostate cancer. Modern genetic and genome-based technologies have provided evidence for the presence of somatic alterations, germline variations, and inactivity of tumor suppressors which not only individualize PCa but also serve as a driving force for prostate carcinogenesis and its progression toward advance incurable stages. The detection technologies applied for the analyses of somatic alterations in PCa revealed the role of certain tumor-suppressor genes (TSGs) and oncogenes. The different patterns including mutations, deletions, and rearrangements of oncogenes lead to uncontrolled growth of malignant cells. For example, a single mutation in the Ras gene converts it into a potent oncogene. In addition, c-Myc, c-ErbB2, and Bcl-2 oncogenes were also detected in prostate cancer (Table 13.3).

Besides, single-nucleotide polymorphisms (SNPs) are the most common type of polymorphisms and occur with a frequency of 1 in 250 base pairs in the entire genome including the promoter region, exonic sequences, intronic sequences, and other noncoding sequences. SNPs may influence promoter activity, DNA and pre-mRNA conformation, mRNA stability, and influence directly or indirectly in phenotypic expression. Researchers have demonstrated the importance of the presence of polymorphic CAG/GGC repeats of AR gene, cell cycle-related genes, and cell-adhesion regulatory genes is related with prostate cancer (Table 13.4), of which the variant polymorphism can increase the risk [40–42]. The biological significance of polyglycine (GGC/N) repeat in the exon 1 is less clear. Hence, one study [43] has proposed that the size of the glycine repeats might increase the PCa risk by using cDNA assay.

Table 13.3 Common somatic genetic changes in PCa

Gene	Chr	Event	Function
ETC genes	7	Fusion to AR-targeted genes	Transcription factors, differentiation
NKX3.1	8p21	Inactivation	Cell growth/differentiation
c-MYC	8q	Genomic amplification	Cell growth/differentiation
PTEN	10q23	Inactivation	Cell growth/metabolism
Rb	13q	LOH, mutation	Cell cycle regulation
P53	17p	LOH, mutation	Cell cycle regulation, apoptosis, DNA damage detection
ETS genes	21	Fusion to AR-targeted genes (TMPRSS2)	Transcription factors, cell growth/differentiation
AR	X	Mutation, genomic amplification	Cell growth/differentiation

Quote from: Koochekpour S (2011) Genetic and epigenetic changes in human prostate cancer. Iran Red Crescent Med J 13: 80–98

Table 13.4 SNPs with strong association with PCAa

Chr	Alleles (–/+)	OR	Function	References
11q13	T/G	1.21	Intergenic sequence	[44, 45]
19q13	A/G	1.37	Androgenic effect	[45, 46]
7q21	T/C	1.19	Membrane trafficking	[45]
10q11	C/T	1.38	Tumor suppression	[44, 45]
17q12	G/A	1.22	Tumor suppression/epithelial differentiation	[44, 45, 46]
Xp11	T/C	1.29	Apoptosis, DNA repair, stress response	[45, 47]
10q26	T/C	1.18	Antiapoptic properties	[44]
6q25	C/T	1.21	Drug detoxification properties	[45]
17q24	T/G	1.2	Intergenic sequence	[45, 46]
3p12	C/T	1.3	Intergenic sequence	[45]

Quote from: Koochekpour S (2011) Genetic and epigenetic changes in human prostate cancer. Iran Red Crescent Med J 13: 80–98

A recent report [44] detected an average of 1,725 tumor-specific point mutations with only 1.5 % found within the coding regions were identified in prostate cancer sample by RNA-seq (Fig. 13.1). Most substitutions were T: A to C: G. One plausible explanation for this finding is that this type of point mutation occurred during RNA editing because RNA editing specifically changes adenosine (A) to inosine (I), which, in turn, is read as guanosine (G)[48].

In high eukaryotes especially in human, many genes can produce multiple isoforms of protein products through alternative splicing (AS) of their mRNAs. It has also been observed that alternative splicing is involved in pathogenesis of many diseases especially in cancer. In prostate cancer, aberrant and alternative splicing generates proteins that influence cell phenotypes and survival of patients. Ren et al. [44] used SpliceMap to detect splice junctions and then applied different criteria to detect different types of AS, including exon skipping, intron retention, and alternative 5′ and 3′ splice sites. Intron 4 retention in KLK3 and exon skipping in AMACR were found in the study (Fig. 13.2).

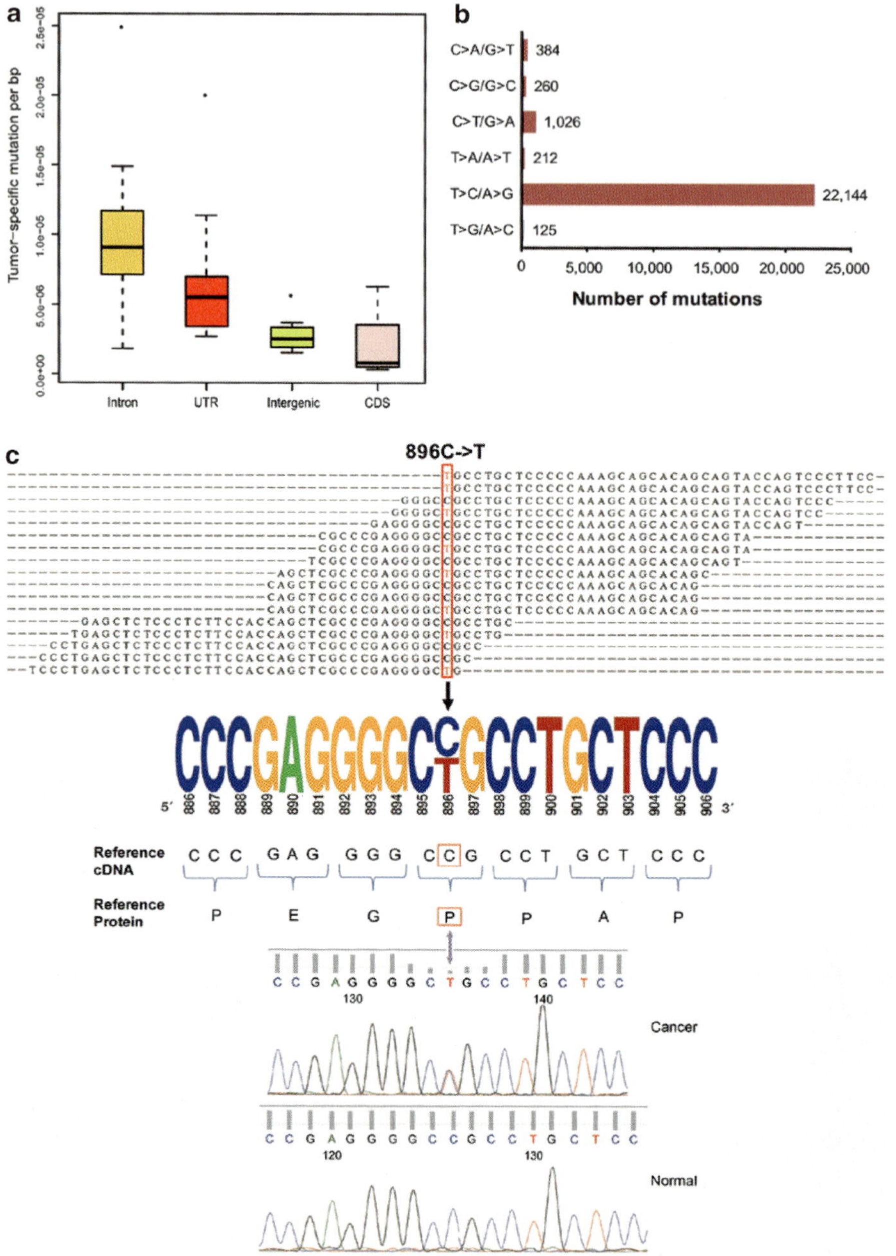

Fig. 13.1 The landscape of somatic mutations in prostate cancers. (**a**)The distribution of somatic mutations among different locations in the genome. (**b**) The frequencies of different substitutions. (**c**) Validation of a somatic mutation in CHAF1A. The mapped reads are shown in the *top panel*. The mutated residue is highlighted by a *red box*. An electropherogram of the Sanger sequencing validation of the mutation and its surrounding nucleotides is shown in the *bottom panel*

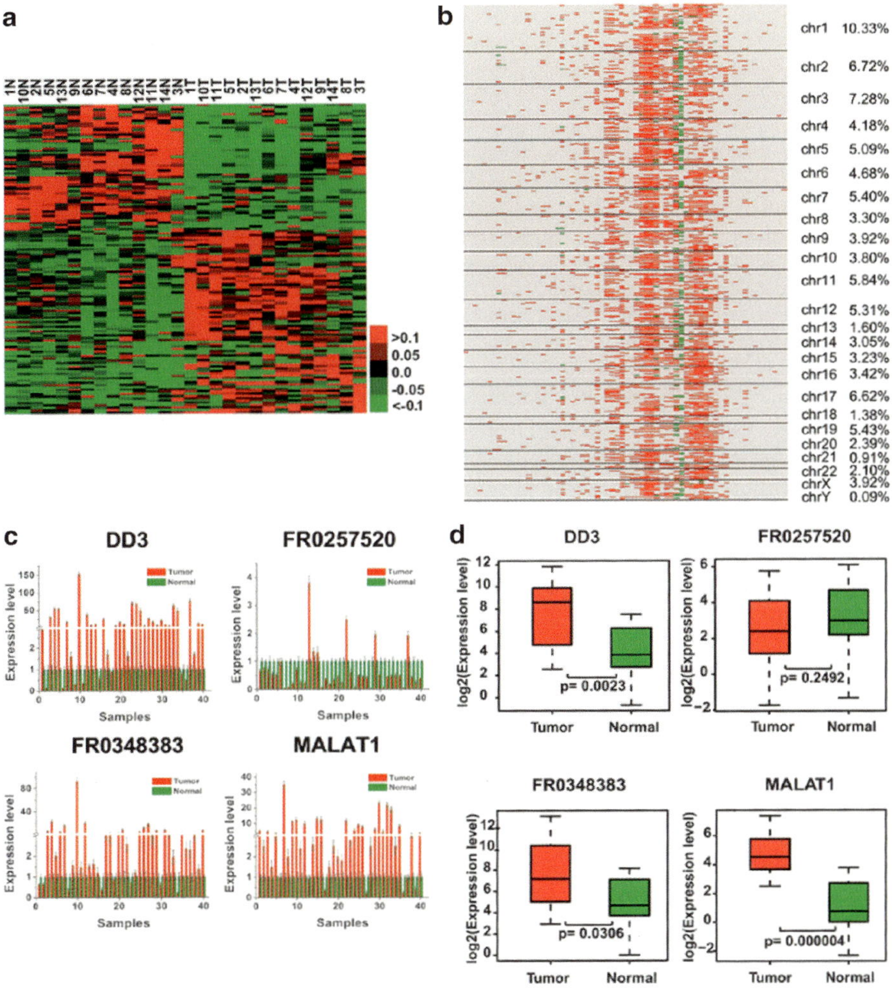

Fig. 13.2 Transcriptional landscape of human long ncRNAs in prostate cancer. (**a**) Supervised hierarchical clustering analysis using 137 long ncRNAs that were consistently upregulated or downregulated in more than 50 % of the prostate cancer samples (≥twofold and FDR ≤0.001). Shades of red and green are used to illustrate whether the expression value is above (*red*) or below (*green*) the mean expression value across all samples (each row in the data was normalized from −1 to +1). (**b**) Correlation heatmap between the expression of long ncRNAs and genes. Rows represent genes aligned according to their chromosomal locations, and the columns represent differentially expressed long ncRNAs. A *red* color indicates a positive correlation, whereas *green bars* represent a negative correlation (absolute correlation coefficient |R| ≥0.85, FDR ≤0.01). (**c**) qRT-PCR assessed the expression levels of DD3, FR0257520, FR0348383, and MALAT1 in the additional set of 40 pairs of prostate cancer and adjacent normal tissues. (**d**) Comparison of the expression levels of DD3, FR0257520, FR0348383, and MALAT1 between prostate cancer and normal tissues by qRT-PCR

In addition, accumulating evidence has indicated that long noncoding RNAs (lncRNAs) or long intergenic noncoding RNAs (lincRNAs) can be regarded as oncogenes or tumor-suppressor genes [26]. Prensner et al. [49] presented the largest RNA-seq analysis and were the first to comprehensively analyze a common epithelial cancer from a large cohort of human tissue samples. They focused on 121 uncharacterized ncRNAs that may have important biological functions in the PCa. Another study reported finding an average of 1,599 known long ncRNA expressing in normal and prostate cancer tissue, in which 137 lncRNAs were proved to be consistently upregulated or downregulated in more than 50 % of prostate cancer by using RNA-seq [44] (Fig. 13.3).

Furthermore, the fusion of AD gene with other genes will cause metastatic or androgen-resistant prostate cancer, which might be explained as (1) AR amplification, (2) a hypersensitive AR resulting from point mutations, (3) promiscuous mutant-AR protein activated by non-androgenic ligands, and (4) AR-polymorphisms changing the response to androgen. In 2005, a study discovered a genetic rearrangement that causes the fusion of two independent genes on chromosome 21: the androgen-regulated gene TMPRSS2 and the ETS transcription factor ERG, a previously known oncogene in sarcoma in prostate cancer [50]. The TMPRSS2-ERG gene fusion has been causally proved to be linked to cancer progression by promoting invasion in vitro, and greatly enhanced prostate cancer development was observed in the condition of overexpression of the fusion gene in mice [51]. Interestingly, Pflueger et al. [52] discovered and characterized seven new cancer-specific gene fusions, two involving the ETS genes ETV1 and ERG and four involving non-ETS genes such as CDKN1A (p21), CD9, and IKBKB (IKK-beta), which are known to present key biological roles in cellular homeostasis or assumed to be critical in tumorigenesis of other tumor entities, as well as the oncogene PIGU and the tumor-suppressor gene RSRC2. These novel fusion genes were all found in prostate cancer harboring the TMPRSS2-ERG gene fusion, but with low frequency. It was also discovered in addition to the well-known fusion between TMPRSS2-ERG, a previously unreported event involving the adjacent genes USP9Y and TTY15 on chromosome Y and an interchromosomal translocation involving genes CTAGE5 and KHDRBS3 in Chinese population with the use of RNA-seq (Fig. 13.4). In the study, the prevalence of TMPRSS2-ERG gene fusion was 19.1 %, which is consistent with previous studies in Asian populations. The incidence of the two novel gene fusions, CTAGE5-KHDRBS3 was 37 % and USP9Y-TTTY15 was 35.2 %, respectively, which was not noted in the population of Caucasian [44]. The difference detected in the fusion genes between Chinese and Caucasian indicates that prostate cancer varies remarkably among races and countries, which helps us step forward in the discovery of the genetic etiology of the disease.

The etiology is the original causes of how the disease is developed under the pressure of genetic and environmental risk factors. RNA-seq analyzes the mechanism of the disease in the transcriptomic level, uncovering the mysterious veil of PCa.

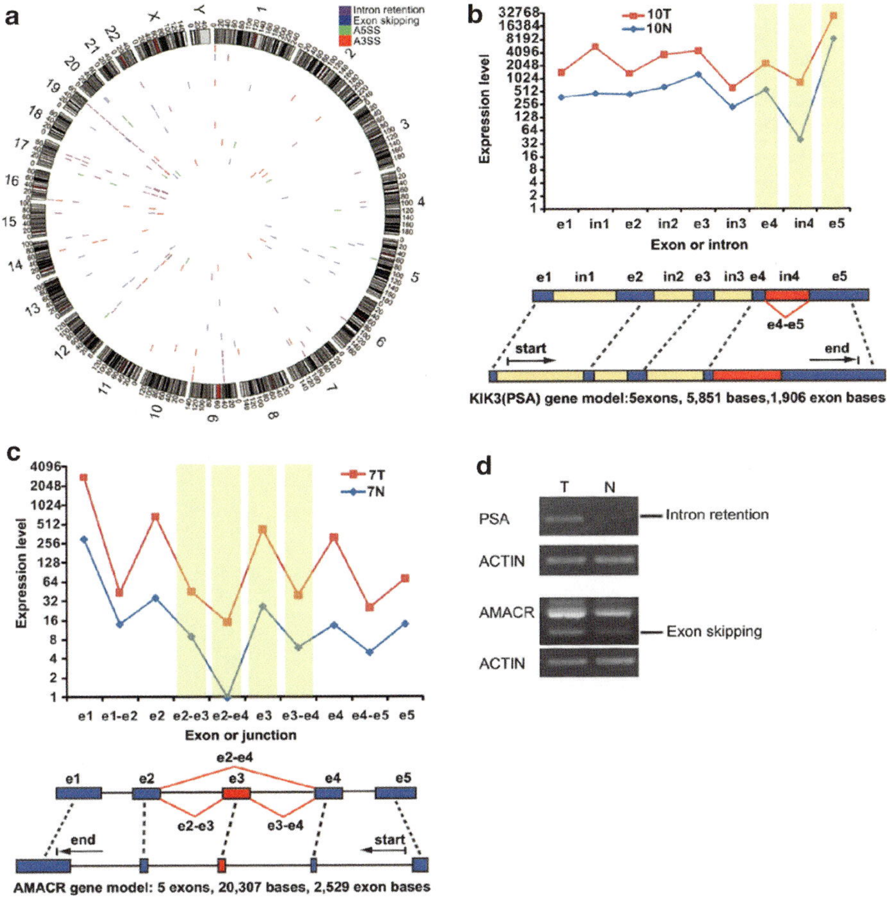

Fig. 13.3 The landscape of alternative splicing in prostate cancer. (**a**) A Circos plot showing the genomic landscape of AS events in the 14 prostate cancer samples discovered by RNA-seq. The *outer ring* shows chromosome ideograms. The *bars along each inner ring* represent AS events in a prostate cancer sample. (**b**) An example of RNA-seq data indicative of intron retention in the KLK3 (PSA) gene. The *line plot* displays the expression of each exon (e1, e2, etc.) and intron (in1, in2, etc.), and alternative expression events are highlighted in *yellow*. (**c**) An example of RNA-seq data indicative of exon skipping in the AMACR gene. The *line plot* displays the expression of each exon (e1, e2, etc.) and exon junction (e1–e2 and others), and alternative expression events are highlighted in *yellow*. (**d**) Validation of KLK3 intron retention and AMACR exon skipping by RT-PCR. A pair of primers was designed to detect only KLK3 intron retention

5.2 Diagnosis

In the aspect of diagnosis, PSA is the best available marker, but it cannot effectively differentiate between PCa and other benign conditions such as BPH, prostatitis, and other non-malignant conditions, which indicates there is no specific threshold for

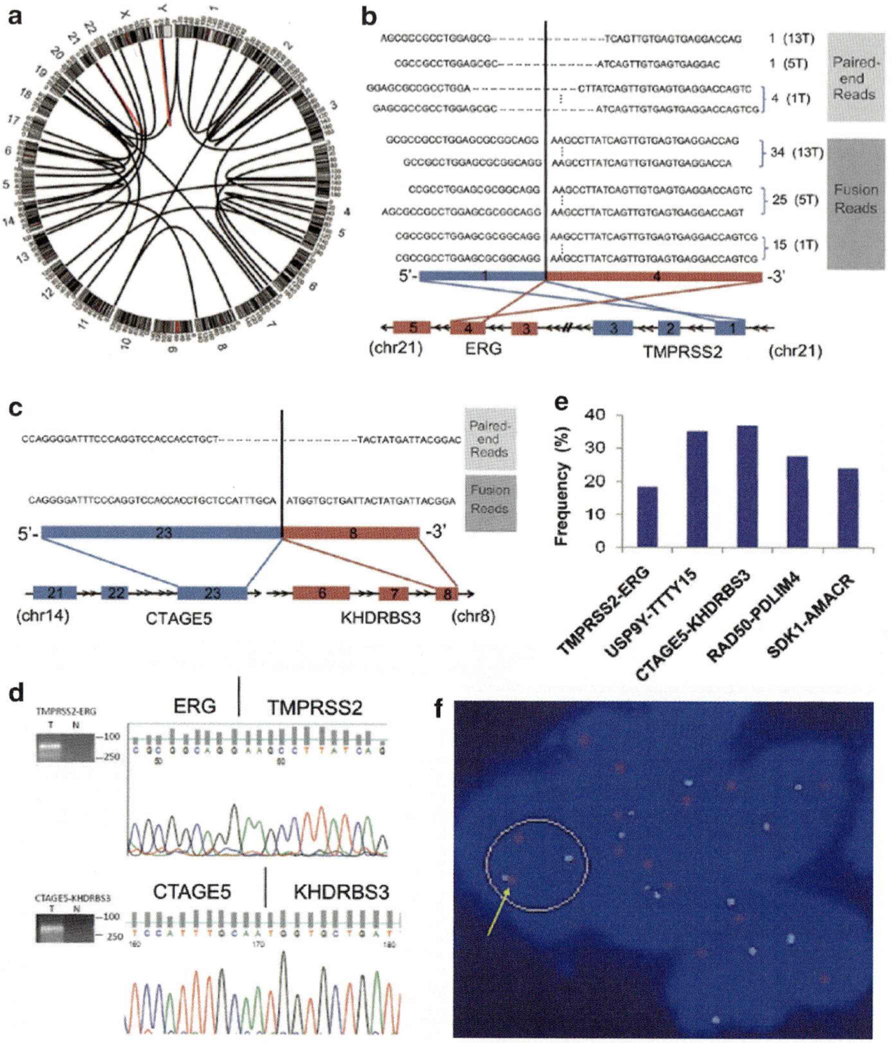

Fig. 13.4 The landscape of gene fusion in prostate cancer. (**a**) A Circos plot of the genomic landscape of gene fusions discovered by RNA-seq in the 14 prostate cancer samples. The *outer ring* shows chromosome ideograms. The gene fusions are shown as *arcs* linking the two genomic loci, each colored according to the frequency with which the gene fusion was found in the 14 prostate cancer samples (*red*=3 and *black*=1). (**b**) TMPRSS2-ERG fusion in three prostate cancers. The TMPRSS2-ERG fusion was between exon 1 of TMPRSS2 (*red*) and exon 4 of ERG (*blue*). The number of reliable pair-end and fusion-spanning reads in each sample is indicated to the right of each read. The sample ID is indicated in *brackets*. (**c**) The CTAGE5-KHDRBS3 fusion in one prostate cancer is revealed by one paired-end and one fusion-spanning read. The CTAGE5-KHDRBS3 fusion was between exon 23 of CTAGE5 (*blue*) and exon 8 of KHDRBS3 (*red*). (**d**) Representative experimental validation of the fusion gene transcript by RT-PCR and Sanger sequencing. (**e**) Prevalence of the TMPRSS2-ERG, USP9Y-TTTY15, CTAGE5-KHDRBS3, RAD50-PDLIM4, and SDK1-AMACR fusions in prostate cancer. (**f**) Interphase FISH on formalin-fixed, paraffin-embedded tissue confirming the fusion of SDK1 and AMACR. Probes for SDK1 (*red*) and AMACR (*green*) demonstrate the fusion of the genomic loci (*yellow arrows*) in cancerous cells

distinguishing insignificant cancers from benign prostate diseases. The suggested use of PSA in the diagnosis emphasizes on its potential role in the early detection of the disease and its use in screening [53]. However, the false positive rate of PSA screening does exist, which leads to the misdiagnosis and expensive approaches such as transrectal prostate biopsy. The Prostate Cancer Prevention Trial (PCPT) demonstrated that 15 % of men with PSA between 0 and 4.0 ng/ml which is considered normal have PCa found on prostate biopsy. And 15 % of these patients have higher Gleason scores [54, 55]. At this point, strategies in PCa early detection rely primarily on surrogate biomarkers that are prostate cancer specific, but not prostate specific. An alternative strategy is to develop clinically robust assays for biomarkers identified through genomic and transcriptomic studies not only cancer specific [56] but also predictive for non-indolent prostate cancer.

Prostate cancer antigen 3 (PCA3), also called DD3, has been shown to be highly overproduced in PCa tissues and metastases as a noncoding RNA produced in the prostate when compared with BPH tissue [57]. Martina et al. analyzed 201 patients' urine samples and found the oversensitivity, specificity, positive predictive value, and negative predictive value for PCA3 were 82 %, 76 %, 67 %, and 87 % as compared to 98 %, 5 %, 40 %, and 83 % respectively for tPSA, indicating that urine PCA3 might be in a greater use in the early diagnosis in patients with PCa than serum PSA. As a noncoding RNA, PCA3 can be also detected by RNA-seq. However, there are no reports on it.

The fusion of TMPRSS2 and ERG loci at the chromosomal level and subsequent overexpression of the TMPRSS2: ERG transcript and truncated ERG protein product is essentially 100 % specific in the tissue of PCa in the reports [58–60]. Several studies have used research-grade RT-PCR-based assays, showing that TMPRSS2: ERG mRNA is also detectable in urine [61–65]. Tomlins et al. [66] have recently reported the results of a clinical-grade, transcription-mediated amplification (TMA) assay for quantifying TMPRSS2: ERG mRNA based on the same technology as the PCA3 assay. Urine TMPRSS2: ERG transcript was measured quantitatively and discovered to be related with indicators of clinically significant PCa including tumor size, Gleason score, and upgrading by prostatectomy in prospectively collected whole urine from multiple cohorts including 218 men undergoing prostatectomy at the University of Michigan and 1,094 men undergoing biopsy at 10 academic and community clinics. They demonstrated that urine TMPRSS2: ERG, in combination with urine PCA3, enhances the specificity of serum PSA for predicting PCa and clinically relevant cancer on biopsy. Although there are lots of reports about the high specificity and positive predictive value for PCa diagnosis by using urine detection of TMPRSS2: ERG, no clinical diagnostic test is available. It is more common to use the combination of urinary PCA3 and TMPRSS2-ERG with serum PSA levels to improve screening compared with PSA alone [67].

Besides, other biomarkers being tested include urinary concentration of b-microseminoprotein (MSMB), the expression of which is decreased by the rs10993994 risk SNP in the MSMB promoter. MSMB levels are decreased in PCa tissue compared to benign prostate disease, and decreased urinary levels have been shown to improve upon urinary PSA, but not serum PSA, for PCa diagnosis [68].

Furthermore, PSA isoform (p2PSA) is proved to be sensitive in the PCa. Jansen et al. [69] measured serum p2PSA, tPSA, fPSA, and BPHA in a total of 756 men from two different prostate centers. And they found p2PSA differed significantly between men with and without PCa. Besides, p2PSA showed a specific additional increase in PCa predictive value when added to a multivariate model consisting of tPSA and fPSA. But p2PSA had limited additional value in identifying men with aggressive PCa (Gleason score ≥7). They indicated that further study was needed to clearly define the definitive value of p2PSA to modify the number of men undergoing prostate biopsy.

Considering the fact that biopsy may arise several complications which affect the basic condition of patients. The early diagnosis of prostate cancer can be achieved through noninvasive methods instead of prostate biopsy, such as serum PSA, digital examination of rectum (DER), and transrectal ultrasound (TREU). But none of them are with high sensitivity and specificity. With the help of RNA-seq, the detections of fusion genes, noncoding RNAs, and PSA isoform in serum and urine can be easily accomplished. These detections can raise the diagnosis accuracy when being combined together, which might be glad tidings for the patients with PCa and can prevent the disease from progressing.

5.3 Treatment

Regularly, surgical or medical castration is highly effective in shrinking tumor burden, decreasing PSA levels, enhancing quality of life, and improving survival. However, a small portion of the patients will eventually experience disease progression despite castration, with a median duration of response of 12–24 months. Following hormonal manipulations, cytotoxic chemotherapy, mitoxantrone, and docetaxel were approved by the US Food and Drug Administration (FDA) to be used in clinical for the treatment of men with CRPC. Although docetaxel plus prednisone extended survival compared to mitoxantrone, the overall benefit was modest with most patients experiencing disease progression within 7 months [70–72]. Thus, novel targeted agents and alternative chemotherapy agents are focused to have effectiveness in the treatment of CRPC.

AR has been and remains the main target for pharmacological treatment of PCa. Recent novel approaches have been developed to target even the lowest levels of androgens by blocking steroid synthesis. For example, as a target drug recently approved by FDA to inhibit cytochrome P450 (CYP) 17 so that the production of androgen is effectively blocked, abiraterone acetate (Zytiga™ Centocor Ortho Biotech, Inc., Philadelphia, PA, USA) is used in combination with prednisone for the treatment of metastatic CRPC [73]. In the study by Attard et al. [74, 75], a decrease in PSA following treatment in 50 % of all men with CRPC was found in the initial Phase I clinical trial.

Another interesting direction of clinical research is the potential to target "actionable" gene fusions in PCa. To date, there are no drugs that target TMPRSS2: ERG specifically. RAF kinase fusions, although rare, are of immediate therapeutic significance given the numerous approved and investigational agents. Palanisamy et al. [21] demonstrated that the RAF kinase fusions were sensitive to sorafenib, an orally active FDA-approved agent that inhibits multiple kinases, including RAF [76].

Besides, Ateeq et al. [77] had proof of principle that PCa with SPINK1 overexpressing can be targeted using cetuximab which is an epidermal growth factor receptor (EGFR) inhibitor.

The choice for the treatment of advanced prostate cancer is varied but with little experience in clinical trials. Typically, systemic hormonal options, such as chronic androgen deprivation therapy (ADT), are initiated when metastatic disease is diagnosed. ADT provides cancer control and palliation for variable periods of time ranging from a few months to several years. For men who progress to the hormone-refractory state, treatment with docetaxel chemotherapy is an option. For the further progressing PCa with metastases, the therapy might depend on target drugs which specifically aim at the disorder points. RNA-seq is the method for the pharmacologists to target the points which are related with the disease, so that the patients will have the expectation of good prognosis, even complete healing.

6 Prospect

Although RNA-seq can be an effective genomic assay, its shortcomings should be cared. As regards the RNA-Seq, we still have to face several critical issues either from a biological or computational point of view. RNA-seq protocols are very sensitive with a high degree of quality control for each step so that the contamination of reagents with RNA-seq and degradation of RNA will affect the results. During the library preparation, strandedness of RNA fragments should be preserved particularly in order to correctly determine the directionality of gene transcription and facilitate the detection of opposing and overlapping transcripts within gen-dense genomic regions. In addition, other aspects of the technique, such as the data analysis, are with great importance which needs to be concerned in the future.

Although facing so much challenges of RNA-seq, recent advances in the techniques have provided researchers with a powerful toolbox for the characterization and quantification of the transcriptome. Using these technological advances, we can build a complete catalogue of transcripts that are derived from genomes ranging from those of simple unicellular organisms to complex mammalian cells as well as in tissues in normal and disease states. In addition, these technologies are bringing us closer to the ability to use RNA measurements for the discovery of the etiology of the diseases, clinical diagnosis, and specific target treatment of the disease with the conception of individualized therapy.

Prostate cancer is a world-concerning disease with a growing incidence both in the developed and developing countries. It is heterogeneous with different causes, mechanisms for different treatment among the races and countries. With the help of RNA-seq which is used in the finding of the etiology, searching the novel diagnosis and prognosis biomarkers, and the choices of the treatment, PCa is becoming clearer and clearer to us. Our ideal goal is that prostate cancer can be diagnosed in the early stage with high sensitivity biomarkers and prevented by individualized therapy with the help of RNA-seq in the future.

Thus, this technology will continue to help us realize the full potential of transcriptomic information as it relates to basic biological questions of differentiation and diversity, as well as its growing impact on the personalization of healthcare.

References

1. Petrosino JF, Highlander S, Luna RA, Gibbs RA, Versalovic J. Metagenomic pyrosequencing and microbial identification. Clin Chem. 2009;55(5):856–66. Epub 2009/03/07.
2. Jemal A, Bray F, Center MM, Ferlay J, Ward E, Forman D. Global cancer statistics. CA Cancer J Clin. 2011;61(2):69–90. Epub 2011/02/08.
3. Lapointe J, Li C, Giacomini CP, Salari K, Huang S, Wang P, et al. Genomic profiling reveals alternative genetic pathways of prostate tumorigenesis. Cancer Res. 2007;67(18):8504–10. Epub 2007/09/19.
4. Mendes A, Scott RJ, Moscato P. Microarrays–identifying molecular portraits for prostate tumors with different Gleason patterns. Methods Mol Med. 2008;141:131–51. Epub 2008/05/06.
5. Harris WP, Mostaghel EA, Nelson PS, Montgomery B. Androgen deprivation therapy: progress in understanding mechanisms of resistance and optimizing androgen depletion. Nat Clin Pract Urol. 2009;6(2):76–85. Epub 2009/02/10.
6. Attar RM, Takimoto CH, Gottardis MM. Castration-resistant prostate cancer: locking up the molecular escape routes. Clin Cancer Res. 2009;15(10):3251–5. Epub 2009/05/19.
7. Ryan CJ, Small EJ. Early versus delayed androgen deprivation for prostate cancer: new fuel for an old debate. J Clin Oncol. 2005;23(32):8225–31. Epub 2005/11/10.
8. Wang Z, Gerstein M, Snyder M. RNA-Seq: a revolutionary tool for transcriptomics. Nat Rev Genet. 2009;10(1):57–63. Epub 2008/11/19.
9. Jacquier A. The complex eukaryotic transcriptome: unexpected pervasive transcription and novel small RNAs. Nat Rev Genet. 2009;10(12):833–44. Epub 2009/11/19.
10. Marguerat S, Bahler J. RNA-seq: from technology to biology. Cell Mol Life Sci: CMLS. 2010;67(4):569–79. Epub 2009/10/28.
11. Costa V, Angelini C, De Feis I, Ciccodicola A. Uncovering the complexity of transcriptomes with RNA-Seq. J Biomed Biotechnol. 2010;2010:853916. Epub 2010/07/14.
12. Gilad Y, Rifkin SA, Bertone P, Gerstein M, White KP. Multi-species microarrays reveal the effect of sequence divergence on gene expression profiles. Genome Res. 2005;15(5):674–80. Epub 2005/05/04.
13. Velculescu VE, Zhang L, Vogelstein B, Kinzler KW. Serial analysis of gene expression. Science. 1995;270(5235):484–7. Epub 1995/10/20.
14. Wilhelm BT, Landry JR. RNA-Seq-quantitative measurement of expression through massively parallel RNA-sequencing. Methods. 2009;48(3):249–57. Epub 2009/04/02.
15. Mortazavi A, Williams BA, McCue K, Schaeffer L, Wold B. Mapping and quantifying mammalian transcriptomes by RNA-Seq. Nat Methods. 2008;5(7):621–8. Epub 2008/06/03.
16. Nagalakshmi U, Wang Z, Waern K, Shou C, Raha D, Gerstein M, et al. The transcriptional landscape of the yeast genome defined by RNA sequencing. Science. 2008;320(5881):1344–9. Epub 2008/05/03.
17. Supper J, Gugenmus C, Wollnik J, Drueke T, Scherf M. Hahn A, et al. Methods: Detecting and visualizing gene fusions; 2012. Epub 2012/10/06.

18. Li H, Wang J, Mor G, Sklar J. A neoplastic gene fusion mimics trans-splicing of RNAs in normal human cells. Science. 2008;321(5894):1357–61. Epub 2008/09/06.
19. Maher CA, Kumar-Sinha C, Cao X, Kalyana-Sundaram S, Han B, Jing X, et al. Transcriptome sequencing to detect gene fusions in cancer. Nature. 2009;458(7234):97–101. Epub 2009/01/13.
20. Berger MF, Levin JZ, Vijayendran K, Sivachenko A, Adiconis X, Maguire J, et al. Integrative analysis of the melanoma transcriptome. Genome Res. 2010;20(4):413–27. Epub 2010/02/25.
21. Palanisamy N, Ateeq B, Kalyana-Sundaram S, Pflueger D, Ramnarayanan K, Shankar S, et al. Rearrangements of the RAF kinase pathway in prostate cancer, gastric cancer and melanoma. Nat Med. 2010;16(7):793–8. Epub 2010/06/08.
22. Zhou JB, Zhang T, Wang BF, Gao HZ, Xu X. Identification of a novel gene fusion RNF213SLC26A11 in chronic myeloid leukemia by RNA-Seq. Mol Med Rep. 2013;7(2):591–7. Epub 2012/11/16.
23. Steidl C, Shah SP, Woolcock BW, Rui L, Kawahara M, Farinha P, et al. MHC class II transactivator CIITA is a recurrent gene fusion partner in lymphoid cancers. Nature. 2011;471(7338):377–81. Epub 2011/03/04.
24. McManus CJ, Duff MO, Eipper-Mains J, Graveley BR. Global analysis of trans-splicing in Drosophila. Proc Natl Acad Sci USA. 2010;107(29):12975–9. Epub 2010/07/10.
25. Metzker ML. Sequencing technologies—the next generation. Nat Rev Genet. 2010;11(1):31–46. Epub 2009/12/10.
26. Prensner JR, Chinnaiyan AM. The emergence of lncRNAs in cancer biology. Cancer Discov. 2011;1(5):391–407. Epub 2011/11/19.
27. Wang X, Liu P, Zhu H, Xu Y, Ma C, Dai X, et al. miR-34a, a microRNA up-regulated in a double transgenic mouse model of Alzheimer's disease, inhibits bcl2 translation. Brain Res Bull. 2009;80(4–5):268–73. Epub 2009/08/18.
28. Beaulieu YB, Kleinman CL, Landry-Voyer AM, Majewski J, Bachand F. Polyadenylation-dependent control of long noncoding RNA expression by the poly(a)-binding protein nuclear 1. PLoS Genet. 2012;8(11):e1003078. Epub 2012/11/21.
29. Flockhart RJ, Webster DE, Qu K, Mascarenhas N, Kovalski J, Kretz M, et al. BRAFV600E remodels the melanocyte transcriptome and induces BANCR to regulate melanoma cell migration. Genome Res. 2012;22(6):1006–14. Epub 2012/05/15.
30. Twine NA, Janitz K, Wilkins MR, Janitz M. Whole transcriptome sequencing reveals gene expression and splicing differences in brain regions affected by Alzheimer's disease. PLoS One. 2011;6(1):e16266. Epub 2011/02/02.
31. Shapiro IM, Cheng AW, Flytzanis NC, Balsamo M, Condeelis JS, Oktay MH, et al. An EMT-driven alternative splicing program occurs in human breast cancer and modulates cellular phenotype. PLoS Genet. 2011;7(8):e1002218. Epub 2011/08/31.
32. Lander ES. Initial impact of the sequencing of the human genome. Nature. 2011;470(7333):187–97. Epub 2011/02/11.
33. Freedman ML, Monteiro AN, Gayther SA, Coetzee GA, Risch A, Plass C, et al. Principles for the post-GWAS functional characterization of cancer risk loci. Nat Genet. 2011;43(6):513–8. Epub 2011/05/27.
34. Shah SP, Kobel M, Senz J, Morin RD, Clarke BA, Wiegand KC, et al. Mutation of FOXL2 in granulosa-cell tumors of the ovary. N Engl J Med. 2009;360(26):2719–29. Epub 2009/06/12.
35. Ha KC, Lalonde E, Li L, Cavallone L, Natrajan R, Lambros MB, et al. Identification of gene fusion transcripts by transcriptome sequencing in BRCA1-mutated breast cancers and cell lines. BMC Med Genomics. 2011;4:75. Epub 2011/10/29.
36. Carraro DM, Ferreira EN, de Campos MG, Puga RD, Abrantes EF, Trape AP, et al. Poly (A)+ transcriptome assessment of ERBB2-induced alterations in breast cell lines. PLoS One. 2011;6(6):e21022. Epub 2011/07/07.
37. Bill-Axelson A, Holmberg L, Ruutu M, Haggman M, Andersson SO, Bratell S, et al. Radical prostatectomy versus watchful waiting in early prostate cancer. N Engl J Med. 2005;352(19):1977–84. Epub 2005/05/13.
38. Pomerantz MM, Freedman ML. Genetics of prostate cancer risk. Mount Sinai J Med, NY. 2010;77(6):643–54. Epub 2010/11/26.

39. Lichtenstein P, Holm NV, Verkasalo PK, Iliadou A, Kaprio J, Koskenvuo M, et al. Environmental and heritable factors in the causation of cancer–analyses of cohorts of twins from Sweden, Denmark, and Finland. N Engl J Med. 2000;343(2):78–85. Epub 2000/07/13.
40. Thomas G, Jacobs KB, Yeager M, Kraft P, Wacholder S, Orr N, et al. Multiple loci identified in a genome-wide association study of prostate cancer. Nat Genet. 2008;40(3):310–5. Epub 2008/02/12.
41. Kote-Jarai Z, Easton DF, Stanford JL, Ostrander EA, Schleutker J, Ingles SA, et al. Multiple novel prostate cancer predisposition loci confirmed by an international study: the PRACTICAL Consortium. Cancer Epidemiol Biomarkers Prev. 2008;17(8):2052–61. Epub 2008/08/19.
42. Zheng SL, Sun J, Wiklund F, Smith S, Stattin P, Li G, et al. Cumulative association of five genetic variants with prostate cancer. N Engl J Med. 2008;358(9):910–9. Epub 2008/01/18.
43. Jenster G, van der Korput HA, Trapman J, Brinkmann AO. Identification of two transcription activation units in the N-terminal domain of the human androgen receptor. J Biol Chem. 1995;270(13):7341–6. Epub 1995/03/31.
44. Ren S, Peng Z, Mao JH, Yu Y, Yin C, Gao X, et al. RNA-seq analysis of prostate cancer in the Chinese population identifies recurrent gene fusions, cancer-associated long noncoding RNAs and aberrant alternative splicings. Cell Res. 2012;22(5):806–21. Epub 2012/02/22.
45. Eeles RA, Kote-Jarai Z, Giles GG, Olama AA, et al. Multiple newly identified loci associated with prostate cancer susceptibility. Nat Genet. 2008;40:316–21.
46. Yeager M, Orr N, Hayes RB, Jacobs KB, Kraft P et al. Genome-wide association study of prostate cancer identifies a second risk locus at 8q24. Nat Genet. 2007;39:645–9.
47. Gudmundsson J, Sulem P, Rafnar T, Bergthorsson JT, Manolescu A, et al. Common sequence variants on 2p15 and Xp11.22 confer susceptibility to prostate cancer. Nat Genet. 2008;40:281–3.
48. Bass BL. RNA editing by adenosine deaminases that act on RNA. Annu Rev Biochem. 2002;71:817–46. Epub 2002/06/05.
49. Prensner JR, Iyer MK, Balbin OA, Dhanasekaran SM, Cao Q, Brenner JC, et al. Transcriptome sequencing across a prostate cancer cohort identifies PCAT-1, an unannotated lincRNA implicated in disease progression. Nat Biotechnol. 2011;29(8):742–9. Epub 2011/08/02.
50. Tomlins SA, Rhodes DR, Perner S, Dhanasekaran SM, Mehra R, Sun XW, et al. Recurrent fusion of TMPRSS2 and ETS transcription factor genes in prostate cancer. Science. 2005;310(5748):644–8. Epub 2005/10/29.
51. Tomlins SA, Laxman B, Varambally S, Cao X, Yu J, Helgeson BE, et al. Role of the TMPRSS2-ERG gene fusion in prostate cancer. Neoplasia. 2008;10(2):177–88. Epub 2008/02/20.
52. Pflueger D, Terry S, Sboner A, Habegger L, Esgueva R, Lin PC, et al. Discovery of non-ETS gene fusions in human prostate cancer using next-generation RNA sequencing. Genome Res. 2011;21(1):56–67. Epub 2010/11/03.
53. Catalona WJ, Smith DS, Ratliff TL, Dodds KM, Coplen DE, Yuan JJ, et al. Measurement of prostate-specific antigen in serum as a screening test for prostate cancer. N Engl J Med. 1991;324(17):1156–61. Epub 1991/04/25.
54. Lucia MS, Darke AK, Goodman PJ, La Rosa FG, Parnes HL, Ford LG, et al. Pathologic characteristics of cancers detected in The Prostate Cancer Prevention Trial: implications for prostate cancer detection and chemoprevention. Cancer Prev Res (Phila). 2008;1(3):167–73. Epub 2009/01/14.
55. Thompson IM, Pauler DK, Goodman PJ, Tangen CM, Lucia MS, Parnes HL, et al. Prevalence of prostate cancer among men with a prostate-specific antigen level < or =4.0 ng per milliliter. N Engl J Med. 2004;350(22):2239–46.
56. Leary RJ, Kinde I, Diehl F, Schmidt K, Clouser C, Duncan C, et al. Development of personalized tumor biomarkers using massively parallel sequencing. Sci Transl Med. 2010;2(20):20ra14. Epub 2010/04/08.
57. de Kok JB, Verhaegh GW, Roelofs RW, Hessels D, Kiemeney LA, Aalders TW, et al. DD3(PCA3), a very sensitive and specific marker to detect prostate tumors. Cancer Res. 2002;62(9):2695–8. Epub 2002/05/01.

58. Tomlins SA, Bjartell A, Chinnaiyan AM, Jenster G, Nam RK, Rubin MA, et al. ETS gene fusions in prostate cancer: from discovery to daily clinical practice. Eur Urol. 2009;56(2):275–86. Epub 2009/05/05.
59. Furusato B, Tan SH, Young D, Dobi A, Sun C, Mohamed AA, et al. ERG oncoprotein expression in prostate cancer: clonal progression of ERG-positive tumor cells and potential for ERG-based stratification. Prostate Cancer Prostatic Dis. 2010;13(3):228–37. Epub 2010/06/30.
60. Park K, Tomlins SA, Mudaliar KM, Chiu YL, Esgueva R, Mehra R, et al. Antibody-based detection of ERG rearrangement-positive prostate cancer. Neoplasia. 2010;12(7):590–8. Epub 2010/07/24.
61. Hessels D, Smit FP, Verhaegh GW, Witjes JA, Cornel EB, Schalken JA. Detection of TMPRSS2-ERG fusion transcripts and prostate cancer antigen 3 in urinary sediments may improve diagnosis of prostate cancer. Clin cancer Res. 2007;13(17):5103–8. Epub 2007/09/06.
62. Laxman B, Morris DS, Yu J, Siddiqui J, Cao J, Mehra R, et al. A first-generation multiplex biomarker analysis of urine for the early detection of prostate cancer. Cancer Res. 2008;68(3):645–9. Epub 2008/02/05.
63. Laxman B, Tomlins SA, Mehra R, Morris DS, Wang L, Helgeson BE, et al. Noninvasive detection of TMPRSS2:ERG fusion transcripts in the urine of men with prostate cancer. Neoplasia. 2006;8(10):885–8. Epub 2006/10/25.
64. Nilsson J, Skog J, Nordstrand A, Baranov V, Mincheva-Nilsson L, Breakefield XO, et al. Prostate cancer-derived urine exosomes: a novel approach to biomarkers for prostate cancer. Br J Cancer. 2009;100(10):1603–7. Epub 2009/04/30.
65. Rostad K, Hellwinkel OJ, Haukaas SA, Halvorsen OJ, Oyan AM, Haese A, et al. TMPRSS2:ERG fusion transcripts in urine from prostate cancer patients correlate with a less favorable prognosis. APMIS. 2009;117(8):575–82. Epub 2009/08/12.
66. Tomlins SA, Aubin SM, Siddiqui J, Lonigro RJ, Sefton-Miller L, Miick S, et al. Urine TMPRSS2:ERG fusion transcript stratifies prostate cancer risk in men with elevated serum PSA. Sci Transl Med. 2011;3(94):94ra72. Epub 2011/08/05.
67. Salami SS, Schmidt F, Laxman B, Regan MM, Rickman DS, Scherr D, et al. Combining urinary detection of TMPRSS2:ERG and PCA3 with serum PSA to predict diagnosis of prostate cancer. Urol Oncol. 2011. Epub 2011/05/24.
68. Whitaker HC, Kote-Jarai Z, Ross-Adams H, Warren AY, Burge J, George A, et al. The rs10993994 risk allele for prostate cancer results in clinically relevant changes in microseminoprotein-beta expression in tissue and urine. PLoS One. 2010;5(10):e13363. Epub 2010/10/23.
69. Jansen FH, van Schaik RH, Kurstjens J, Horninger W, Klocker H, Bektic J, et al. Prostate-specific antigen (PSA) isoform p2PSA in combination with total PSA and free PSA improves diagnostic accuracy in prostate cancer detection. Eur Urol. 2010;57(6):921–7. Epub 2010/03/02.
70. Tannock IF, de Wit R, Berry WR, Horti J, Pluzanska A, Chi KN, et al. Docetaxel plus prednisone or mitoxantrone plus prednisone for advanced prostate cancer. N Engl J Med. 2004;351(15):1502–12. Epub 2004/10/08.
71. Petrylak DP, Tangen CM, Hussain MH, Lara Jr PN, Jones JA, Taplin ME, et al. Docetaxel and estramustine compared with mitoxantrone and prednisone for advanced refractory prostate cancer. N Engl J Med. 2004;351(15):1513–20. Epub 2004/10/08.
72. Armstrong AJ, George DJ. Optimizing the use of docetaxel in men with castration-resistant metastatic prostate cancer. Prostate Cancer Prostatic Dis. 2010;13(2):108–16. Epub 2010/01/13.
73. Barrie SE, Potter GA, Goddard PM, Haynes BP, Dowsett M, Jarman M. Pharmacology of novel steroidal inhibitors of cytochrome P450(17) alpha (17 alpha-hydroxylase/C17-20 lyase). J Steroid Biochem Mol Biol. 1994;50(5–6):267–73. Epub 1994/09/01.
74. Attard G, Swennenhuis JF, Olmos D, Reid AH, Vickers E, A'Hern R, et al. Characterization of ERG, AR and PTEN gene status in circulating tumor cells from patients with castration-resistant prostate cancer. Cancer Res. 2009;69(7):2912–8. Epub 2009/04/03.
75. Attard G, Reid AH, Yap TA, Raynaud F, Dowsett M, Settatree S, et al. Phase I clinical trial of a selective inhibitor of CYP17, abiraterone acetate, confirms that castration-resistant prostate

cancer commonly remains hormone driven. Journal Clin Oncol. 2008;26(28):4563–71. Epub 2008/07/23.
76. Wilhelm SM, Adnane L, Newell P, Villanueva A, Llovet JM, Lynch M. Preclinical overview of sorafenib, a multikinase inhibitor that targets both Raf and VEGF and PDGF receptor tyrosine kinase signaling. Mol Cancer Ther. 2008;7(10):3129–40. Epub 2008/10/15.
77. Ateeq B, Tomlins SA, Laxman B, Asangani IA, Cao Q, Cao X, et al. Therapeutic targeting of SPINK1-positive prostate cancer. Sci Transl Med. 2011;3(72):72ra17. Epub 2011/03/04.

Chapter 14
Next-Generation Sequencing for High-Throughput RNA Interference Screens

Toby M. Ward, Anna-Maria Jegg, and Elizabeth Iorns

Abstract Ribonucleic acid interference (RNAi) screening has emerged as an indispensable genetic research tool, allowing determination of phenotypic effects after silencing entire suites of genes. As the catalog of fully sequenced genomes and transcriptomes grows, production of small interfering/short-hairpin RNA libraries that target every gene in a particular cell, tissue, or organism is achievable, allowing high-throughput "genome-wide" RNAi screening. This technology has been embraced by cancer biologists and has been used to analyze a myriad of phenotypic effects of genetic loss of function in human cancers.

A basic RNAi screening scheme includes silencing of a panel of genes in a cell population, followed by identification of a phenotypic change (in cancer research, this might include cell death, cell invasion, sensitivity to therapeutics). Upon identification of cells exhibiting the phenotype of interest, it is necessary to determine which specific shRNAs are responsible. Previously, this process was laborious, requiring tedious DNA extraction, PCR amplification, and individual cloning and sequencing of PCR amplicons to determine the specific shRNA(s) harbored by cells.

With the advent of next-generation sequencing (NGS), identification of individual shRNAs harbored by cells has been revolutionized. NGS allows rapid and specific identification of shRNA oligomers present in the cell(s) of interest and requires minimal amounts of source material. This chapter will describe the use of NGS in RNAi screens with a focus on cancer biology and provide resources for those interested in pursuing NGS-powered RNAi screens.

T.M. Ward, Ph.D. (✉)
Stanford Cancer Institute, Stanford University School of Medicine, Palo Alto, CA, USA
e-mail: tward1@stanford.edu

A.-M. Jegg, Ph.D.
University of Miami Miller School of Medicine, Miami, FL, USA

E. Iorns, Ph.D.
Science Exchange, Inc., Palo Alto, CA, USA

W. Wu and H. Choudhry (eds.), *Next Generation Sequencing in Cancer Research: Volume 1: Decoding the Cancer Genome*, DOI 10.1007/978-1-4614-7645-0_14,

Keywords RNAi screen • Next-generation sequencing • Massively parallel sequencing • In vivo screen • High-throughput screen • shRNA • Loss-of-function screen

1 Introduction

RNA interference (RNAi) is a powerful technique to silence gene expression in eukaryotic cells and has enjoyed intense research focus since its discovery. The mechanism of endogenous cellular RNAi has been reviewed [1, 2] and relies on small double-stranded RNAs (dsRNAs) with sequence complementarity to target mRNAs. RNAi by cellular small interfering RNAs (siRNAs), microRNAs (miRNAs), and PIWI-interacting RNAs (piRNAs) leads to either mRNA transcript degradation or inhibition of translation [3], thereby reducing or silencing expression of the encoded protein(s). In addition to physiological RNAi to control gene expression at the posttranscriptional level, delivery of chemically synthesized siRNAs or expression of vectored short hairpin (shRNAs) to knock down expression of target genes is now a standard tool in molecular biology. The ability to selectively and specifically silence any gene of interest using synthesized siRNA oligomers and vectored shRNA has revolutionized the field of functional genetics.

RNA interference has proven invaluable in cell-based functional genetic screening, where assembled libraries of different siRNAs or shRNAs targeting hundreds or thousands of genes can be used [4–6]. The ability to silence gene expression in such a broad and robust manner allows investigators to screen for a cellular phenotype of interest (i.e., cell transformation, drug resistance/susceptibility) and then to determine which genes, when silenced, are responsible for the observed phenotypes. Together with the ever-expanding catalog of fully sequenced genomes and transcriptomes, which facilitates the design of siRNAs and shRNAs targeting every predicted mRNA in a particular organism, tissue, or cell, RNA interference screening has been and will continue to be an immensely powerful and commonly used genetic tool.

In cell-culture assays, RNAi screens are often performed using siRNAs in a microplate-based approach, where target cells are transfected in an arrayed format (e.g., in 96- or 384-well plates), with each well containing a known siRNA. Thus, it is relatively simple to identify the specific siRNAs that resulted in a particular phenotype after transfection and performance of the selection screen [9]. In contrast, pooled multiplex approaches for RNAi screening are often useful in settings where plate-based assays are not possible (e.g., in vivo RNAi screens) or when using cells which are not amenable to efficient transfection in culture (i.e., primary human cells, quiescent cells, and certain cell lines). In these settings, pools consisting of hundreds or thousands of individual shRNAs encoded by appropriately pseudotyped lentiviral delivery vectors have emerged as the tool of choice for gene silencing. Use of such shRNA libraries allows for large-scale loss-of-function screens in numerous cell types (including primary and nondividing cells) and in assays that require non-microplate-based methods (such as in vivo screens) [9]. Moreover, the stable integration of lentiviral vector shRNA-encoding DNA into the target cell genome allows for the use of longer experimental timelines (weeks to months) than does transfection of short-lived siRNA (days), further enabled by the inclusion of drug-resistance markers for stable cell selection [7]. The use of expression vectors also allows employment of chemically inducible promoters to drive expression of shRNAs, allowing investigators to control the timing of shRNA expression in target cell populations.

A major bottleneck in the process of pooled shRNA screening is deconvolution of the resulting shRNA pools that are enriched or depleted after the selective screen, in other words the identification of specific shRNAs (and by association their target genes) that are responsible for an observed phenotype. For instance, in a setting where hundreds or thousands of individual shRNA sequences are introduced into a population of cells, identifying single shRNAs that are responsible for a particular phenotype is a daunting task. Moreover, the scale of large-scale screens, shRNA library density, and gene knockdown efficiency all impact on the rates of false-negative and false-positive "hits" detected. The development and availability of massively parallel next-generation sequencing (NGS) has revolutionized the ability to identify the presence and representation of individual shRNAs within screened cell populations [7]. Using NGS technology, it is possible to identify and quantify individual shRNAs from cells at the termination of the phenotypic screening assay, with enhance dynamic range and sensitivity over other methods. In addition, the use of next-generation sequencing for deconvolution has the immense advantage of facilitating both positive and negative selection RNAi screens.

2 NGS in Positive Selection RNAi Screens

The most straightforward RNAi screening design in cancer biology is introduction of a silencing library into target cells, followed by selection of cells that gain an appreciable phenotype of interest, such as increased proliferation rate, enhanced invasion, drug resistance, or altered morphology. In this type of positive selection or "enrichment" screen, the shRNAs responsible for the observed phenotypes are maintained in the cells of interest, which are then isolated, followed by harvesting of cellular DNA and sequencing or hybridization for shRNA identification [7]. Figure 14.1 provides a workflow for a basic cell-culture positive selection "enrichment" RNAi screen. As discussed in detail later in this chapter, identification of shRNAs from positive selection screens is possible using standard PCR, cloning, and sequencing; however, this method is laborious and expensive. Alternatively, next-generation sequencing of harvested genomic DNA from cells after phenotypic screening provides a method to reliably quantify the shRNAs present in the population of cells exhibiting the desired phenotype.

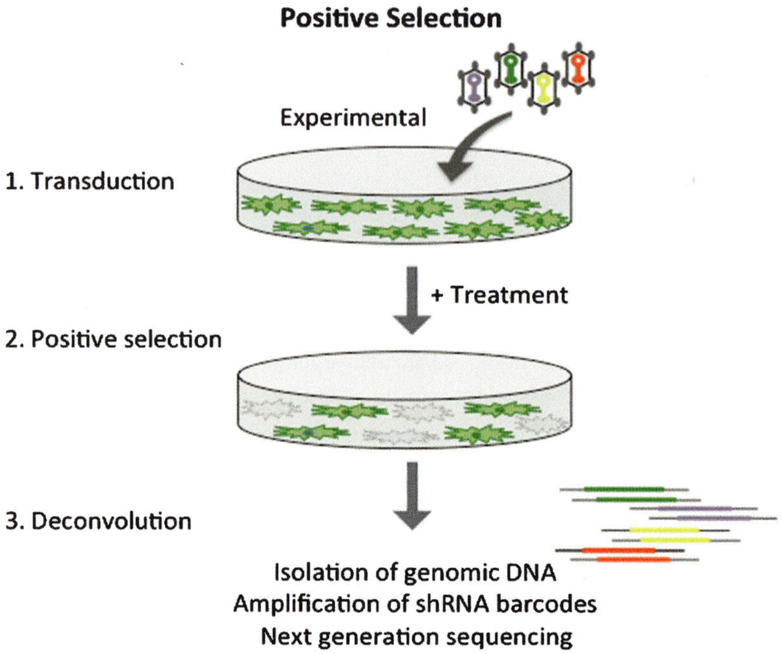

Fig. 14.1 Workflow for a basic cell-culture positive selection shRNA screen using NGS. Following transduction of target cells with viral particles containing the library of shRNA sequences, cells that gain an appreciable phenotype of interest are selected, and shRNAs responsible for the observed phenotypes are deconvoluted via amplification of shRNA barcodes and next-generation sequencing

3 Examples of Positive Selection RNAi Screens in Cancer Research

In cancer biology, pooled positive selection shRNA screens have been successfully used to identify genes that mediate drug resistance [8], to identify novel tumor suppressors [9, 10], and to identify genes involved in cancer cell invasion [11]. As shRNA libraries become more widely available, it is expected that such screens will increasingly be of use in oncology research, to further understanding of malignancy and to identify novel targets for clinical intervention. In this section, several published journal articles describing the use of pooled shRNA screens in cancer research will be described.

To study cancer cell resistance to chemotherapy, Burgess et al. [8] used a retroviral shRNA library (the "Cancer 1,000" library composed of approximately 2,300 individual shRNAs targeting 1,000 murine transcripts) to identify genes that, when silenced, conferred resistance to doxorubicin in mouse lymphoma cells. After identification of hits from an initial cell-culture screen, the authors validated the resistance-promoting effects of shRNA targeting the TOP2A gene in vivo. In this positive selection screen, the authors used PCR amplification of shRNAs that had integrated into the genomes of doxorubicin-resistant cells, followed by cloning and sequencing of the resulting PCR amplicons to determine shRNA identities.

A seminal paper on the use of pooled shRNA screening to identify novel tumor suppressors was published in 2005 [10]. Westbrook et al. used partially transformed human mammary epithelial cells in a soft agar assay to identify a panel of 25 candidate tumor suppressor genes. An shRNA library targeting approximately 9,000 human transcripts was retrovirally transduced into this cell population, and cells that were able to grow in soft agar were selected for DNA harvesting and shRNA identification. Using a combination of barcode array hybridization and PCR/cloning/sequencing for deconvolution, the authors successfully identified the RE1-silencing transcription factor (REST) gene as a novel tumor suppressor, a result which was validated in human cancer samples. Importantly, this screening strategy was further validated by its identification of the canonical tumor suppressor phosphatase and tensin homolog (PTEN) and transforming growth factor beta-receptor 2 (TGFBR2).

As illustrated in these 2 studies, positive selection/enrichment shRNA screens using smaller libraries are amenable to deconvolution using a standard PCR/cloning/sequencing approach. However, when using massively multiplex shRNA libraries (tens or hundreds of thousands of total shRNAs) in large pool formats (>10,000 shRNAs per pool), this deconvolution technique may become prohibitively expensive or laborious. Next-generation sequencing represents a method to circumvent and improve on this deconvolution method (discussed in detail later in this chapter).

An example of an NGS-powered enrichment shRNA screen was performed by Iorns et al. and published in 2012 [9]. In this study, partially transformed human breast epithelial cells transduced with the pGIPZ shRNA library (composed of ~70,000 shRNAs targeting 16,000 human transcripts) were used to identify novel candidate breast tumor suppressors. Briefly, cells were transduced using 7 pools of ~10,000 shRNAs and then used in an in vivo tumor formation assay via injection into the mammary fat pads of female mice. After tumor formation, genomic DNA was extracted from the xenografts, followed by next-generation sequencing to identify enriched shRNAs. Using this methodology, the authors were able to identify leukemia inhibitory factor receptor (LIFR) as a novel breast tumor suppressor, as well as the canonical tumor suppressors p53 and Max dimerization protein (MNT). In addition to NGS for deconvolution, the authors also performed PCR/cloning/sequencing on a subset of resulting tumor genomic DNA, a complementary method which confirmed screen hits identified by NGS.

4 NGS in Negative Selection RNAi Screens

A major advantage of next-generation sequencing in large-scale pooled shRNA screening is the ability to perform negative selection (or "dropout") screens, in which the phenotype of interest causes loss or dropout of cells (and thus loss of the responsible shRNAs) as opposed to enrichment [7]. In particular, this methodology

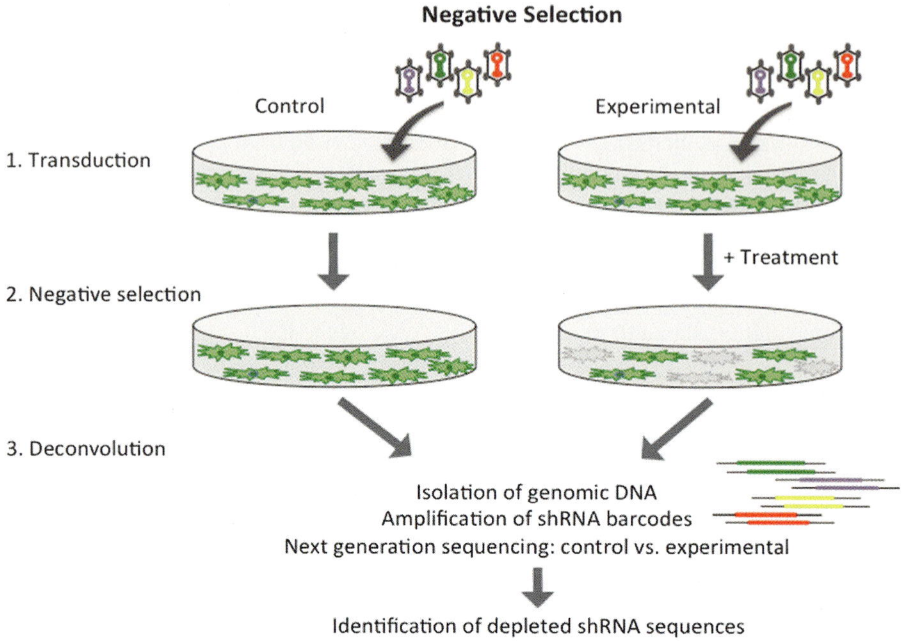

Fig. 14.2 Workflow for a basic cell-culture negative selection shRNA screen using NGS. Control and experimental cell populations are transduced with the shRNA library, and experimental cells are placed under selective pressure (i.e., drug treatment). The control population of cells harboring the entire transduced shRNA library is sequenced to determine shRNA representation before and after selection. Comparison of the relative representation of shRNAs in the starting reference pool of cells to those remaining in the cell pool after negative selection reveals which shRNAs have been lost or decreased

is well suited for shRNA screens attempting to identify gene targets that, when silenced, inhibit cancer cell growth, survival, or other malignant phenotypes, or result in enhanced drug sensitivity. While potentially more complicated than positive selection enrichment screens, negative selection screens have proved to be a powerful tool in cancer research and are potentially more relevant for target identification/therapeutic evaluation for the aforementioned reasons.

A general scheme for a negative selection screen is provided in Fig. 14.2; for this method, a reference population of cells harboring the entire transduced shRNA library is sequenced shortly after transduction (before negative selection) to determine the shRNA representation in the starting cell pool. After the appropriate length of time to screen for the desired phenotype, a second experimental population of cells is sequenced to determine shRNA representation. By comparing the relative representation of shRNAs in the starting reference pool of cells to those remaining in the cell pool after negative selection, it is possible to determine which shRNAs have been lost or decreased, i.e., "dropped out." For more information on the design and use of negative selection screens, see [7, 12].

5 Examples of Negative Selection RNAi Screens in Cancer Research

Negative selection screens powered by next-generation screening are invaluable for the identification of genes that, when silenced, confer sensitivity to existing therapeutic agents. Cells rendered sensitive to drugs upon gene silencing will die, enter growth arrest, or suffer inhibited proliferation, resulting in decreased representation of the shRNAs in the treated cells compared to the initial (control) population. In addition, negative selection screens are useful for identification of novel targets for therapeutic intervention in cancer cells, i.e., genes that, when silenced, result in cancer cell death, quiescence, etc., are candidate for drug development targets.

An example of a negative selection shRNA screen in cancer research was published by Silva et al. in 2008 [13]. In this study, genes essential to cancer cell growth and survival were identified via lentiviral integration into target cell populations, followed by cultured growth in vitro for 1 or 2 weeks. Using an shRNA library that was grouped into pools of ~6,000, ~10,000, and ~20,000 shRNAs per pool, the authors tested their screening strategy at different coverage densities and found that a pool of ~20,000 shRNAs was as effective as the smaller pools for identifying true hits. Hits from the initial screens were then validated by transducing several cancer cell types with multiple shRNAs targeting the candidate essential transcripts. This screening strategy identified a panel of essential genes in cancer cell lines, including genes involved in the cell cycle, mitotic checkpoint proteins, etc. In this study, microarray hybridization of shRNA barcodes was employed for deconvolution of the shRNA screen data; this group of investigators published further on the use of this multiplex dropout screening methodology [14].

In a study to identify genes that cooperate with Ras mutation to drive cancer cell proliferation and survival, Luo et al. used shRNA negative selection screening in Ras-mutant and Ras-wild-type cells [15]. An shRNA library was employed in 6 pools of ~16,000 shRNAs each and was transduced into the DLD-1 colon cancer cell line harboring either mutant or wild-type Ras proto-oncogene. The shRNA library used consisted of 74,905 retroviral shRNAs targeting 32,293 unique human transcripts. Using microarray hybridization for deconvolution of the screen data, the authors identified and validated shRNAs targeting components of the proteasomal degradation pathway and the anaphase-promoting complex, revealing genes that may be targets for the treatment of Ras-mutant cancers.

In an example of a next-generation sequencing powered multiplex shRNA screen, Zuber et al. identified potential therapeutic target genes in acute myeloid leukemia (AML) using an shRNA library consisting of 1,094 unique shRNAs targeting 243 murine transcripts [16]. After transducing the library as a single pool into mouse AML model cells, the investigators used next-generation sequencing to deconvolute the shRNAs that were depleted after 14 days of culture in vitro. Using this screen and follow-up experiments, the BRD4 gene was successfully identified and validated as a novel critical mediator of AML disease maintenance.

6 Approaches for Identification of shRNAs in Pooled RNAi Screens

To facilitate amplification and identification of shRNA sequences from cells after screening, also known as deconvolution, shRNA constructs in commercial libraries typically contain defined flanking sequences termed "barcodes," which allow identification of individual shRNAs using array hybridization (discussed below). In addition, the constant regions surrounding the shRNA cassette in vectored libraries can serve as a template for PCR amplification of shRNA, facilitating cloning/sequencing deconvolution approaches or next-generation sequencing. While not the focus of this chapter, the cloning/sequencing and microarray hybridization approaches are described below.

6.1 shRNA Amplification, Cloning, and Sanger Sequencing

For positive selection (enrichment) screens, standard PCR amplification, subcloning of PCR products, and sequencing to detect individual shRNAs are possible [9, 17]. By using the sequences flanking the ends of the shRNA constructs, which are conserved across the library, PCR amplification can be used to amplify the entire remaining collection of shRNAs following an enrichment screen. Resulting PCR amplicons can then be cloned into plasmid vectors, isolated from bacteria, and sequenced to reveal shRNAs present in the selected population of cells. However, concerns over the cost, limited scalability, and labor intensiveness of this methodology are prohibitive for many laboratories. In addition, this method is not readily amenable to negative selection screens.

6.2 Microarray Hybridization

Microarray chip hybridization-based detection of shRNAs present in cell populations after a pooled phenotypic screen has been used successfully in both positive and negative selection screens, as described above. In this method, the unique barcodes present in each shRNA construct in the library used are PCR amplified, and amplicons are fluorescently labeled and then hybridized to a custom oligonucleotide microarray [18]. Comparison of microarray-detected shRNAs recovered from each population of cells (i.e., drug treated vs. untreated) allows for assessment of enrichment or dropout of individual shRNAs. One limitation of microarray-based deconvolution is the requirement for a custom array designed for the shRNA library being used. In addition, microarray detection has been shown to have a narrower dynamic range than next-generation sequencing [7], due to its potential for saturation.

6.3 Next-Generation Sequencing

The use of next-generation sequencing for deconvolution of shRNA screening is advantageous for several reasons: custom microarrays engineered toward the specific library are unnecessary, NGS is particularly scalable to individual experimental needs, and NGS provides an improved dynamic range as well as increased sensitivity [7]. Moreover, NGS has the added benefits of use for determination of lentiviral integration sites as well as determination of library composition and fidelity, as discussed below. To use next-generation sequencing to identify the frequency of each shRNA construct in cell populations, PCR primers complementary to constant regions found in all shRNA constructs are used to amplify the shRNA target sequence specific to each individual shRNA construct. The PCR primers also encompass p5 and p7 sequences that allow sequence capture and sequencing by synthesis on the Illumina GAIIx platform. There are four steps involved: (1) barcode recovery (genomic DNA is extracted from shRNA bearing cells. shRNA sequences integrated into genomic DNA are then recovered by PCR amplification), (2) qPCR DNA quantification (this quantifies the products from step 1 to allow accurate NGS library concentrations to be inferred), (3) barcode sequencing (massively parallel sequencing is used to sequence denatured shRNA-seq libraries to identify and quantify the presence and number of specific shRNA constructs), and (4) data analysis (shRNA-seq data can be mapped to reference shRNA libraries using shALIGN [7]). The shALIGN program outputs the total number of short reads from step 3 matching to each library hairpin. Statistical analysis can then be conducted using shRNA-seq (a novel R package) [7]. This allows the identification of significantly over- or underrepresented shRNAs which may regulate the phenotype of interest that was screened for.

7 Identification of shRNA Genomic Integration Sites

As mentioned, it is often advantageous to use retro- or lentiviral delivery of shRNA-encoding constructs into eukaryotic cells, yielding integration into the cellular genome. However, genomic integration of lentiviral vectors containing promoter sequences can impact expression of genes proximal and distal to the integration site(s), as well as resulting in insertional mutagenesis, gene truncation, silencing, etc. Therefore, it is advantageous to enable determination of the genomic loci in which the delivered shRNA constructs have integrated, for instance, to validate that a phenotype elicited by a particular shRNA construct is not driven by insertional off-target effects. Recently, several groups have described the use of NGS to determine sites of genomic integration of transposons and lentiviruses [19–21]; these methods are directly applicable to shRNA screens where vectors containing known sequences (barcodes) were used. In this manner, it is possible not only to use NGS for deconvolution of the shRNA screen itself but also to determine the specific shRNA integration sites within cells.

8 shRNA Libraries

Concomitant with the increase in the number of laboratories employing RNAi screens for a multitude of uses, there are now several commercially available shRNA libraries. These libraries can be tailored to an investigator's particular interests, i.e., to silence genes known to be involved in a certain cellular pathway or process and genes encoding druggable targets, or larger libraries targeting the entire genome/transcriptome of an organism can be used. In addition, several libraries are currently available either as bacterial stocks, in which the end user produces the lentiviral particles for transduction, or as preprepared lentiviral stocks. With regard to cancer research, there are currently several libraries available from both academic and commercial resources. Table 14.1 contains a list and description of several available shRNA libraries targeting murine and human mRNA transcripts, which have been used in the study of cancer.

9 NGS to Analyze shRNA Library Fidelity

A major concern in the development and maintenance of large shRNA libraries is ensuring that the entire catalog of shRNAs is represented each time the library is reconstituted and employed. For example, reproduction of a plasmid-based library in bacteria yields the potential for inadvertent loss of shRNAs, due to stochastic events, differences in culture conditions, etc. Thus, surveillance of large shRNA libraries is necessary to ensure proper coverage of the intended target group of transcripts/genes and to minimize variation in shRNA representation. In addition to its role in identifying shRNAs after RNAi screens, next-generation sequencing is also useful for library surveillance. As described in [7, 22], plasmid pools isolated from bacterial cultures can be sequenced using NGS technology to determine the relative shRNA representation within a given library prep. While it is unlikely that researchers need to sequence an entire shRNA library each time it is to be used, it is important to ensure library fidelity after large-scale production in bacteria.

10 Limitations

As with any RNAi experiment, one of the most pressing concerns in correctly interpreting the experimental data from library-based screens is the possibility of off-target effects, which arise when shRNAs impact expression of a gene(s) other than its intended cognate mRNA [23]. To control for off-target effects, validation of screen "hits" using multiple shRNAs targeting a given mRNA of interest is often used. An additional control is forced expression of shRNA-resistant codon-swapped cDNA coding for the silenced protein, to rescue (reverse) the phenotype.

Table 14.1 Human and murine targeting shRNA libraries, including the source company, species specificity, number of shRNAs present, and number of transcripts targeted

Source	Species	Coverage	Vector system	Comments	Link
Thermo Scientific	Human	Up to 18,205 genes (95,703 shRNAs)	GIPZ/TRIPZ (inducible)	Available as sub-libraries GFP/RFP markers	www.thermoscientificbio.com/rnai-and-custom-rna-synthesis/shrna/
	Mouse	Up to 22,345 genes (10,1919 shRNAs)	GIPZ/TRIPZ (inducible)	Available as sub-libraries GFP/RFP markers	
Sigma Aldrich MISSION®	Human	Up to 20,018 genes (129,695 shRNAs)	pLKO.1	Available as sub-libraries	www.sigmaaldrich.com
	Mouse	Up to 21,171 genes (118,062 shRNAs)	pLKO.1	Available as sub-libraries	
System Biosciences GeneNet™	Human	Up to 38,500 genes (200,000 shRNAs)	pSIH1-H1/pSIF-H1	Available as sub-libraries Various markers (Puro, GFP)	www.systembio.com/rnai-libraries/
	Mouse	Up to 34,000 genes (150,000 shRNAs)	pSIH1-H1/pSIF-H1	Available as sub-libraries Various markers (Puro, RFP)	
Cellecta (DECIPHER project)	Human/mouse/rat	Up to 10,000 genes (55,000 shRNAs)	pRSI	Available as sub-libraries DECIPHER libraries are available under a MTA at no charge	www.cellecta.com/products-services/pooled-shRNA-libraries/ www.decipherproject.net

While silencing screens in cell culture have been used to identify novel candidate therapeutic targets, it is important to note that si/shRNA-mediated silencing is qualitatively distinct from the biological activities of small molecule drugs [24], i.e., proteins inhibited by pharmaceuticals may be impaired for activity but are still present in cells. To wit, the use of dominant negative isoforms of targeted proteins may be useful as a complementary method to assist in validating hits from shRNA screens. For more information on the limitations and concerns around the design of large-scale RNAi screens, see Kaelin 2012 [24].

11 Conclusions

Large-scale, high-throughput RNAi screening has emerged as a powerful functional genetics tool and has been successfully employed in cancer research using both in vitro and in vivo experimental methods. The advent of next-generation sequencing, its increasing availability to researchers, and its continuing downward cost trajectory have substantially impacted the field of high-throughput RNAi screening. Together with the availability of vectored massively multiplex shRNA libraries, this technology enables cancer biologists to perform large, robust screening assays to study the effects of genetic loss of function in almost any human cell type. As discussed, NGS offers several advantages for the deconvolution of data from shRNA screens, including applicability to any shRNA library, scalability, sensitivity, and potentially reduced costs. As shRNA libraries become increasingly diverse to cover increased numbers of transcripts, as well as increasingly dense (i.e., more shRNAs per gene) to improve specificity and to reduce false-negative/false-positive detection, NGS will continue to be an invaluable tool for efficient deconvolution and interpretation of the resulting data.

References

1. Carthew RW. Molecular biology. A new RNA dimension to genome control. Science. 2006;313(5785):305–6. Epub 2006/07/22.
2. Tomari Y, Zamore PD. Perspective: machines for RNAi. Genes Dev. 2005;19(5):517–29. Epub 2005/03/03.
3. Fire A, Xu S, Montgomery MK, Kostas SA, Driver SE, Mello CC. Potent and specific genetic interference by double-stranded RNA in Caenorhabditis elegans. Nature. 1998;391(6669):806–11. Epub 1998/03/05.
4. Boutros M, Ahringer J. The art and design of genetic screens: RNA interference. Nat Rev Genet. 2008;9(7):554–66. Epub 2008/06/04.
5. Cullen LM, Arndt GM. Genome-wide screening for gene function using RNAi in mammalian cells. Immunol Cell Biol. 2005;83(3):217–23. Epub 2005/05/10.
6. Fuchs F, Boutros M. Cellular phenotyping by RNAi. Brief Funct Genomic Proteomic. 2006;5(1):52–6. Epub 2006/06/14.
7. Sims D, Mendes-Pereira AM, Frankum J, Burgess D, Cerone MA, Lombardelli C, et al. High-throughput RNA interference screening using pooled shRNA libraries and next generation sequencing. Genome Biol. 2011;12(10):R104. Epub 2011/10/25.

8. Burgess DJ, Doles J, Zender L, Xue W, Ma B, McCombie WR, et al. Topoisomerase levels determine chemotherapy response in vitro and in vivo. Proc Natl Acad Sci U S A. 2008;105(26):9053–8. Epub 2008/06/25.
9. Iorns E, Ward TM, Dean S, Jegg A, Thomas D, Murugaesu N, et al. Whole genome in vivo RNAi screening identifies the leukemia inhibitory factor receptor as a novel breast tumor suppressor. Breast Cancer Res Treat. 2012;135(1):79–91. Epub 2012/04/27.
10. Westbrook TF, Martin ES, Schlabach MR, Leng Y, Liang AC, Feng B, et al. A genetic screen for candidate tumor suppressors identifies REST. Cell. 2005;121(6):837–48. Epub 2005/06/18.
11. Lin G, Aranda V, Muthuswamy SK, Tonks NK. Identification of PTPN23 as a novel regulator of cell invasion in mammary epithelial cells from a loss-of-function screen of the "PTP-ome". Genes Dev. 2011;25(13):1412–25. Epub 2011/07/05.
12. Campeau E, Gobeil S. RNA interference in mammals: behind the screen. Brief Funct Genomics. 2011;10(4):215–26. Epub 2011/07/28.
13. Silva JM, Marran K, Parker JS, Silva J, Golding M, Schlabach MR, et al. Profiling essential genes in human mammary cells by multiplex RNAi screening. Science. 2008;319(5863):617–20. Epub 2008/02/02.
14. Schlabach MR, Luo J, Solimini NL, Hu G, Xu Q, Li MZ, et al. Cancer proliferation gene discovery through functional genomics. Science. 2008;319(5863):620–4. Epub 2008/02/02.
15. Luo J, Emanuele MJ, Li D, Creighton CJ, Schlabach MR, Westbrook TF, et al. A genome-wide RNAi screen identifies multiple synthetic lethal interactions with the Ras oncogene. Cell. 2009;137(5):835–48. Epub 2009/06/06.
16. Zuber J, McJunkin K, Fellmann C, Dow LE, Taylor MJ, Hannon GJ, et al. Toolkit for evaluating genes required for proliferation and survival using tetracycline-regulated RNAi. Nat Biotechnol. 2011;29(1):79–83. Epub 2010/12/07.
17. Douma S, Van Laar T, Zevenhoven J, Meuwissen R, Van Garderen E, Peeper DS. Suppression of anoikis and induction of metastasis by the neurotrophic receptor TrkB. Nature. 2004;430(7003):1034–9. Epub 2004/08/27.
18. Brummelkamp TR, Bernards R. New tools for functional mammalian cancer genetics. Nat Rev Cancer. 2003;3(10):781–9. Epub 2003/10/23.
19. Brady T, Roth SL, Malani N, Wang GP, Berry CC, Leboulch P, et al. A method to sequence and quantify DNA integration for monitoring outcome in gene therapy. Nucleic Acids Res. 2011;39(11):e72. Epub 2011/03/19.
20. Ustek D, Sirma S, Gumus E, Arikan M, Cakiris A, Abaci N, et al. A genome-wide analysis of lentivector integration sites using targeted sequence capture and next generation sequencing technology. Infect Genet Evol. 2012;12(7):1349–54. Epub 2012/05/23.
21. Duncavage EJ, Magrini V, Becker N, Armstrong JR, Demeter RT, Wylie T, et al. Hybrid capture and next-generation sequencing identify viral integration sites from formalin-fixed, paraffin-embedded tissue. J Mol Diagn. 2011;13(3):325–33. Epub 2011/04/19.
22. Bassik MC, Lebbink RJ, Churchman LS, Ingolia NT, Patena W, LeProust EM, et al. Rapid creation and quantitative monitoring of high coverage shRNA libraries. Nat Methods. 2009; 6(6):443–5. Epub 2009/05/19.
23. Sigoillot FD, Lyman S, Huckins JF, Adamson B, Chung E, Quattrochi B, et al. A bioinformatics method identifies prominent off-targeted transcripts in RNAi screens. Nat Methods. 2012;9(4):363–6. Epub 2012/02/22.
24. Kaelin Jr WG. Molecular biology. Use and abuse of RNAi to study mammalian gene function. Science. 2012;337(6093):421–2. Epub 2012/07/28.

Chapter 15
Assessment of Mapping and SNP-Detection Algorithms for Next-Generation Sequencing Data in Cancer Genomics

Weixin Wang, Feng Xu, and Junwen Wang

Abstract The rapid development of next-generation sequencing (NGS) technology provides a new chance to extend the scale and resolution of genomic research. How to efficiently map millions of short reads to the reference genome and how to make accurate variant call are two major challenges in NGS analysis. In this chapter, we review current software for aligning short reads and detecting single-nucleotide polymorphisms (SNPs) and extensively evaluate their performance on normal and cancer samples from the Cancer Genome Atlas project and trio's data from the 1000 Genomes Project. We find that Burrows–Wheeler transform-based aligners are proven to be the most suitable for Illumina platform, and NovoalignCS shows the best overall performance for SOLiD data. We also demonstrate FaSD as the most reliable SNP caller compared with several state-of-the-art programs. Furthermore, NGS shows significantly lower coverage and poorer SNP-calling performance in

W. Wang • F. Xu
Department of Biochemistry, LKS Faculty of Medicine, The University of Hong Kong, Hong Kong, China

Shenzhen Institute of Research and Innovation, The University of Hong Kong, Shenzhen, China

J. Wang (✉)
Department of Biochemistry, LKS Faculty of Medicine, The University of Hong Kong, Hong Kong, China

Shenzhen Institute of Research and Innovation, The University of Hong Kong, Shenzhen, China

LKS Faculty of Medicine, Centre for Genomic Sciences, The University of Hong Kong, Hong Kong, China
e-mail: junwen@hku.hk

W. Wu and H. Choudhry (eds.), *Next Generation Sequencing in Cancer Research: Volume 1: Decoding the Cancer Genome*, DOI 10.1007/978-1-4614-7645-0_15,

the CpG island, promoter, and 5′UTR regions of the human genome. We show that both high GC-content and low repetitive elements are the causes of lower coverage in the promoter regions.

Keywords Next-generation sequencing • Alignment • SNP • Genotype • Cancer

1 Introduction

The Sanger method [1] has been applied in many large-scale sequencing projects in the past three decades and was treated as a "gold standard" because of its high accuracy in targeted sequencing [2]. Next-generation sequencing (NGS), as a high-throughput method, generates three to four orders of magnitude more sequences in less time and with considerably less cost in comparison with the Sanger method [2–4]. Meanwhile, the advent of NGS has significantly promoted the sequence-based genomic research [5], such as metagenomics, epigenetics, gene expression, RNA splicing [6, 7], and whole genome association studies [8]. In the recent several years, the first-generation Sanger sequencing has already given way to NGS platforms, such as the Roche/454 GS FLX, the Illumina/Solexa Genome Analyzer, and the Applied Biosystems SOLiD System. Recently, sequencing technology even extends to the third generation, including Helicos Biosciences' single-molecule sequencing [9] and Pacific Biosciences' single-molecule, real-time technology [10]. Here we primarily focus on the bioinformatics applications involved in two major parts of the NGS analysis pipeline: raw reads alignment and single-nucleotide polymorphisms (SNPs) detection.

2 Bioinformatics in Cancer Genome Research

2.1 Alignment

2.1.1 Performance Assessment of Alignment Software

After the raw reads are generated, the first step is to align them to the reference genome (if available) or de novo assemble them (if reference unavailable) for further analysis. With lots of reliable reference genomes in hand, for example, the human reference genome is maintained and improved by the Genome Reference Consortium (GRC), the alignment procedure can be immediately implemented.

Due to the tremendous volume of reads and the huge size of the human genome, speed and memory are the two major concerns in mapping NGS reads. Classic algorithms like BLAST [11] and BLAT [12] can align NGS reads to the reference very precisely, but usually take a few days even on computer grids, not to mention personal computers. Another challenge is how to pick true hit if there are multiple putative alignments. Normally, aligners will record all possible locations with the specified tags or pick a location arbitrarily.

Furthermore, since the sequenced genome is more or less different from the reference genome and there are unavoidable errors in base calling, alignment algorithms should be sensitive enough for genomic variant detection but robust enough for sequencing errors [7]. So far, several short-read alignment programs have been developed. Here are the lists of a brief review of the popular programs:

BFAST (http://sourceforge.net/apps/mediawiki/bfast/): It supports both Illumina and ABI SOLiD platforms, as well as many other NGS technologies (454, Helicos). It performs a full Smith–Waterman alignment [13, 14] which particularly emphasizes on alignment sensitivity.

Bowtie-v1(http://bowtie-bio.sourceforge.net/): It uses a Burrows–Wheeler transform to create a permanent, reusable index of the genome and supports MAQ-like and SOAP-like alignment policies [15].

BWA (http://bio-bwa.sourceforge.net/): It uses a Burrows–Wheeler transform to create an index of the genome. It implements two algorithms, BWA-short and BWA–SW. The former is suitable for query sequences shorter than 200 bp and the latter for longer sequences up to around 100 kbp. Both algorithms do gapped alignment [16, 17].

MAQ (Mapping and Assembly with Qualities, http://maq.sourceforge.net/): It is based on a "space seeding indexing" strategy. It can efficiently winnow the candidate locations within the reference [18].

MOM (Maximum Oligonucleotide Mapping, http://mom.csbc.vcu.edu/): It establishes a seed hash table to perform exact match between reference sequences and short reads, designed to better handle the higher frequency of sequencing errors near the 3′ and 5′ ends of Illumina reads [19].

MOSAIK (http://bioinformatics.bc.edu/marthlab/Mosaik): It is based on a banded Smith–Waterman algorithm seeded by results from a k-mer hashing scheme. It supports read with a wide range of sizes.

Novoalign (http://www.novocraft.com/): It supports gapped alignment of single-end and paired-end reads from Illumina GA I and II and from the new Helicos Heliscope Genome Analyzer and uses base qualities at all steps in the alignment. The program includes adapter trimming, base quality calibration, Bi-Seq alignment, and option to report multiple alignments per read.

RazerS (http://www.seqan.de/projects/razers.html): No read length limit. Hamming or edit distance mapping with configurable error rates. It has configurable and predictable sensitivity (run time/sensitivity trade-off). It supports paired-end read mapping [20].

RMAP (http://rulai.cshl.edu/rmap/): It can map reads with or without error probability information (quality scores) and supports paired-end reads or bisulfite-treated reads mapping. There is no limitation on read widths or number of mismatches [21, 22].

SeqMap (http://biogibbs.stanford.edu/~jiangh/SeqMap/): It is written in ANSI C++. While doing the mapping, SNPs as well as insertions/deletions of the nucleotide bases in the sequences can be tolerated and furthermore detected. By default, it outputs alignment in the Eland format [23].

SHRiMP (http://compbio.cs.toronto.edu/shrimp/): It supports letter space and color space reads and allows paired-end alignment [24].

Slider (http://www.bcgsc.ca/platform/bioinfo/software/slider): It uses the "probability" files instead of the sequence files as input for alignment to a reference sequence or a set of reference sequences [25].

SOAP-v2 (http://soap.genomics.org.cn/): It is efficient in time and space through the usage of Burrows–Wheeler transform (BWT) compression index of the reference genome. It is an updated version of SOAP for short oligonucleotide alignment. The new program is fast and accurate for huge amounts of short reads generated by Illumina/Solexa Genome Analyzer [26].

ZOOM (http://www.bioinformaticssolutions.com/products/zoom/): It has extreme high sensitivity for a wide range of read length and mismatches by a spaced seed strategy. It is capable of mapping 15× coverage of a human genome in 1 day [27].

Based on the algorithms, these programs can be classified into three categories [16, 28]. The first category is based on hashing table and can be further divided into two subcategories, either hashing the reads then using the reference genome to scan the hash table, such as RMAP [21, 22], MAQ [18], ZOOM [27], SeqMap [23], SHRiMP [24] (for the updated version 2, it hashes the genome [29]), and RazerS [20], or hashing the reference genome then using reads to scan the hash table, such as MOM [19], Novoalign, MOSAIK, and BFAST [13]. The second category of programs, such as Bowtie-v1 [15], BWA [16, 17], and SOAP-v2 [26], are based on the Burrows–Wheeler transform (BWT) [30]. These methods basically use the FM index data structure, proposed by Ferragina and Manzini, which introduces the concept that a suffix array is much more efficient if it is created from the BWT sequence, rather than from the original sequence [31]. The FM index retains the suffix array's ability for quick pattern search and is generally smaller than the input genome size [32]. Slider [25] is a representative aligner of the third category which is implemented by merge sorting the reference subsequences and read sequences.

Based on the assessment of 2×13,326,195 paired-end reads (SRR018643) sequenced on Illumina Genome Analyzer II from the untreated primary glioblastoma multiforme (GBM) sample (SRS004141) in a male (TCGA-06-0188) [33], BWT-based aligners beat the methods in the other two categories [34]. The previous assessment showed that Bowtie-v1 finished the job within two and a half hours, with over 67.5 % reads aligned and with a peak memory consumption of 2.9GB. Meanwhile, BWA needed 88 % more time to do the alignment but with 5 % more reads aligned (72.99 %) when the same 2-seed-mismatch MAQ-like policy was applied [34]. MAQ reached a better sensitivity of 72.0 %, but used 39 h for alignment, which is consistent with the previous comparison between MAQ and Bowtie on reads of 76 bp [15]. Due to the underlying Needleman–Wunsch (Novoalign) and Smith–Waterman (SHRiMP) exact search algorithms, they showed excellent sensitivity. SHRiMP had a sensitivity of 81.2 %, which was nearly 20 % higher than Bowtie, but it took 100-folds as much time as Bowtie did.

These eight programs were then tested on Illumina single-end data from the same GBM sample (15,578,118 single-end reads (SRR018725) with length of 76 bp)

Table 15.1 Performance assessment of eight NGS mapping tools on Illumina single-end sequencing data of SRR018725

Program	Version	Index time (h:m:s)	Peak memory footprint (gigabyte)	Alignment time (h:m:s)	Peak memory footprint (gigabyte)	Reads aligned (%)
Bowtie-v1	0.12.7	3:43:36	5.5	1:04:58	2.2	65.96
BWA	0.5.8c	1:46:42	1.5	4:21:04	2.3	56.39
SOAP-v2	2.20	1:45:54	2.3	4:35:24	5.3	61.17
RMAP	2.0.5	N/A	N/A	6:39:54	6.0	54.48
ZOOM	1.5.0	N/A	N/A	3:17:39	5.2	59.10
MAQ	0.7.1	0:01:56	0.34	15:19:43	7.3	67.58
Novoalign	2.07.06	0:06:28	13.5	71:43:07	13.1	73.32
SHRiMP	2.1.0	4:08:13	12	520:45:13	12	75.30

Table 15.2 Performance assessment of eight NGS mapping tools on Illumina paired-end sequencing data of SRR018658

Program	Version	Index time (h:m:s)	Peak memory footprint (gigabyte)	Alignment time (h:m:s)	Peak memory footprint (gigabyte)	Reads aligned (%)
Bowtie-v1	0.12.7	3:43:36	5.5	2:37:09	2.9	50.01
BWA	0.5.8c	1:46:42	1.5	6:58:12	5.0	57.22
SOAP-v2	2.20	1:45:54	2.3	10:25:11	6.8	49.75
RMAP	2.0.5	N/A	N/A	10:21:67	10.0	45.18
ZOOM	1.5.0	N/A	N/A	7:12:23	10.2	50.42
MAQ	0.7.1	0:01:56	0.34	39:07:24	8.1	63.18
Novoalign	2.07.06	0:06:28	13.5	138:45:32	13.1	68.45
SHRiMP	2.1.0	4:08:13	12	1051:41:15	12.0	71.23

in order to check whether aligners will perform differently on single-end data. Similar trends were observed for each aligner on running time, memory consumption, and sensitivity (Table 15.1). To test whether the aligners perform the same in normal sample, we ran aligners on 2×13,716,752 paired-end reads (SRR018658) with length of 76 bp each from blood-derived normal sample (SRS004142) of the same male (TCGA-06-0188) sequenced on Illumina Genome Analyzer II. The relative ranking of software on three measurements stays the same in both cancer and normal samples, indicating that the performances of the aligners are not affected by sample heterogeneity (Table 15.2). Aligners are also evaluated on the ABI SOLiD System 3.0 platform. One million single-end reads (SRR030482) with length of 50 bp from the serous cystadenocarcinoma sample (SRS004260) were used, and NovoalignCS showed the best alignment capability [34]. This previous study showed on average they have a lower proportion of mapped reads compared with the Illumina data. Though SHRiMP has extremely high sensitivity in SOLiD data, its alignment time is 600-folds more than Bowtie's.

2.1.2 Sequencing Depth, CpG Islands, Repetitive Elements, and Genomic Coverage

By mapping and pooling 13 runs of the GBM sample to the reference human genome hg18(NCBI build 36.1), we studied the relationship between sequencing depth and genome coverage [34]. It is obvious that the percentage of genome covered increases with the increase of sequencing depth. At 1× sequencing depth (1×=3.0 gigabases of human genome), less than 50 % of the genome was covered at least once, and less than 20 % was covered at least twice. At 10× depth, nearly 90 % of whole genome was covered, and 83 % was covered at least twice.

Moreover, the distribution of NGS coverage on human genome is uneven. CpG island regions showed a significant lower coverage than the whole genome and gene regions (both *p*-values less than 2.2e–16, by one-tailed *t*-test after 1,000 times bootstrap). At 10× depth, only 50 % of CpG islands were covered at least once, compared to 90 % for the whole genome, and the corresponding numbers were 20 and 50 %, respectively, at 1× depth [34]. With the fact that CpG islands are in 74 % of upstream promoters and 40 % of the downstream promoters of mammalian genes [35], the promoter and 5′UTR regions, which are important for regulatory roles of the genome, are also undercovered by the NGS compared with whole genome background (both *p*-values less than 2.2e–16, by one-tailed *t*-test after 1,000 times bootstrap). At 10× depth, only 83 % promoter and 76 % 5′UTR regions were covered at least once, compared to 90 % for whole genome, and the corresponding numbers were 42, 40, and 50 %, respectively, at 1× depth [34]. In these tests, CpG island regions were retrieved from CpG island table in CpG island track for hg18 through the UCSC genome browser; 5′UTR and gene regions were retrieved from RefGene table in RefSeq genes track for hg18 through UCSC genome browser. Promoters for each gene were defined as the region 5 kb upstream/downtream of the transcriptional start site of the gene.

Furthermore, we drew the histogram for the relationship between GC-content and window coverage under 10× depth by picking 10,000 windows with 1 kb length each from human genome and computing their GC-content and coverage (Fig. 15.1), whose pattern is consistent with the previous study [36]. On the basis of that histogram, the CpG island, promoter, and 5′UTR regions have average GC-contents of 68.6, 57.7, and 51.1 %, which are higher than the peak at GC-content of 45 %, so the GC-content and sequence coverage are inversely related. Specifically, we investigated all human promoters annotated in RefSeq, ranked them by GC-content, and picked 200 promoters for both top-ranked and bottom-ranked promoters each. It is demonstrated that the GC-rich promoters have significantly lower coverage than the GC-poor promoters (top, 0.10±0.10; bottom, 0.92±0.13; *p*-value, 1.15e–222, Fig. 15.2, by one-tailed *t*-test).

Besides region's GC-content, it is also reported that the repetitive elements in DNA sequences create ambiguities in alignment in NGS analysis [5, 37], which may cause the low NGS coverage in repetitive elements-enriched region. We ranked the promoters by the proportion of the promoter sequence covered by repetitive elements then did similar test. However, the *t*-test showed a significant higher

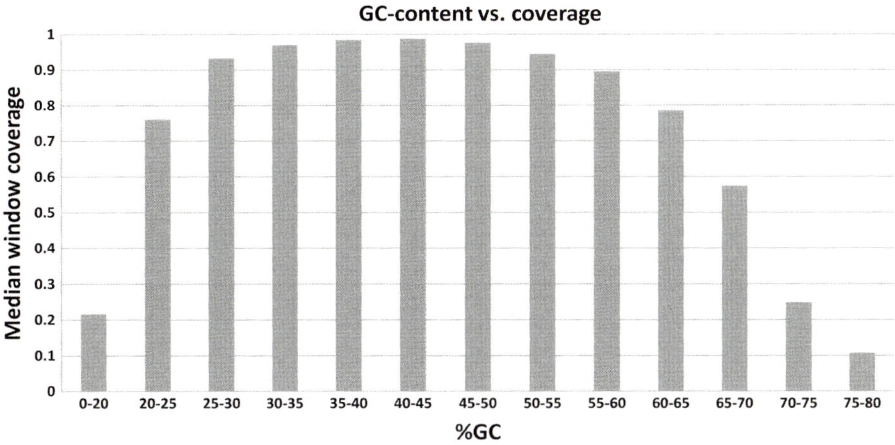

Fig. 15.1 The histograms of median per window NGS coverage categorized by GC-content bins under the condition of 10× sequencing depth. If GC-content is around 45 %, window coverage reach maximum

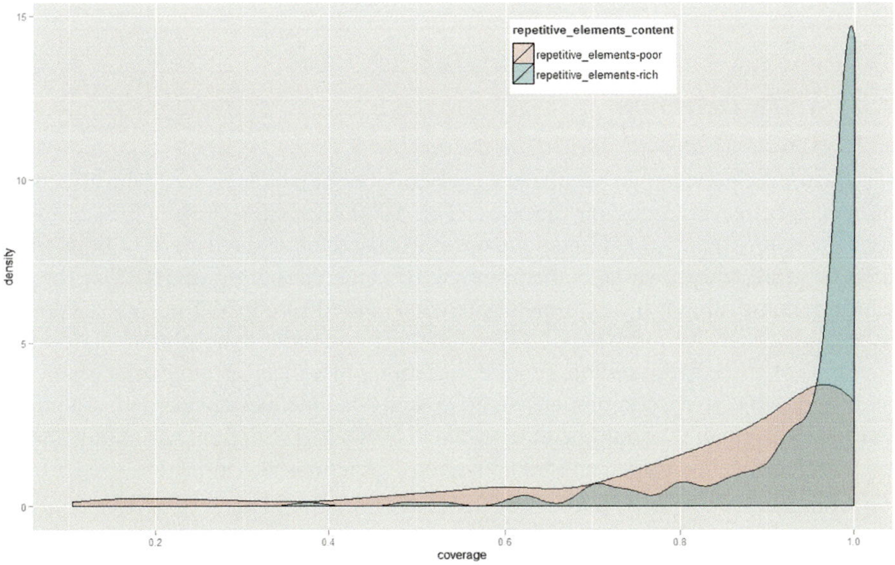

Fig. 15.2 Density histogram of NGS coverage for 200 GC-rich (*green*) and 200 GC-poor promoters (*red*)

NGS coverage for repetitive elements-enriched promoters (top, 0.94 ± 0.10; bottom, 0.83 ± 0.21; *p*-value, 7.33e–12, Fig. 15.3, by one-tailed *t*-test), which contradicts previous reports. The results indicated that the relatively higher GC-content and lower repetitive elements are the major causes for the lower coverage in regulatory regions.

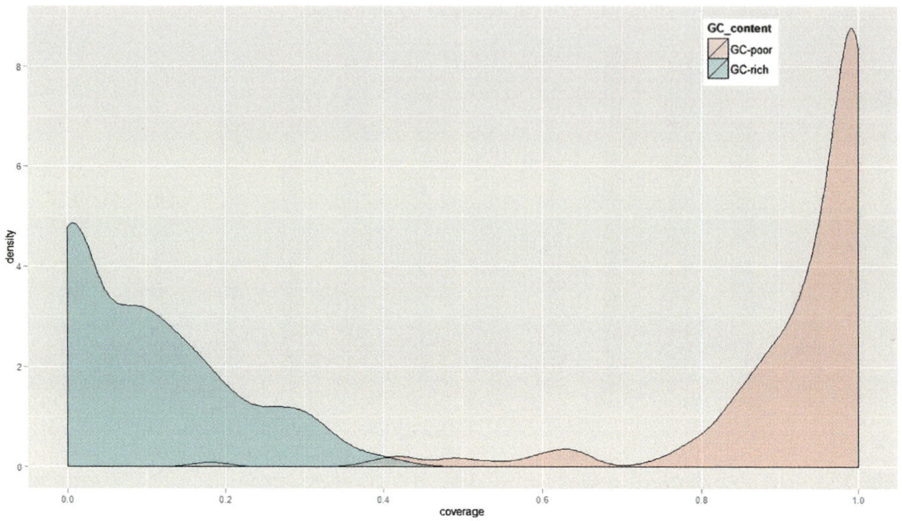

Fig. 15.3 Density histogram of NGS coverage for 200 repetitive elements-rich (*green*) and 200 repetitive elements-poor promoters (*red*)

2.2 SNPs Detection

It is becoming prevalent to sequence a large number of people to study the effects of human genetic variations on diseases. The 1000 Genomes Project [38] aims to sequence more than 1,000 human genomes to characterize the pattern of both common and rare genetic variants (http://www.1000genomes.org/), and TCGA (http://cancergenome.nih.gov/) has been sequencing paired tumor and normal tissue of many individuals for more than 20 types of cancer in order to accelerate our understanding of the corresponding genetic variations of cancer at molecular level. To realize the objectives, comprehensive NGS analysis pipelines are needed. An effective analysis pipeline should be able to detect SNPs and structural variations: insertions, deletions, inversions, and translocations. Deletions and insertions longer than 1 kb are usually defined as copy number variations (CNVs) [39], while insertions or deletions less than 1 kb are called indels. Among them, SNPs are thought to be the most common class of genetic variations because they have been reported to influence transcriptional regulation [40], alternative splicing [41], noncoding RNA regulation [42], and protein coding [43]. A large proportion of SNPs have been identified in the Human Genome Project [44, 45] and the International HapMap Project [46, 47], and their biological functions and roles have been studied extensively in various linkage analyses and genome-wide association studies (GWAS) [48]. Advances of NGS enable us to detect SNPs in individual or population with dramatically lower cost and less time compared with traditional hybridization-based or enzyme-based genotyping methods.

Many applications have been developed to detect SNPs from NGS data in recent years, including Pyrobayes, PolyBayes, MAQ, SOAPsnp, Varscan, SNVMix2, SeqEM, Atlas-SNP2, SliderII, FaSD, Bcftools, and GATK. Pyrobayes and PolyBayes recalibrate base calling from raw data and then implement a Bayesian approach that incorporates prior information with population mutation rates to detect SNPs [49, 50]. MAQ calls SNPs based on a Bayesian statistical model, which incorporates several features, including mapping qualities which measure the confidence that a read actually comes from the position it is aligned to, error probabilities from the raw sequence quality scores, sampling of the two haplotypes, and an empirical model for correlated errors at a site [18]. SOAPsnp is also based on the Bayes' theorem. It recalibrates the sequencing quality score to calculate the likelihood of genotype for each position with existing conversion matrix and then combines the prior probability to infer the true genotype in each position [51]. VarScan utilizes the uniquely mapped reads and then combines polymorphism events occurred in one locus and detected in multiple reads into unique SNPs/indels. For each predicted variant, VarScan determines the overall coverage, as well as the number of supporting reads, average base quality, and number of strands observed for each allele [52]. SNVMix2 combines three binomial mixture models to model allelic counts, nucleotide, and mapping qualities of the reads and infers SNPs and model parameters with expectation maximization (EM) algorithm [53]. SeqEM estimates parameters underlying the posterior probabilities with an adaptive approach rather than specifying an arbitrarily prior, which is implemented by using the EM algorithm to numerically maximize the observed data likelihood with respect to genotype frequencies and the nucleotide-read error rate [54]. Atlas-SNP2 is basically similar to SOAPsnp, but it infers systematic errors of base substitutions on single reads by fitting training datasets using a logistic regression model, which identified read sequence-related covariates to the base quality score [55]. For shallow coverage data, SliderII [56] is capable of utilizing information about known SNPs of a target genome as priors, to enhance its capability of detecting these known SNPs and novel SNPs and mutations in their vicinity. FaSD [57] combines an alternative score and a geometric mean of a mutation probability of reads to create an FaSD score to measure the polymorphism probability that a certain locus is an SNP location and to determine its corresponding genotype. Other SNP callers used in NGS analysis are integrated into pipelines, such as Bcftools [58] in SAMtools and UnifiedGenotyper in GATK [59]. Both Bcftools and GATK use Bayesian likelihood to infer the posterior probability of a locus being an SNP and to call the genotype.

2.2.1 Performance Assessment of SNP Callers

Performance Evaluation Benchmarked by SNP Array

In previous comparison [57], we selected five representative tools, SOAPsnp, MAQ, FaSD, Bcftools, and GATK, to do the performance test on datasets derived from the same GBM sample and its corresponding blood normal sample mentioned above

(10× aligned by Bowtie or BWA, respectively). With Affymetrix SNP array 6.0 and Illumina SNP Humanhap550 Genotyping BeadChip array from the same samples as gold standards, the area under the receiver operating characteristic curve (AUC), a widely accepted measurement of a classifier's performance without assigning specific cutoffs, was computed. For AUC calculations, the reference homozygote genotype was assigned as 0 and all other genotypes as 1. For SOAPsnp and MAQ, the Phred-scaled likelihood that the genotype is identical to the reference, which is also called "SNP quality," was assigned as the predictor. For FaSD, the FaSD score is used as the predictor. GATKs and Bcftools generate SNP calls in the VCF format, and the number in QUAL field can be used as predictor. FaSD had the largest AUC compared with the other software for all the categories of both the tumor and normal datasets, no matter which array was used as the benchmark. Because of the similar underlying algorithms, MAQ and SOAPsnp showed a nearly identical AUC performance. Here the performance of GATK and Bcftools in the lowest depth category [4, 5] is more than 20 % worse than the above software, which indicates their limitation for low-depth data in the AUC measurement.

Genotype concordance is also an important measure to compare SNP-calling tools and the SNP arrays with each other [60]. Looking at the normal dataset with either Bowtie or BWA as the aligner [57], the genotypes called by both arrays were very similar with concordance rates of more than 0.95 (0.997 for Bowtie and 0.957 for BWA), which indicates the high reliability of their usage as the gold standard. SOAPsnp and MAQ, again, showed high concordances of 0.997 with either Bowtie or BWA as the aligner. By comparing the results from the SNP-calling programs with those from the two arrays, we found that FaSD and Bcftools were the best methods. FaSD is better than Bcftools when Illumina array was used as the benchmark by a concordance of 0.882 for FaSD versus 0.865 for Bcftools when aligned by Bowtie and 0.833 versus 0.674 when aligned by BWA [57].

Performance Evaluation Benchmarked by High-Depth MAQ Call Set

The evaluations benchmarked on the SNP arrays are biased towards probes' loci because both Illumina and Affymetrix SNP arrays only contain a small portion of known SNPs (906,600 SNP markers for Affymetrix SNP array 6.0 and 555,000 for Illumina's Sentrix HumanHap550 Genotyping BeadChip), which is less than one tenth of true SNPs in human genome. To avoid this bias, we evaluated the performance of the software using sequencing data of chromosomes 21 and 22 from a Yoruba individual (NA19240) with average 35× whole genome depth from the 1000 Genome Project. Because MAQ has been proven to be highly accurate for high-depth data [18], we used the results of MAQ SNP calling on the complete dataset as our benchmark. For each chromosome, 10 % reads of the high-depth data were randomly sampled four times to form four test sets. MAQ, SOAPsnp, FaSD, GATK, and Bcftools were then used to call SNPs on these test sets, and these results were compared with the benchmark.

On the chromosome 21, Bcftools outperformed other software in 4–5 and 6–10 depth categories, while GATK was the best tool in 11–15 and 16–20 categories.

Table 15.3 The comparison of AUC of different tools on different sequencing depth categories

		4_5	6_10	11_15	16_20
21	FaSD	0.726±0.001	0.783±0.005	0.774±0.003	0.778±0.006
	SOAPsnp	0.689±0.001	0.638±0.003	0.599±0.002	0.578±0.007
	MAQ	0.698±0.001	0.639±0.003	0.604±0.004	0.589±0.006
	GATK	0.785±0.002	0.804±0.013	0.814±0.012	0.840±0.013
	Bcftools	0.807±0.005	0.840±0.011	0.784±0.020	0.791±0.010
22	FaSD	0.814±0.001	0.797±0.001	0.818±0.006	0.796±0.009
	SOAPsnp	0.699±0.001	0.686±0.001	0.637±0.010	0.627±0.006
	MAQ	0.699±0.001	0.686±0.001	0.640±0.010	0.632±0.005
	GATK	0.824±0.003	0.892±0.010	0.781±0.042	0.842±0.054
	Bcftools	0.822±0.007	0.817±0.024	0.764±0.092	0.739±0.102

The first number in each cell is the average AUC; the second number is the corresponding standard deviation. The result of MAQ in high sequencing depth dataset is applied as benchmark. The dataset is sampled from MAQ alignment of chromosomes 21 and 22 from NA19240. The average depth of test dataset in chromosomes 21 and 22 is 4×

However, the performance of FaSD was only a little worse than the above two tools and better than MAQ and SOAPsnp. On the chromosome 22, GATK and Bcftools outperformed FaSD in the 4–5 and 6–10 categories, while FaSD could beat other software in the 11–15 categories. Overall, GATK was the best tools for the SNPs that were not covered by SNP arrays in both chromosomes 21 and 22 (Table 15.3).

The majority of loci (>99.5 %) in the human genome have the genotype of AA, whose inclusion might overwhelm any differences, that is, the overall genotype concordance won't differ much among the various caller methods. Therefore, we used non-reference concordances [59] to assess the quality of SNP calling by excluding the concordant AA genotypes. In this measurement, FaSD had the highest non-reference concordance with the benchmark in chromosome 21 [57]. For chromosome 22, GATK had the best non-reference concordance with the benchmark. However, the non-reference concordances of both FaSD and Bcftools were only around 0.5 and 2.5 % lower than GATK, respectively [57].

Performance on Trio's Data

Sequencing data from multiple individuals in one family can improve variant discovery with the help of the heredity information and augmentation of the sequencing depth. Among the above tools, FaSD, GATK, and Bcftools have the function to utilize this kind of trio's data. We took one high-depth Yoruba trio's data from pilot 2 project of 1000 Genome Project (NA19238 (23×, chromosome 21 coverage), NA19239 (29×), NA19240 (35×)) and randomly picked 10 % of each individual's alignment on chromosome 21 and merged them together. We repeated the procedure for four times to form four pooled samples. Similarly, we used MAQ to call SNPs from full original chromosome 21 alignment file of each individual as gold standard. Compared with the individual call set in Table 15.3, we found the AUCs of trio's call set in chromosome 21 of NA19240 were significantly improved in the last two depth categories, for all three software (Table 15.4).

Table 15.4 GATK and Bcftools' performance of chromosome 21 on trio's data

		4_5	6_10	11_15	16_20
19238	FaSD	0.755±0.003	0.833±0.007	0.836±0.006	0.817±0.001
	GATK	0.779±0.003	0.860±0.003	0.881±0.004	0.814±0.001
	Bcftools	0.759±0.010	0.833±0.008	0.858±0.008	0.784±0.003
19239	FaSD	0.782±0.002	0.815±0.004	0.858±0.004	0.824±0.002
	GATK	0.760±0.002	0.830±0.004	0.890±0.004	0.852±0.001
	Bcftools	0.732±0.004	0.801±0.006	0.856±0.004	0.811±0.006
19240	FaSD	0.839±0.001	0.814±0.002	0.851±0.001	0.833±0.001
	GATK	0.765±0.001	0.793±0.003	0.842±0.002	0.843±0.002
	Bcftools	0.734±0.004	0.743±0.006	0.801±0.008	0.809±0.004

The first number in each cell is the mean of AUC, and the second number is the corresponding standard deviation in the depth category. The average depth of this merged dataset is 8.7×

Table 15.5 AUC performance for SNP callers in different regions

Region	MAQ	SOAPsnp	FaSD	GATK	Bcftools
CpG island	0.8429	0.8379	0.8723	0.7443	0.6441
Promoter	0.8865	0.8798	0.9015	0.7628	0.6719
5′UTR	0.8690	0.8527	0.9172	0.7477	0.6585
Genome	0.8872	0.8866	0.9646	0.7780	0.6952

2.2.2 Poorer SNP-Calling Performance in the Regulatory Region

Taken the fact that NGS has poorer coverage in regulatory region into account, we investigated whether SNP-calling performance follows that pattern. Affymetrix SNP probes with high-quality genotyping result were picked by discarding the probes with a confidence score above 0.018, which result in 583,891 high-quality probes. Among them, 98 % (575,765/583,891) were covered by 10× depth sequencing data (the 30GB GBM data aligned by Bowtie in 2.1.2), and they formed the final test set. In this test set, 711 SNP probes are located in CpG island, while 3169 in promoter and 1099 in 5′UTR region. No matter which tool was applied, AUC performance in CpG island related regions was significantly lower than the whole genome background (p-value <0.01 by bootstrap, Table 15.5).

2.2.3 Total Number of SNPs Called

Though MAQ, SOAPsnp, FaSD, GATK, and Bcftools have been already tested in above experiments, there is still a doubt on how many SNPs can be called for each software with default cutoff. Previous study [57] showed that FaSD called 69,768 SNPs in chromosome 21 and 78,240 SNPs in chromosome 22. The numbers are close to the SNPs called by the high-depth MAQ benchmark, which were 78,679 and 68,017, respectively. SOAPsnp and MAQ called 25–40 % more SNPs than the benchmark, whereas Bcftools and GATK called 40–50 % less SNPs than the benchmark.

Table 15.6 Numbers of SNPs called by each program from normal and tumor datasets at the whole genome scale aligned by Bowtie

Dataset\software	Bcftools	GATK	FaSD	MAQ	SOAPsnp
Normal	1,446,800	4,103,233	8,709,478	15,769,193	15,790,046
Tumor	1,391,247	4,387,713	11,382,453	16,244,577	16,259,369

The average depth of both normal and tumor datasets was 10×. Each tool's parameters are set as default

Table 15.7 Numbers of SNPs called by each program from normal and tumor datasets at the whole genome scale aligned by BWA

Dataset\software	Bcftools	GATK	FaSD	MAQ	SOAPsnp
Normal	1,954,900	3,296,263	13,757,146	18,196,240	18,197,348
Tumor	1,798,365	3,500,104	15,331,923	17,068,827	17,063,211

The average depth of both normal and tumor datasets was 10×. Each tool's parameters are set as default

The number of SNPs called by each tool in 10× depth GBM and normal data aligned by both Bowtie and BWA were also compared. Bowtie-v1 only assigns two values to mapping quality (MQ) field in alignment SAM files, 0 or 255, and this ignorance resulted in a huge inconsistency between the numbers of SNPs of data prepared by different aligners. As shown in Tables 15.6 and 15.7, GATK, which is sensitive to MQ, reported 25 % more SNPs in both tumor and normal samples when aligned by Bowtie. On the contrary, Bcftools and FaSD reported 30 % more SNPs when samples are aligned by BWA. MAQ and SOAPsnp showed a slightly variation in number of SNPs between samples aligned by Bowtie and BWA.

3 Summary

Following a mature NGS analysis pipeline (Fig. 15.4), we assessed 8 popular NGS aligner and 5 SNP caller. In this assessment, we found that Bowtie and NovoalignCS are the best tools for mapping short reads generated by Illumina and SOLiD platforms, respectively. Meanwhile, we should admit that a lot of aligners can run in multi-thread mode in practical, and hardware limitation may not be the barrier for biology groups. For example, the SHRiMP2 paper compares SHRiMP2, BFAST, BWA, and Bowtie's performance on artificial data for different variation cases while utilizing an 8-core 3.0 Ghz Intel Xeon machine with 16 Gb RAM [29]. For that case, SHRiMP2 showed an acceptable speed (20-folds slower than Bowtie) with significantly higher precision and recall rate. Thus, if the users would like to pay more attention to the reads, which contain relatively higher proportion of polymorphic loci, and parallel computation is affordable, this Smith–Waterman string matching algorithm-based aligner should be the first choice. Then by assessing SNP callers on cancer data in TCGA and trio's data in 1000 Genome Project, FaSD has been

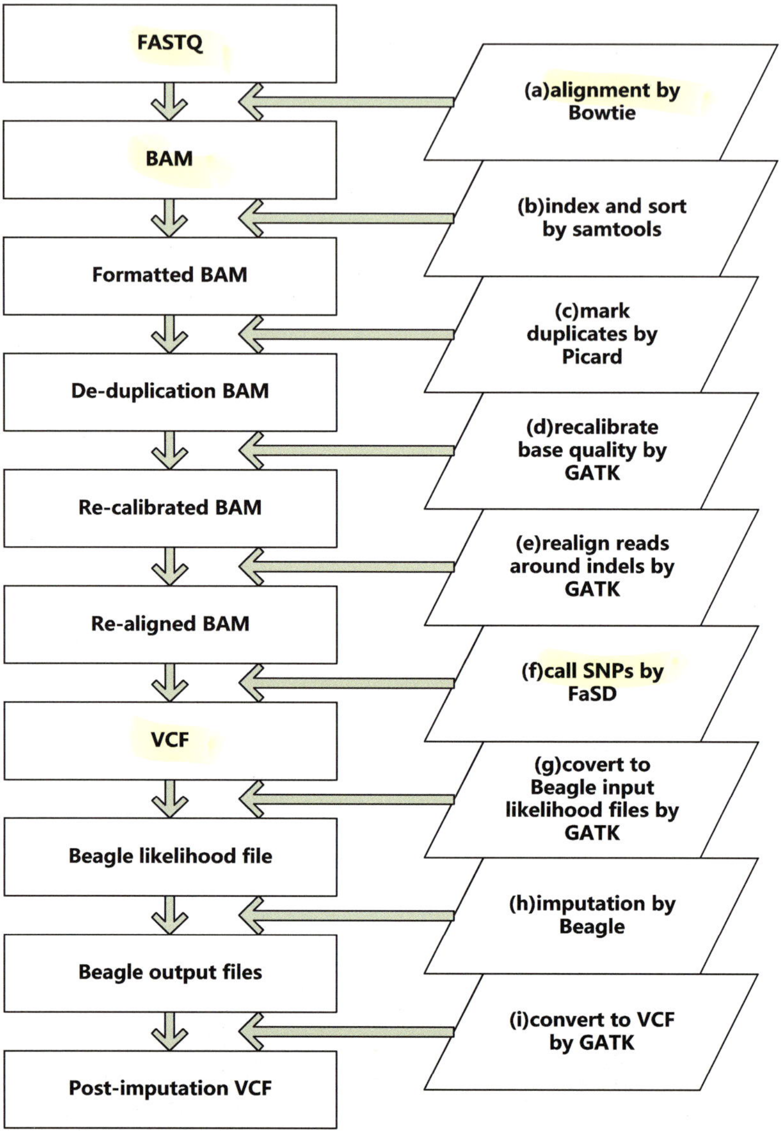

Fig. 15.4 The example of analysis pipeline for Illumina pair-end reads of pooled samples. Several of them may need to be modified for customized situations, such as single-end reads, data from individual, and data generated by other platforms

proven to be the best tool for SNPs discovery. Furthermore, we found that CpG-rich regions, such as promoter and 5′UTR, where regulatory elements are usually located, are poorly covered by the NGS platforms. For example, to cover 50 % of genomic region at least one depth, we need only onefold of the whole genome. However, to cover CpG island regions with the same criteria, we need tenfolds of the whole genome. Similar to mapping coverage, the CpG island, promoter, and 5′UTR regions have significantly lower SNP-calling performance than the genome and gene body regions no matter which SNP caller we used. This discovery raises the concerns for NGS technology, particularly when the corresponding studies focus on regulatory elements. NGS experiments for studying these regions should have higher sequencing depth than the normal genomic region.

References

1. Sanger F, Nicklen S, Coulson AR. DNA sequencing with chain-terminating inhibitors. Proc Natl Acad Sci U S A. 1977;74(12):5463–7.
2. Bonetta L. Genome sequencing in the fast lane. Nat Methods. 2006;3(2):141–7.
3. von Bubnoff A. Next-generation sequencing: the race is on. Cell. 2008;132(5):721–3.
4. Schuster SC. Next-generation sequencing transforms today's biology. Nat Methods. 2008;5(1):16–8.
5. Flicek P, Birney E. Sense from sequence reads: methods for alignment and assembly. Nat Methods. 2009;6 Suppl 11:S6–S12.
6. Mardis ER. The impact of next-generation sequencing technology on genetics. Trends Genet. 2008;24(3):133–41. Review.
7. Mardis ER. Next-generation DNA, sequencing methods. Ann Rev Genomics Hum Genet. 2008;9:387–402.
8. Li MJ, Sham PC, Wang J. Genetic variant representation, annotation and prioritization in the post-GWAS era. Cell Res. [Letter Research Support, Non-U.S. Gov't]. 2012;22(10):1505–8.
9. Pushkarev D, Neff NF, Quake SR. Single-molecule sequencing of an individual human genome. Nat Biotechnol. 2009;27(9):847–50.
10. Eid J, Fehr A, Gray J, Luong K, Lyle J, Otto G, et al. Real-time DNA sequencing from single polymerase molecules. Science. 2009;323(5910):133–8.
11. Altschul SF, Gish W, Miller W, Myers EW, Lipman DJ. Basic local alignment search tool. J Mol Biol. 1990;215(3):403–10.
12. Kent WJ. BLAT—the BLAST-like alignment tool. Genome Res. 2002;12(4):656–64.
13. Homer N, Merriman B, Nelson SF. BFAST: an alignment tool for large scale genome resequencing. Plos One. 2009;4(11):A95–A106.
14. Homer N, Merriman B, Nelson SF. Local alignment of two-base encoded DNA sequence. BMC Bioinformatics. 2009;10:175.
15. Langmead B, Trapnell C, Pop M, Salzberg SL. Ultrafast and memory-efficient alignment of short DNA sequences to the human genome. Genome Biol. 2009;10(3):R25.
16. Li H, Durbin R. Fast and accurate short read alignment with Burrows-Wheeler transform. Bioinformatics. 2009;25(14):1754–60.
17. Li H, Durbin R. Fast and accurate long-read alignment with Burrows-Wheeler transform. Bioinformatics. 2010;26(5):589–95.
18. Li H, Ruan J, Durbin R. Mapping short DNA sequencing reads and calling variants using mapping quality scores. Genome Res. 2008;18(11):1851–8.
19. Eaves HL, Gao Y. MOM: maximum oligonucleotide mapping. Bioinformatics. 2009;25(7):969–70.
20. Weese D, Emde AK, Rausch T, Doring A, Reinert K. RazerS-fast read mapping with sensitivity control. Genome Res. 2009;19(9):1646–54.
21. Smith AD, Chung WY, Hodges E, Kendall J, Hannon G, Hicks J, et al. Updates to the RMAP short-read mapping software. Bioinformatics. 2009;25(21):2841–2.
22. Smith AD, Xuan ZY, Zhang MQ. Using quality scores and longer reads improves accuracy of Solexa read mapping. BMC Bioinformatics. 2008;9:128.

23. Jiang H, Wong WH. SeqMap: mapping massive amount of oligonucleotides to the genome. Bioinformatics. 2008;24(20):2395–6.
24. Rumble SM, Lacroute P, Dalca AV, Fiume M, Sidow A, Brudno M. SHRiMP: accurate mapping of short color-space reads. Plos Comput Biol. 2009;5(5):e1000386.
25. Malhis N, Butterfield YSN, Ester M, Jones SJM. Slider-maximum use of probability information for alignment of short sequence reads and SNP detection. Bioinformatics. 2009;25(1):6–13.
26. Li RQ, Yu C, Li YR, Lam TW, Yiu SM, Kristiansen K, et al. SOAP2: an improved ultrafast tool for short read alignment. Bioinformatics. 2009;25(15):1966–7.
27. Lin H, Zhang Z, Zhang MQ, Ma B, Li M. ZOOM! Zillions of oligos mapped. Bioinformatics. [Research Support, N.I.H., Extramural Research Support, Non-U.S. Gov't]. 2008;24(21):2431–7.
28. Li H, Homer N. A survey of sequence alignment algorithms for next-generation sequencing. Brief Bioinform. 2010;11(5):473–83.
29. David M, Dzamba M, Lister D, Ilie L, Brudno M. SHRiMP2: sensitive yet practical SHort Read Mapping. Bioinformatics. 2011;27(7):1011–2.
30. Basti G, Perrone AL. A fast hybrid block-sorting algorithm for the lossless interferometric data compression. P Soc Photo-Opt Ins. 2003;5103:92–100.
31. Ferragina P, Manzini G. Opportunistic data structures with applications. Ann Ieee Symp Found. 2000:390–8.
32. Graf S, Nielsen FGG, Kurtz S, Huynen MA, Birney E, Stunnenberg H, et al. Optimized design and assessment of whole genome tiling arrays. Bioinformatics. 2007;23(13):I195–204.
33. Network TCGAR. Comprehensive genomic characterization defines human glioblastoma genes and core pathways. Nature. 2008;455(7216):1061–8.
34. Wang WX, Wei Z, Lam TW, Wang JW. Next generation sequencing has lower sequence coverage and poorer SNP-detection capability in the regulatory regions. Sci Rep. 2011;1:55.
35. Wang JW, Ungar LH, Tseng H, Hannenhalli S. MetaProm: a neural network based meta-predictor for alternative human promoter prediction. Bmc Genomics. 2007;8:374.
36. Harismendy O, Ng PC, Strausberg RL, Wang X, Stockwell TB, Beeson KY, et al. Evaluation of next generation sequencing platforms for population targeted sequencing studies. Genome Biol. [Evaluation Studies Research Support, N.I.H., Extramural]. 2009;10(3):R32.
37. Treangen TJ, Salzberg SL. Repetitive DNA and next-generation sequencing: computational challenges and solutions. Nat Rev Genet. 2012;13(1):36–46.
38. Genomes Project C. A map of human genome variation from population-scale sequencing. Nature. 2010;467(7319):1061–73.
39. Freeman JL, Perry GH, Feuk L, Redon R, McCarroll SA, Altshuler DM, et al. Copy number variation: new insights in genome diversity. Genome Res. 2006;16(8):949–61.
40. Kim BC, Kim WY, Park D, Chung WH, Shin KS, Bhak J. SNP@Promoter: a database of human SNPs (single nucleotide polymorphisms) within the putative promoter regions. BMC Bioinformatics. 2008;9 Suppl 1:S2.
41. Yang JO, Kim WY, Bhak J. ssSNPTarget: genome-wide splice-site single nucleotide polymorphism database. Hum Mutat. 2009;30(12):E1010–20.
42. Hariharan M, Scaria V, Brahmachari SK. dbSMR: a novel resource of genome-wide SNPs affecting microRNA mediated regulation. BMC Bioinformatics. 2009;10:108.
43. Ng PC, Henikoff S. Predicting the effects of amino acid substitutions on protein function. Ann Rev Genomics Hum Genet. 2006;7:61–80.
44. Venter JC, Adams MD, Myers EW, Li PW, Mural RJ, Sutton GG, et al. The sequence of the human genome. Science. 2001;291(5507):1304–51.
45. Lander ES, Linton LM, Birren B, Nusbaum C, Zody MC, Baldwin J, et al. Initial sequencing and analysis of the human genome. Nature. 2001;409(6822):860–921.
46. Frazer KA, Ballinger DG, Cox DR, Hinds DA, Stuve LL, Gibbs RA, et al. A second generation human haplotype map of over 3.1 million SNPs. Nature. 2007;449(7164):851–61.
47. Altshuler D, Brooks LD, Chakravarti A, Collins FS, Daly MJ, Donnelly P, et al. A haplotype map of the human genome. Nature. 2005;437(7063):1299–320.
48. Li MLJ, Wang PW, Liu XR, Lim EL, Wang ZY, Yeager M, et al. GWASdb: a database for human genetic variants identified by genome-wide association studies. Nucleic Acids Res. 2012;40(D1):D1047–54.

49. Quinlan AR, Stewart DA, Stromberg MP, Marth GT. Pyrobayes: an improved base caller for SNP discovery in pyrosequences. Nat Methods. 2008;5(2):179–81.
50. Marth GT, Korf I, Yandell MD, Yeh RT, Gu ZJ, Zakeri H, et al. A general approach to single-nucleotide polymorphism discovery. Nat Genet. 1999;23(4):452–6.
51. Li RQ, Li YR, Fang XD, Yang HM, Wang J, Kristiansen K, et al. SNP detection for massively parallel whole-genome resequencing. Genome Res. 2009;19(6):1124–32.
52. Koboldt DC, Chen K, Wylie T, Larson DE, McLellan MD, Mardis ER, et al. VarScan: variant detection in massively parallel sequencing of individual and pooled samples. Bioinformatics. 2009;25(17):2283–5.
53. Goya R, Sun MG, Morin RD, Leung G, Ha G, Wiegand KC, et al. SNVMix: predicting single nucleotide variants from next-generation sequencing of tumors. Bioinformatics. 2010;26(6): 730–6.
54. Martin ER, Kinnamon DD, Schmidt MA, Powell EH, Zuchner S, Morris RW. SeqEM: an adaptive genotype-calling approach for next-generation sequencing studies. Bioinformatics. 2010;26(22):2803–10.
55. Shen YF, Wan ZZ, Coarfa C, Drabek R, Chen L, Ostrowski EA, et al. A SNP discovery method to assess variant allele probability from next-generation resequencing data. Genome Res. 2010;20(2):273–80.
56. Malhis N, Jones SJ. High quality SNP calling using Illumina data at shallow coverage. Bioinformatics. 2010;26(8):1029–35.
57. Xu F, Wang W, Wang P, Jun Li M, Chung Sham P, Wang J. A fast and accurate SNP detection algorithm for next-generation sequencing data. Nat Commun. 2012;3:1258.
58. Li H, Handsaker B, Wysoker A, Fennell T, Ruan J, Homer N, et al. The sequence alignment/map format and SAMtools. Bioinformatics. 2009;25(16):2078–9.
59. McKenna A, Hanna M, Banks E, Sivachenko A, Cibulskis K, Kernytsky A, et al. The genome analysis toolkit: a MapReduce framework for analyzing next-generation DNA sequencing data. Genome Res. 2010;20(9):1297–303.
60. DePristo MA, Banks E, Poplin R, Garimella KV, Maguire JR, Hartl C, et al. A framework for variation discovery and genotyping using next-generation DNA sequencing data. Nat Genet. 2011;43(5):491–8.

Chapter 16
Applications of Very Low-Coverage Sequencing in Cancer Genomics: Copy Number, Virus Detection and Survival

Henry M. Wood

Abstract Since the advent of next-generation sequencing technology, much technological progress has been directed to increasing the depth of sequence coverage, either by improvements in chemistry or hardware to ever-increasing read numbers and lengths or by targeted capture of specific regions.

However there have also been developments in using lower-depth sequencing, mostly by utilising the aligned position of each read. Counting the frequency of aligned reads across the genome allows genomic copy number patterns to be measured. This technique has been shown to work using small amounts of DNA from fixed pathology archive samples and is amenable to high levels of multiplexing. A slight modification also allows viral infection in HPV-positive tumours to be detected, genotyped and quantified. Analysis of a number of tumours has allowed a sophisticated normalisation method to be developed which takes into account the amount of normal cells infiltrating a tumour sample and any tumour aneuploidy. This enables a more accurate definition of genomic gain and loss. This normalisation was applied to a cohort of early stage lung squamous cell carcinomas. No individual regions of gain or loss were associated with known clinical features such as stage, grade or length of survival; however, a pan-genomic index was developed which was linked to survival. This index is based on the total amount of genomic gain and loss seen in a tumour and the level of within-tumour heterogeneity, with loss and heterogeneity being linked to poorer survival.

Keywords Copy number • Next-generation sequencing • Lung cancer • Survival • Heterogeneity

H.M. Wood (✉)
Leeds Institute of Molecular Medicine, University of Leeds, Leeds LS9 7TF, UK
e-mail: h.m.wood@leeds.ac.uk

W. Wu and H. Choudhry (eds.), *Next Generation Sequencing in Cancer Research: Volume 1: Decoding the Cancer Genome*, DOI 10.1007/978-1-4614-7645-0_16,

1 Introduction

Since the development of next-generation sequencing technology, much genomic research effort has been split between two different complementary approaches. The larger genome centres have pushed forward whole genome sequencing, and recently technological developments such as the Illumina HiSeq and the advent of companies such as Complete Genomics have made this research accessible to laboratories with a more modest budget. In the meantime, many developments have been made in the field of targeted sequencing, either by the sequencing of PCR amplicons or custom gene capture or whole exome capture kits. This enables researchers to focus their efforts on a small subset of the genome at a much reduced cost. With both whole genome sequencing and targeted sequencing, the goal has always been to increase coverage. The higher the coverage, the more reliable any mutation detection will be.

Quite early in the history of this technology, a landmark paper [1] was published from which it became apparent that it was not necessary to use the sequence produced to search for point mutations, but that huge amounts of information could be generated by analysing the positions in the genome where those sequences aligned and, by implication, the structure of the genome. The authors used tens of millions of paired whole genome sequencing reads to find genomic breakpoints in lung tumour cell lines by looking for paired reads in which the pairs aligned to different genomic regions. They also examined copy number by counting how many reads fell into short "windows" along the genome and observing how the number of reads per window varied in proportion to the number of copies per cell of that region of the genome. Both these approaches are very important in the field of cancer research, as many tumours show substantial genomic rearrangements, some of which are known to be important in the oncogenic process [2–4].

This approach to detecting copy number changes using sequence read depth is known by various terms, but consensus in places such as the discussion forum http://seqanswers.com/ [5] seems to be moving towards "CNV-seq". It has been further developed by the implementation of various mathematical algorithms to improve breakpoint detection [6–8]. However, there was still the assumption that many millions of reads were used per sample, thus making the cost of analysing large numbers of samples prohibitive. All the published work at this point had also used optimal DNA samples, either as cell lines or fresh-frozen clinical samples.

A big step towards making this technology more usable in the clinical research community was the demonstration that Illumina libraries could be made from formalin-fixed paraffin-embedded (FFPE) samples [9]. Another was the adaptation of tagging procedures originally developed for amplicon sequencing [10] to whole genome library preparation to enable copy number detection by sequencing multiple libraries at very low coverage together on one lane of a sequencer [11]. As few as 1 million reads were shown to be sufficient to produce data comparable with array platforms, and the technique was shown to work with as little as 5 ng DNA from FFPE samples. Hospital archives typically contain vast repositories of biopsy and surgical samples from tumours and precancerous lesions, so these advances allowed modern genomic analyses to be performed with large, pre-existing sample sets. It should be stressed that since 1 million reads represent very low coverage, only one read for every 3 Kb of the human genome, the actual sequence cannot be used for mutational analysis at base pair level. It is only the position of each read, as determined by alignment, that can be used. The technique can be thought of as analogous to array data, with each genomic window being equivalent to one array probe, but using sequencing has a number of advantages. Firstly, sequencing appears to be more tolerant of smaller DNA amounts and the use of FFPE material. Arrays typically require microgram amounts of good-quality DNA. Using FFPE material with arrays often requires extra preprocessing steps [12]. When DNA quantities are reduced, it is often necessary to introduce a whole genome amplification step [13], which might introduce bias into the copy number detected. Nanogram amounts of DNA from FFPE material might not produce a library suitable for sequencing a genomic sample to 30× coverage, due to introduced sequence errors and PCR duplications, but as long as enough reads are produced which uniquely align to the genome, then the data is usually more than enough for copy number studies. Sequencing using tagged libraries also offers highly tailored resolution according to the research needs and budget available. The libraries can be reused a number of times and pooled at different levels of multiplexing. This allows the researcher to choose between a high-resolution experiment with a higher cost per sample and a low-resolution experiment with a correspondingly reduced cost. It could be envisaged that a large number of samples are pooled for a low-resolution scan and then the more interesting samples re-sequenced at higher coverage for a higher-resolution study. No extra DNA is needed for this second experiment, as the sequencing library can be reused.

2 Applications of Low-Coverage Sequencing in Clinical Samples

This technique has already been used in a number of studies. As a proof of principle study, tumour clonality and progression were demonstrated by examining sequential FFPE biopsies and surgical samples of a patient with recurrent multifocal oral cancer [11]. Between 55 and 270 ng DNA were used for 7 samples taken between 2007 and 2009. All samples showed some aberrations in common, most noticeable a gain of the p arm of chromosome 10. All samples taken from 2009 showed an additional 8 Mb gain in the distal portion of 9p.

A cohort of 44 FFPE head and neck squamous cell carcinomas were examined to investigate the link between human papillomavirus (HPV) status and genomic copy number [14]. Genomic copy number was calculated as before, by aligning the reads to the human genome and counting the number of reads falling into a number of genomic windows. HPV status was determined by realigning the same reads against all known HPV subtypes. This detection method showed good concordance with the immunohistochemistry techniques which are currently considered the "gold standard" for HPV detection [15]. Detection by sequencing has the advantage over other methods in that it provides a numerical estimation of the number of virus genomes per cell and the ability to distinguish between any number of viral species and subtypes. Indeed, one sample was found to harbour a rare subtype (HPV-61) which is undetected by standard diagnostic means. Looking at the genomic copy number results, it was shown that HPV-positive tumours showed a simpler pattern of aberrations, with less intra-tumour heterogeneity. There were also differences in the regions altered, most noticeably that 3p loss and 3q gain were seen in most of the HPV-negative tumours, but that only 3q gain was seen in most of the HPV-positive tumours. By taking into account the viral load of the samples, it was found that the HPV-positive samples with a low (1–5 copies per cell) viral load were more similar to the HPV-negative samples and that the samples with high viral load exhibited the HPV-positive karyotype patterns most strongly. These differences could be important. It has been suggested that treatment of HPV-positive patients be de-escalated due to their increased survival rates and to avoid the unpleasant side effects of treatment [16]. If it were to be shown from sequence data that HPV-positive patients could be further stratified into high-load HPV-typical genomes and low-load tobacco/alcohol-typical genomes, then more personalised treatments could be offered.

3 Normalisation of Copy Number

Over the course of analysing a number of samples, it became apparent that normalising samples according to mean read count was not satisfactory. At its most simple, the ratio of tumour and normal samples was normalised to 1:1, and gain or loss was called once this ratio crossed some arbitrary threshold, say 1.2:1 and 0.8:1, respectively. This failed due to the different ploidy and tumour cell content of each sample. For example, a sample with 100 % tumour cells such as a cell line will have 1:1 ratio for normal regions and 1.5:1 ratio for a gain of one copy, but a surgical sample with only 50 % tumour cells will go from 1:1 ratio for normal regions to a 1.25:1 ratio for regions of gain and 0.75:1 ratio for regions of loss. Ploidy has a similar effect. A diploid sample with a gain of one copy will go from a 1:1 ratio to a 1.5:1 ratio, whereas a triploid sample with a gain of one copy will go from a 1:1 ratio to a 1:1.33 ratio and a tetraploid sample will go from 1:1 to 1.25:1.

Combining the effects of ploidy and tumour cell content is a complicated process. There have been array-based methods such as CGHcall [17] which look to address these issues; however it was decided to develop a mathematical method specifically for sequencing data. This is because sequence data has some advantages over array data in that the data is a raw count. This allows the copy number to be directly proportional to the read count and means it cannot be saturated in the case of very high copy number and can achieve zero in the case of complete loss, which is not the case with arrays. A Bioconductor package called CNAnorm [18] was developed to address these issues. There are other copy number packages designed to work with sequence data; see the SEQanswers wiki (http://seqanswers.com/wiki/SEQanswers) [19] for an up-to-date list. At the time of writing, most of these are purely breakpoint-calling algorithms, which do not take into account normalisation. Several of these packages are designed to look for germ line mutations rather than the aberrations seen in cancer samples. Many of them expect higher coverage data and break down with only 1 million reads per sample. CNAnorm is specifically designed to work with the very low-coverage data obtained from multiplexing. It is not a breakpoint-calling method, but a normalisation technique, although it does call up the DNAcopy [20] algorithm from Bioconductor to call breakpoints. It can be used for germ line mutations, but it is mainly designed for the somatic effects seen in tumours. It first smooths the copy number data and looks for common copy number states across the genome. It then uses a semi-automatic process to determine which of these states corresponds to which copy number and what level of normal cell contamination is present. By doing this it ensures that all samples are properly normalised, so that analysis of gain and loss is not compromised by tumour cell content and ploidy. It makes a best guess as to the tumour cell content and underlying ploidy, but the user is able to alter this if other information such as FISH or a pathology report suggests otherwise.

4 CNV-seq to Study Survival in Lung Cancer

The third published account of using multiplex CNV-seq on FFPE samples [21] is the one which will take up the bulk of this chapter. Sequence data was produced from 89 early stage lung squamous cell carcinoma samples for which survival data was available. Strict admission criteria were enforced so that all patients had received similar treatment, no patients had any other cancer in the previous 5 years, and no patients whose death could be attributed to complications arising from curative surgery were used. The purpose of this study was to examine whether any copy number patterns could be observed which were predictive of survival, with the hope that eventually such patterns could be used to improve the staging and grading process which is currently used to plan treatment. The strict admission criteria ensured that noise was reduced and that no patterns seen to affect survival were in fact a proxy for other contributing factors such as tumour subtype or late versus early stage disease.

Historically, non-small cell lung cancer (NSCLC) has been treated as a single disease comprising three distinct major histological subtypes, adenocarcinoma (ADC), squamous cell carcinoma (SCC) and large cell carcinoma (LCC). It is only recently that it has been proposed that the subtypes should be treated differently [22] and, as such, the study of the individual subtypes is not as advanced as it might be. This study represents one of the largest genomic analyses to be published comprising purely the squamous subtype.

It was initially hoped that by analysing enough samples from a clinically homogenous group of patients, a few regions of the genome would emerge which would be linked to either good or poor survival. Then it would be simply a matter of finding the minimal region of gain or loss and identifying the genes present in that region to look for potential prognostic markers. This approach is very common for copy number studies. Indeed a PubMed search using the terms "copy number" will generally reveal at least one new study of this type every week for various tumour types.

Sequence data was obtained for 89 patients, with an average of around 1 million reads per patient. Reads were aligned to the human genome using BWA [23] and then split into equally sized genomic windows averaging 300 reads each. A normal control sample was constructed from a pool of 20 normal British individuals downloaded from the 1000 Genomes Project [24]. Examples of karyograms plotted from these data are shown in Fig. 16.1. The ratio of tumour to normal reads in each window was normalised according to mean read number per window and then again later by CNAnorm. A number of methods were then used to try and identify a genomic region or pattern of regions which associated with survival. Initially, frequency plots were drawn by pooling the data from patients with poor or good survival, or other clinical features, and counting how many patients exhibited gain or loss in any one region. A more sophisticated version of the same approach was to use the Bioconductor package KC-SMARTR [25] which looks for statistically different regions of gain and loss in copy number data. An alternative approach used was to draw heatmaps and use clustering algorithms to try and identify subgroups within the patients which could then be correlated to clinical features. Neither approach was able to split the patients into any meaningful groupings. The frequency plot/KC-SMARTR method showed that the patients' genomes appeared to gain and lose most regions seemingly at random and that a few regions such as 3p loss and 3q gain were almost universal. The clustering didn't split the patients into any distinct subgroups. To test the effectiveness of these methods, the lung samples were compared to another group of head and neck SCC patients. Both the locus-by-locus approach of KC-SMARTR and the clustering methods were able to easily split the samples into either lung or head and neck groups, showing that the lung SCC genomes did appear to have a very homogenous appearance distinct from that of other cancers.

Following the failure of these attempts to find individual genomic loci linked to survival, it was decided to adopt a whole genome method of analysis. This approach had previously been shown to work in breast cancer [26], when three distinct patterns of genomic damage ("simplex", "sawtooth" and "firestorm") were observed, and a genomic index, the "F"-stat, was calculated which linked total genomic damage to survival. These exact methods did not work on the lung SCC samples. The F-stat did not link to survival, and the genomes did not easily fit into the three classifications, mostly being in a continuum between "simplex" and "sawtooth". However, by analysing the data using a similar whole genomic approach, two new measures were calculated, which together had a significant effect on survival.

Fig. 16.1 (continued) (**b**) shows mostly gain and loss of entire chromosomes; (**c**) shows some chromosomal arm-size events and some smaller events; (**d**) similar in general to (**c**) but with a few local regions that are heavily altered; (**e**) almost every chromosomal arm demonstrates regions of loss and gain. For all samples show gain in 3q, either of the whole arm or localised to the distal portion

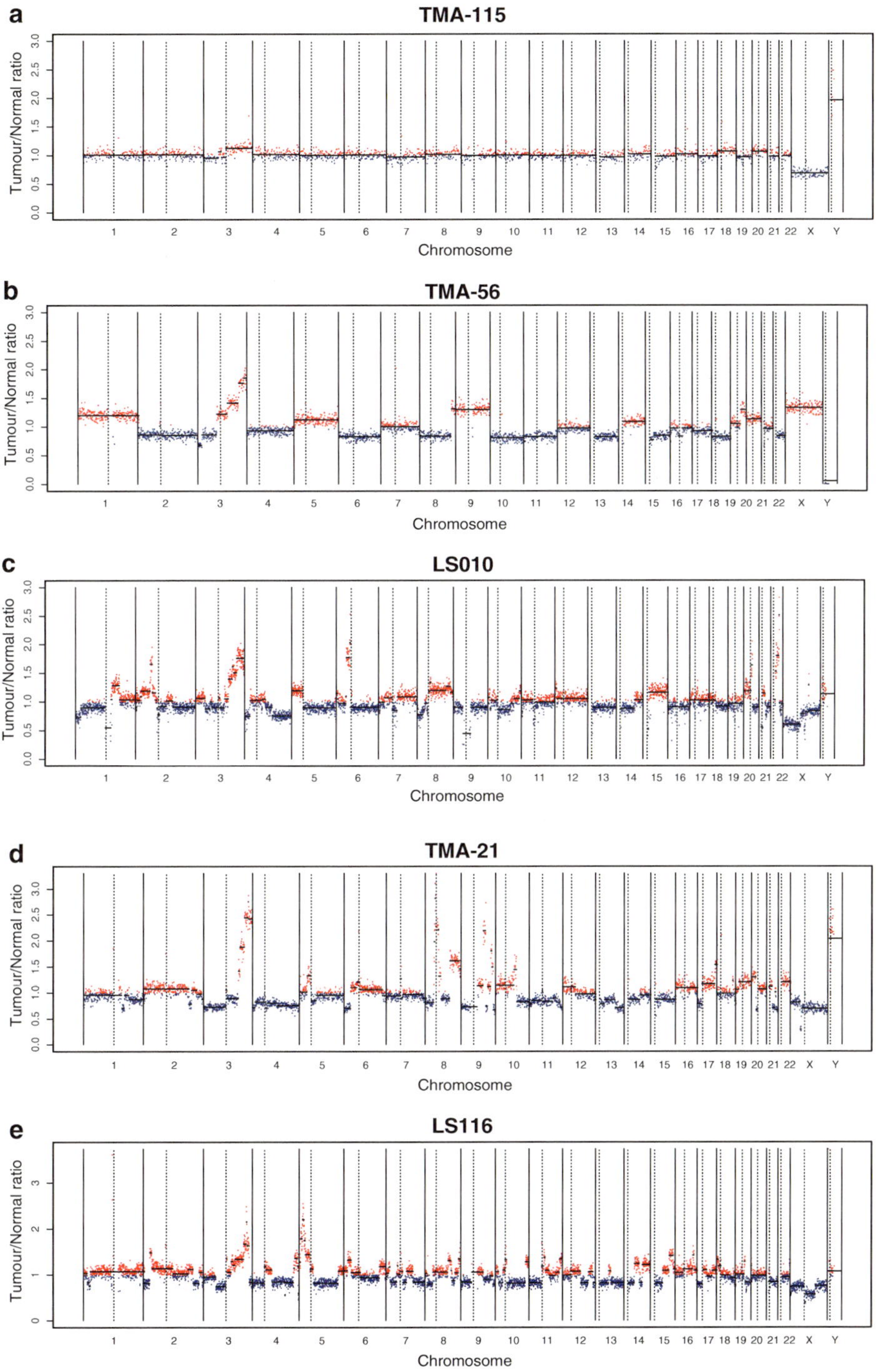

Fig. 16.1 Examples of karyograms from lung SCC patients showing some of the range of genomic damage seen between patients. *Red dots* are above the "normal" value, indicative of gain, whilst *blue dots* are below this value, indicating loss. (**a**) Shows a genome with very few aberrations;

The approach emerged from the realisation that the definitions of normal copy number, gain and loss were not always easy to make. When defining normal, assumptions have to be made. Sometimes, normal is assumed to be diploid, but this seems unreasonable if the vast majority of the genome is triploid or tetraploid. Normal could be assumed to be either the median or mean of all the copy number data. This is not ideal if a sample contains much more gain than loss, or more loss than gain, as it will skew the mean or median away from what is actually the normal. Once a normal has been decided, then gain and loss need to be defined. The most common method is to call gain and loss when the copy number goes past some threshold. CNAnorm allows a more sophisticated method by taking into account tumour cell content and aneuploidy. However all these methods break down if the tumour is highly heterogenous, containing many clones, each with their own copy number patterns.

Examples of the problems faced when normalising genomes and calling gain and loss are shown in Fig. 16.2. Examples 1–4 show some of the problems of defining what is normal. Example 1 shows a largely diploid genome with a simple gain. Mean, median and mode are all around 2, so they would all be acceptable for defining what is normal, as would the assumption of a diploid genome. Example 2 is similar but with more regions of gain. The extra gains have pushed both mean and median to 3 copies, even though very little of the genome is triploid. Most of the genome has 2 copies, with 4 copies being the next most common. For this sample, using the mean or median as the normal would result in most of the genome being called as loss. Example 3 is mostly triploid with a region of loss. Mean, median or mode will all give the normal as 3 copies. Assuming the diploid state to be normal will call most of this genome as gain. It is feasible to argue that a largely triploid sample has gained most of the genome, but when looking for disease-causing variations, it is not helpful to call the entire genome as a gain. Example 4 is largely triploid with regions of gain. The mean is a value between 3 and 4 copies. Using this value would result in the entire genome being called as either gain or loss. The median and mode are both at 3 copies. Example 5 is a very heterogenous sample. Using mean, median or mode results in a normal value of 3.5 copies of each genomic region. There is no meaningful way of calling gain or loss, because only a fraction of the cells exhibit any of the features seen. Imposing a threshold will clearly result in large numbers of false positives and false negatives.

Fig. 16.2 Examples of copy number plots from heavily stylised genomes showing how different normalisation strategies would affect calling of gain and loss and how the GH measure is calculated. For each plot, the copy number is on the *left*, and a density plot of the copy number is on the *right*. Each data point represents one window in a fictional genome on the x-axis, with absolute copy number on the y-axis. The mean copy number is shown as a *dotted line*, the median as a *dashed line* and the most common as a *solid line*. All data points with a copy number less than the mode are coloured *blue*, and all those with a copy number greater than the mode are coloured *red*. The G-stat is simply the proportion of the data points which are below the modal value (i.e. *blue*). The two highest peaks of the density plot, representing the two most common copy number states, are labelled with *arrows*. The H-stat is the relative height of these two peaks, as the height of the second highest divided by the height of the highest. The GH stat is $G \times (1 - H)$

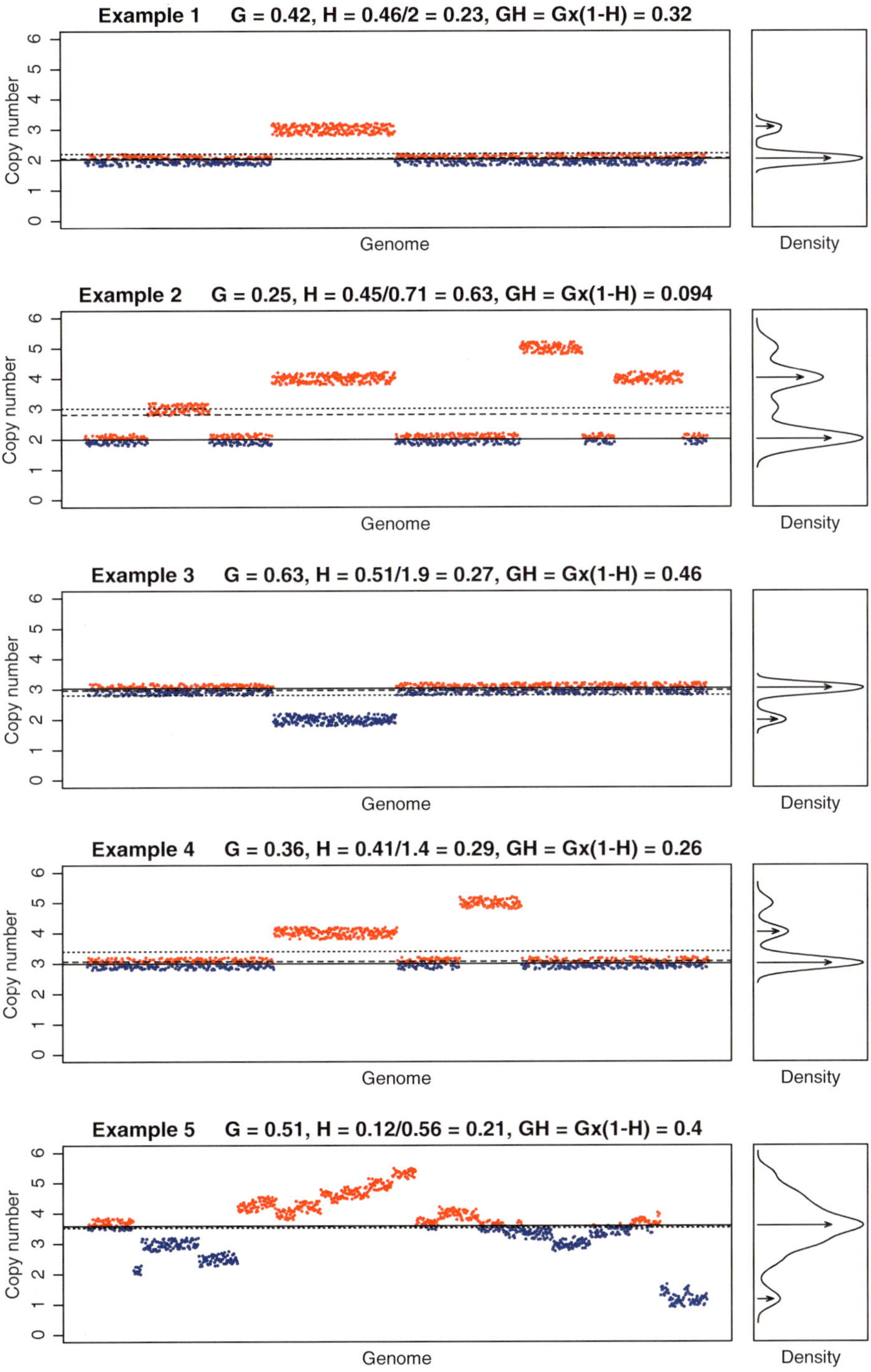
Example 1 G = 0.42, H = 0.46/2 = 0.23, GH = Gx(1-H) = 0.32
Copy number
0 1 2 3 4 5 6
Genome
Density
Example 2 G = 0.25, H = 0.45/0.71 = 0.63, GH = Gx(1-H) = 0.094
Copy number
0 1 2 3 4 5 6
Genome
Density
Example 3 G = 0.63, H = 0.51/1.9 = 0.27, GH = Gx(1-H) = 0.46
Copy number
0 1 2 3 4 5 6
Genome
Density
Example 4 G = 0.36, H = 0.41/1.4 = 0.29, GH = Gx(1-H) = 0.26
Copy number
0 1 2 3 4 5 6
Genome
Density
Example 5 G = 0.51, H = 0.12/0.56 = 0.21, GH = Gx(1-H) = 0.4
Copy number
0 1 2 3 4 5 6
Genome
Density

Whilst examining the genomes of the series of lung SCC series, most of the samples exhibited some degree of the heterogeneity demonstrated by example 5. This made calling regions of gain and loss for these samples quite problematical. However, it was noticed that the more problematical samples tended to exhibit poorer survival. It was also seen that those samples where a large amount of loss was seen, relative to the most common state, such as example 3, also fared worse.

It was decided to consider the most common copy number state as normal, and to make no calling of gain and loss, just measure the distribution of data away from this normal value. In a completely normal sample, even though the entire genome is normal, half of the data will be above the normal value and half below it, just due to the random distribution of reads. A tumour might be expected to have its own characteristic proportion of the genome above and below the normal value. Similarly, a homogenous tumour such as a cell line would be expected to have well-defined copy number states, and a highly heterogenous tumour would have a mixture of copy number distributions. Two measures were developed to report the proportion of the genome above and below the normal value and the amount of homogeneity of the copy number patterns.

These measures were calculated by using data produced during the operation of CNAnorm. One of the ways CNAnorm can attempt to normalise the copy number data across the genome is to produce a density plot of all the smoothed genome-wide copy number states. A normal sample will produce one peak; a sample with a mixture of normal and heterozygous deletion will produce two peaks. A sample with gains and losses will produce more peaks. The first measure calculated (the "G"-stat) is a measure of genomic loss. The peaks are examined, and the highest peak, representing the greatest proportion of the genome, is deemed to be the normal. The G-stat is the proportion of the genome which is below that value. A normal sample, or one with equal amounts of gain and loss, will have a G-stat of 0.5. A sample with more loss than gain will have a high G-stat, and a sample with more gain than loss will have a low G-stat.

The second measure (the "H"-stat) is a measure of homogeneity and complexity of genomic damage. It is the relative heights of the two highest peaks calculated by CNAnorm, representing the proportion of the genome in the most common abnormal copy number state compared with the proportion of the genome which is normal. A homogenous tumour with a simple pattern of gain and loss will have relatively well-defined peaks and therefore a relatively high H-stat. A heterogenous tumour, or one with a complex pattern of gains and losses, will have a large number of small peaks, or indistinct peaks, merging into each other. This will produce a low H-stat.

Individually, the G- and H-stats each showed a nonsignificant effect on survival, but by combining them by multiplying $G \times (1-H)$, a combined GH index was produced which had a significant effect on survival (examples of GH calculation are shown in Fig. 16.2). The samples were simply split into two halves according to whether they were above or below the median GH index. Kaplan-Meier estimates [27] were performed and compared using a log rank test [28]. Samples with a low GH index showed significantly better survival ($p=0.003$) than those with a higher

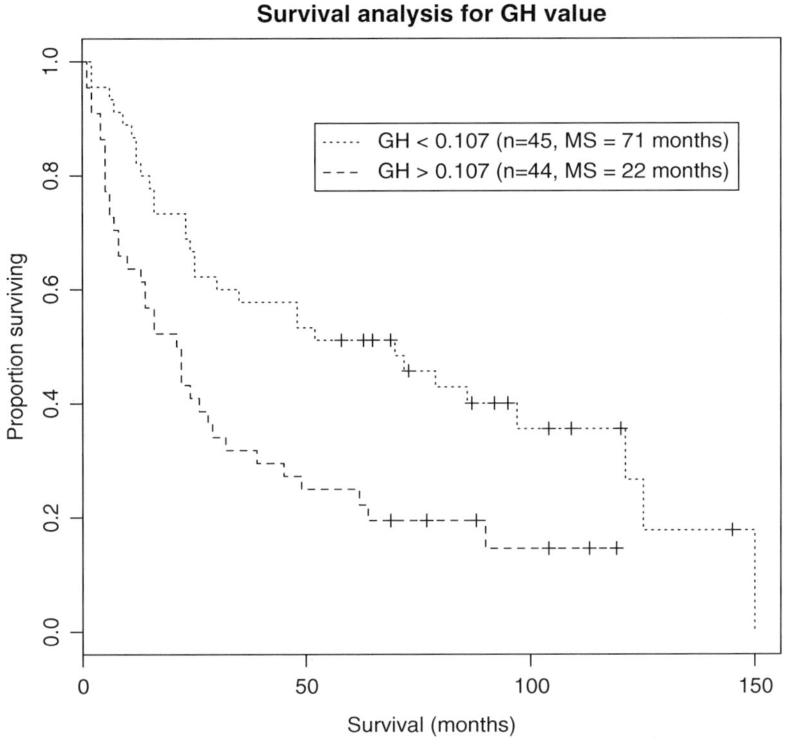

Fig. 16.3 Kaplan-Meier plot showing survival for patients with different GH values. The *dotted line* shows patients with a GH value less than the median value of 0.107 (i.e. relatively little loss and more homogenous genomes). The *dashed line* shows patients with a GH value above 0.107 (relatively more loss and more heterogenous genomes)

value, with a mean survival of 71 months compared to 22 months (Fig. 16.3). Further analysis demonstrated that neither G nor H correlated with any known clinical features, such as stage or grade, but that they were separate, independent prognostic markers. This was confirmed by multivariate analysis [29] which compared the GH index with the effect of age, stage, grade and other clinical features. GH emerged as having an independent, significant effect on survival ($p = 0.04$).

How this pan-genomic measure is able to predict survival in lung SCC patients and why the more traditional, locus-by-locus approach does not requires further study. Historically lung SCC patients have been studied alongside other NSCLC patients. Other studies which find regions apparently affecting survival might in fact only be discovering a region which is able to separate the different subtypes with their respectively different outcomes. Conversely, it might be the case that what is currently defined as SCC might in future be further subdivided into a greater number of subtypes and that within these subtypes, genomic regions do act as a marker for survival.

Better stratification of patients enables more personalised treatment and monitoring. Therefore any measure, such as the GH index, which can add information on top of the current TNM staging procedure, is to be welcomed.

It may be that this index is only valuable for lung SCC patients. Indeed the F-stat shown by Hicks et al. [26] to correlate with survival in breast cancer did not show the same pattern in these patients. Even if that is the case, hopefully the kind of pan-genomic thinking behind this approach and that of Hicks will become more widespread, especially in those tumour types where the locus-by-locus approach has so far proved unsuccessful. Too many studies simply look for regions of gain and loss, as defined by some arbitrary threshold. This is not a problem if gain and loss are easy to call, but when the effects of tumour cell content, different ploidies and clonal heterogeneity are ignored, then such calling can become increasingly meaningless.

How exactly G and H affect survival is currently speculation. G is a measure of genomic loss, with more loss being linked to worse prognosis. Tumours with more loss tend to be characterised by more regions of loss of heterozygosity, exposing tumour suppressor genes to haplo-insufficiency. It is likely that this will have a different effect than generating a tumour by gain or increased expression of oncogenes. Possibly loss is less disruptive than gain, so the cells are more normal and able to survive for longer in situ. H is a measure of homogeneity and complexity of genomic damage. Patients with simpler, more homogenous tumours tend to survive for longer. This could be because heterogenous tumours have more clonal variety so are more likely to be preadapted to withstand treatment or relocation to a distant metastatic site.

5 Future Developments

Future work can be divided into two broad categories: what is possible technologically at the moment and what is likely to be possible in the future. Finding validation data for the GH measure falls into the first category. This has not been possible so far for logistical reasons. Finding enough samples with clinically similar features has proved very challenging. Another study of similar size has recently been published trying to link copy number to survival in lung SCC [30], but these samples represent a mixture of early, late and metastatic disease, with treatment regimes unrecorded.

Other work in this field will most likely be driven by further advances in technology. All of the work presented here used data produced on an Illumina GAII machine. Producing copy number data using this method is approximately equivalent in price and quality to using an array. With the arrival of higher-throughput machines such as the Illumina HiSeq, it is possible to process many more samples for the same amount of money or to get much more data per sample. In a very basic way, this will make studies similar to those shown here to be much easier even for modestly funded laboratories.

This increase in throughput might also make it possible to better investigate what really lies beneath the rather simplistic G and H measures. G and H are derived from the mixtures of copy number states seen in DNA taken from large numbers of cells in a tumour sample. Recently methods have been published allowing copy number to be measured from large numbers of single cells [31]. The samples are separated into single cells by flow sorting. The DNA is then subjected to whole genome amplification before copy number for several cells is measured in a multiplexed HiSeq lane. If this approach were to be used on samples with various G and H scores, then their true nature could be revealed, with possible insights as to why they affect survival.

6 Conclusions

The rise of new sequencing technology is revolutionising cancer research. Whilst most research effort is focused on fully characterising genomes, exomes or regions at base pair level, there is much to be gained from very low-coverage sequencing. The position of the sequence reads, rather than the sequence itself, can be used to measure copy number and viral load across the entire genome. This method is flexible and suitable for use with small or FFPE samples. Using this technique, it has been possible to generate a genomic index linked to survival in lung SCC patients. This index does not relate to any individual region of the genome, but is a measure of the total amount of genomic gain, loss and clonal heterogeneity. It is perhaps unlikely that this measure will become part of a standard diagnostic or prognostic test in the near future, but it is hoped that it will lead to better understanding of the genomic factors which influence the progression of this disease. This in turn may lead to a better stratification of patients and more personalised treatment.

References

1. Campbell PJ, Stephens PJ, Pleasance ED, O'Meara S, Li H, Santarius T, et al. Identification of somatically acquired rearrangements in cancer using genome-wide massively parallel paired-end sequencing. Nat Genet. 2008;40(6):722–9.
2. Kallioniemi A, Kallioniemi OP, Sudar D, Rutovitz D, Gray JW, Waldman F, et al. Comparative genomic hybridization for molecular cytogenetic analysis of solid tumors. Science. [Comparative Study Research Support, Non-U.S. Gov't Research Support, U.S. Gov't, Non--P.H.S. Research Support, U.S. Gov't, P.H.S.]. 1992;258(5083):818–21.
3. Albertson DG, Collins C, McCormick F, Gray JW. Chromosome aberrations in solid tumors. Nat Genet. [Research Support, Non-U.S. Gov't Research Support, U.S. Gov't, P.H.S. Review]. 2003;34(4):369–76.
4. Rabbitts TH. Chromosomal translocations in human cancer. Nature. 1994;372(6502):143–9. Review.
5. Li JW, Schmieder R, Ward RM, Delenick J, Olivares EC, Mittelman D. SEQanswers: an open access community for collaboratively decoding genomes. Bioinformatics. 2012;28(9):1272–3.
6. Chiang DY, Getz G, Jaffe DB, O'Kelly MJ, Zhao X, Carter SL, et al. High-resolution mapping of copy-number alterations with massively parallel sequencing. Nat Methods. 2009;6(1):99–103.

7. Xie C, Tammi MT. CNV-seq, a new method to detect copy number variation using high-throughput sequencing. BMC Bioinformatics. 2009;10:80.
8. Yoon S, Xuan Z, Makarov V, Ye K, Sebat J. Sensitive and accurate detection of copy number variants using read depth of coverage. Genome Res. 2009;19(9):1586–92.
9. Schweiger MR, Kerick M, Timmermann B, Albrecht MW, Borodina T, Parkhomchuk D, et al. Genome-wide massively parallel sequencing of formaldehyde fixed-paraffin embedded (FFPE) tumor tissues for copy-number- and mutation-analysis. PLoS One. 2009;4(5):e5548.
10. Craig DW, Pearson JV, Szelinger S, Sekar A, Redman M, Corneveaux JJ, et al. Identification of genetic variants using bar-coded multiplexed sequencing. Nat Methods. 2008;5(10):887–93.
11. Wood HM, Belvedere O, Conway C, Daly C, Chalkley R, Bickerdike M, et al. Using next-generation sequencing for high resolution multiplex analysis of copy number variation from nanogram quantities of DNA from formalin-fixed paraffin-embedded specimens. Nucleic Acids Res. 2010;38(14):e151.
12. Hostetter G, Kim SY, Savage S, Gooden GC, Barrett M, Zhang J, et al. Random DNA fragmentation allows detection of single-copy, single-exon alterations of copy number by oligonucleotide array CGH in clinical FFPE samples. Nucleic Acids Res. 2010;38(2):e9.
13. Little SE, Vuononvirta R, Reis-Filho JS, Natrajan R, Iravani M, Fenwick K, et al. Array CGH using whole genome amplification of fresh-frozen and formalin-fixed, paraffin-embedded tumor DNA. Genomics. 2006;87(2):298–306.
14. Conway C, Chalkley R, High A, Maclennan K, Berri S, Chengot P, et al. Next-generation sequencing for simultaneous determination of human papillomavirus load, subtype, and associated genomic copy number changes in tumors. J Mol Diagn. 2012;14(2):104–11.
15. Smeets SJ, Hesselink AT, Speel EJ, Haesevoets A, Snijders PJ, Pawlita M, et al. A novel algorithm for reliable detection of human papillomavirus in paraffin embedded head and neck cancer specimen. Int J Cancer. 2007;121(11):2465–72.
16. Leemans CR, Braakhuis BJ, Brakenhoff RH. The molecular biology of head and neck cancer. Nat Rev Cancer. [Research Support, Non-U.S. Gov't Review]. 2011;11(1):9–22.
17. van de Wiel MA, Kim KI, Vosse SJ, van Wieringen WN, Wilting SM, Ylstra B. CGHcall: calling aberrations for array CGH tumor profiles. Bioinformatics. 2007;23(7):4.
18. Gusnanto A, Wood HM, Pawitan Y, Rabbitts P, Berri S. Correcting for cancer genome size and tumour cell content enables better estimation of copy number alterations from next-generation sequence data. Bioinformatics. [Research Support, Non-U.S. Gov't]. 2012;28(1):40–7.
19. Li JW, Robison K, Martin M, Sjodin A, Usadel B, Young M, et al. The SEQanswers wiki: a wiki database of tools for high-throughput sequencing analysis. Nucleic Acids Res. [Research Support, Non-U.S. Gov't]. 2012 Jan;40(Database issue):D1313–7.
20. Venkatraman ES, Olshen AB. A faster circular binary segmentation algorithm for the analysis of array CGH data. Bioinformatics. 2007;23(6):657–63.
21. Belvedere O, Berri S, Chalkley R, Conway C, Barbone F, Pisa F, et al. A computational index derived from whole-genome copy number analysis is a novel tool for prognosis in early stage lung squamous cell carcinoma. Genomics. [Research Support, Non-U.S. Gov't]. 2012;99(1):18–24.
22. Gazdar AF. Should we continue to use the term non-small-cell lung cancer? Ann Oncol. 2010;21 Suppl 7:vii225–vii9.
23. Li H, Durbin R. Fast and accurate short read alignment with Burrows-Wheeler transform. Bioinformatics. 2009;25(14):60.
24. Durbin RM, Abecasis GR, Altshuler DL, Auton A, Brooks LD, Gibbs RA, et al. A map of human genome variation from population-scale sequencing. Nature. [Research Support, N.I.H., Extramural Research Support, N.I.H., Intramural Research Support, Non-U.S. Gov't Research Support, U.S. Gov't, Non-P.H.S.]. 2010;467(7319):1061–73.
25. de Ronde JJ, Klijn C, Velds A, Holstege H, Reinders MJ, Jonkers J, et al. KC-SMARTR: an R package for detection of statistically significant aberrations in multi-experiment aCGH data. BMC Res Notes. 2010;3:298.

26. Hicks J, Krasnitz A, Lakshmi B, Navin NE, Riggs M, Leibu E, et al. Novel patterns of genome rearrangement and their association with survival in breast cancer. Genome research. [Research Support, N.I.H., Extramural Research Support, Non-U.S. Gov't Research Support, U.S. Gov't, Non-P.H.S.]. 2006;16(12):1465–79.
27. Kaplan EL, Meier P. Nonparametric estimation from incomplete observations. J Am Stat Assoc. 1958;53(282):457–81.
28. Miller RG, Gong G, MuÒoz A. Survival analysis. New York: Wiley; 1981.
29. Selvin S. Statistical analysis of epidemiologic data. Oxford: Oxford University Press; 2004. http://dx.doi.org/10.1093/acprof:oso/9780195172805.001.0001.
30. Lockwood WW, Wilson IM, Coe BP, Chari R, Pikor LA, Thu KL, et al. Divergent genomic and epigenomic landscapes of lung cancer subtypes underscore the selection of different oncogenic pathways during tumor development. PLoS One. 2012;7(5):e37775.
31. Baslan T, Kendall J, Rodgers L, Cox H, Riggs M, Stepansky A, et al. Genome-wide copy number analysis of single cells. Nat Protoc. [Research Support, Non-U.S. Gov't Research Support, U.S. Gov't, Non-P.H.S.]. 2012;7(6):1024–41.

Chapter 17
Tumor Phylogenetics in the NGS Era: Strategies, Challenges, and Future Prospects

Ayshwarya Subramanian, Stanley Shackney, and Russell Schwartz

Abstract Tumor phylogenetics is a strategy for interpreting the evolution of tumors using computer algorithms for phylogenetics, i.e., the inference of evolutionary trees. The approach takes advantage of a large body of phylogenetic theory and algorithms, developed primarily for inferring evolution among species, to interpret complex tumor data sets as evidence for evolutionary processes. The result is a tumor phylogeny, or phylogenetic tree, a reconstruction of the sequences of mutations that cells within a tumor or class of tumors accumulate over the course of their progression. The goals of finding such trees are to better interpret heterogeneity within and among tumors, identify and classify tumor subtypes, learn markers of progression for key steps in tumor evolution, and enable predictive modeling of likely tumor progression steps that may ultimately assist in diagnosis and treatment. With the rise of whole-genome sequencing data, the need for sophisticated models and algorithms that can make sense of such data has never been more acute. In this chapter, we cover the fundamentals of reconstructing tumor phylogenies with a special focus on next-generation sequencing data and discuss recent research, current trends, and challenges and opportunities currently facing the field.

A. Subramanian
Department of Biological Sciences, Carnegie Mellon University, Pittsburgh, PA 15213, USA
e-mail: ayshwarya@cmu.edu

S. Shackney
Intelligent Oncotherapeutics, LLP, Pittsburgh, PA 15243, USA
e-mail: sshackney@verizon.net

R. Schwartz (✉)
Department of Biological Sciences, Carnegie Mellon University, Pittsburgh, PA 15213, USA

Lane Center for Computational Biology, Carnegie Mellon University, Pittsburgh, PA 15213 USA
e-mail: russells@andrew.cmu.edu

W. Wu and H. Choudhry (eds.), *Next Generation Sequencing in Cancer Research: Volume 1: Decoding the Cancer Genome*, DOI 10.1007/978-1-4614-7645-0_17,

Keywords Phylogenetics • Evolution • Oncogenetics trees • Cancer progression • Tumor heterogeneity

1 Introduction

Cancer research has made tremendous progress in understanding the basic biology of tumors. One of the key insights that has informed work in this area is the recognition that a tumor is an evolutionary system, in which individual cells undergo a process of rapid mutation and selection leading to a progression in phenotypes and, typically, aggressiveness of the tumor [1]. As the data available for understanding tumors has grown in size and complexity, largely due to next-generation sequencing (NGS), it has increasingly become apparent that an understanding of the process of tumor evolution is necessary to make sense of these data. A more detailed understanding of the common pathways of tumor evolution could not only shed light on the basic mechanistic events driving tumor development but also provide powerful evidence to guide treatment strategies and predict progression. Phylogenetic algorithms have provided one powerful set of tools for drawing meaningful inferences from complex tumor data sets in terms of models of tumor evolution. In this chapter, we provide background on the key principles and methods of tumor phylogenetics and survey seminal results in the field. We then describe the emerging intersection between NGS and tumor phylogenetics. We conclude by considering implications of past work, current challenges, and possibilities for the future of tumor phylogenetics in the NGS era.

2 Principles of Tumor Phylogenetics

2.1 Understanding Tumors as Evolutionary Systems

Tumor evolution is generally understood to be a process of progressive acquisition of genetic or epigenetic abnormalities [2]. Our understanding of the details of this process has become more nuanced over time alongside our ability to profile tumors at the genetic level. Early mathematical models [3] in the field suggested that tumor progression was driven by a series of DNA mutations, an insight that helped lay the basis for the more nuanced two-hit hypothesis [4] of the requirement of pairs of mutations for tumor formation and later more sophisticated models of accumulated genetic abnormalities [5]. More recent tumor progression models have been increasingly informed by population genetic theory. A defining characteristic of tumors is hypermutability [6], providing a source of high diversity in genotype and phenotype, which may be reinforced by external environmental or hereditary factors [7], but is now generally understood that the tumor genome evolves following the principles of Darwinian selection [2, 8]. A framework for understanding these selective pressures was developed by Hanahan and Weinberg [9], who sought to categorize the specific functions for which tumor evolution selects. Work continues on more detailed mathematical models of tumors as evolutionary systems (cf., [10–12]).

An important consequence of this process of rapid evolution and selection is high heterogeneity both within and between tumors. Tumors are not homogeneous masses of cells but rather contain many distinct cell populations representing different stages or directions of evolution within single tumors [13, 14]. Debate exists within the field [15] as to whether this diversity primarily reflects evolution within tumors themselves (the clonal evolution hypothesis) or evolution within a separate population of progenitor stemlike cells (the cancer stem cell hypothesis). At the same time, the variability of this process from patient to patient also produces high heterogeneity between tumors. Genomic technologies have revealed, however, that even clinically indistinguishable tumors may show very different gene expression profiles [16–18] potentially predictive of different prognoses or responses to treatment [19, 20]. This recognition of distinct molecular mechanisms underlying diverse subtypes of tumors has in turn helped spawn the notion of targeted therapeutics [21], drugs that target defining abnormalities of specific tumor subtypes and, in the process, more selectively target tumor cells with reduced toxicity for healthy tissue (cf., [22]).

Much of the research challenge in oncology today concerns the difficulty of finding meaningful information to guide the development of diagnostics and therapeutics in the face of the enormous diversity within and between tumors. Identifying the meaningful features defining subtypes of tumors is ultimately a question of characterizing major pathways of evolution by which these subtypes develop. Distinguishing mutations causal for tumor progression that may make good therapeutic targets (known as *drivers*) from those that result at random from tumor hypermutability (known as *passengers*) is ultimately a question of distinguishing those mutations under selective pressure from those that are not. These are extremely challenging problems for a field that is suddenly awash in data, data that is complicated by the enormous diversity cell-to-cell and tumor-to-tumor as a direct result of these same evolutionary processes we seek to understand.

2.2 Phylogenetics Basics

Phylogenetics provides one answer to the problem of how the field of cancer research can draw meaningful inferences of the underlying evolutionary process of cancer from the large, highly heterogeneous data sets that confront the field. To help frame that discussion, we first provide a brief introduction to the field of phylogenetics, including the basic concepts and terminology needed for the remainder of this chapter. Readers interested in a more detailed grounding in phylogenetics can refer to any of many texts on the topic, such as [23, 24].

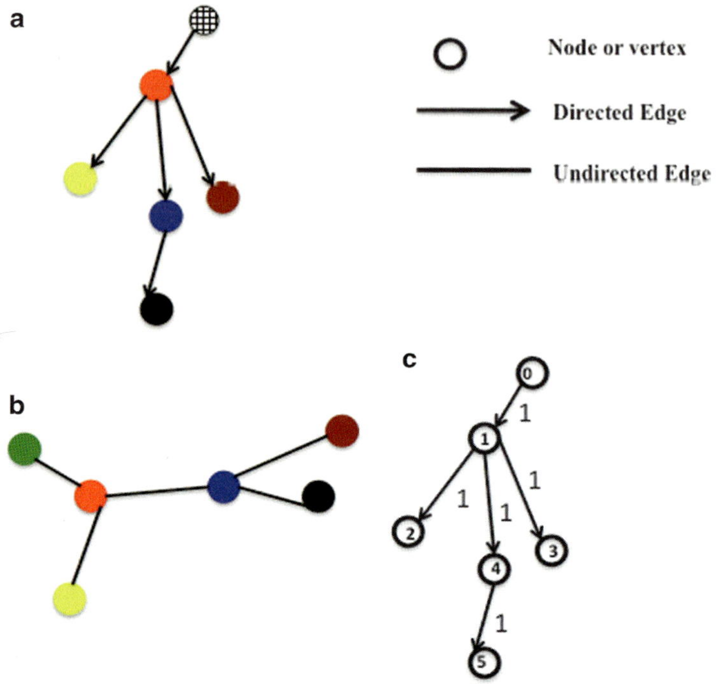

Fig. 17.1 Examples of phylogenies: A phylogeny is an acyclic graph consisting of nodes or vertices connected by edges. A phylogeny can have one of two forms: (**a**) rooted, where the checkered node is the root and all nodes below it are children, and (**b**) unrooted, where there is no root and hence no direction in the phylogeny. The edges may be weighted or unweighted. (**c**) An example of a weighted phylogeny where each edge has weight 1

A phylogeny is a representation of the evolutionary history of a set of species or organisms and their common ancestors. At the most basic level, a phylogeny is a tree, defined by a set of *nodes* (typically representing the species, or *taxa*, under study) and a set of *edges* connecting pairs of nodes (typically representing ancestral relationships between nearest relatives in the tree). Figure 17.1 provides examples of hypothetical phylogenies. A tree may have a single node known as the *root*, identified as the common ancestor of all other nodes in the tree. For any pair of nodes sharing an edge, the node that is higher in the tree (closer to the root) is called a *parent* node and that lower in the tree (farther from the root) is called the *child*. Nodes with no children are known as *leaves* and commonly represent observed members of a species.

Phylogenetics is a branch of computational biology that arose to infer phylogenies for collections of organisms. In order to infer a phylogeny, three components are required: (1) data describing evolutionary differences between the taxa under consideration, (2) an underlying model of evolution, and (3) an algorithm, or computational procedure, for building a tree given the data and model of evolution. Many variations exist on all three components of the phylogenetic inference process.

At a high level, phylogeny inference approaches are generally split into two categories depending on the kind of data they use: distance-based, where data is assumed to describe a measure of distance between or similarity of taxa and the goal is generally to find a tree consistent with pairwise distance data, or character-based, where data is assumed to be a discrete array of evolutionary features characterizing the taxa (e.g., specific DNA bases found in some conserved region of all taxa) and the goal is generally to find a tree describing specific evolutionary changes (mutations) along tree edges, explaining how the observed taxa might have evolved from a common ancestor.

Models of evolution typically follow from the type of data available. The simplest use the principle of parsimony, i.e., that simpler trees are more likely to be correct. For distance-based trees, parsimony leads to a model called minimum evolution [25], in which one seeks to minimize the total amount of evolutionary distance in the tree. In the character-based case, it is known as maximum parsimony, in which one seeks to minimize numbers or weights of mutations across the tree [26]. Parsimony models are often deprecated in favor of probabilistic models capturing more detailed quantitative representations of a model of evolution. Most commonly used are maximum-likelihood (ML) models, in which one seeks a tree that maximizes the probability of the tree having produced the observed data [27]. A more sophisticated alternative is Bayesian modeling [28], which uses a similar sort of probabilistic model but samples over possible ranges of model parameters to provide a more nuanced picture of uncertainty in the phylogeny inference.

Just as models must be designed to suit the available data, algorithms are determined by both data and models. Phylogenetic inference in all common models is computationally intractable, meaning that there are no known computer algorithms that can reliably find the most plausible phylogeny for a given data set and evolutionary model. In practice, then, a variety of heuristic methods are usually used. For distance-based phylogenetics, more tractable simplifications of the major models are generally used, such as the neighbor-joining approximation to minimum evolution [29]. Character-based models commonly use variants of Markov chain Monte Carlo (MCMC) methods, an approach for randomly searching sets of possible trees [23]. Trade-offs are often required between model realism and tractability, with ML often favored as a good compromise between more realistic but computationally difficult Bayesian models and more tractable but less realistic parsimony models. Many software packages are available for the major phylogeny algorithms, however, that make it easy to access effective phylogenetics without needing a deep understanding of the algorithmic theory behind it. Popular codes include PHYLIP [30], PAUP [31], RAxML [32], and MRBAYES [33].

2.3 Adapting Phylogenetic Methods to Tumor Evolution

The application of phylogenetics to tumor data is a promising approach to detailing the multiple interacting events underlying tumor progression. It relies on the key

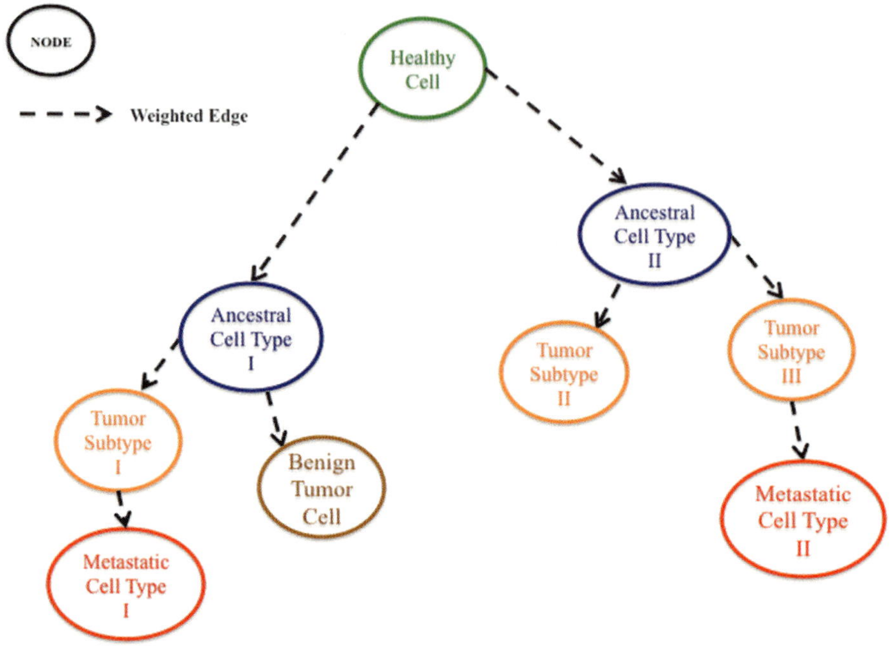

Fig. 17.2 A tumor phylogeny represents a model of possible trajectories of evolution through discrete stages of progression as a normal healthy tissue transforms into malignant tissue with increasing degrees of aggression. Here, the ovals represent nodes of a rooted phylogeny and the dashed arrows represent weighted edges. The weights can be changes in mutations or copy number variations or structural variations. The green circle, which is the root, represents a healthy cell and it can acquire genomic changes to attain either ancestral cell type I or II. Ancestral cell type I can further mutate to either acquire the status of tumor subtype I or become a benign tumor cell. Tumor subtype I can in turn mutate to acquire an aggressive metastatic cell type I. Similarly, ancestral cell type II can acquire distinct mutations to become either tumor subtype II or III. Tumor subtype III is more aggressive and can mutate to become metastatic cell type II

observations that cancer is at its heart an evolutionary phenomenon and a tumor is an evolving system, suggesting that computational methods for reconstructing evolutionary systems should in principle provide a way to reconstruct processes of tumor evolution (Fig. 17.2). The field of tumor phylogenetics was introduced by Desper et al. in 1999 in a pioneering paper [34] setting out the concept of tumor phylogenies, also called oncogenetic trees. Since this seminal work, methods for tumor phylogenetics have been extended to many data sources, evolutionary models, and algorithms. Figure 17.3 summarizes the basic steps of a generic tumor phylogeny pipeline, and Table 17.1 summarizes the major contributions in the field. Figure 17.4 provides a simple example of the possible output of such a tumor phylogeny

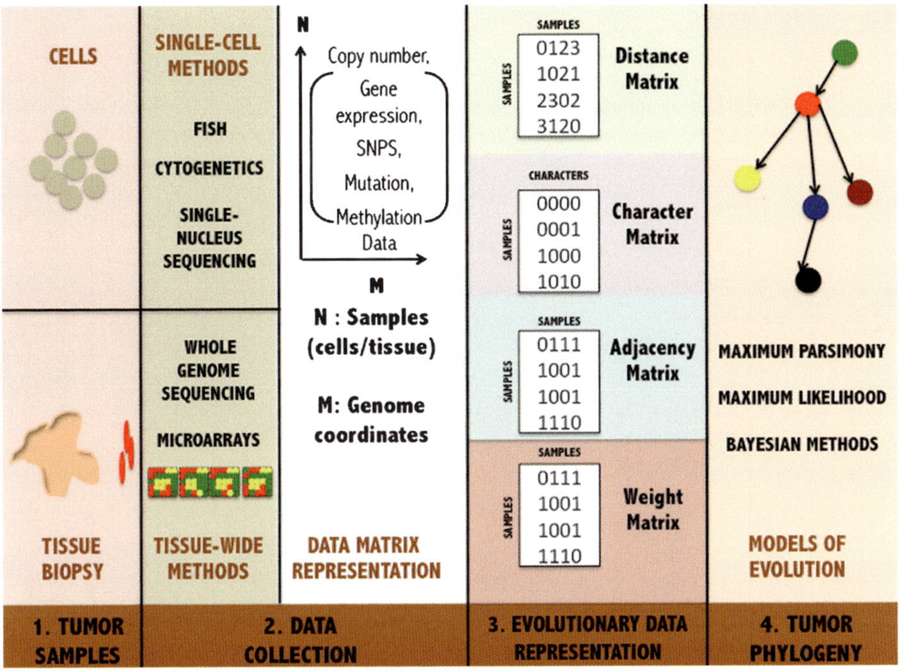

Fig. 17.3 A general tumor phylogeny pipeline. In *Step 1*, tumor samples are collected from patients. The samples are organized either at the single-cell resolution or the whole tissue resolution as biopsies or blood samples. In *Step 2*, evolutionary data is obtained from the raw tumor tissue sample using either single-cell based methods like FISH or single-nucleus sequencing or tissue-wide methods like microarray experiments or next-generation sequencing. In *Step 3*, the evolutionary data is processed into a phylogeny data matrix. The data format may vary depending on whether distance-based or character-based methods are used for phylogenetics. In *Step 4*, a tumor phylogeny inference algorithm supporting the underlying model of evolution is applied to generate the tumor phylogeny

inference, in this case an unrooted neighbor-joining tumor phylogeny built from a set of primary breast tumor samples with multiple sections from each of several individual biopsies [64].

The remainder of this chapter will survey the range of methods and research efforts in the field of tumor phylogenetics. It will begin with a historical look at the field, covering the spectrum of methods that emerged in the pre-genomic era. It will then examine how high-throughput sequencing has begun to change the practice of tumor phylogenetics. Finally, it will look to the future, examining the challenges for tumor phylogenetics in an era of ever greater availability of sequencing data as well as the opportunities this situation creates for tumor phylogenetics in the broader cancer research agenda.

Table 17.1 Summary of major contributions in the field of tumor phylogenetics

Approach no.	Tumor data retrieval technology	Molecular data type	Evolutionary data type	Model of evolution	Phylogeny algorithm	References
1.	Tissue wide: CGH	Chromosome copy number values	Matrix of weights (joint probabilities)	Oncogenetic tree model	Chow-Liu	[34, 35, 37, 44]
2.	Tissue wide: CGH	Chromosome copy number values	Matrix of weights (joint probabilities)	Minimum evolution	Fitch and neighbor (PHYLIP)	[36]
3.	Tissue wide: CGH	Chromosome break-point data	Matrix of patterns of correlations with noise correction	Oncogenetic tree model	Graph model	[41]
4.	Tissue wide: microarray	Gene expression data	Distance matrix	Minimum evolution	Neighbor-joining, weighted least squares, FastME	[46]
5.	Tissue wide: microarray	Gene expression data	Distance matrix	Minimum evolution	Neighbor-joining, weighted least squares, FastME	[46]
6.	Single cell: cytogenetic data	Copy number, structural rearrangement	Matrix of statistical dependencies	Maximum-likelihood estimation	Variant of Felsenstein 1981	[43]
7.	Tissue wide: CGH, loss of heterozygosity (LOH) scores	Copy number	Character-based feature matrix	K-mutagenic trees mixture model	Mtreemix	[65, 66]
8.	Tissue wide	Microsatellite data	Matrix of probabilistic weights	Minimum evolution	Maximum weight branching algorithm, distance-based	[38]
9.	Single cell: FISH	Copy number	Character-based	Minimum evolution	Minimum spanning tree inference	[59]
10.	Tissue wide: microarray	Gene expression	Character-based using polarity assessment	Maximum parsimony	Heuristic-based (PHYLIP)	[56]

11.	Tissue wide: CGH	Chromosome copy number values	Matrix of pairwise correlations	Oncogenetic tree model with error correction	Chow-Liu	[39]
12.	Tissue wide: microarray	Gene expression	Distance-based	Maximum-likelihood Bayesian model	Minimum spanning tree inference	[69]
13.	Regionally sectioned tissue wide: SPP	Copy number	Distance-based	Minimum evolution	Neighbor joining	[64]
14.	Tissue wide: microarray	Differential gene expression	Distance-based	Minimum evolution	Neighbor joining	[54]
15.	Tissue wide: aCGH	Copy number	Character-based	Maximum parsimony	Heuristic-based (PAUP)	[74]
16.	Single-cell genome sequencing	Copy number	Distance-based	Minimum evolution	Neighbor joining	[91]
17.	Genome and exome sequencing of regional sections	Mutation profile, copy number	Character-based	Based on clonal ordering	Empirically derived	[14]
18.	Deep sequencing of reads	Copy number	Character-based	Graph theoretical	Allelic and connectivity graphs	[88]

Note: Lists of the various tumor phylogenetic approaches sorted by categories of tumor data retrieval technology, the molecular data type, the evolutionary data types, the model of evolution, the phylogeny algorithms employed, and the corresponding publications

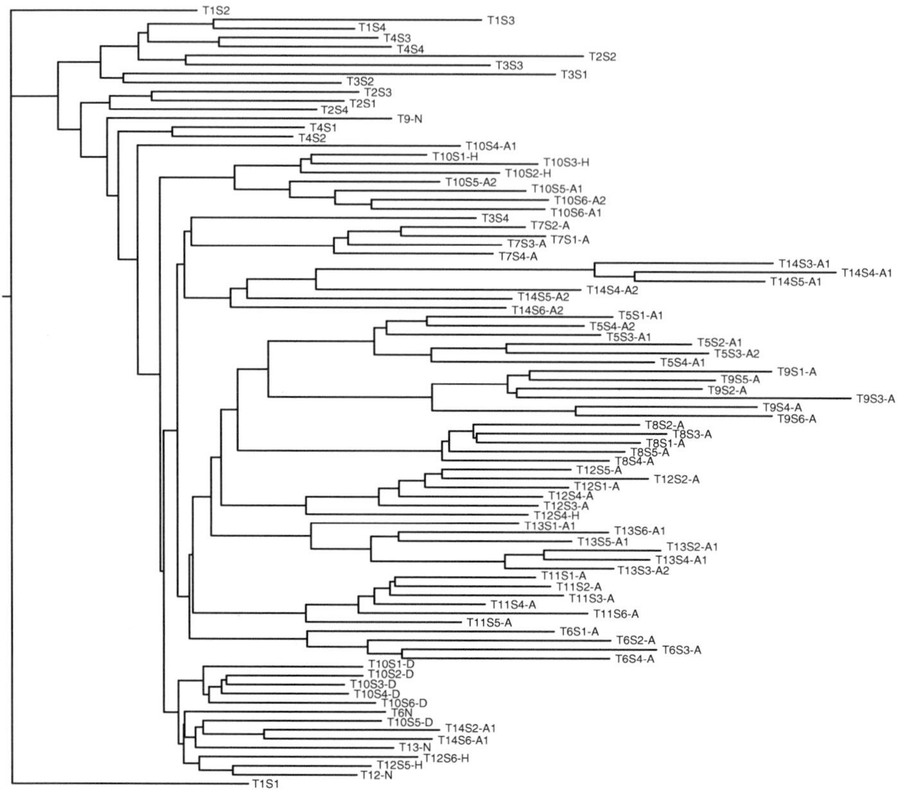

Fig. 17.4 An example tumor phylogeny. The data consisted of log DNA copy number ratios from aCGH experiments on 87 tumor sections derived from 14 primary ductal breast tumors [64]. Phylogenies were built using the phylogeny reconstruction software PHYLIP to build an unrooted neighbor-joining tree using Euclidean distances between log copy number vectors. In the tree, leaves are labeled by section (S) number and tumor (T) sample number (e.g., T1S1 indicates section 1 from tumor 1). N stands for normal and A stands for aneuploidy based on additional information on ploidy available in the primary reference for the data [64]

3 Tumor Phylogenetics in the Pre-NGS Era

3.1 Phylogenetics at the Tissue Level

The first approaches to tumor phylogenetics relied on a model of treating distinct tumors as species and seeking a tree among the tumors. Although tumors in distinct patients are not literally descended from common ancestors, they nonetheless can be explained by descent from common ancestors provided they undergo similar pathways of evolution. By observing how different tumors group into a tree, one can in principle identify tumors with similar molecular bases, identify sequences of events not apparent from examination of individual tumors, and project early decision points in the evolution of tumors.

The earliest approaches depended on pre-genomic methods for assaying tumor state. The original work of Desper et al. [34], for example, used comparative genomic hybridization (CGH) data probing copy number gains and losses of large genomic regions preselected for their relevance to tumor progression. Similar coarse-grained karyotyping approaches were employed by a variety of other studies on tumor phylogenetics [35–37]. More sophisticated, but still essentially pre-genomic, approaches emerged to allow a more nuanced portrait of tumor evolution, for example, through the use of microsatellite typing to detect allelic imbalances inaccessible to CGH methods [38]. Other studies have also explored how multiple heterogeneous types of marker could be combined in a single tumor phylogenetic profile, e.g., combining point mutations of target genes, epigenetic methylation markers, and microsatellite instability into a unified model of progression [39].

Several computational approaches were attempted for analyzing these pre-genomic data types. The earliest work on oncogenetic trees [34, 35] used a custom character-based likelihood model and adapted a classic combinatorial optimization algorithm [40] to find trees under this model. A number of theoretical results later improved on this approach, for example, by extending it to error-prone data [41, 42] and by developing more sophisticated probabilistic [43] and combinatorial [44] frameworks in which to model patterns of correlations among frequent mutations. An alternative approach using distance-based methods was soon developed to account for dependencies between mutational events and to bring to bear many sophisticated algorithms already available for distance-based phylogeny inference [35, 36, 45].

A significant advance in tumor phylogenetics came with the appearance of whole-genome data sets, predominantly microarrays, which allowed much larger marker sets as well as de novo discovery of significant progression markers. Expression microarrays were adopted for this task [46], with phylogenetic methods providing a generalization of clustering methods and then being applied for tumor classification [47–49] to give a finer-scale hierarchical classification as well as predictions of early stages along progression of distinct subtypes. A variety of other similar array data types have been used for studies of tumor classification, including SNP arrays [50], methylation arrays [51], and aCGH arrays [52].

The move to whole-genome data sets had important implications for algorithms for phylogenetics, primarily by encouraging a shift to distance-based methods better able to deal with the much larger numbers of markers available [46]. Additional algorithmic work has aimed to adapt phylogenetic approaches more specifically to tumor development. One novel problem is proper normalization of evolutionary distances, since large changes in global expression patterns do not necessarily correspond to large numbers of mutations, a problem partially addressed by measuring evolutionary distance by changes at the level of inferred expression modules rather than genes [53] and by establishing novel tumor evolutionary distance scales based on degrees of differentiation [54]. Another problem has been rooting tumor phylogenies, addressed through novel strategies for outgrouping, a common phylogenetic technique for determining the root of a tree by adding a species distant from those under study [55, 56].

3.2 *Phylogenetics at the Cellular Level*

An alternative approach to tumor phylogenetics arose from a somewhat different conception of the problem focused on building phylogenies of single tumors based on cell-to-cell heterogeneity within single tumors. The major source of such data for this strategy has been fluorescence in situ hybridization (FISH), generally used to probe copy numbers of specific genes, genomic regions, or chromosomes of interest. Such data can provide a detailed and accurate profile of copy number variations cell-to-cell in a tumor, allowing one to reconstruct profiles of fine steps in tumor development that would not be seen at the population resolution. Such data is, however, limited to small numbers of markers per cell, ranging from two in the earliest studies [13, 57] to eight probes per cell in the most recent [58].

Superficially, the algorithmic problem of phylogenetics from FISH data is similar to that of phylogenetics on other forms of discrete copy number data. Algorithms for this kind of data used similar forms of character-based phylogenetics to those of pre-genomic copy number studies on whole tumors, although with special handling of some kinds of evolutionary mechanism characteristic of tumor evolution, such as aneuploidy [59, 60]. In addition, the cell-level resolution resulted in treating inference of phylogenies within single patients as a separate problem from synthesizing phylogenies from distinct tumors into a model of common trends across a population. While the former was solved by phylogenetic tree-building algorithms, the latter required a simple variant of ideas from two other forms of common phylogenetic algorithm: supertree algorithms [61], which join simple trees into more complex models, and consensus tree algorithms [62], which identify common features among a set of trees.

These single-cell approaches provide a way around a major challenge facing tissue-level methods: turning intratumor heterogeneity from a confounding factor into a valuable source of information about likely pathways of fine-scale progression. They are, however, severely limited by the difficulty of probing more than a few markers of state in single cells. Such a limitation precludes discovery of novel markers and limits the complexity of models one can build, a significant disadvantage relative to the genomic methods available for tissue-level analysis.

3.3 *Bridging the Gap Between Tissue and Cell Data*

Tissue-level and cell-level analyses each bring advantages and disadvantages, prompting a search for new methods capable of bringing the benefits of both. Some experimental strategies were developed to better control for intratumor heterogeneity in phylogenetic studies. One strategy is to use cultured cell lines rather than primary tumors [63], providing a more uniform set of primary data from which to perform phylogenetics, although at the cost of having data that may be poorly representative of the diversity and heterogeneity found within and between primary tumors. Two important steps were made in an influential paper that advanced the importance of tumor heterogeneity to understanding progression [64]: more precise subsampling of biopsies by microdissection and the use of cell sorting to subdivide tumor cell populations by ploidy, each allowing generation of more uniform cell populations for phylogenetic analysis.

An alternative approach to bridging cell- and tissue-level analyses was computational inference of uniform populations from heterogeneous samples. A theoretical basis for such analysis was developed based on earlier pre-genomic assays of tumors by computationally modeling tumor data as products not of pure evolutionary trees but as mixtures of distinct samples from trees, known as mutagenic tree models [65–67], suitable for relatively small numbers of markers. These methods have been applied, for example, to reconstruct progression pathways from karyotyping data on meningiomas [68]. Mixture modeling emerged as a computational strategy for genome-scale analysis, in which one attempts to explain multiple raw array data sets as mixtures of cell types [69, 70] to be inferred computationally, an approach similar to one previously used to control for contamination of tumor data by normal cells [71] and applied for discovery of likely sites of origin of tumors [72, 73]. Further algorithmic advances adapted these initially distance-based methods into character-based by segmenting inferred mixture components [74, 75], enabling both phylogenetic inference and discovery of phylogenetically informative markers from whole-genome array data.

These hybrid approaches represented a step forward for tumor phylogenetic studies, enabling in principle the simultaneous study of whole-genome marker sets and cellular-scale evolutionary processes. In practice, however, highly noisy data sets, limited amounts of data, and uncertainty introduced at every stage of computational processing hinder the discovery of detailed, accurate models of tumor phylogenetics. In the face of high heterogeneity both within and between tumors, new advances were needed in data generation and data processing to build accurate and detailed models of tumor evolution.

3.4 *Profiling Tumors by NGS*

Next-generation sequencing offers the promise of massive throughput data, high resolution, and extensive detail. Large amounts of NGS data have been gathered in a series of major patient studies [76, 77], while continuing improvements in sequencing technology are making it easier and cheaper to incorporate NGS data into more patient studies. NGS data provides a way to mitigate several challenges facing earlier studies by allowing typing and discovery of markers at arbitrary resolution, providing more accurate quantitation of copy numbers, and allowing measurement of many kinds of data (copy number, mutation, structural rearrangement, and epigenetic) from the same basic technology. At the same time, NGS data create substantial new problems for phylogenetic analysis due to computational and statistical challenges of handling the much larger marker sets NGS produces and the difficulty of interpreting some forms of NGS data (e.g., structural rearrangements) phylogenetically.

Despite these challenges, phylogenetics from NGS data is not in principle substantially different from that with pre-NGS genomic data types. Additional preprocessing is, however, required. Typically, a first step is mapping NGS data to a reference genome. Data spanning different kinds of variation will then require custom preprocessing. The most common approach for dealing with NGS copy number data has been to bin the DNA sequence data into windows and then determine copy numbers for the windows. Several algorithms have been published that delve deeper into the nuances of segmenting the genome data [78, 79]. For structural rearrangement information, paired-end sequencing is commonly needed to reveal sources of novel chimeric sequences [80–82]. Processing mutation data follows the model of haplotype inference in population data where after alignment to the genome, genotypes are inferred and variants identified [83, 84]. RNA-seq follows a similar approach of measuring counts at individual base locations to estimate fine-scale expression levels [85–87].

4 Phylogenetics on Whole Tumor Sequencing

Surprisingly, few NGS tumor phylogenetic studies have yet been published given the amount of data now available. NGS has, however, already enabled some important new directions in tumor phylogenetics. For example, NGS has made it possible to reconstruct rearrangement phylogenies [88], revealing an incredibly complex and diverse landscape of rearrangements across tumor types that was invisible to prior genomic methods. As NGS has helped reveal the extent of genomic aberrations, it has also made apparent the need to deal with intratumoral heterogeneity. Advances in profiling tumor state from the pre-NGS era have thus been adapted and extended for NGS data. Regional sectioning and ploidy profiling techniques similar to those that began to reveal tumor heterogeneity at the whole-genome level in the pre-NGS era [65] have been developed for NGS studies. An important recent study in NGS tumor phylogenetics [14] showed some of the power of these methods through NGS studies of regionally sectioned renal carcinomas, revealing a complex landscape characterized by high intratumor heterogeneity. Clonal ordering [89], a simple model of tumor evolution, makes it possible to reconstruct evidence for multiclonal progression pathways from such data. Such regional profiling studies provide substantial power to characterize heterogeneity beyond that visible at the whole tumor level but nonetheless provide only a partial solution to the problem of separating heterogeneous signals in even small subregions of tumors. Single-cell studies [57, 90] have shown that far greater heterogeneity is present than can be resolved by even a fine-grained microdissection.

A key advance for tumor studies then has been the development of single-cell sequencing. By sequencing individual cells from single tumors, it becomes possible to systematically assay large numbers of markers of varying types within single tumors. The first approaches to this task made it possible to reconstruct likely tumor phylogenies largely from relatively coarse-grained data on aneuploidy of individual cells [91]. Further studies have extended the range and quality of the data, for example, by allowing one to recover nearly complete spectra of point mutations from which it is possible to reconstruct detailed models of progression of single tumors [92, 93]. Single-cell sequencing technology has been applied in other contexts also to single-cell expression profiling [94], although not yet to our knowledge to tumor phylogenetics.

Despite the huge promise of single-cell sequencing, there remain substantial technological gaps before we can truly characterize tumor genomes accurately at the single-cell level. A key problem is the need for single-cell genome amplification, which introduces biases, both systematic and random, in measured copy numbers, and potentially nonspecific amplification of contaminants and introduction of chimeric sequences that might be confused with true translocations [95]. Technological improvements such as multidisplacement amplification (MDA) [96, 97] are reducing the systematic bias in amplification over early whole-genome amplification (WGA) approaches [91], but cannot overcome the inherent randomness of the process. Furthermore, adequate scalable computational methods are still lacking for key problems of interpreting single-cell NGS data, such as reconstructing detailed genomic rearrangement phylogenies.

5 Challenges and Future Directions

As this chapter has argued, oncogenesis is at its heart an evolutionary process and accurate characterization of common tumor progression pathways through phylogenetics provides a powerful method for characterizing this process and its many variations. Tumor phylogenetics is a young field but has advanced dramatically since its development due to improvements in models and algorithms for phylogenetic reconstruction of tumor data and advances in technologies for gathering the data itself. Perhaps no development has been more promising in this direction than the availability of fast, inexpensive sequencing through NGS. Phylogenetics, in turn, provides a key tool for making sense of the incredible complexity NGS tumor data is beginning to reveal. Nonetheless, we remain far from being able to truly reconstruct how even a single tumor has progressed, much less building a collective profile of the major pathways of progression across tumor populations and translating that knowledge into improved patient outcomes. In this final section, we survey some of the major challenges remaining at present and prospects for overcoming them.

5.1 *Tumor Heterogeneity and Single-Cell Sequencing*

The progress and challenges in tumor phylogenetics let us identify some features we would ideally want in a technology for profiling tumor state. Given the high heterogeneity between tumors, it should be able to produce a complete profile of potential mutation events without bias. It should further be able to accommodate heterogeneous forms of data, including copy number variations, single nucleotide changes, structural rearrangements, and epigenetic modification, as well as ancillary reporters, such as changes in gene expression. Given heterogeneity within tumors, it should be able to distinguish these variations at the level of single cells. Furthermore, it should have high enough throughput and low enough cost to be able to assay large numbers of cells in large numbers of patients.

As we have seen, NGS is approaching this ideal data source, particularly with the advent of single-cell NGS, but nonetheless faces substantial technical challenges. At present, high genome coverage is achievable, leading to relatively thorough and accurate typing of point mutations [92, 93]. In principle, similar approaches could provide acceptable data with current technologies for epigenetic modifications. Accurate quantization remains a problem, however, a challenge for studying copy number variations in tumors. Furthermore, reconstruction of rearrangements from tumor data is still in its early stages, and it remains unclear what magnitude or quality of data it will require at the single-cell level. Scales of studies still remain orders of magnitude smaller than would be ideal, with major studies either typically profiling hundreds of patients at a tumor-wide level [77] or profiling on the order of a hundred cells in individual patients [91–93]. Detailed models of major progression pathways and accurate statistics for identifying common features of major progression pathways are likely to require at least the order of hundreds of cells each for hundreds of patients, something so far only achieved with pre-genomic FISH technologies [13, 58]. Furthermore, the amount of data needed can be expected to increase in proportion to the rarity of the subtype or mutation being studied. Meeting this challenge will require the development of next-generation sequencing technologies that can handle small amounts of starting material to yield high coverage and large multiplexing capabilities to meet time, demands, and cost. Thus, we can conclude that faster and cheaper single-cell sequencing and more accurate quantization are likely to pay large dividends for tumor phylogeny approaches.

5.2 Computational Challenge of Tumor Phylogenetics

Another area of challenge for tumor phylogenetics comes in the computational processing of NGS data. Accurately assembling genomes remains a challenging computational problem in general, and all the more so for tumor genomes, which are likely to have undergone large rearrangements relative to available reference genomes [77, 97]. Further error comes from the assumptions in the noise models accompanying variant calling methods. While large numbers of reads can overcome inaccuracy due to random sequencing errors, more data will not overcome systematic biases or misassembly leading to incorrect inference of variations. Better models of errors in sequencing and assembly are an ongoing and active area of research [98]. Compounding this challenge is the fact that error models can differ greatly from technology to technology and will likely need to evolve along with new technologies for sequencing.

Given the limitations of the available technologies, especially for single-cell sequencing, computational models are likely to play a role for some time in controlling for biases in sampling, sequencing, and reconstruction of tumor state from NGS data. Mixture model methods developed for array technologies [69, 73] can be expected to have continued value in more accurately reconstructing profiles of tumor heterogeneity from NGS data. NGS presents substantial new challenges, though, for example, in scaling methods to handle large numbers of markers, dealing with challenging computational problems arising from mixtures of rearrangements [99] and developing methods to integrate heterogeneous forms of data (e.g., simultaneous point mutation, copy number, and epigenetic sequencing data).

Advances are further needed in the phylogenetic inference step itself. Tumor phylogenetics depends on the identification of robust markers of progression [74, 75], a problem that becomes more challenging the more data one examines. Figuring out how to separate drivers of oncogenesis from passenger mutations [100] is a challenging problem for the field of tumor biology as a whole but also central to characterizing pathways of tumor progression. Many promising approaches have been explored to attempt to reduce the scope of the problem. For copy number data, this can be approached as a kind of segmentation problem, in which one can improve statistical power by grouping nearby markers on the genome [75, 101]. For point mutation data, other groupings, such as by gene or pathway, can provide similar help in simplifying the statistical problem [102]. Nonetheless, the problem remains unsolved.

Models of evolution remain a further obstacle. Tumor phylogeny studies to date have generally modeled tumor evolution using standard generic phylogenetic algorithms, such as neighbor joining [36, 64] or maximum parsimony [74]. Some single-cell studies have used custom but still quite simplified models [46, 59]. Accurate tumor phylogenies, however, depend on having an accurate quantitative model of how tumors specifically evolve. Although tumors are indeed evolutionary systems, they are systems that behave quite differently in some ways than evolving systems of organisms, with very high mutation rates and strong selective pressures that distinguish them from typical species evolution problems. Furthermore, since damage to the cell replication machinery is a hallmark of tumor development [9], models of evolutionary events in tumor cells will be very different from those in individual organisms. We know a great deal about some of the specific mechanisms by which tumors evolve [102], and a variety of mathematical models have been developed by the field [10, 12, 103–106] that may be useful for more accurately judging the plausibility of various tumor phylogeny scenarios. Nonetheless, we do not have good models for how different tumors differ in their propensities for these various events, much less the detailed quantitative models of likelihoods of possible mutational events that would ideally be used for phylogenetic inference. Furthermore, this gap in quantitative models represents a difficult chicken-and-egg problem for the field, as we will only learn these models by studying accurate tumor phylogenies. Computational approaches can in principle allow one to solve such problems by iteratively cycling between better models from which we can learn phylogenies and better phylogenies from which we can learn models [59], but this process is likely to require a better qualitative understanding of the basic mechanisms of oncogenesis, better data, and algorithms capable of making use of such data.

A final but substantial challenge is to algorithms for tumor phylogenetics. Phylogenetics is a challenging computational problem even in classic species scenarios and standard algorithms will not scale to the volumes of data NGS is making available, especially for the character-based approaches needed to reconstruct detailed events along evolutionary pathways [74]. The important role of genomic rearrangements in tumor development represents a major challenge in itself [77, 97]. Furthermore, as improved models and new kinds of data are developed, new algorithms are likely to be needed to fit to those data. New approaches to phylogenetics will be needed to integrate heterogeneous data sources available through NGS technologies. Experimental validation of inferred phylogenies is likewise a difficult problem with no obvious solutions.

6 Conclusions

Phylogenetics provides a crucial set of tools for making sense of the evolutionary processes that underlie tumor development in the face of the flood of data that NGS has unleashed, but major challenges remain to realize its potential. Cancer biology, like much of biological and medical research, has transformed in a span of a few years to a data-driven field dependent on computational algorithms for managing data and drawing meaningful inferences from it. Phylogenetic theory provides a framework and a set of models and algorithms well suited to understanding what is at its heart an evolutionary system. This framework has advanced alongside technologies for probing tumor state, helping us to assemble profiles of the common ways tumors develop across patient populations. With the NGS era, these computation methods have never been more necessary. However, major challenges confront both sequencing technologies and phylogenetic inferences if we are to truly reconstruct in detail the common evolutionary trajectories that underlie tumor development across cancers, subtypes, and patients.

References

1. Cahill DP, Kinzler KW, Vogelstein B, Lengauer C. Genetic instability and Darwinian selection in tumours. Trends Cell Biol. 1999;9(12):M57–60.
2. Nowell PC. The clonal evolution of tumor cell populations. Science. 1976;194:23–8.
3. Nordling C. A new theory on cancer-inducing mechanism. Br J Cancer. 1953;7(1):68–72.
4. Knudson A. Mutation and cancer: statistical study of retinoblastoma. Proc Natl Acad Sci USA. 1971;68(4):820–3.
5. Fearon ER, Vogelstein B. A genetic model for colorectal tumorigenesis. Cell. 1991;61: 759–67.
6. Strauss BS. Hypermutability in carcinogenesis. Genetics. 1998;148(4):1619–26. Review.

7. Danaei G, Vander-Hoorn S, Lopez AD, Murray CJ, Ezzati M, Comparative Risk Assessment collaborating group (Cancers). Causes of cancer in the world: comparative risk assessment of nine behavioural and environmental risk factors. Lancet. 2005;366(9499):1784–93.
8. Anderson AR, Weaver AM, Cummings PT, Quaranta V. Tumor morphology and phenotypic evolution driven by selective pressure from the microenvironment. Cell. 2006;127(5):905–15.
9. Hanahan D, Weinberg RA. Hallmarks of cancer: the next generation. Cell. 2011;144(5):646–74. Review.
10. Iwasa Y, Michor F. Evolutionary dynamics of intratumor heterogeneity. PLoS One. 2011;6(3):e17866.
11. Martins FC, De S, Almendro V, Gönen M, Park SY, Blum JL, et al. Evolutionary pathways in BRCA1-associated breast tumors. Cancer Discov. 2012;2(6):503–11.
12. Cheng Y-K, Beroukhim R, Levine RL, Mellinghoff IK, Holland EC, Michor F. A mathematical methodology for determining the temporal order of pathway alterations arising during gliomagenesis. PLoS Comput Biol. 2012;8(1):e1002337.
13. Janocko LE, Brown KA, Smith CA, Gu LP, Pollice AA, Singh SG, et al. Distinctive patterns of Her-2/neu, c-myc, and cyclin D1 gene amplification by fluorescence in situ hybridization in primary human breast cancers. Cytometry. 2001;46(3):136–49.
14. Gerlinger M, Rowan AJ, Horswell S, Larkin J, Endesfelder D, Gronroos E, et al. Intratumor heterogeneity and branched evolution revealed by multiregion sequencing. N Engl J Med. 2012;366(10):883–92.
15. Campbell LL, Polyak K. Breast tumor heterogeneity: cancer stem cells or clonal evolution? Cell Cycle. 2007;6(19):2332–8. Review.
16. Perou CM, Sorlie T, Eisen MB, Rijn M, Rees CA, Pollack JR, et al. Molecular portraits of human breast tumors. Nature. 2000;406:747–52.
17. Perou CM. Molecular stratification of triple-negative breast cancers. Oncologist. 2011;16 Suppl 1:61–70.
18. Sorlie T, Perrou CM, Tibshirani R, Aas T, Geisler S, Johnsen H, et al. Gene expression profiles of breast carcinomas distinguish tumor subclasses with clinical implications. Proc Natl Acad Sci USA. 2001;98:10869–74.
19. Golub TR, Slonim DK, Tamayo P, Huard C, Gaasenbeek M, Mesirov JP, et al. Molecular classification of cancer: class discovery and class prediction by gene expression monitoring. Science. 1999;286(5439):531–7.
20. Sotiriou C, Neo SY, McShane LM, Korn EL, Long PM, Jazaeri A, et al. Breast cancer classification and prognosis based on gene expression profiles from a population-based study. Proc Natl Acad Sci USA. 2003;100(18):10393–8.
21. Pegram MD, Konecny G, Slamon DJ. The molecular and cellular biology of HER2/neu gene amplification/overexpression and the clinical development of herceptin (trastuzumab) therapy for breast cancer. Cancer Treat Res. 2000;103:57–75. Review.
22. Hait WN, Hambley TW. Targeted cancer therapeutics. Cancer Res. 2009;69(4):1263–7. discussion 1267.
23. Felsenstein J. Inferring phylogenies. Sunderland, MA: Sinauer Associates; 2004.
24. Semple C, Steel M. Phylogenetics. Oxford: Oxford University Press; 2003.
25. Kidd KK, Sgaramella-Zonta LA. Phylogenetic analysis: concepts and methods. Am J Hum Genet. 1971;23:235–52.
26. Cavalli-Sforza LL, Edwards AW. Phylogenetic analysis. Models and estimation procedures. Am J Hum Genet. 1967;19(3 Pt 1):233–57.
27. Edwards AWF, Cavalli-Sforza LL. Reconstruction of evolutionary trees. In: Heywood VH, McNeill J, editors. Phenetic and phylogenetic classification. London: Systematics Association pub. no. 6; 1964. p. 67–76.
28. Huelsenbeck JP, Ronquist F, Nielsen R, Bollback JP. Bayesian inference of phylogeny and its impact on evolutionary biology. Science. 2001;294(5550):2310–4.
29. Saitou N, Nei M. The neighbor-joining method: a new method for reconstructing phylogenetic trees. Mol Biol Evol. 1987;4(4):406–25.
30. Felsenstein J. PHYLIP—Phylogeny inference package (version 3.2). Cladistics. 1989;5:164–6.

31. Swofford DL. PAUP*. Phylogenetic analysis using parsimony (*and other methods). Sunderland, MA: Version 4. Sinauer Associates; 2003.
32. Stamatakis A, Ott M, Ludwig T. RAxML-OMP: an efficient program for phylogenetic inference on SMPs. In: Proceedings of 8th international conference on Parallel Computing Technologies (PaCT2005), lecture notes in computer science, 3506:288–302. Springer Verlag.
33. Huelsenbeck JP, Ronquist F. MRBAYES: Bayesian inference of phylogenetic trees. Bioinformatics. 2001;17(8):754–5.
34. Desper R, Jiang F, Kallioniemi OP, Moch H, Papadimitriou CH, Schäffer AA. Inferring tree models for oncogenesis from comparative genome hybridization data. J Comput Biol. 1999;6(1):37–51.
35. Jiang F, Desper R, Papadimitriou CH, Schäffer AA, Kallioniemi OP, Richter J, et al. Construction of evolutionary tree models for renal cell carcinoma from comparative genomic hybridization data. Cancer Res. 2000;60(22):6503–9.
36. Desper R, Jiang F, Kallioniemi OP, Moch H, Papadimitriou CH, Schäffer AA. Distance-based reconstruction of tree models for oncogenesis. J Comput Biol. 2000;7(6):789–803.
37. Huang Z, Desper R, Schäffer AA, Yin Z, Li X, Yao K. Construction of tree models for pathogenesis of nasopharyngeal carcinoma. Genes Chromosomes Cancer. 2004;40(4):307–15.
38. Chen L, Nordlander C, Behboudi A, Olsson B, Levan KK. Deriving evolutionary tree models of the oncogenesis of endometrial adenocarcinoma. Int J Cancer. 2007;120(2):292–6.
39. Sweeney C, Boucher KM, Samowitz WS, Wolff RK, Albertsen H, Curtin K, et al. Oncogenetic tree model of somatic mutations and DNA methylation in colon tumors. Genes Chromosomes Cancer. 2009;48(1):1–9.
40. Chow CK, Liu CN. Approximating discrete probability distributions with dependence trees. IEEE Trans Inform Theor. 1968;14(3):462–7.
41. Szabo A, Boucher K. Estimating an oncogenetic tree when false negatives and positives are present. Math Biosci. 2002;176(2):219–36.
42. Yamamoto K, Oba S, Ishii S. Considering false negatives in mixture oncogenetic trees model for chromosomal alteration data. Genome Informatics Workshop (GIW2006). 2006. p. 134.
43. von Heydebreck A, Gunawan B, Füzesi L. Maximum likelihood estimation of oncogenetic tree models. Biostatistics. 2004;5(4):545–56.
44. Radmacher MD, Simon R, Desper R, Taetle R, Schäffer AA, Nelson MA. Graph models of oncogenesis with an application to melanoma. J Theor Biol. 2001;212(4):535–48.
45. Desper R, Gascuel O. Fast and accurate phylogeny reconstruction algorithms based on the minimum-evolution principle. J Comput Biol. 2002;9(5):687–705.
46. Desper R, Khan J, Schäffer AA. Tumor classification using phylogenetic methods on expression data. J Theor Biol. 2004;228(4):477–96.
47. Weigelt B, Glas AM, Wessels LF, Witteveen AT, Peterse JL, van't Veer LJ. Gene expression profiles of primary breast tumors maintained in distant metastases. Proc Natl Acad Sci USA. 2003;100(26):15901–5.
48. Quackenbush J. Microarray analysis and tumor classification. N Engl J Med. 2006; 354(23):2463–72. Review.
49. Perez-Diez A, Morgun A, Shulzhenko N. Microarrays for cancer diagnosis and classification. Adv Exp Med Biol. 2007;593:74–85. Review.
50. Dutt A, Beroukhim R. Single nucleotide polymorphism array analysis of cancer. Curr Opin Oncol. 2007;19(1):43–9. Review.
51. Zheng S, Zhao Z. DNA methylation profiling distinguishes three clusters of breast cancer cell lines. Chem Biodivers. 2012;9(5):848–56.
52. O'Hagan RC, Brennan CW, Strahs A, Zhang X, Kannan K, Donovan M, et al. Array comparative genome hybridization for tumor classification and gene discovery in mouse models of malignant melanoma. Cancer Res. 2003;63(17):5352–6.
53. Park Y, Shackney S, Schwartz R. Network-based inference of cancer progression from microarray data. IEEE/ACM Trans Comput Biol Bioinform. 2009;6(2):200–12.
54. Riester M, Stephan-Otto Attolini C, Downey RJ, Singer S, Michor F. A differentiation-based phylogeny of cancer subtypes. PLoS Comput Biol. 2010;6(5):e1000777.

55. Abu-Asab M, Chaouchi M, Amri H. Evolutionary medicine: a meaningful connection between omics, disease, and treatment. Proteomics Clin Appl. 2008;2(2):122–34.
56. Abu-Asab MS, Chaouchi M, Amri H. Phylogenetic modeling of heterogeneous gene-expression microarray data from cancerous specimens. OMICS. 2008;12(3):183–99.
57. Ryott M, Wangsa D, Heselmeyer-Haddad K, Lindholm J, Elmberger G, Auer G, et al. EGFR protein overexpression and gene copy number increases in oral tongue squamous cell carcinoma. Eur J Cancer. 2009;45(9):1700–8.
58. Heselmeyer-Haddad K, Garcia LYB, Bradley A, Melendez CO, Lee W, Christensen R, et al. Single-cell genetic analysis of ductal carcinoma in situ and invasive breast cancer reveals enormous tumor heterogeneity, yet conserved genomic imbalances and gain of MYC during progression. Am J Pathol. 2012;181(5):1807–22.
59. Pennington G, Smith CA, Shackney S, Schwartz R. Expectation-maximization method for reconstructing tumor phylogenies from single-cell data. Comput Syst Bioinformatics Conf. 2006:371–80.
60. Pennington G, Smith CA, Shackney S, Schwartz R. Reconstructing tumor phylogenies from heterogeneous single-cell data. J Bioinform Comput Biol. 2007;5(2a):407–27.
61. Bininda-Emonds ORP, Gittleman JL, Steel MA. The (super) tree of life: procedures, problems, and prospects. Annu Rev Ecol Syst. 2002;33:265–89.
62. Gordon AD. Consensus supertrees: the synthesis of rooted trees containing overlapping sets of leaves. J Classif. 1986;3:31–9.
63. Nugoli M, Chuchana P, Vendrell J, Orsetti B, Ursule L, Nguyen C, et al. Genetic variability in MCF-7 sublines: evidence of rapid genomic and RNA expression profile modifications. BMC Cancer. 2003;3:13.
64. Navin N, Krasnitz A, Rodgers L, Cook K, Meth J, Kendall J, et al. Inferring tumor progression from genomic heterogeneity. Genome Res. 2010;20(1):68–80.
65. Beerenwinkel N, Rahnenführer J, Däumer M, Hoffmann D, Kaiser R, Selbig J, et al. Learning multiple evolutionary pathways from cross-sectional data. J Comput Biol. 2005;12(6):584–98.
66. Beerenwinkel N, Rahnenführer J, Kaiser R, Hoffmann D, Selbig J, Lengauer T. Mtreemix: a software package for learning and using mixture models of mutagenetic trees. Bioinformatics. 2005;21(9):2106–7.
67. Yin J, Beerenwinkel N, Rahnenführer J, Lengauer T. Model selection for mixtures of mutagenetic trees. Stat Appl Genet Mol Biol. 2006;5:Article17.
68. Ketter R, Urbschat S, Henn W, Feiden W, Beerenwinkel N, Lengauer T, et al. Application of oncogenetic trees mixtures as a biostatistical model of the clonal cytogenetic evolution of meningiomas. Int J Cancer. 2007;121(7):1473–80.
69. Schwartz R, Shackney SE. Applying unmixing to gene expression data for tumor phylogeny inference. BMC Bioinformatics. 2010;11:42.
70. Tolliver D, Tsourakakis C, Subramanian A, Shackney S, Schwartz R. Robust unmixing of tumor states in array comparative genomic hybridization data. Bioinformatics. 2010;26(12):i106–14.
71. Etzioni R, Hawley S, Billheimer D, True LD, Knudsen B. Analyzing patterns of staining in immunohistochemical studies: application to a study of prostate cancer recurrence. Cancer Epidemiol Biomarkers Prev. 2005;14(5):1040–6.
72. Quon G, The YW, Chan E, Brudno M, Hughes T, Morris Q. A mixture model for the evolution of gene expression in non-homogenous datasets. In: Proceedings of the Neural Information Processing Systems Conference (NIPS) 2008. 2008.
73. Quon G, Morris Q. ISOLATE: a computational strategy for identifying the primary origin of cancers using high-throughput sequencing. Bioinformatics. 2009;25(21):2882–9.
74. Subramanian A, Shackney SE, Schwartz RS. Inference of tumor phylogenies from genomic assays on heterogeneous samples. J Biomed Biotechnol. 2012;2012:797812.
75. Subramanian A, Shackney SE, Schwartz RS. Novel multi-sample scheme for inferring phylogenetic markers from whole genome tumor profiles. In: Proceedings of the 8th international conference on Bioinformatics Research and Applications (ISBRA'12).
76. Cancer Genome Atlas Research Network. Comprehensive genomic characterization defines human glioblastoma genes and core pathways. Nature. 2008;455(7216):1061–8.

77. Campbell PJ, Stephens PJ, Pleasance ED, O'Meara S, Li H, Santarius T, et al. Identification of somatically acquired rearrangements in cancer using genome-wide massively parallel paired-end sequencing. Nat Genet. 2008;40(6):722–9.
78. Kim TM, Luquette LJ, Xi R, Park PJ. rSW-seq: algorithm for detection of copy number alterations in deep sequencing data. BMC Bioinformatics. 2010;11:432.
79. Xie C, Tammi MT. CNV-seq, a new method to detect copy number variation using high-throughput sequencing. BMC Bioinformatics. 2009;10:80.
80. Medvedev P, Stanciu M, Brudno M. Computational methods for discovering structural variation with next-generation sequencing. Nat Methods. 2009;6(11 Suppl):S13–20.
81. Zhang ZD, Du J, Lam H, Abyzov A, Urban AE, Snyder M, et al. Identification of genomic indels and structural variations using split reads. BMC Genomics. 2011;12:375.
82. Neuman JA, Isakov O, Shomron N. Analysis of insertion-deletion from deep-sequencing data: software evaluation for optimal detection. Brief Bioinform. 2012;14(1):46–55.
83. Elsharawy A, Forster M, Schracke N, Keller A, Thomsen I, Petersen BS, et al. Improving mapping and SNP-calling performance in multiplexed targeted next-generation sequencing. BMC Genomics. 2012;13(1):417.
84. Adams MD, Veigl ML, Wang Z, Molyneux N, Sun S, Guda K, et al. Global mutational profiling of formalin-fixed human colon cancers from a pathology archive. Mod Pathol. 2012;25(12):1599–608.
85. Li H, Handsaker B, Wysoker A, Fennell T, Ruan J, Homer N, et al. The Sequence Alignment/Map format and SAMtools. Bioinformatics. 2009;25(16):2078–9.
86. Marçais G, Kingsford C. A fast, lock-free approach for efficient parallel counting of occurrences of k-mers. Bioinformatics. 2011;27(6):764–70.
87. Melsted P, Pritchard JK. Efficient counting of k-mers in DNA sequences using a Bloom filter. BMC Bioinformatics. 2011;12:333.
88. Greenman CD, Pleasance ED, Newman S, Yang F, Fu B, Nik-Zainal S, et al. Estimation of rearrangement phylogeny for cancer genomes. Genome Res. 2012;22(2):346–61.
89. Merlo LM, Pepper JW, Reid BJ, Maley CC. Cancer as an evolutionary and ecological process. Nat Rev Cancer. 2006;6(12):924–35.
90. Ried T, Liyanage M, du Manoir S, Heselmeyer K, Auer G, Macville M, et al. Tumor cytogenetics revisited: comparative genomic hybridization and spectral karyotyping. J Mol Med. 1997;75(11–12):801–14.
91. Navin N, Kendall J, Troge J, Andrews P, Rodgers L, McIndoo J, et al. Tumour evolution inferred by single-cell sequencing. Nature. 2011;472(7341):90–4.
92. Hou Y, Song L, Zhu P, Zhang B, Tao Y, Xu X, et al. Single-cell exome sequencing and monoclonal evolution of a JAK2-negative myeloproliferative neoplasm. Cell. 2012;148(5):873–85.
93. Xu X, Hou Y, Yin X, Bao L, Tang A, Song L, et al. Single-cell exome sequencing reveals single-nucleotide mutation characteristics of a kidney tumor. Cell. 2012;148(5):886–95.
94. Tang F, Barbacioru C, Wang Y, Nordman E, Lee C, Xu N, et al. mRNA-Seq whole-transcriptome analysis of a single cell. Nat Methods. 2009;6(5):377–82.
95. Yilmaz S, Singh AK. Single cell genome sequencing. Curr Opin Biotechnol. 2012;23(3):437–43.
96. Dean FB, Hosono S, Fang L, Wu X, Faruqi AF, Bray-Ward P, et al. Comprehensive human genome amplification using multiple displacement amplification. Proc Natl Acad Sci USA. 2002;99(8):5261–6.
97. Stephens PJ, McBride DJ, Lin ML, Varela I, Pleasance ED, Simpson JT, et al. Complex landscapes of somatic rearrangement in human breast cancer genomes. Nature. 2009;462(7276):1005–10.
98. Yang X, Chockalingam SP, Aluru S. A survey of error-correction methods for next-generation sequencing. Brief Bioinform. 2012;14(1):56–66.
99. Mahmoody A, Kahn CL, Raphael BJ. Reconstructing genome mixtures from partial adjacencies. BMC Bioinformatics. 2012;13 Suppl 19:S9. doi:10.1186/1471-2105-13-S19-S9.

100. Bozic I, Antal T, Ohtsuki H, Carter H, Kim D, Chen S, et al. Accumulation of driver and passenger mutations during tumor progression. Proc Natl Acad Sci USA. 2010;107(43): 18545–50. doi:10.1073/pnas.1010978107.
101. Olshen AB, Venkatraman ES, Lucito R, Wigler M. Circular binary segmentation for the analysis of array-based DNA copy number data. Biostatistics. 2004;5(4):557–72.
102. Navin NE, Hicks J. Tracing the tumor lineage. Mol Oncol. 2011;4(3):267–83. doi:10.1016/j.molonc.2010.04.010. Erratum in: Mol Oncol. 2011 Jun; 5(3):302.
103. Vandin F, Upfal E, Raphael BJ. Finding driver pathways in cancer: models and algorithms. Algorithms Mol Biol. 2012;7(1):23.
104. Durrett R, Foo J, Leder K, Mayberry J, Michor F. Intratumor heterogeneity in evolutionary models of tumor progression. Genetics. 2011;188(2):461–77.
105. Podlaha O, Riester M, De S, Michor F. Evolution of the cancer genome. Trends Genet. 2012;28(4):155–63.
106. Gerstung M, Eriksson N, Lin J, Vogelstein B, Beerenwinkel N. The temporal order of genetic and pathway alterations in tumorigenesis. PLoS One. 2011;6(11):e27136.

Chapter 18
Impact and Challenges in Assessing Tumor Purity by Next-Generation Sequencing

Xiaoping Su, Gabriel G. Malouf, and Francisco J. Esteva

Abstract Cancer research is receiving a tremendous boost from the technological advancements known as next-generation sequencing, which have enabled researchers to identify gene mutations that are highly recurrent in several different subtypes of cancer and to discover new subtypes of cancer based on a profile of gene mutations. The ability to identify and then target the specific genetic profile of each tumor is the promise of developing personalized cancer treatments. Much research effort is needed before that promise can be realized. One area of research supporting this effort is estimating tumor purity and heterogeneity. Tumors removed from patients contain a mixture of normal and tumor cells. As a result, analyses of mutations related to cancer must include a determination of how much the tumor's genome sequence differs from that of the normal matched tissue as a result of mutation. We explore tumor purity estimation in this chapter and discuss a new software named PurityEst that uses the major tumor clones as estimation of tumor purity.

Keywords Tumor purity • Tumor heterogeneity • NGS • Purity estimation • Genetic alteration • Somatic mutation

X. Su, Ph.D. (✉)
Department of Bioinformatics and Computational Biology, The University of Texas MD Anderson Cancer Center, Unit 1410, Houston, TX 77230-1402, USA
e-mail: xsu1@mdanderson.org

G.G. Malouf, M.D., M.Sc.
Department of Medical Oncology, Groupe Hospitalier Pitie-Salpetriere
University Pierre and Marie Curie (Paris VI), Paris, France
e-mail: gabriel.malouf@psl.aphp.fr

F.J. Esteva, M.D., Ph.D.
Division of Cancer Medicine, Department of Breast Medical Oncology,
The University of Texas MD Anderson Cancer Center, Houston, TX 77230, USA
e-mail: fjesteva@mdanderson.org

W. Wu and H. Choudhry (eds.), *Next Generation Sequencing in Cancer Research: Volume 1: Decoding the Cancer Genome*, DOI 10.1007/978-1-4614-7645-0_18,

1 Introduction

Cancer refers to at least 100 forms of disease, bearing different histologies and arising in different organs. Cancer has many etiologies, with specific causes related to each cancer type, even though predisposing factors for tumor development are unknown for the majority of cases. Cancer can be secondary to viral infections, such as that associated with the hepatitis B or C virus (i.e., hepatocellular carcinoma) [1]; it can be hereditary, which is the case for BRAC1 and BRCA2 mutations that predispose individuals to breast cancer [2]; and it can also be influenced by the environment, as may be the case for lung cancer in smokers. Beyond etiology, the evolution of a normal cell toward a cancerous one involves different genetic and epigenetic alterations that confer selective advantages which allow a tumor to grow and escape apoptosis [3]. These alterations have been thoroughly investigated over the last few decades, leading to the emergence of an ideal personalized medicine model, in which patients will be offered "a la carte" treatment, according to a defined tumor genetic alteration [4]. For example, patients with breast cancer characterized by the overexpression of estrogen and progesterone receptors and HER2 amplifications benefit from hormonal therapy and anti-HER2 blockade (i.e., Herceptin) [5–7]. However, beyond those molecular markers, we currently lack genetic data to inform the routine assignment of personalized treatment for most individuals with breast cancer [8]. For other tumor subtypes, such as renal cell carcinoma, pancreatic cancer, or liver cancer, no reliable marker has been validated for clinical use, and we have limited understanding of the full spectrum of the genetic alterations associated with these cancer types.

Historically, the study of human hereditary cancer syndromes has provided important insights that improve our understanding of common, nonhereditary malignancies. This has been true for genes that are vital to cellular processes such as metabolism, cell division, cell repair, and apoptosis. An important example is the von Hippel-Lindau (VHL) hereditary syndrome, which is characterized by germline mutations of the *VHL* gene [9]. Of note, the VHL gene is frequently inactivated in sporadic clear-cell renal cell carcinoma [9, 10]. Another example is the Li-Fraumeni syndrome, which is an extremely rare autosomal dominant hereditary disorder linked to germline mutations of the p53 tumor suppressor gene that lead to increased susceptibility to different types of cancer [11]. The p53 gene is a master regulator of various signaling pathways and thus is seen as the guardian of the genome. Its inactivation leads to the loss of tumor-defensive functions, including DNA repair, apoptosis, and cell cycle arrest. Over time, researchers have discovered p53 mutations in almost every type of cancer, at rates varying from 10 % in hematological malignancies to almost 100 % in squamous cell carcinoma [12, 13].

Notable progress in identifying mutations in different oncogenes or tumor suppressor genes has led to a prolific amount of literature analyzing the role of those alterations as prognostic markers for recurrence-free survival and overall survival of patients with specific tumor subtypes [14–16]. The role of gene mutations as markers predictive of patient response to chemotherapy has also been investigated [15].

More recently, mutations in selected genes have been used to develop inhibitors that can target cancer cells; this is the case for the BRAF V600E mutation, which occurs in 60 % of melanomas. Vemurafenib, a BRAF enzyme inhibitor, was shown to work exclusively in patients with this mutation and has been approved for the treatment of melanoma [17]. Of note, this drug paradoxically stimulates normal BRAF and may promote tumor proliferation [18]. The result of this research is that all patients with advanced melanoma are now systematically tested for the BRAF V600E mutation. Another type of mutation for which routine screening has been implemented is the EGFR mutations, which are used as markers to predict a patient's response to EGFR inhibitors in non-small cell lung cancer [19]. In brief, genomics has entered the arena of clinical medicine.

All these discoveries and treatment improvements were based on a hypothesis-driven model and involved the analysis of a handful of mutations in a specific cancer type. They occurred before the advent of several sequencing platforms collectively known as next-generation sequencing (NGS), which have been developed over the last 6 years. In this chapter, we will briefly describe the impressive discoveries that have been achieved since the implementation of NGS techniques for simultaneously analyzing the genetics of multiple tumor subtypes. A major focus will be the issues of tumor purity and heterogeneity in the interpretation of NGS data.

2 Contribution of Next-Generation Sequencing in Understanding Genetic Alterations in Different Tumor Subtypes

Compared with Sanger sequencing, NGS enables the detection of genomic alterations in cancer cells in an unbiased way, much faster, and with higher sensitivity. Thus, the 23,000 genes of the human genome can be sequenced and processed in less than 2 weeks, allowing for the identification of somatic mutations, translocations, and gene copy number alterations.

The first whole genome sequencing of a tumor was performed on a sample of cytogenetically normal acute myeloid leukemia [20]. Since then, many cancer subtypes have been sequenced at both the DNA and RNA levels. In a relatively short time, NGS technologies have allowed researchers to identify highly recurrent mutations in several cancer subtypes, as is the case for PBRM1 in clear-cell renal cell carcinoma [21]. NGS has also allowed for the determination of recurrent translocations, leading to the identification of new cancer subtypes, an example of which is bone sarcoma that is defined by BCOR-CCNB3 gene fusion [22].

The landscape of recurrent genetic alterations for major cancer subtypes was recently reported by the Cancer Genome Atlas Network [12, 23–25], which revealed that each tumor subtype has a unique mutation spectrum. For instance, a mean of 360 exonic mutations has been reported in squamous cell lung cancer [12]. These studies have the potential to guide drug development by calling attention to the deep analysis of recurrent mutations, as is the case for the NFE2L2 and KEAP1 gene mutations that were discovered in 34 % of squamous cell lung tumors [12]. Indeed, the NFF2L2 and KEAP1 genes belong to the antioxidant pathway, and KEAP1, along with CUL3, plays a key role in the degradation of NFF2L2 by ubiquitination. A mutation of NFF2L2 or KEAP1 leads to the translocation of NFF2L2 in the nucleus, which, by binding to the antioxidant response element in the upstream promoter regions of antioxidative genes, activates them, favoring tumor proliferation. This provides a rationale for the development of several NFF2L2 inhibitors. In clear-cell renal cell carcinomas, recurrent mutations of histone remodeling enzymes such as PBRM1, UTX, and SETD2 have been frequently found to open avenues for developing novel agents that target the epigenome [21, 26].

3 Emerging Issues of Cancer Heterogeneity and Tumor Purity in Interpreting Next-Generation Sequencing Data

Concomitantly with the publication of the landscape of genetic mutations in different cancer subtypes, a few studies used NGS to analyze tumor heterogeneity within different areas of a tumor taken from the same patient [27–29]. Those reports found that some tumors are homogeneous, with high concordance between the mutations found in different areas of the tumor, as shown recently for hepatocellular carcinoma [28]. The majority of the tumors analyzed, however, showed heterogeneity of mutations in different areas of the tumor [27]. Indeed, Gerlinger et al. [27] evaluated the intra-tumoral mutation landscape using exome sequencing of primary renal cell carcinomas and associated metastatic sites. The authors revealed high rates (63–69 %) of differences in the somatic mutations in separate sections of a single tumor. Interestingly, they also found phylogenetic relationships between tumor regions [27]. Notably, they performed gene expression profiling in different areas of the primary renal tumor and metastatic tissue and found that different areas of the tumor may harbor distinct gene expression signatures, with some associated with the clear-cell A signature (good prognosis) and some with the clear-cell B signature (poor prognosis). Those findings require validation in a large set of patients; thus, their clinical utility in predicting patients' outcomes and optimizing patients' treatments remains to be established.

4 Tumor Heterogeneity and Tumor Purity as Two Sides of the Same Coin

In addition to cancer cells, the tumor microenvironment is composed of lymphocytes, macrophages, and mainly cancer-associated fibroblasts (CAF), referred to as cancer stroma [30–34]. Thus, analyses of gene mutations in tumor cells must account for the genes associated with a mixture of cell types within the sample. As part of that mixture, CAF can represent different proportions of tumor bulk that exist within regions of the primary tumor as well as between tumors belonging to different patients. Interestingly, CAF are particularly abundant in the stroma of desmoplastic cancers (i.e., pancreatic cancer, cholangiocarcinoma), where they promote tumor invasion and metastasis [30–34]. The expression of the CAF gene may also differ in distinct cancer subtypes [35]. Hence, in addition to assessing tumor heterogeneity, analyzing the primary tumors from patient samples requires the estimation of tumor purity, especially for desmoplastic cancers. This is the only way to be able to identify the mutations that drive carcinogenesis. Low tumor purity can result in underestimating the number of mutations, especially if the sequencing coverage depth is low.

We suggest that tumor heterogeneity and tumor purity are two sides of the same coin. Epithelial cells within the same tumor can be phenotypically and genetically distinct. Short of having to conduct single-cell sequencing on the different components of each tumor, it is crucial to develop algorithms to better estimate tumor purity, as this can have major implications in tumor biomarker development and personalized medicine. It is currently impossible to sequence a tumor sample that does not have a component of stromal cells, even after macrodissection. In the case of a "monogenomic" tumor, as defined by Navin et al. [36], increasing the coverage depth of cancer samples may be enough to compensate for issues of tumor purity. Conversely, in the case of "polygenomic" tumors, increasing the coverage depth may not be enough, and distinguishing tumor purity from minor clone subpopulations may be an issue. Determining how to best handle these issues represents an ongoing challenge. PurityEst, an algorithm that was developed to better estimate tumor purity by NGS data [37], will be discussed in detail below.

Although not the topic of this chapter, heterogeneity within the same tumor also may confound the interpretation of NGS data. Indeed, emerging data highly suggest that tumor heterogeneity may hamper the development of drug biomarkers, as well as drug discovery. Thus, a deeper understanding of both tumor purity and heterogeneity may facilitate a better description of the natural development of a tumor [38].

5 PurityEst Algorithm for Tumor Purity Estimation

Because tumor tissues usually consist of a mixture of multiple tumor clones and normal cells including fibroblasts and infiltrating lymphocytes, the observed magnitude of copy number changes is diminished, which is the basis of tumor clone purity estimation using SNP array data [39–43]. Gusnanto et al. [44] converted mapped reads to DNA copy number ratios between tumor and normal genomes for purity estimation. In principle, NGS provides an alternative to methods based on DNA copy numbers, which is to use mutant allele frequencies in the heterozygous loci with somatic mutations in a tumor mixture.

Tumor purity is an estimation of how much of the tumor's genome sequence is different from that of the normal matched tissue as a result of mutation. A sample of only tumor cells (containing no normal tissue) should show a mean frequency of 0.5 for mutant alleles at heterozygous loci with somatic mutations. The contamination of tumor tissue with normal tissue, the level of which is sample dependent, affects mutant allele frequencies and indeed decreases their relative fraction for both homozygote and heterozygote somatic mutations.

PurityEst [37] implements a simple approach to the purity estimation problem. PurityEst assumes that the tumor tissue can be largely approximated by a mixture of a normal clone and a tumor clone and infers the tumor purity from the allelic differential representation of heterozygous loci with somatic mutations in a tumor sample with a matched normal tissue sample. PurityEst gives a purity estimate from somatic mutations in each chromosome and takes a robust average of the chromosome-wide estimates to be the purity estimate of the tumor tissue.

Assume that the tumor sample contains one type of tumor cells contaminated with normal cells so that the proportion of tumor cells is $\gamma \in [0,1]$ ("tumor purity") and the proportion of normal cells is $1-\gamma$ ("normal contamination"). Tumor purity γ is the difference between the expected mutant allele frequency (equal to 0.5 without reference allele bias) without normal tissue contamination and the mean observed mutant allele frequency in heterozygous loci with somatic mutations in a tumor sample with matched normal tissue. The expected mutant allele frequency of heterozygous loci with somatic mutations without normal tissue contamination is usually less than 0.5 because of reference allele bias and could be approximated with the mean SNP allele frequency in heterozygous loci. The SNP allele frequencies in a tumor sample are defined as $X = (x_1, \ldots, x_n)$, where $x_i = \frac{B_i}{A_i + B_i}$, A_i indicates the wild allele count, B_i the mutant allele count in SNP heterozygous loci, n is the total number of SNP heterozygous loci, and the summations include all the SNP heterozygous loci attributed to germline mutations. The mutant allele frequencies in a tumor sample are defined as $Y = (y_1, \ldots, y_m)$, where $y_i = \frac{B_i}{A_i + B_i}$, A_i indicates the wild allele count, B_i the mutant allele count in heterozygous loci with somatic mutations, and m is the total number of heterozygous loci with somatic mutations. Thus, the tumor purity γ is defined as the ratio of the allele frequency means:

$$\gamma = \frac{\frac{1}{m}\sum_{i=1}^{m} y_i}{\frac{1}{n}\sum_{i=1}^{n} x_i} = \frac{\mu_y}{\mu_x}$$

In order to minimize the effect of individual allele frequencies on the allele frequency means by the loci with low allele count, both the SNP allele frequency mean μ_x and the mutant allele frequency mean μ_y were estimated by $\frac{\sum_i B_i}{\sum_i (A_i + B_i)}$. Note that the set of somatic mutations is assumed to be mutually exclusive with the set of germline mutations and the later set is usually much greater than the former.

Theoretically, the expected value of μ_x is 0.5. However, empirical data showed that the mean value is typically slightly lower than 0.5, which suggests that different alleles are not equally represented with the current sequencing technology. Hence, PurityEst uses the computation of the empirical value of μ_x to correct for this representational bias.

The above formulation does not explicitly consider the effects of copy number gains and losses in tumor genomes. In principle, mutant alleles can be greatly amplified in tumors which can bias the tumor purity estimation. However, based on our empirical observations, the biases appear to affect only a small fraction of the tumor genome. Therefore, to minimize the effect of such biases, PurityEst estimates tumor purity γ_i separately for each autosomal chromosome with a threshold of 20 somatic mutations per chromosome, where $i = 1,\ldots,22$, and obtains a final estimate from robustly averaging γ_i, excluding the outliers that might be caused by factors such as aneuploidy. The tumor purity is estimated by

$$\bar{\gamma} = \frac{1}{n}\sum_{i=1}^{n}\gamma_i$$

$$\mathrm{SE}_{\bar{\gamma}} = \frac{S}{\sqrt{n}}$$

where n is the number of autosomal chromosomes with purity estimation excluding the outliers and $\mathrm{SE}_{\bar{\gamma}}$ is the estimated error from the sample standard deviation.

PurityEst also implements the "extreme studentized deviate" (ESD) multiple-outlier procedure [45] to remove the outliers, which can accurately identify either single or multiple outliers and will be less susceptible to masking problem. For this purpose, a reasonable upper bound must first be decided for the number of outliers present in a dataset. In our experience, a reasonable upper bound for the number of possible outliers is [n/5], where [n/5] is the largest integer $\leq n/5$. The following multiple-outlier procedure [45] achieves this goal:

Suppose we have $x_1,\ldots, x_n \sim N(\mu, \sigma^2)$ for a large majority of the sample points, we suspect that we may have as many as k outliers, where $k = [n/5]$ and $[n/5]$ is the largest integer $\leq n/5$. We wish to have a type I error rate of α to test the hypothesis H_0: there is no outliers versus H_1: there are between 1 and k outliers and would like to use a decision rule that can specifically identify the outliers. For this purpose:

1. The ESD statistic is computed based on the full sample $\max_{i=1,\ldots,n} |x_i - \bar{x}| / s$ where $\bar{x}$ is the sample mean and s is sample standard deviation. This statistic is denoted by $\mathrm{ESD}^{(n)}$ and the most outlying data point by $x^{(n)}$.
2. $x^{(n)}$ is removed from the sample and the mean, standard deviation, and ESD statistic are computed from the remaining $n-1$ data points. The ESD statistic from the reduced sample is denoted by $\mathrm{ESD}^{(n-1)}$.
3. The most outlying sample point is removed and the ESD statistic is recomputed until k ESD statistics have been computed which are denoted by

$ESD^{(n)}, ESD^{(n-1)}, \ldots, ESD^{(n-k+1)}$ based on the original sample of size n and successively reduced samples of size $n-1, \ldots, n-k+1$. The most outlying values identified at each of the k steps are denoted by $x^{(n)}, x^{(n-1)}, \ldots, x^{(n-k+1)}$.

4. The critical values corresponding to the ESD statistics are $ESD_{n,1-\alpha}, ESD_{n-1,1-\alpha}, \ldots, ESD_{n-k+1,1-\alpha}$.
5. The following decision rule for the outlier decision will then be used:

   If $ESD^{(n-k+1)} > ESD_{n-k+1,1-\alpha}$, then we declare the k values $x^{(n)}, x^{(n-1)}, \ldots, x^{(n-k+1)}$ as outliers.

   else if $ESD^{(n-k+2)} > ESD_{n-k+2,1-\alpha}$, then we declare the $k-1$ values $x^{(n)}, x^{(n-1)}, \ldots, x^{(n-k+2)}$ as outliers.

   ...

   else if $ESD^{(n)} > ESD_{n,1-\alpha}$, then we declare one outlier, $x^{(n)}$.

   else if $ESD^{(n)} \le ESD_{n,1-\alpha}$, then we declare no outlier is present.

   Thus, we have the ability to declare either $0, 1, \ldots,$ or k sample points as outliers.

Note that all k outlier test statistics $ESD^{(n)}, ESD^{(n-1)}, \ldots, ESD^{(n-k+1)}$ must be computed regardless of whether any specific test statistic (e.g., $ESD^{(n)}$) is significant or not. This procedure has good power to either declare no outliers or detect from 1 up to k outliers with little susceptibility to masking effects, unless the true number of outliers is larger than k.

6 Application of PurityEst

We used a publically available dataset of seven matched prostate tumor samples [46] to test PurityEst algorithm. The sequencing data was generated with Illumina GAII, containing 7 matched prostate cancer samples with 76nt paired-end (PE) reads. We used a reference (GRCh37/hg19) sequence-guided alignment and assembly tool, MOSAIK [47], to map the paired-end 76nt reads, and analyzed the resulting alignments and detected the putative single-nucleotide variations (SNVs) using the Bayesian model-based software GigaBayes [47] that enables efficient analysis of billions of aligned short-read sequences. We also filtered out all known SNVs based on the public database UCSC dbSNP 135. We then determined the somatic status of each SNV by comparing the genotypes between matched normal tissue and tumor samples. Both wild and mutant allele counts at each heterozygous locus of both SNPs and somatic mutations were also generated by GigaBayes to be used in tumor purity estimation.

We calculated the genomic coverage of the tissue samples calculated in terms of the mean depth of each base across the whole genome by using uniquely aligned reads, for which the mapping criteria were relatively stringent, with a maximum of 5 % mismatch allowed. We found the tumor samples to have a mean genomic coverage ranging from 29.5 to 35.8 and the matched normal tissue samples to have a mean coverage ranging from 18.8 to 34.9. For each matched prostate tumor sample,

Table 18.1 Tumor purity estimate for each autosomal chromosome and the outlier detection in sample PR-1701 by PurityEst

Chrom	SNP allele frequency	Mutant allele frequency	Tumor purity estimate	Chrom outlier status
1	0.4966	0.2568	0.5170	No
2	0.4838	0.2558	0.5288	No
3	0.4844	0.2493	0.5147	No
4	0.4848	0.2458	0.5070	No
5	0.4863	0.2387	0.4908	No
6	0.4770	0.2537	0.5319	No
7	0.4808	0.2544	0.5290	No
8	0.4959	0.2574	0.5191	No
9	0.4826	0.2612	0.5414	No
10	0.4783	0.1840	0.3847	Yes
11	0.4842	0.2489	0.5140	No
12	0.4859	0.2497	0.5139	No
13	0.4848	0.2580	0.5321	No
14	0.4818	0.2390	0.4960	No
15	0.4793	0.2410	0.5028	No
16	0.4714	0.2219	0.4707	No
17	0.4768	0.2757	0.5782	No
18	0.4980	0.2747	0.5516	No
19	0.4830	0.2868	0.5939	No
20	0.4896	0.2732	0.5580	No
21	0.4722	0.2434	0.5155	No
22	0.4905	0.2729	0.5563	No

Chrom autosomal chromosome

we used PurityEst to estimate the tumor purity level and its error bar by bootstrap sampling for each autosomal chromosome, performed the outlier detection, and removed the outliers in the overall tumor purity estimation for each matched prostate sample if any outliers were detected in a specific sample (Table 18.1 and Fig. 18.1a). The purity level estimate from chromosome 10 was found to be substantially lower than other chromosomes. It is not clear what caused this outlier, as no major copy number gain, and only minor fraction of losses were observed from the chromosome.

Figure 18.1b (Table 18.2) showed the PurityEst estimates along with the estimates obtained by using ABSOLUTE [42] and reported by Berger et al. [46], who used copy number changes derived from SNP array data of the same seven prostate cancer samples. The correlation coefficient between the two kinds of estimates is 0.91, demonstrating that PurityEst estimates are consistent with those from DNA copy number data.

PurityEst can be used to estimate tumor purity based on mutant allele fractions in a mixture of a tumor clone and a normal clone. Multiple factors such as coverage, copy number changes, and representational bias can potentially affect the purity

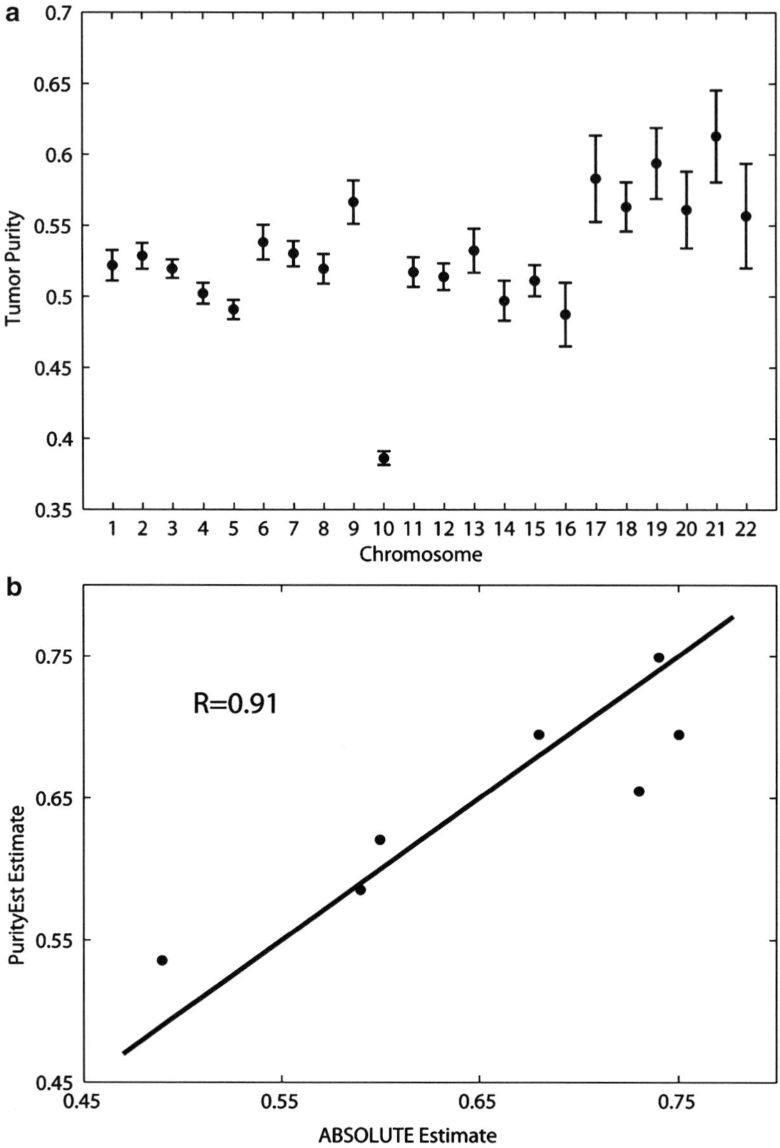

Fig. 18.1 (**a**) Tumor purity estimations by PurityEst, for each autosomal chromosome in sample PR-1701. The error bars are estimated by bootstrap sampling. Chromosome 10 is an outlier. The mean purity level excluding the outlier is 0.535. (**b**) Scatter plot of tumor purity estimates from ABSOLUTE and PurityEst. The correlation coefficient between the two kinds of estimates is 0.91 from seven patient samples

estimation. PurityEst can handle some, but not all of these effects. When adequate coverage is available, this method may be extended to model multiple mutant allele fractions that reflect intra-tumor heterogeneity. PurityEst provides a simple and effective solution for tumor purity estimation in cancer studies.

Table 18.2 Tumor purity estimate for seven matched prostate tumor samples, by both ABSOLUTE and PurityEst

Sample ID	ABSOLUTE	PurityEst
PR-0580T	0.73	0.6549
PR-0581T	0.6	0.6208
PR-1701T	0.49	0.5356
PR-1783T	0.75	0.6946
PR-2832T	0.59	0.5855
PR-3027T	0.74	0.7491
PR-3043T	0.68	0.6947

Sample ID sample identification number

7 Personalized Medicine Challenges

Historically, much effort has been exerted to select patients according to specific clinical (i.e., lymph node involvement, metastasis), pathological (i.e., microvascular invasion), and genomic features (i.e., HER2 amplification). These efforts have led to the design of several clinical trials and have facilitated the approval of several drugs as adjuvant therapy or to treat metastases. The overwhelming amount of genetic data available today makes data interpretation very challenging, and we add to that the difficulties arising from needing to assess tumor purity and tumor heterogeneity. From this perspective, many questions remain to be answered.

References

1. Spangenberg HC, Thimme R, Blum HE. Targeted therapy for hepatocellular carcinoma. Nat Rev Gastroenterol Hepatol. 2009;6(7):423–32.
2. Bishop DT. BRCA1 and BRCA2 and breast cancer incidence: a review. Ann Oncol. 1999;10 Suppl 6:113–9.
3. Hanahan D, Weinberg RA. Hallmarks of cancer: the next generation. Cell. 2011;144(5):646–74.
4. Hamburg MA, Collins FS. The path to personalized medicine. N Engl J Med. 2010; 363(4):301–4.
5. Romond EH, Perez EA, Bryant J, Suman VJ, Geyer Jr CE, Davidson NE, et al. Trastuzumab plus adjuvant chemotherapy for operable HER2-positive breast cancer. N Engl J Med. 2005;353(16):1673–84.
6. Esteva FJ, Guo H, Zhang S, Santa-Maria C, Stone S, Lanchbury JS, et al. PTEN, PIK3CA, p-AKT, and p-p70S6K Status. Association with trastuzumab response and survival in patients with HER2-positive metastatic breast cancer. Am J Pathol. 2010;177(4):1647–56.
7. Esteva FJ. Monoclonal antibodies, small molecules, and vaccines in the treatment of breast cancer. Oncologist. 2004;9 Suppl 3:4–9.
8. Esteva FJ, Sahin AA, Cristofanilli M, Arun B, Hortobagyi GN. Molecular prognostic factors for breast cancer metastasis and survival. Semin Radiat Oncol. 2002;12(4):319–28.
9. Maher ER, Neumann HP, Richard S. von Hippel-Lindau disease: a clinical and scientific review. Eur J Hum Genet. 2011;19(6):617–23.
10. Moore LE, Nickerson ML, Brennan P, Toro JR, Jaeger E, Rinsky J, et al. Von Hippel-Lindau (VHL) inactivation in sporadic clear cell renal cancer: associations with germline VHL polymorphisms and etiologic risk factors. PLoS Genet. 2011;7(10):e1002312.
11. Varley JM. Germline TP53 mutations and Li-Fraumeni syndrome. Hum Mutat. 2003;21(3): 313–20.

12. Hammerman PS, Hayes DN, Wilkerson MD, Schultz N, Bose R, Chu A, et al. Comprehensive genomic characterization of squamous cell lung cancers. Nature. 2012;489(7417):519–25.
13. Peller S, Rotter V. TP53 in hematological cancer: low incidence of mutations with significant clinical relevance. Hum Mutat. 2003;21(3):277–84.
14. Osborne C, Wilson P, Tripathy D. Oncogenes and tumor suppressor genes in breast cancer: potential diagnostic and therapeutic applications. Oncologist. 2004;9(4):361–77.
15. Ross JS, Fletcher JA. The HER-2/neu oncogene in breast cancer: prognostic factor, predictive factor, and target for therapy. Oncologist. 1998;3(4):237–52.
16. Esteller M, Gonzalez S, Risques RA, Marcuello E, Mangues R, Germa JR, et al. K-ras and p16 aberrations confer poor prognosis in human colorectal cancer. J Clin Oncol. 2001;19(2): 299–304.
17. Chapman PB, Hauschild A, Robert C, Haanen JB, Ascierto P, Larkin J, et al. Improved survival with vemurafenib in melanoma with BRAF V600E mutation. N Engl J Med. 2011;364(26):2507–16.
18. Hatzivassiliou G, Song K, Yen I, Brandhuber BJ, Anderson DJ, Alvarado R, et al. RAF inhibitors prime wild-type RAF to activate the MAPK pathway and enhance growth. Nature. 2010; 464(7287):431–5.
19. Pirker R, Herth FJ, Kerr KM, Filipits M, Taron M, Gandara D, et al. Consensus for EGFR mutation testing in non-small cell lung cancer: results from a European workshop. J Thorac Oncol. 2010;5(10):1706–13.
20. Ley TJ, Mardis ER, Ding L, Fulton B, McLellan MD, Chen K, et al. DNA sequencing of a cytogenetically normal acute myeloid leukaemia genome. Nature. 2008;456(7218):66–72.
21. Varela I, Tarpey P, Raine K, Huang D, Ong CK, Stephens P, et al. Exome sequencing identifies frequent mutation of the SWI/SNF complex gene PBRM1 in renal carcinoma. Nature. 2011;469(7331):539–42.
22. Pierron G, Tirode F, Lucchesi C, Reynaud S, Ballet S, Cohen-Gogo S, et al. A new subtype of bone sarcoma defined by BCOR-CCNB3 gene fusion. Nat Genet. 2012;44(4):461–6.
23. Comprehensive molecular portraits of human breast tumours. Nature. 2012;490(7418):61–70.
24. Comprehensive molecular characterization of human colon and rectal cancer. Nature. 2012;487(7407):330–7.
25. Integrated genomic analyses of ovarian carcinoma. Nature. 2011;474(7353):609–15.
26. Dalgliesh GL, Furge K, Greenman C, Chen L, Bignell G, Butler A, et al. Systematic sequencing of renal carcinoma reveals inactivation of histone modifying genes. Nature. 2010;463(7279):360–3.
27. Gerlinger M, Rowan AJ, Horswell S, Larkin J, Endesfelder D, Gronroos E, et al. Intratumor heterogeneity and branched evolution revealed by multiregion sequencing. N Engl J Med. 2012;366(10):883–92.
28. Huang J, Deng Q, Wang Q, Li KY, Dai JH, Li N, et al. Exome sequencing of hepatitis B virus-associated hepatocellular carcinoma. Nat Genet. 2012;44(10):1117–21.
29. Swanton C. Intratumor heterogeneity: evolution through space and time. Cancer Res. 2012; 72(19):4875–82.
30. Erez N, Truitt M, Olson P, Arron ST, Hanahan D. Cancer-associated fibroblasts are activated in incipient neoplasia to orchestrate tumor-promoting inflammation in an NF-kappaB-dependent manner. Cancer Cell. 2010;17(2):135–47.
31. Hwang RF, Moore T, Arumugam T, Ramachandran V, Amos KD, Rivera A, et al. Cancer-associated stromal fibroblasts promote pancreatic tumor progression. Cancer Res. 2008;68(3): 918–26.
32. Karagiannis GS, Poutahidis T, Erdman SE, Kirsch R, Riddell RH, Diamandis EP. Cancer-associated fibroblasts drive the progression of metastasis through both paracrine and mechanical pressure on cancer tissue. Mol Cancer Res. 2012;10(11):1403–18.
33. Liao D, Luo Y, Markowitz D, Xiang R, Reisfeld RA. Cancer associated fibroblasts promote tumor growth and metastasis by modulating the tumor immune microenvironment in a 4T1 murine breast cancer model. PLoS One. 2009;4(11):e7965.

34. Pazolli E, Alspach E, Milczarek A, Prior J, Piwnica-Worms D, Stewart SA. Chromatin remodeling underlies the senescence-associated secretory phenotype of tumor stromal fibroblasts that supports cancer progression. Cancer Res. 2012;72(9):2251–61.
35. Tchou J, Kossenkov AV, Chang L, Satija C, Herlyn M, Showe LC, et al. Human breast cancer associated fibroblasts exhibit subtype specific gene expression profiles. BMC Med Genomics. 2012;5:39.
36. Navin N, Krasnitz A, Rodgers L, Cook K, Meth J, Kendall J, et al. Inferring tumor progression from genomic heterogeneity. Genome Res. 2010;20(1):68–80.
37. Su X, Zhang L, Zhang J, Meric-Bernstam F, Weinstein JN. PurityEst: estimating purity of human tumor samples using next-generation sequencing data. Bioinformatics. 2012;28(17):2265–6.
38. Nik-Zainal S, Van Loo P, Wedge DC, Alexandrov LB, Greenman CD, Lau KW, et al. The life history of 21 breast cancers. Cell. 2012;149(5):994–1007.
39. Sun W, Wright FA, Tang Z, Nordgard SH, Van Loo P, Yu T, et al. Integrated study of copy number states and genotype calls using high-density SNP arrays. Nucleic Acids Res. 2009;37(16):5365–77.
40. Bengtsson H, Neuvial P, Speed TP. TumorBoost: normalization of allele-specific tumor copy numbers from a single pair of tumor-normal genotyping microarrays. BMC Bioinformatics. 2010;11:245.
41. Van Loo P, Nordgard SH, Lingjaerde OC, Russnes HG, Rye IH, Sun W, et al. Allele-specific copy number analysis of tumors. Proc Natl Acad Sci USA. 2010;107(39):16910–5.
42. Carter SL, Cibulskis K, Helman E, McKenna A, Shen H, Zack T, et al. Absolute quantification of somatic DNA alterations in human cancer. Nat Biotechnol. 2012;30(5):413–21.
43. Yu G, Zhang B, Bova GS, Xu J, Shih IM, Wang Y. BACOM: in silico detection of genomic deletion types and correction of normal cell contamination in copy number data. Bioinformatics. 2011;27(11):1473–80.
44. Gusnanto A, Wood HM, Pawitan Y, Rabbitts P, Berri S. Correcting for cancer genome size and tumor cell content enables better estimation of copy number alterations from next-generation sequence data. Bioinformatics. 2011;28(1):40–7.
45. Rosner B. Percentage points for a generalized ESD many outlier procedure. Technometrics. 1983;25(2):165–72.
46. Berger MF, Lawrence MS, Demichelis F, Drier Y, Cibulskis K, Sivachenko AY, et al. The genomic complexity of primary human prostate cancer. Nature. 2011;470(7333):214–20.
47. Hillier LW, Marth GT, Quinlan AR, Dooling D, Fewell G, Barnett D, et al. Whole-genome sequencing and variant discovery in C. elegans. Nat Methods. 2008;5(2):183–8.

Index

W. Wu and H. Choudhry (eds.), *Next Generation Sequencing in Cancer Research: Volume 1: Decoding the Cancer Genome*, DOI 10.1007/978-1-4614-7645-0,

Q

R

V

W

X

Z

Printed by Printforce, the Netherlands

Next Generation Sequencing in Cancer Research